AF529886

Drainage Engineering: Principles and Practices

Dr. S.K. Gupta

Born (1949), Dr. Gupta is Fellow of the National Academy of Agricultural Sciences, Indian National Academy of Engineering and Indian Society of Agricultural Engineering. He obtained B.Sc. (Agricultural Engineering & Technology) in 1970 and M.Sc. (Agricultural Engineering) in 1976 from PAU, Ludhiana and Ph. D (Civil Engineering) from JNTU, Hyderabad in 1984. Since his joining CSSRI in 1971, besides holding scientific positions, he has been Head, Division of Drainage and Water Management, Head, Indo-Dutch Network Project, Head, Division of Irrigation and Drainage Engineering and Project Coordinator (AICRP on Management of Salt-Affected Soils and Use of Saline Water in Agriculture). After superannuation, Dr. Gupta has been working as Emeritus Scientist (ICAR) and INAE Distinguished Professor at the same institute. Currently he is an independent Researcher. Dr. Gupta has been engaged in conducting researches in subsurface drainage, irrigation water management, leaching, use of saline water in agriculture, disaster/environmental management and mathematical modelling. He has published around 150 research papers in high impact factor journals and published 20 books. Dr. Gupta has been consultant to WAPCOS-Louis Berger, MoWR, MoRD, UP-Dutch Tube Well Project, ActionAid International, Synergics Hydro India-Oromia Water Works Enterprise, Ethiopia and others. He has been bestowed with Rafi Ahmad Kidwai Award by ICAR. Besides he is recipient of several awards from NAAS, ISAE, IE (India), CBIP, MoWR among other organizations. He is the Chief Editor of Journal of Water Management of the Indian Society of Water Management.

Drainage Engineering: Principles and Practices

S.K. Gupta

Ex-Head Irrigation and Drainage Engineering

Independent Researcher

Published by
SCIENTIFIC PUBLISHERS (INDIA)

Jodhpur –

5 A, New Pali Road
P.O. Box 91
Jodhpur - 342 001 INDIA

Delhi –

4806/24, Ansari Road
Daryaganj
New Delhi - 110 002 INDIA

ISBN : 978-93-88172-02-8

Visit the Scientific Publishers (India) website at
http://www.scientificpub.com
Printed in India

DR R.C. SRIVASTAVA Dr Rajendra Prasad Central Agricultural University
Vice-Chancellor Pusa (Samastipur)- Bihar

FOREWORD

Projections of the future food grains requirement to maintain marginal self-sufficiency achieved so far vary widely. A higher figure pegs the need at 367 million tons by 2025 and 581 million tons by 2050 probably taking into account the need of farm animals, which is often overlooked. With no scope to expand the net sown area hovering around 140+ million hectare (M ha) for the last two decades or so, the additional food grains production must come from increase in the cropping intensity, expansion of irrigation and introduction of scientific ways and means in agriculture production programmes. Although India´s accomplishments so far have been impressive, but the future challenge is to get equal or even higher increase from diminishing per capita arable land, diminishing irrigation water resources and expanding abiotic and biotic stresses. For example, our irrigated lands are under stress due to age-old nemesis of abiotic stresses of water logging and soil salinization. It is assessed that currently 6.73 M ha area has already gone out of cultivation due to the twin problems. It is likely to exacerbate as our irrigated lands lack appropriate and adequate drainage. An assessment made by CSSRI, a premier organization working on reclamation and management of salt affected soils, reveals that by 2050, extent of salt affected soils may increase to 16.2 M ha. It is likely to adversely impact agricultural production besides having serious socio-economic and environmental implications. Even this estimate is likely to be surpassed as assessed by TERI, New Delhi because of anticipated climate change. The projected climate aberrations may trigger natural disasters and sea level changes that may result in submergence of large coastal lands rendering them waterlogged saline.

The prescription of land drainage to manage water logging and soil salinity has been known to the mankind since the beginning of agriculture. Irrigation and drainage go together has been repeated umpteen times in various reports since the middle of 19th century. Yet the axiom has rarely been employed in practice in the irrigation systems designed and implemented in India. It is a matter of great satisfaction that drainage technologies are now being increasingly employed to reclaim waterlogged salt affected soils. More than 60,000 hectares of waterlogged saline lands have been reclaimed in Haryana, Rajasthan, Gujarat, Andhra Pradesh, Maharashtra and Karnataka with the intervention of subsurface drainage technology. The documented success stories under varied agro-climatic conditions have the potential of up-scaling. While crops yields in all cases increased by from more than 20% to 250%, cropping intensity showed an upward trend going up to 200% on lands lying barren before the intervention. Besides, the reclamation has remarkable favorable impact on capital formation, factor productivity, socio-economic changes in the society and proved to be an instrument of integrated rural development. The immense environmental benefits make this intervention a win-win proposition.

Land drainage is an engineering intervention that requires inputs from skilled personnel having background of engineering and agriculture. Most State Agricultural Universities train agricultural engineering students in the field of Irrigation and Drainage Engineering at graduate and post graduate level. Realizing the constraint of limited literature in the field of land drainage, I am happy to note that Dr. S.K. Gupta has ventured to prepare a book titled ′Drainage Engineering: Principles and Practices`. I believe that it should go a long way in meeting the demand of teachers and students to have a text book in this field. It may not be out of place to mention that Dr. Gupta has been associated with land drainage since 1980 and has actively participated in many experimental, pilot projects and large scale drainage projects in India and abroad. I am sure that this publication will have a good blend of theory and practical experiences on land drainage in India.

I had the chance to have a glimpse at the contents of this book. The whole text is divided into 11 well-defined chapters. All important issues such as extent, nature and distribution of waterlogged salt affected lands, drainage investigations, surface drainage, subsurface drainage design, materials and construction, alternate drainage systems such as mole, interceptor, vertical and bio drainage have been included. Separate chapters on reclamation of saline and alkali soils, acid soils and management of poor quality groundwater are included. Socio-economic and environmental issues have been discussed in a separate chapter. All the chapters are designed based on latest Dean′s committee recommended course content for land drainage.

I congratulate the author for his commendable efforts in compiling the current advanced theory and practical techniques on diverse aspects of agricultural land drainage for reclamation of waterlogged saline lands. I am convinced that the theory and practical experiences compiled in this textbook will help the teachers and students to have a better teaching/learning experience. Besides the teachers and students, researchers and field officers will find this book a handy reference to design and construct the drainage systems. All the information needed for this purpose has been included in this volume. I am sure that this publication will adore the shelves of teachers, researchers, field practitioners, and policy planners.

R.C. Srivastava

PREFACE

Current state of the land and water resources has been worrying all the stakeholders considering precarious nature of the food security achieved so far. The concerns have multiplied in view of the recently enacted 'Food Security Bill" that makes it imperative to ensure food for all. As a result, India needs to produce significantly more food grains to meet the requirements of its burgeoning population. On the other hand, several problems ail our land and water resources impacting the production from these resources. Twin problems of water logging and soil salinity have emerged as the major abiotic stresses in irrigated lands that have worried the Ministry of Water Resources, Ministry of Agriculture and others organizations connected with natural resources management. The problems have been a cause of worry mainly because expansion of irrigation has been our best bet to increase agricultural production. It is assessed that if the 'Business As Usual' approach continues, area under water logging and soil salinity by 2050 may treble to 20.0 million ha from the current level of 6.73 million ha.

Artificial land drainage is a prime requirement for sustainable irrigated agriculture and to reclaim waterlogged salt affected soils. More so because natural drainage is inadequate and is being further restricted through infrastructural development projects. Non-scientific methods of irrigated agriculture are compounding the situation. It is assessed that by 2050 India may need to spend Rs. 4000/- billion on artificial land drainage to ensure food and nutritional security of the teeming millions. Such a huge investment requires scientific and technical support to the participating organizations in effectively implementing appropriately chosen interventions.

It is heartening to note that 'Artificial Land Drainage' is being taught in all the Agricultural Engineering colleges of India. The course work is regularly updated by the ICAR in the overall guidance of the Dean's Committee. This textbook has been prepared strictly as per the course adopted by the colleges on the recommendations of ICAR. To give it a practical touch and the shape of a manual on land drainage, explanation of all issues is illustrated through the collation of Indian experiences on subsurface drainage. With support from global experiences, guidelines have been framed and included in the publication which may prove helpful in implementing large-scale drainage projects in India.

This book has been divided into 11 chapters. Chapter 1 mainly includes the general introduction to water logging, a brief introduction to soil salinity and kinds of drainage methods. Surface drainage methods, an essential intervention in monsoon climatic regions and as supplement to the subsurface drainage, are included in Chapter 2. Drainage investigations needed to diagnose the problems and to arrive at realistic drainage design parameters are included in Chapter 3 followed by drainage design procedures in Chapter 4. While drainage materials are the subject matter of Chapter 5, drainage construction procedures and methodologies to monitor and evaluate completed projects are included in Chapter 6. Role of special drainage techniques such as mole, interceptor, vertical and bio-drainage have been included in Chapter 7. Since artificial drainage is mainly directed towards reclamation of chemically degraded soils, reclamation of salt affected soils and acid soils are included in Chapters 8 and 9. A major criticism of drainage intervention stems from the fact that a good amount of byproducts in the nature of poor quality water is generated, which needs to be disposed of. Its eco-friendly reuse and several disposal techniques form the subject matter of Chapter 10. Finally, cost calculations, socio-economic and environmental issues associated with drainage projects have been included in Chapter 11. Clearly, each and every aspect of surface and subsurface drainage to reclaim waterlogged salt affected lands for sustainable agriculture has been covered.

It is now emerging that land drainage, as a water management strategy, will impact future agriculture more than anything else. It calls for all out efforts to popularize the technologies. New literature on land drainage needs to be developed and published. The easily accessible literature will provide much needed support to the students who are likely to man the key projects in future. Current genre of officers of the line departments who are being entrusted with this enormously important task also feel frustrated as there is a general lack of self-explanatory literature on the subject. This book is designed to fillup this gap.

This publication is a fully revised version of the final report of the Emeritus Scientist Project submitted to the ICAR by the author. Additions and deletions have made to make it a textbook for the students of agricultural and civil engineering. I take this opportunity to express my sincere thanks and gratitude to Dr S. Ayyappan, Secretary, Department of Agricultural Research and Education and Director General, Indian Council of Agricultural Research (ICAR) for my selection as Emeritus Scientist and providing me an opportunity to work on this project. I express my deep gratitude to Dr Arvind Kumar, DDG (Education), Dr A.K. Singh, former DDG (NRM) and Dr A.K. Sikka, DDG (NRM) for providing full support to the project at all stages. My heartfelt thanks are due to Dr D. K. Sharma, Director, CSSRI, Karnal, Dr Kusumakar Sharma, ADG, Agricultural Education Division, Dr P.S. Minhas, former ADG (Soil and Water) and Dr J.C. Dagar, former ADG (A &AF), New Delhi for their consistent support and guidance. Their guidance has been a source of inspiration. The facilities provided by Dr D.K. Sharma at CSSRI have been beyond expectation and I would like to convey my sincere thanks for this support. The excellent cooperation received from Dr S.K. Ambast, PC, AICRP on Management of Salt Affected Soils and Use of Saline Water in Agriculture is thankfully acknowledged. Special thanks are due to the staff of computer center, administration and finance sections of CSSRI for extending full support to the project.

The current publication is prepared to serve as a textbook of Drainage Engineering in Agricultural and Civil Engineering colleges. Appropriate theoretical and objective type questions have been framed to provide students a feel of what they can expect in their examinations. As such, it can prove to be a handy sourcebook for the competitive examinations. Besides, the book should serve as a handy training material in specialized training programs for the officers of the line departments such as water resources, agriculture, irrigation, private enterprises and non-government organizations.

The author at the end expresses his heartfelt thanks to his wife and other family members as they ungrudgingly allowed me to snatch their quality time while preparing the textbook. Without their support, it might not have been possible to update the textbook to its current state.

The author will highly appreciate the feedback and contributions in the form of case studies or research results that might help to enrich the future versions of this textbook.

S.K. Gupta
Ex-Head, Irrigation and Drainage Engineering
Independent Researcher

CONTENTS

Foreword v
Preface vii

1. WATER LOGGING – SOIL SALINITY AND DRAINAGE 1-27

Soil and Water Related Problems 1
Irrigation, Drainage and Associated Problems 2
Water Logging 2
Brief History and Extent of Water Logging 7
Impacts of Water Logging 8
Soil Salinization 13
Drainage Problems in India 16
Prevention, Management and Reclamation 18
Artificial Drainage 19
Types of Drainage Systems 20
Realization of Drainage Needs 23
Global and Indian Scene 24
Questions (Theory) 25
Questions (Objective) 26

2. SURFACE DRAINAGE METHODS 28-70

Definition 28
Surface Drainage Requirement and Crops 29
Factors Affecting Design of Surface Drainage 30
Design of Surface Drainage Systems 31
Drainage Coefficient 32
Empirical Formulae for Collector/Main Drains Design 41
Drainage Components and Field Practices 45
Drainage systems 47
Design of Field Drains 51

Outlet 61
Construction 61
Maintenance of Open Drains 62
Field Assessment of Benefits of Surface Drainage 64
Non-conventional Drainage Systems 65
Questions (Theory) 67
Questions (Objective) 68

3. DRAINAGE INVESTIGATIONS 71-129

Introduction 71
Stages in Drainage Investigations 72
Reconnaissance Survey 74
Topographic Survey 74
Soil and Geological Survey 77
Photogrammetric Survey 86
Groundwater Surveys 86
Groundwater Investigations 87
Water Transmission Characteristics 95
Hydraulic Conductivity 99
Laboratory Methods 101
Field Methods 104
Drainable Pore Space 114
Subsurface Drainage Coefficient 120
Collation and Application of Data 124
Questions (Theory) 124
Questions (Objective) 126

4. DESIGN OF SUBSURFACE DRAINAGE SYSTEMS 130-186

Introduction 130
Layout 131
Drainage Criteria 134
Drain Depth 136
Spacing of Lateral Drains 141
Design Equations (Steady State) 142
Selection of Steady State Equation 156
Unsteady State Equations 157

Fluctuating Water Table 166
Recommendations on Drain Spacing 169
Interceptor Drainage for Sloping Lands 169
Capacity/Size of Drains 171
Drainage of Paddy Fields 178
Questions (Theory) 179
Questions (Objective) 180

5. DRAINAGE MATERIALS 187-224

Introduction 187
Materials for Drain Pipes 187
Testing Flexible Pipes 194
Tests on Fittings 195
Load on Conduits 197
Envelopes 203
Need of an Envelope 205
Parameters to Assess the Need of an Envelope 206
Entrance Head Loss and Resistance 208
Criteria for Granular Envelope 211
Design and Selection of Synthetic Envelope 214
Tests on Synthetic Envelope Materials 219
Organic Envelopes 219
Questions (Theory) 220
Questions (Objective) 220

6. CONSTRUCTION AND EVALUATION OF SYSTEMS 225-277

Introduction 225
Pre-installation activities 225
Construction Methods 226
Manual Construction 226
Mechanical Installation 229
Land Drainage for Peat and Muck Soils 234
Drainage Structures 235
Maintenance of Pipe Drainage Systems 247
Inspection and Quality Control 251
Tender Document 252

	Organizational Set-up	253
	Monitoring and Evaluation	257
	Drain Outflow Method to Evaluate Soil Parameters	262
	Crop Cutting Experiments	268
	Measurement of Sediment Deposits	270
	Water and Salt Balance	272
	Questions (Theory)	275
	Questions (Objective)	275
7.	**SPECIAL DRAINAGE TECHNOLOGIES**	**278-314**
	Introduction	278
	Mole Drainage	278
	Mole Drainage: Indian Experiences	281
	Failures and Remedies	282
	Drainage for Seepage Interception	283
	On-farm Ponds for Seepage Interception	287
	Vertical Drainage	288
	Theory of Flow to Drainage Wells	291
	Bio-Drainage	300
	Suitability of Drainage Methods	308
	Integrated Drainage Systems	308
	Special Activities to Improve Drainage Effectiveness	309
	Questions (Theory)	311
	Questions (Objective)	312
8.	**SALT AFFECTED SOILS: RECLAMATION AND MANAGEMENT**	**315-355**
	Terminology	315
	Salt Affected Soils	321
	Formation of Salt Affected Soils	325
	Extent and Distribution of Salt Affected Soils	326
	Reclamation of Alkali Soils	328
	Determination of Gypsum Requirement (GR)	330
	Alkali tolerance and Crop Selection	335
	The Package (Alkali Lands)	338
	Phyto-remediation of Alkali Lands	338

Reclamation of Saline Lands 339
Cropping in Saline Soils under Reclamation 346
Land Modifications and Planting Management 346
The Package (Saline Lands) 350
Reclamation of Saline-Alkali Soils 350
Questions (Theory) 351
Questions (Objective) 352

9. RECLAMATION AND MANAGEMENT OF ACID SOILS 356-381

Introduction 356
Definition, Characterization and Formation of Acid Soils 356
Kinds of Soil Acidity 358
Extent and Distribution 359
Causes of Low Productivity in Acid Soils 359
Management of Acid Soils 361
Liming 361
Industrial wastes as Amendments 370
Engineering and Agronomic Measures 373
Crop Selection: Acid Tolerant Crops 374
Special Cases of Acid Soils 376
Questions (Theory) 379
Questions (Objective) 379

10. CHARACTERIZATION AND MANAGEMENT OF SALINE WATER 382-435

Introduction 382
Origin of Salts in Irrigation Water 382
Extent and Distribution of Saline Water 383
Chemical Characteristics to Characterize Water Quality 385
Salinity Hazard: Electrical Conductivity (EC) 387
Sodium/Alkali Hazard 389
Toxicity Hazard 394
Combination of Parameters to Assess Water Quality 396
Infiltration Problems 398
Soil Salinization 399
Water and Salt Balance of the Root Zone 403

Leaching Requirement 405
Leaching Fraction and Soil Salinization 409
Management Strategies for Saline/Alkali Water 414
Chemical Amendments 422
Water Quality Guidelines for Irrigation 424
Disposal of Saline Drainage Effluents 426
Management of Toxic Water 427
Management of Waste Water 428
Questions (Theory) 429
Questions (Objective) 430

11. ECONOMIC ANALYSIS AND SOCIO-ENVIRONMENTAL ISSUES 436-464

Costs 436
Investment on Machinery 439
Benefits 440
Economic Analysis 443
Evaluation Indices 444
Calculation of Indices 447
Case Studies of Economic Analysis 449
Comparison of Drainage Systems 452
Cost Recovery 452
Indirect Benefits 453
Environmental Impact Assessment 455
Farmers' Participation 457
Land Drainage: Current Status and Issues 459
Questions (Theory) 461
Questions (Objective) 462

Annexures I - VI

Annexure I: Bill of Quantities for Surface and Subsurface Drainage Systems Construction 465
Annexure II: Ground Water Balance 470
Annexure III: Irrigation Water Salinity Limits for Selected Crops 475
Annexure IV: Farmer's Consent /Agreement 478
Annexure V: MOA For Farmer Drainage Societies (FDS) 479

Annexure VI: Handing over the Block to Drainage Society 480
Glossary 481
References 500
Answers to Objective Type Questions 514
Conversion Tables 519
Subject index 526

Chapter 1

WATER LOGGING – SOIL SALINITY AND DRAINAGE

SOIL AND WATER RELATED PROBLEMS

Soil is nature's gift to mankind. It is one of the basic resources to sustain plant and animal life. Crop productivity is sustained by the top fertile soil. Whereas it may take around a century to build-up one cm of precious top soil, it is being eroded on an average @16.4 t/ha annually (Singh *et al.,* 1992). Besides erosion, soil is also being degraded in several other ways. The quantitative estimates of the problem vary largely. According to oft quoted estimate made by the Ministry of Agriculture (1984-85), 175 million hectare (M ha) out of a total land area of 304 M ha for which records are available, are subjected to a variety of environment problems. Following the methodology proposed by Oldoman, Sehgal and Abrol (1994) assessed that about 188 M ha or 57% of the geographical area of the country is under the processes of land degradation (Table 1.1). Recent estimate shows that problems are expanding and cover more than 200 M ha (Table 1.2).

Table 1.1 Extent of soil degradation in India

Degradation type	Area (M ha)	% of total geographical area
Erosion		
Water	148.9	45.3
Wind	13.5	4.1
Chemical degradation (Fertility/soil salinity)	13.8	4.2
Physical degradation (water logging)	11.6	3.5
Total	187.7	57.1

Naturally occurring poor quality water in arid, semiarid and sub-humid coastal regions is another problem causing environmental concerns. The extent of the poor quality groundwater ranges from 30 to 84% of the total groundwater resource of some states. In fact, these are the areas that are most prone to water logging and soil salinity upon introduction of irrigation. Pollution of surface and groundwater resources through diffuse sources (say agriculture) as well as through point sources (say disposal of domestic and industrial effluents) has also assumed serious dimensions. Water inadequacy in times to come may not only result from inadequate resources but because of impaired quality as well.

Table 1.2 Soil problems and their extent (M ha)

Problem	**Area (M ha)**	
	Ministry of Agriculture estimate (1984-85)	**Recent estimate**
Serious water and wind erosion	150.0	172*
Shifting cultivation	3.0	3.0
Water logging	6.0	10.0**
Saline-alkali soils	7.0	8.6***
Diara lands	2.5	2.5
Other cultivable lands fit for cultivation	6.6	6.6
Total	175.1	202.7

* Singh *et al.* (1992); ** Bowonder and Ravi (1984); *** Gupta (2005)

IRRIGATION, DRAINAGE AND ASSOCIATED PROBLEMS

Irrigation is an essential input to increase agricultural productivity and to sustain it without major year to year variations in the productivity. Often irrigation systems are constructed with the assumption that drainage is either not needed or it can be introduced at a later stage whenever required. It is contrary to the common dictum that irrigation and drainage go together. In fact drainage constitutes a subset of water resources system and is crucial to sustain irrigated agriculture. Neglecting drainage has resulted in several environmental problems and large tracts in the irrigation commands have gone out of cultivation due to spread of water logging and soil salinity often referred as wet deserts. Currently, water logging and soil salinity are major constraints to sustainable agricultural production in India. Both the problems are encountered either individually or together in irrigation commands and even in rain fed situations.

WATER LOGGING

In its simplest form, water logging connotes restricted drainage, whether surface or subsurface. It refers to a condition of the land (soil) in which water stagnates on the land surface or water table comes within or very near to the root zone resulting in severe aeration problems. As a result, crop yields decline below the normal levels. In extreme cases the land is taken out of cultivation because it is no more economical to plough the land.

Water logging has been defined by various workers and organization in different ways although the basic premise is same, as indicated in the opening lines of the previous paragraph. Clayton defined a land *'as waterlogged when the water table is within ± 45.75 cm of the ground surface'*. It may be noted that this definition takes into account both surface and subsurface water logging although +45.75 cm above the ground surface seems quite high. Usually, surface water logging connotes shallow submergence, and higher depths of submergence often associated with floods. The definition was first adopted in India, by the Drainage Board, in 1917. A modification to this definition was made in 1925 in the state of Punjab, when *'a land was designated as waterlogged, if it is thrown permanently out of cultivation due to rise in sub-soil water and is recorded as banjar in the revenue records`*. The water table in the month of June should be within 1.25 m of the surface. A further modification in the definition was made to account for the yield of crops. It was decided to term *'a land to be waterlogged if the yield is reduced to 25% of the optimum value, due to rise in sub-soil water'*.

A special committee of the Central Board of Irrigation and Power has defined water logging as follows:

An area is said to be waterlogged when the water table rises to a level that the soil pores in the root zone of a crop become saturated resulting in restriction of the normal circulation of air, decline in the level of oxygen and increase in the level of carbon dioxide.

The National Commission on Agriculture (1976) adopted the definition given by Central Board of Irrigation and Power and defined water logging as *' excess water in the root zone due to shallow water table restricting the normal aeration of the crop roots'*. From practical point of view the Commission categorized the area as water logged when the water table was within 1.5 m of the soil surface (Table 1.3). An Experts' Group, set up by the Ministry of Water Resources in October, 1986 suggested that water logging conditions prevail when the water table rises to within 2m depth from the ground surface in the month of November and April. This definition included the time of the year during which the water table needs to be considered, water table being dynamic in nature (Table 1.3). It may be noted that a shallow water table in pre-monsoon months may trigger soil salinization while a shallow water table in the post-monsoon months may hamper timely completion of field operations for the winter (*rabi*) season crops. Many workers suggest that criterion of water table within 2 m in the month of November can be relaxed unless it becomes detrimental to tillage for the sowing of *rabi* crops.

Table 1.3 The criteria adopted by different agencies to define water logging

Water logging	National Commission on Agriculture (1976)	Working Group (1991)
Water logged/critical	Water table < 1.5 m	Water table < 2 m
Potentially water logged	–	Water table 2-3 m
Safe area	–	Water table > 3 m

According to FAO (FAO, 1973), *'waterlogged areas are those where soils are temporarily saturated or where the water table is too shallow such that capillary rise of groundwater encroaches upon the root zone and may even reach the soil surface'*. It would affect the root proliferation and have adverse impact on growth and yield of plants (Fig. 1.1).

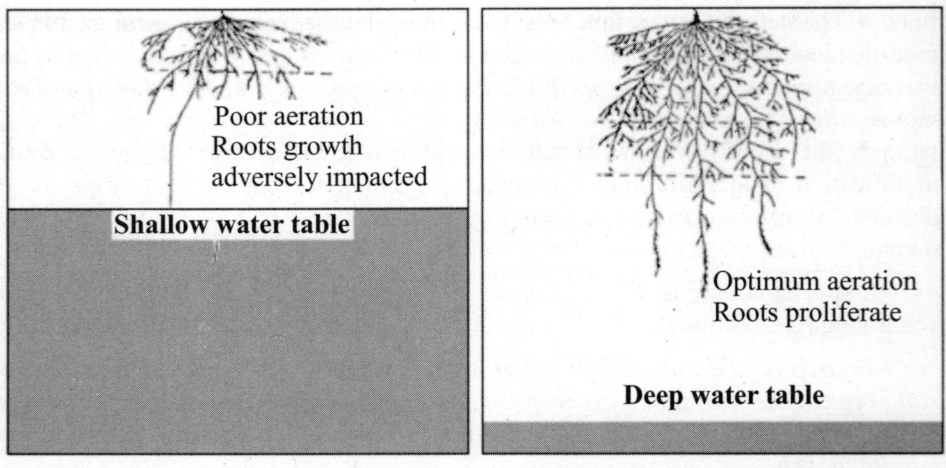

Fig. 1.1 Impact of depth to water table on root distribution and proliferation

Surface Stagnation

Surface stagnation of water is a very common problem in the monsoon climatic conditions and is experienced in almost all the irrigation commands (Fig. 1.2). Although common in humid regions, analysis of rainfall data for various stations in the country revealed that such a possibility cannot be ruled out even in semiarid or arid climatic conditions. With introduction of irrigation, degree of water stagnation and the periods for which water stagnates on the land surface got a sea change. It is because the capacity of the soil to absorb water reduces due to rise in the water table.

Fig. 1.2 Surface stagnation of rainwater and farmers response to save crop

Causes of surface stagnation

- Construction of roads, bridges, railway lines, buildings and undertaking other developmental activities resulting in choking of natural drainage
- Poor natural drainage as a consequence of unfavourable topography
- Submergence of agricultural lands due to spilling of rivers
- Heavy rain storms coupled with poor natural drainage

- Rise in water table in the commands adversely affecting rainwater absorption
- Inadequate maintenance of existing drainage systems and outlets
- Formation of a hard layer at plough depth in rice-wheat rotation
- Poor on-farm water management resulting in poor application efficiencies and increased deep percolation losses
- Increasing encroachments of wetlands for developmental activities
- Deforestation and poor upkeep of watersheds
- Tidal ingress, cyclones, etc. in coastal regions

To avoid or mitigate the problem, surface drainage improvements are needed whenever an irrigation project is planned.

Areas prone to surface water logging

Surface stagnation and shallow water table continue to be a major stress factors in irrigation commands in monsoons. Although, most agricultural lands in monsoon climatic conditions are prone to short-term surface stagnation, the most affected areas are as follows:

- As per the reports of the CWC, on an average 8.0 M ha area suffers from floods every year out of which at least 3.7 M ha area is under crops. While floods are more devastating and a recurring phenomenon, short-term water stagnation is of more common occurrence in humid, sub-humid and even in semi-arid regions especially during the *kharif* season. Excepting arid regions, one day rain storms may exceed 125 mm practically all over the country. It results in reduced crops yield.
- Introduction of irrigation through inter-basin transfer of water has caused rise in water table in large areas of several irrigation commands. As a result, rainwater absorption into the soil is much less. As such, degree and duration of water stagnation will be much higher in areas with shallow water table.
- A plough sole/hard pan develops in lands under continuous rice-wheat cropping system. The plough layer impedes the vertical movement of water in the soil profile. Thus, these areas are prone to surface stagnation.
- In alkali (sodic) lands under reclamation, since only top 15 cm of the surface is initially reclaimed, lower layers continue to have high exchangeable sodium percentage (ESP) even after several years of reclamation. As a result of impeded vertical movement these lands get submerged following rainfall.
- Irrigation with high SAR saline or alkali waters with high RSC reduce infiltration rate of the irrigated lands because of structural breakdown of the soils. As a result, these lands are prone to surface stagnation following rain events of higher magnitude.

- In general, drainage systems provided have inadequate capacities because of several assumptions made and thumb rules adopted. These systems often fail to cater to the needs at critical junctures.
- Poor upkeep and maintenance of the drainage systems and submergence of natural outlets causes back flows resulting in submergence problems in larger areas.
- Besides, changing land use patterns and poor on-farm water management (field-to-field irrigation, flood irrigation, improper land leveling) favour short-term stagnation of water.

Shallow Water Table

The problem is associated with groundwater levels rising and remaining in the root zone for periods enough to adversely affect aeration and crop productivity (Fig. 1.3). While rise in water table is imminent upon introduction of irrigation, several factors together determine the rate at which water table rises in a given setting. Soil salinity accompanies and is often termed as a twin to the problem of water logging. Amongst the various factors that lead to water logging and soil salinity in irrigation commands, some are common to most irrigation commands while few others are site specific and are applicable only under specific situations. The idea to split these into two groups is to emphasize that both general and local factor should be identified for the menace of water logging in irrigation commands.

Common factors

- Leakage and seepage from the irrigation networks
- Poor on-farm water management resulting in excessive deep percolation losses
- Inadequate internal and artificial drainage
- Poor maintenance of irrigation and drainage systems
- Change in cropping pattern in favour of water guzzling crops such as rice and sugarcane
- Stakeholder′s lack of knowledge of irrigated agriculture
- Water pricing, almost free supply does not provide any incentive for implementing water management
- Lack of institutional support for pursuing science-based agriculture

Site-specific factors relevant to particular set-up

- Excess water release in the command as in Sharda Sahayak irrigation command in Uttar Pradesh and Ukai-Kakrapar project in Gujarat
- Lack of irrigation during night in many irrigation commands
- Nature of the soils
- Shrinking area under agriculture due to industrialization

- Filling of the Ghaggar depressions as in the case of Indira Gandhi *Nahar Pariyojana* (IGNP)
- Structural inability of the delivery system
- Sea water ingress and intrusion in coastal regions

Amongst the various factors, inadequate drainage is ranked as the number one factor in causing water logging in irrigation commands.

Fig. 1.3 A view of shallow water table in Western Jamuna canal command

Perched Water Table

When the deep percolation is impeded by a poorly permeable layer, perched water table conditions may develop. Its occurrence is temporary as the impeded water will continue to seep through to deeper layers or drain laterally. It may, however, exist long enough to cause a serious problem of excess water in the soil. Perched water table is detected by drilling a borehole into (but not through) the impeding layer.

BRIEF HISTORY AND EXTENT OF WATER LOGGING

Water logging and soil salinity has charted a parallel course with development of irrigation in India and elsewhere. The problems in India came into prominence with the opening of several canal systems by the British. By 1880, some 10,680 km of canals (main and distributaries) existed in the North West Province including 6215 km in Punjab. A recorded incidence of water logging/soil salinity from village Munak in the Western Jamuna Canal command in Haryana dates back to middle of the 19th century, 1855 to be precise. The appearance of problems so early after introduction of irrigation in 1840, suggested a practical remedy that drainage should accompany irrigation. By 1864, *usar* (moderate to severe alkalinity) was well known in Faizabad, Behraich, Lucknow, Sitapur and Hardoi of United Provinces and more reports were pouring in with the Government of India. In 1876, another alarm was raised about the disastrous spread of *reh* in parts of Aligarh, Meerut and throughout the Kali *Nadi* Valley of Uttar Pradesh. Through a petition, David Robarts of *pargana* Sikandra Rao, Aligarh district, requested the Board of Revenue for relief due to appearance of saline efflorescence and deterioration of his land as a result of the introduction of canal irrigation. Similar situation was reported in the Nira Valley Canal Project in 1889, only 5 years after the opening of the canal. The situation was no better in minor irrigation commands as reports of water logging and soil salinity came from tank irrigated areas of the Madras Presidency. In the state of Punjab visual soil surveys, locally termed as *Thur girdawari,* were initiated to demarcate areas damaged by salinity (*thur*) or high moisture content (*sem*) in the root zone.

Even in this century, spectacular rise in water table has been reported in irrigated lands such that many areas that used to be under dry farming few years back became unfit for cultivation of any upland crop during the rainy season. Projects, commissioned in the post-independence period, also developed serious water logging and soil salinity problems. Water table rose at an annual rate of 0.6 m/annum during 1975-2000 at Bhuna (Fatehabad, Haryana) in Bhakra canal command. Consistent rise in water table is reported in the command area of Jui Lift Irrigation Canal in Haryana, ranging from 0.3 to 0.8 m per annum (Sharma and Bhargava, 1982). This situation emerged in spite of the fact that only limited quantities of water were introduced through the lift irrigation scheme. In southwestern Punjab water table is rising due to limited/non-extraction of groundwater because of its brackish/saline quality and more availability of canal water for domestic and irrigation purposes. Within last 17 years the water table in Kuttianwali village in Lambi block of Mukatsar district has risen about 10 m and has come within 2 m of the soil surface (Fig. 1.4A). The water table isobaths map reveal that the water table in the district is in the range of 2-7 m below ground level (Fig. 1.4B). The water logging developed so fast in Tawa project in Madhya Pradesh that within few years of opening of the canal, farmers formed "*Mitti Bachao Abhiyan Samities*" (Save the Soil Committees). The water table in an area of about 100,000 ha got raised at an alarming rate (about 1 m per annum) in the Phase I of the Indira Gandhi *Nahar Pariyojana* (Hooja *et al.,* 1994). Recent experiences also reveal an inter link between the introduction of surface irrigation and spread of water logging and soil salinity in India.

Extent

The precise estimates of the extent of water logging are unavailable. Several attempts have been made to estimate the waterlogged areas by different agencies at different points of time (Table 1.4). The extent varies from 1.61 (Working Group, 1991) to 8.53 M ha (Ministry of Agriculture, 1984-85). The data reported by National Commission on Agriculture (1976) and Ministry of Agriculture probably covered both surface stagnation of water and rise in water table. A report of the planning Commission has mentioned that during 2008-09, 8.7 M ha area was affected in various states (Table 1.4). Apparently, there are large year to year variations as in the case of Bihar. Recently, Indian Council of Agricultural Research (2009) estimated that soils afflicted with physical problems mostly due to water logging are 5.57 M ha although the distribution of area in various states differed from the earlier estimates of similar magnitude. Tata Energy Research Institute (TERI, 1997) projected a 3-fold increase in water logging from 12.7 to 39.6 M ha during 1997-2047.

IMPACTS OF WATER LOGGING

Water logging is known to adversely impact agriculture and environment in a number of ways (Table 1.5). Three major physico-chemical changes that manifest upon water logging are:

- Lack of aeration in the root zone resulting in accumulation of CO_2 at the expense of O_2
- Reduced availability of plant nutrients
- Production of toxic elements to adversely impact crop yields

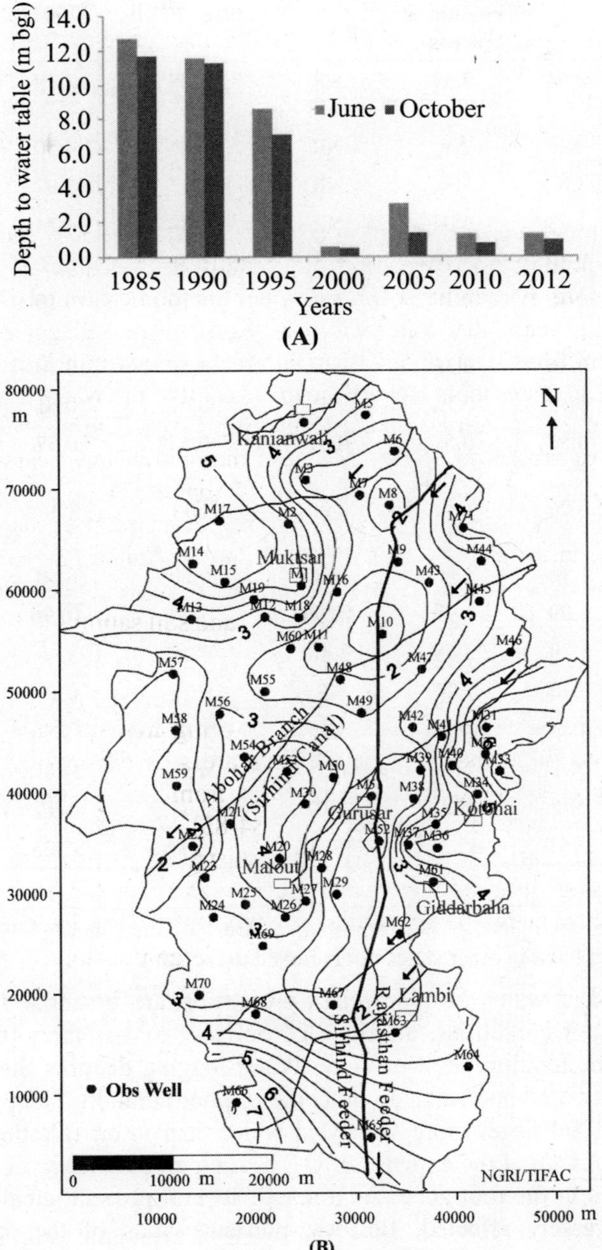

Fig. 1.4 The rising water table in Kuttianwali village in Lambi block (A) (data source Sharma *et al.*, 2013) and groundwater levels in Mukatsar district Punjab (B) (Gurunadharao, 2016)

Table 1.4 Extent and distribution of water logged areas in India (lakh ha)

State	Water logged area						
	Canal Command	Total[1] (including non-command areas)	Irrigation Commission (1972)	National Commission on Agriculture (1976)	Ministry of Agriculture (1994)	Working Group (1991)	Planning Commission (2013)[§]
Andhra Pradesh	2.66	3.39	NR	3.39	3.39	2.66	1.08
Assam	NR	NR	NR	NR	4.50	NR	8.88
Bihar	3.63	3.63	NR	1.17	7.07	3.63	57.4
Gujarat	1.73	4.84	NR	4.84	4.84	0.89	0.24
Haryana	2.30	2.75	6.5	6.20	6.20	2.30	0.49
Jammu and Kashmir	NR	NR	NR	0.10	0.10	0.02	0.85
Karnataka	0.36	0.36	0.07	0.10	0.10	0.25	0.18
Kerala	0.12	0.12	NR	0.61	0.61	0.12	0.08
Madhya Pradesh	0.57	0.57	0.57	0.57	0.57	0.04	0.0
Maharashtra & Goa	0.06	1.11	0.28	1.11	1.11	0.06	1.13
Odisha	1.96	1.96	NR	0.60	0.60	1.96	4.02
Punjab	1.99	1.99	10.90	10.90	10.90	2.00	1.18
Rajasthan	1.79	3.48	3.48	3.48	3.48	1.80	1.18
Tamil Nadu	0.18	1.28	NR	0.18	0.18	0.02	1.19
Uttar Pradesh	4.55	19.81	8.15	8.10	19.80	0.35	8.65
West Bengal	NA	NA	18.5	18.50	21.80	NR	0.20
Delhi	NR	NR	NR	0.01	0.01	NR	0.07
Total	21.90	45.29	48.40	59.86	85.26	16.10	87.03
Total (M ha)	2.19	4.53	4.84	5.99	8.53	1.61	8.70

Source: [1]Singh, 1992; NR = Not reported, § Area affected during 2008-09, Here total includes area affected in other states not included in the table

Both air and water in appropriate proportion are essential for plant growth. Except for winter conditions, an air-filled porosity of less than 10% is considered inadequate from aeration point of view. Water logging disturbs the delicate balance between these two components, an outcome of poor aeration. Since gases in general, diffuse about 1000 times more slowly in water than in air (Grable, 1966), there is accumulation of CO_2 at the expense of O_2, causing an imbalance in the proportion of these two gases in the root zone. As a result, several physiological processes of the plants are adversely affected. Besides, nutrient status of the soils in impacted affecting the availability of many plant nutrients. It can be ascribed to the differences in the chemistry and microbiology of submerged lands as compared to non-submerged lands (Ponnamperuma, 1972). Nitrogen is easily lost to the atmosphere as a gas because it is readily denitrified. Non-availability of nutrients to the plants is also ascribed to leaching of plant nutrients. The anaerobic (reducing) environment results in changes that can result in elemental deficiencies (reduced uptake) or toxicities (excess). The important inorganic and organic compounds that are produced in submerged soils are iron, sulfides and organic acids.

Besides what is stated in the foregoing paragraph, lower temperature in waterlogged lands (larger heat capacity of waterlogged soils), compaction of the soils, delay in farm operations and increasing incidences of diseases also affect crop yield in waterlogged lands. The damage is more if water logging occurs in the early stages or at sensitive growth stages, on hot days and in saline/alkali soils.

Table 1.5 Adverse impacts of water logging on agricultural and environment

Agriculture	Socio-economic	Natural	Regional
Reduced agricultural production	Accentuates socio-economic disparity	Damage to soil health	Higher expenditure on health services
Limited scope of crop diversification	Migration from rural to urban areas	Damage to eco-system health	Higher expenditure on infrastructure maintenance
Reduced cropping intensity	Higher gender disparity		Landscape impairment
Impairment of product quality	Deterioration in law and order situation		

Soil Aeration

Surface stagnation is visually observable and the degree of its severity judged by the duration of water stagnation, depth of water stagnation, time of water stagnation and the crop growth. Its status can as well be measured in the following two ways.

Oxygen Diffusion Rate

Oxygen Diffusion Rate (ODR) is used to assess the aeration status of the root zone or oxygen deficiency. ODR characterizes the soil O_2 conditions and determines the rate at which oxygen can be replenished if used by respiring plant roots or replaced by water. ODR decreases with increasing duration of water stagnation and increases with increasing duration after the water stagnation ends. Values of ODR may continue to be low even after water stagnation has been removed because soils remain saturated for some time. As such, the adverse effect may continue even after the water stagnation is not visible.

ODR is measured with an oxygen diffusion rate meter. The measurements are commonly made at 10 cm depth by inserting an electrode (Fig. 1.5). The readings are taken at 5-10 locations to determine the representative ODR value (Sharma *et al.*, 2002). For adequate aeration values of ODR in the range of 20-35 $*10^{-8}$ g/cm^2-min are desirable (Stolzy and Letey, 1964). According to Stolzy and Letey (1964), minimum ODR values of the order of 40 to 80$*10^{-8}$ g/cm^2/min should ensure good germination and emergence of seeds. If ODR values fall below 20$*$ 10^{-8} g/cm^2-min, roots of many plants fail to grow.

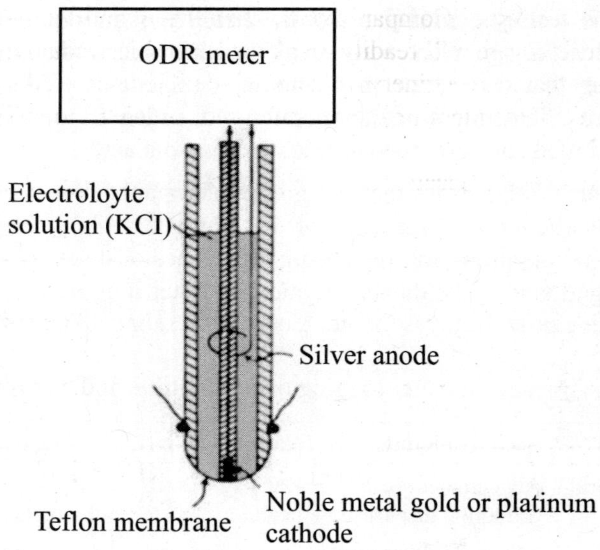

Fig. 1.5. Schematic view of the electrode of an ODR meter

Soil redox potential

Soil redox potential (Eh) is a semi-quantitative indirect measurement of *in-situ* soil oxygen status. Generally, aerobic (oxidized) soils having an Eh value greater than +300-350 mV allow the roots to efficiently carry out their functions. On the contrary, anaerobic (reduced) soils have values less than +300, severity of water logging increases with decreasing values (Fig. 1.6). Under similar conditions, the values of Eh are lower at higher pH indicating that the adverse effects will be more pronounced in alkali than a normal soil. On prolonged water logging, Eh values may attain negative values indicating occurrence of reducing reactions (loss of electrons; more positive ion results). When Eh is negative, phytotoxic concentrations of several elements such as Fe, Mn and Al in the root zone appear impacting the crop productivity. For example, in Western Australia, high to toxic concentrations of Al, Mn, and Fe were observed in shoot tissues of wheat after water logging of acidic soils. Although redox measurements tend to be less stable and less reproducible in an oxidizing environment, nevertheless, even in well aerated soils, there is a close relationship between the soil redox potential and oxygen concentration in the soil atmosphere.

Soil Workability

Water content is the chief factor that determines the strength and the deformation characteristics of cohesive soils. The unconfined compressive strength of a soil is exponentially related with the moisture content. Thus, with increasing soil moisture, soils lose their load bearing capacity. With high moisture content, the soil is susceptible to plastic deformation and consolidation. As a result, dry soils can support heavy loads because of rigidity of the individual soil particles and the high internal friction of the soil mass. In order to manipulate the soil without compaction or puddling, the moisture content of the soil must be below some critical level, depending upon the soil type. Compared with sand, clays have a narrow range of water content in which tillage will readily break the soil. A direct consequence of not being able to employ heavy machinery on waterlogged lands is to delay the sowing/harvesting operations. Timeliness of the operations significantly affects the yield of crops particularly in humid regions and in soils with high clay content. Even the quality of the crop is affected because of delay in harvesting.

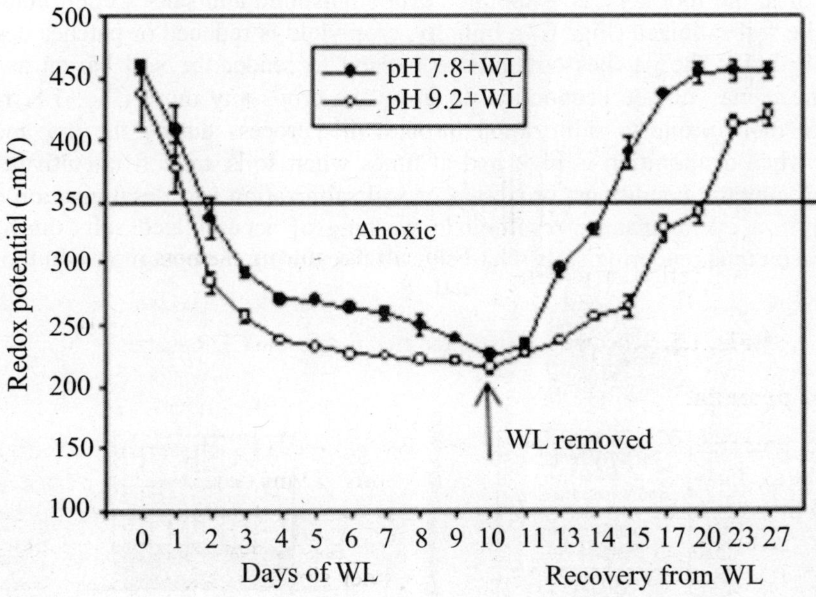

Fig. 1.6 Variation of redox potential in waterlogged soils at 2 pH levels (WL stands for water logging)

SOIL SALINIZATION

Soil salinity problems in irrigation commands develop whenever soil and hydrological conditions favour accumulation of soluble salts in the root zone. The main causes of salt accumulation are:

- Weathering of rocks and the salts brought down from upstream to the plains by rivers and subsequent deposition along with alluvial materials.
- Secondary soil salinization through capillary rise from subsoil salt beds or from shallow brackish groundwater (Fig. 1.7).
- Lack of natural leaching due to topographic situation, especially in arid and semiarid regions.
- Indiscriminate use of irrigation water of variable qualities including poor quality groundwater.
- Ingress of sea water along the coast.
- Salt laden sand blown by sea winds.

While several of these issues are taken up in a subsequent chapter (Chapter 8), secondary soil salinization through capillary rise is illustrated through Fig. 1.7. It may be seen that as long as water table is deeper, water does move up but remains within the soil profile below the root zone. As soon as water table reaches 2 m or so depth, it is drawn up to the soil surface through soil capillaries. In heavy textured soils this movement may begin early even when the water is relatively deeper but in light textured soils it begins at relatively shallower depths. Water reaching the soil surface or in the root zone is evaporated/evapotranspired and salts are left behind to render the soil salinized (Fig. 1.7). Initially, crop yield is reduced or patches develop (Fig. 1.8, left). The patches continue to expand to render the soil barren as after sometime it may not be economical to grow the crops any more (Fig. 1.8, right). Soils are more prone to salinization through this process during the dry and hot months when evaporation is high and at times when soils are left uncultivated. In monsoon climatic conditions, processes of soil salinization and desalinization occur together in a cyclic manner resulting in washing of accumulated salts during the rainy season thus rendering soils relatively salt free during the post monsoon months.

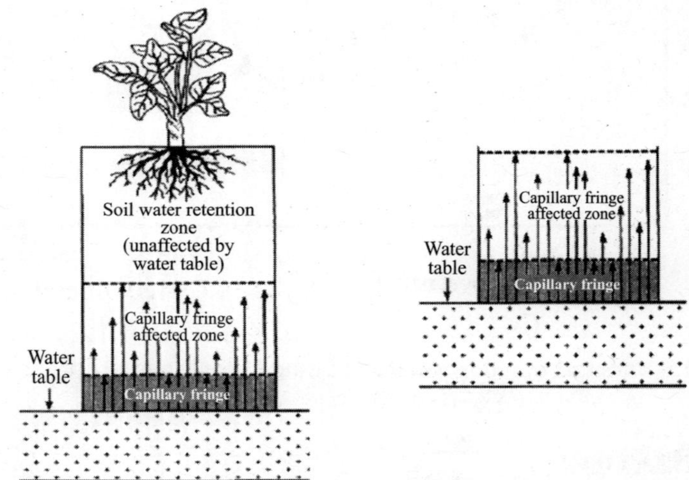

Fig. 1.7 Capillary transport of water and salts under shallow water table

Other impacts

Waterlogged soils have higher heat capacity and therefore, are relatively colder than dry soils. Since all biological processes are strongly influenced by soil temperature, crop growth in water logged soils, therefore, starts later and is slower than dry soils.

The excess water and the resulting continuously wet root zone can lead to some serious and fatal diseases of the root and stem (Poysa *et al.*, 1987).

It is now apparent that the adverse effect of temporary water logging depends not only on plant species but also on the physiological stage of growth, the time and duration of water logging, the light intensity, temperature and the fertility of the soil. Injury to crops is particularly severe when the crops are flooded on hot days. The damaging effects are generally more if the water logging conditions are accompanied by salinity/alkalinity. It may be noted that no single factor affects the crop yield in a water logged soil. Generally, it is a combination of poor aeration, reduced uptake of plant nutrients, imbalance between the uptake of various ions and toxicity effects of various elements/cations.

Fig. 1.8 Development of soil salinity patches (left) and a completely barren saline soil (right)

Yield Losses Due to Water Logging/Soil Salinity

Both the surface stagnation and shallow water table reduce the yield. The yield of several crops reduced with surface stagnation, the loss increasing with increasing duration of water stagnation (Table 1.6). Similarly, wheat yield reduced from an optimum value of 3.8 t/ha at a water table of 120 cm below the ground surface to less than 1 t/ha with water table at or less than 70 cm below the ground level (Anonymous, 2002). The yield loss in soybean was 45% at a water table depth of 45 cm compared to 80 cm below the ground surface (Fig. 1.9). It may be mentioned here that differentiation must be made whether the yield losses are due to water stagnation (land submergence) or due to shallow water table because the strategies for the reclamation and management of the two groups of soil are quite different.

Table 1.6 Yield losses in various crops due to water stagnation of various durations

Crop	**Yield losses over drained (%) with water stagnation (days)**			
	1	**2**	**4**	**6**
Sorghum	3	11	16	20
Pearl millet	6	15	22	27
Pigeon pea	4	14	18	21
Wheat	8	17	27	39
Barley	4	7	13	25
Mustard	8	16	22	29
Berseem (seed)	2	21	35	48
Sunflower	13	19	26	30

Yield losses are relative to drained plots

The crops yield losses due to water logging and soil salinity in Chambal command are depicted in Table 1.7. The data shows average yield losses of about 31 to 59%, highest being in the mustard crop (RAJAD, 1995, Table 1.7). It may be added that the yield losses in several fields were as much as 100% i.e. the lands were not cultivated because of severe water logging or soil salinity.

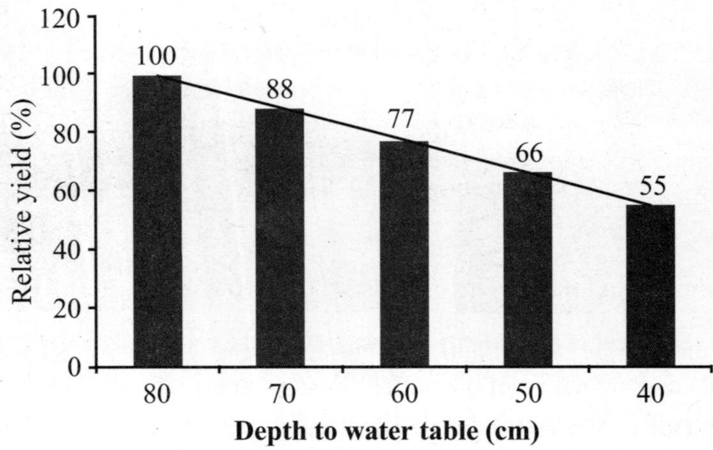

Fig. 1.9 The effect of depth to water table on soybean yield in Chambal command, Rajasthan

Table 1.7 Yield losses due to water logging and soil salinity in RAJAD project, Rajasthan

Crop	Crop yield (t/ha)		Yield loss due to water logging/soil salinity (%)
	Saline / waterlogged	Non-saline / non waterlogged	
Wheat	17.0	34.0	50
Mustard	6.1	14.7	59
Paddy	17.0	29.0	41
Soybean	10.2	21.8	53
Berseem	360.0	520.0	31

DRAINAGE PROBLEMS IN INDIA

Considering India's wide range of climatic, physiographic and geo-hydrological conditions, difficulty in differentiating between rainfall and irrigation induced water logging and primary and secondary salinization (i.e. irrigation induced salinity), there are immense difficulties in pinpointing the kind and degree of drainage problems (Singh, P. http://citeseerx.ist.psu.edu/viewdoc/download). Nonetheless, Central Water Commission characterized drainage problems of India into five main geographical zones:

Northwest India

This zone includes the states of Punjab, Haryana, northwestern Rajasthan and western Uttar Pradesh having semiarid to arid climate. Excess runoff, excess irrigation water and/or salts cannot be efficiently evacuated due to the lack of natural outlets. It leads to the twin problems of water logging and salinity.

Central Peninsular India

This zone covers the states/parts of the states of Madhya Pradesh, Maharashtra, Karnataka and western Andhra Pradesh. The zone consists of hard rock area generally with shallow soil cover and undulating topography. Soils are heavy with low permeability. These soils are suitable for dry land farming. If irrigated without proper drainage, water logging and soil salinity problems are inevitable to develop. The problems could also be of local nature.

Eastern Plains and Deltas

This zone comprises the states/parts of states of Uttar Pradesh (eastern), Bihar, West Bengal, Odisha, Andhra Pradesh and Tamil Nadu. The problem of water logging arises mainly due to high rainfall during monsoon season. High water stage persist in the main drains due to locking of the outfall. In some areas introduction of irrigation has aggravated water logging. Soil salinity is not a major concern because of high rainfall.

Coastal Areas of Western India

In Gujarat, large areas along the coast are affected by severe salinity primarily of marine origin. Introduction of irrigation can further aggravate the problem. Problem gets accentuated because of deficient salt leaching due to low rainfall; poor natural soil permeability and rising water tables in irrigated lands.

Alkali Lands of Western Gangetic Plains

Alkali soils are predominantly found in the states of Haryana, Punjab, Uttar Pradesh and parts of Bihar. The alkali soils have extremely low permeability. Therefore, surface water logging is common in this region. At few places, where alkali groundwater containing carbonates/bicarbonates is used for irrigation, the chances of alkali development are quite favourable.

Rainfed Areas

About 80 M ha of the cropped land of India is likely to remain unirrigated and as a consequence suffer from vagaries of nature. On the contrary, climate and kinds of soils encountered in many of these regions cause excess water problem during monsoon season followed by excess salts during the irrigation and dry season (non-monsoon season). It is especially true of the coastal areas and areas irrigated by saline/alkali groundwater. While lands in the plains of Bihar and Uttar Pradesh are uncultivable during monsoons due to excess water, flat lands in south-western Haryana and south-western Punjab, despite a low annual rainfall, get waterlogged due to sudden rains. As such, these lands need rainwater harvesting technology to manage water and salt problems.

Familiarization with Drainage Problems in Punjab and Haryana

The problems of water logging and soil salinization in the two states are being reported since the middle of the 19th century. Both the states suffer from problems of surface water stagnation as well as shallow water table. The surface stagnation problem manifests from deficient natural drainage and having the largest percentage of irrigated areas. Surface congestion is a common scene in the states during the monsoon season because the existing drainage systems are incapable of draining surface runoff sufficiently fast, especially following heavy rain storms. Non-construction of the field drains by the stakeholders is reflected in large pockets of standing water in agricultural fields impacting production and productivity. Spread of irrigation led to a steady rise in water table in almost all irrigation commands. The rising water table is now limited to poor quality groundwater regions and most critical areas in the two states occur along the canal networks.

From drainage point of view, Haryana is drained by River Yamuna and River Ghaggar, the two drainage basins are referred to as the Yamuna basin and the Ghaggar basin respectively. The former has the most elaborate drainage network yet the system fails to drain the areas fast enough to save the crops from adverse impacts of water logging. Though less severe than in the Yamuna Zone, flooding in the Ghaggar basin is exacerbated by cross-Ghaggar inflows in the districts of Kaithal, Fatehabad and Sirsa. The remaining area called the internal drainage basin poses the most severe problems as the topography is saucer shaped. Even surface drainage water has to be removed using pumps. The groundwater conditions also pose problem resulting in widespread water logging and salinity in the central portion and irrigation commands. It is estimated that nearly 50,000 ha in Haryana is in need of subsurface drainage to control water logging and soil salinization.

Southwest Punjab is the most affected region of Punjab where water logging and soil salinity problems are acute. A report of the Planning Commission revealed that in the Sikhwala village in Muktsar District, no crop could be sown for the last 15 years due to water logging. Similarly, in Shajrana village of Fazilka District, more than 160 ha of agricultural land on the west of Bikaner Canal have been waterlogged for more than two decades. The Government of Punjab has indicated that about 2 lakh ha of fertile agricultural land is water logged impacting more than 200,000 farmers who have lost their primary income source from agriculture.

Concerted efforts are being made by various departments to solve the drainage problems and reclaim waterlogged salt affected lands. Lining of canal networks, construction of new surface collector drains, subsurface drainage, irrigation water management including switchover from surface to pressurized irrigation, application of bio-drainage technology, saline aquaculture in waterlogged areas are a few steps that are being adopted. It seems that all these steps are being taken in piecemeal fashion. For any fruitful long-term solution, there is a need to integrate all these solutions region-wise and as per the master drainage plans. In case modifications in drainage plans are needed they must be incorporated on priority in both the states.

PREVENTION, MANAGEMENT AND RECLAMATION

A distinction has to be made between prevention, management and reclamation although the distinction sometimes gets blurred as few interventions cut across the three categories. Nonetheless, techniques and strategies need to be developed and categorized accordingly.

Prevention, as the term suggests, is to prevent the problem to develop in the first place. Although, it might not be possible to prevent the problem to develop for all times to come, strategies can be put in place to delay the problem as much as possible. To prevent rise in water table, the basic principle of prevention is to minimize the accretions to the groundwater and maximize the withdrawals from the aquifer. Besides, proper identification of the source of 'excess water' can help to avoid the problem to develop (Bos and Boers, 1994). For example:

- If irrigation water is the prime cause of water logging, practices that improve on-farm water use efficiency might work to prevent the problem or in some cases even solve the problem.
- If runoff from the adjoining areas is the cause of water logging, a simple interception drain might prevent the problem.
- If the seepage from the canal network or groundwater inflow from adjoining areas is the cause of water logging, groundwater can be intercepted by a series of tube wells or bio-interceptors to prevent the problem from developing
- If an area is partially inundated due to insufficient discharge capacity of a natural or manmade stream or outlet, renovation of the stream/outlet can prevent or solve the drainage problem.

Management of the problem is a way to manage the affected land without major interventions. It includes application of cheap cost-effective non-structural, agronomical or other interventions. The basic principle is to minimize the adverse effect of the problems on crops and or on soil/water resources. Improved tillage practices, crop selection largely on the basis of their tolerance to water logging and salinity or their water requirement, land forming, skipping/delaying irrigation, switch over to improved irrigation techniques, application of additional doses of plant nutrients and mulching are some of the management practices that can be adopted under different situations. Management strategies help to buy time before major interventions are implemented or funds for these interventions are arranged.

Interventions to bring back the soils to their normal or near normal level of productivity and quality fall in the domain of reclamation. Drainage improvement, chemical management and other strategies are used to reclaim the lands. Normally, no one kind of strategy will be effective and one should try to exploit the synergies of prevention, management and reclamation to delay or minimize investments on reclamation.

ARTIFICIAL DRAINAGE

In general parlance, the terms 'drainage', 'land drainage', 'agricultural drainage' and 'field drainage' are used as synonyms although agricultural drainage is more common to differentiate it from urban drainage. Agricultural drainage is defined *'as the removal and disposal of excess water from agricultural lands'*. Excess water may result from one or a combination of more than one water sources such as precipitation, snow melt, irrigation excess, seepage from water sources, seepage from surrounding areas including overland flow, artesian flow from aquifers, spillage from rivers, natural channels or drains and tidal inundation in coastal areas. It also includes the excess water that may be purposely applied to agricultural land such as water for leaching salt affected soils or for temperature control. Need for artificial drainage therefore, is obvious to supplement the natural drainage. The adequacy of drainage is judged by the fact that excess water is removed efficiently within a desirable period without any damage to the crops or the land resource.

Since agricultural drainage besides removal of water is also necessary to remove salts from the soil, definition of drainage has been refined by the International Commission on Irrigation and Drainage (ICID, 1979). According to ICID (1979), land drainage is defined as follows:

'Land drainage is the removal of excess surface and subsurface water from the land to enhance crop growth, including the removal of soluble salts from the soil'.

The above definition is most commonly applied in practice.

Benefits of Land Drainage

The positive impacts of land drainage are reflected in the improved physico-chemical environment of the root zone. Specifically, the following positive changes are observed.

- Deeper water table results in improved soil aeration resulting in an optimum mix of air and water
- Specific heat of water being higher, dry soils are warmer with relatively higher temperature in the root zone
- Soil structure is maintained as tillage operations are carried out under optimum moisture conditions
- Length of the growing season is more because soils dry earlier making early cultivation possible
- Quality of crops is better with timely harvest of crops. Timely harvest also allows ample time for field preparation for the next crop
- Lesser run-off because of greater storage capacity of the soil to absorb water
- No or reduced production of toxic substances which otherwise may retard plant growth
- Helps in increasing cropping intensity
- Improved biological quality of soils with profuse development of micro-organisms
- Facilitates leaching of salts resulting in reduced electrical conductivity of the soil profile
- Enhanced availability of plant nutrients

TYPES OF DRAINAGE SYSTEMS

Most agricultural lands have some kind of natural surface/subsurface drainage. When this natural drainage is inadequate or is rendered as such by man's interference, some kind of artificial drainage is needed to bridge the gap between the existing natural drainage and the required degree of drainage. Agricultural drainage is broadly categorized into five categories, namely, surface drainage, subsurface drainage, vertical drainage, interceptor drainage and bio-drainage (Fig. 1.10). Some people categorize subsurface drainage as horizontal (tile or pipe) drainage because the tiles/pipes are laid parallel to the ground surface as against the vertical drainage where pipes or drain material is placed perpendicular to the ground surface. Subsurface pipe drainage is also categorized as relief drainage to differentiate it from interceptor drainage. Vertical drainage is also categorized as a sub-group of subsurface drainage. In that case, even bio-drainage can be grouped as a sub-group of subsurface drainage. On the contrary, many workers classify vertical drainage as a class in itself because the nature of investigations and design techniques are entirely different than horizontal drainage. Another kind of drainage is interceptor drainage, which many times is included as a part of subsurface drainage but many times is designated as a separate kind of drainage system. Recently, bio-interceptor drainage has been introduced although it can be categorized as a sub-group of bio-drainage.

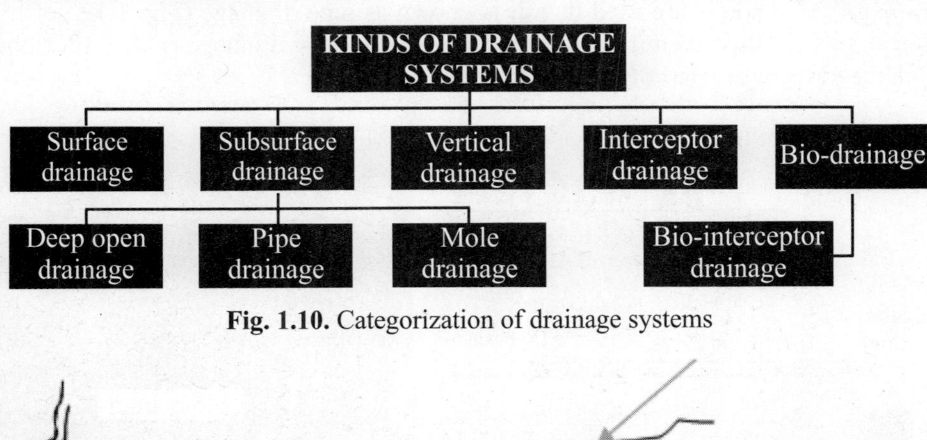

Fig. 1.10. Categorization of drainage systems

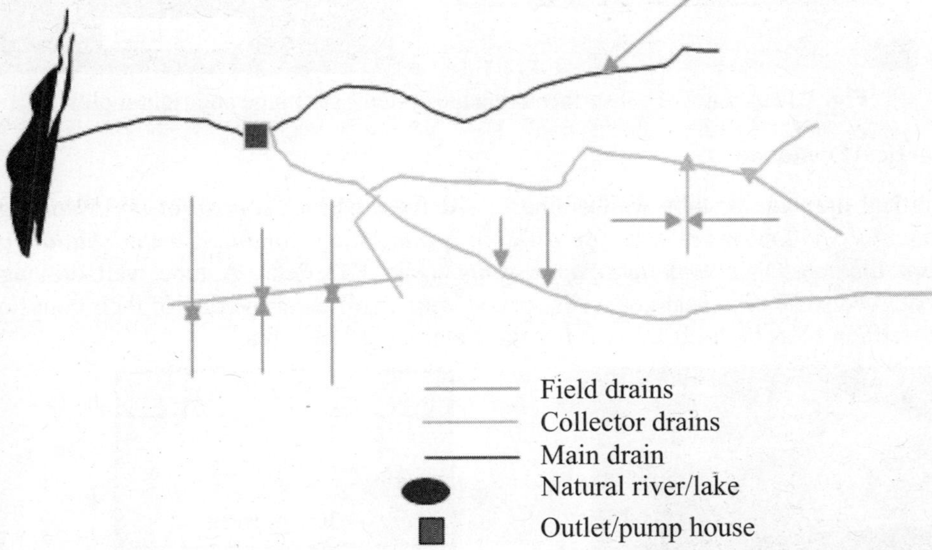

Fig. 1.11 Schematic view of a surface drainage system

Surface Drainage

Surface drainage problems occur in flat or nearly flat areas, in areas having uneven land surfaces with depressions or ridges preventing natural runoff and in the areas where there is no outlet. It is defined as the *'process of removal of water from the land surface in time to prevent damage to crops and to remove ponding water on the soil surface using surface drains without causing soil erosion'* (ASAE, 1979). The amount of water to be removed is expressed in units of mm/d. The system designed to improve surface drainage of an area is known as surface drainage system. A surface drainage system consisting of field drains, collector and main drains is shown in Fig. 1.11.

Subsurface Drainage

'Removal of excess groundwater to control the groundwater table or to remove excess salts in time to prevent damage to crops or to prevent build-up of salts in the root zone using water as the medium of salt transport is known as subsurface drainage'. If conduits are used then it is known as pipe drainage (Fig. 1.12, left). If no conduits are used then it is known as pipeless or mole drainage (Fig. 1.12, right). Both the systems are part of subsurface drainage system.

Fig. 1.12 A view of subsurface drainage systems (left pipe and right mole)

Vertical Drainage

Vertical drainage or tube well drainage is defined as the *'control of an existing or potential shallow water table or artesian groundwater condition using shallow or deep tube wells or sometimes open wells'* (Fig. 1.13, left). A tube well drainage system consists of a group of wells spaced with a sufficient overlap of their cones of depression so as to uniformly control the water table in an area.

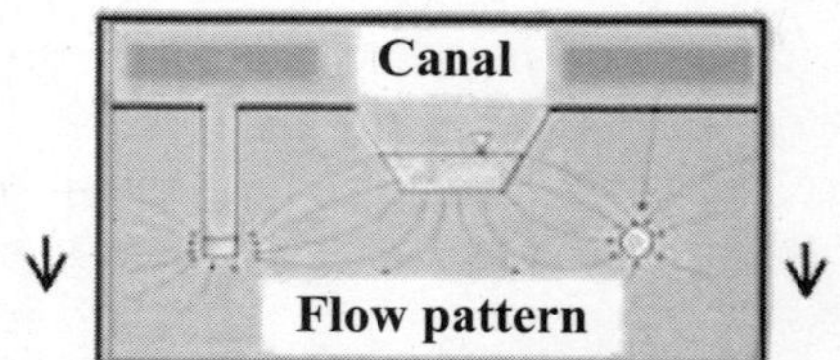

Fig. 1.13 Vertical drainage (left) and interceptor drainage for intercepting canal seepage (right)

Interceptor Drainage

The interceptor drainage intercepts the subsurface flow before it re-surfaces on the earth surface. Normally it is used on long slopes and on shallow permeable surface soils overlying relatively impermeable soils. It is also used to intercept the seepage from water distribution systems to prevent development of water logging in the areas adjoining these systems (Fig. 1.13, right).

Bio-drainage

'The process of removal of excess groundwater using plants (a biological medium) is known as bio-drainage' (Fig. 1.14, left). Bio-interceptor drainage system is another version of this kind of drainage used to intercept canal seepage (Fig. 1.14, right).

Fig. 1.14 Bio (left) and bio-interceptor drainage (right)

REALIZATION OF DRAINAGE NEEDS

The need of drainage of agricultural lands has been realized since time immemorial. In *Narada Smriti* (3150 B.C.), the importance of irrigation and drainage has been vividly mentioned. It says *'No grain is ever produced without water, but too much water tends to spoil the grain'*. Inundation is as injurious to growth as dearth of water. It is known that the Egyptians and the Babylonians practiced drainage in pre-historic times. One of the early recorded evidence of drainage is by Roman Cato' in 2 B.C. References to irrigation wells and tanks in India appear in many religious texts. The tradition of worshipping tanks, which must have a great bearing on man's life and activity then, continues even today. It is reasonably expected that these tanks must have been used for the dual purpose of irrigation and drainage to save human and animal lives and improve agricultural productivity. The need of drainage in the present context in India was realized as early as 1865 when the Punjab Government reported development of soil salinity to the then Governor General. It was emphasized in the reports of Drainage Board that irrigation and drainage should go together. The Royal Commission on Agriculture (1928) mentioned, 'Now lessons have been learnt and in all future projects, drainage will form an essential component of the irrigation projects'. It however, remained on paper and irrigation projects were commissioned without proper provision of drainage even after independence. The expediency with which irrigation potential was to be created diverted the whole attention of the planners to provide the basic input for production i.e. water.

Although drainage component was not lost sight of, adequate provisions were not made in most irrigation projects probably because of the following reasons:

- Drainage component was either neglected or as a token, inadequate provisions were made to minimize the cost of the irrigation projects so as to keep the cost-benefit ratio of the irrigation project at a respectable level.
- Funds at the disposal of planners were not adequate to include both irrigation and drainage.
- The benefits of drainage were not as eloquently documented as that of irrigation. Although irrigation and drainage go together was known, its appreciation was lacking both amongst the planners and the farmers.
- Besides these points, it was the opinion of many experts that drainage may not be immediately needed. A promise of providing drainage was there but it was not implemented till the adverse effects appeared to an extent that a hue and cry was raised by the farmers.

That irrigation and drainage go together was religiously emphasized by the Second Irrigation Commission (1972) and the National Commission on Agriculture (1976). In the same year, the first Irrigation Minister's conference was held and it was made abundantly clear that henceforth, no new irrigation project will be sanctioned which does not have components of drainage and command area development. The Standing Committee of Agriculture (*Lok Sabha* Secretariat, 1996) of the 11th *Lok Sabha* of India, on the undesirable neglect of the agricultural drainage in the irrigated areas of India may also be mentioned. It appears that lately the importance of drainage has been recognized and now drainage is getting high priority in the modernization programmes of the existing irrigation projects (Gupta, 2005, Gupta, 2014). These days, besides the 'need to remove the excess water' drainage is strongly influenced by the concern for the environment. The design and operation of all drainage systems therefore, not only ensure sustainable agriculture in the drained area but must minimize the pollution of surface and groundwater resource from irrigation return flow or drainage effluent (Bos and Boers, 1994). Chakraborty *et al.* (2012) have assessed that about Rs. 4000 billion have to be spent on subsurface drainage by 2050 to sustain irrigated agriculture.

GLOBAL AND INDIAN SCENE

There is no substitute for irrigated agriculture to ensure food security of the increasing billions on the earth. Since sustainability of irrigated hinges upon drainage; drainage has a major role to play here. Even rainfed agriculture in temperate regions and humid tropics needs drainage to control shallow groundwater and manage runoff resulting from heavy rainfall. Out of total world crop land of 1,500 M ha, 1,110 M ha is rainfed while only 390 M ha is irrigated. Most of the rainfed cropped land (90%) is undrained. It is estimated that nearly 250-300 M ha is in immediate need of drainage improvement. Out of an irrigated land of 390 M ha, 190 M ha is provided with some kind of drainage infrastructure, out of which only 60 M ha is known to have adequate drainage. Area covered under drainage in few selected countries is given in Table 1.8.

Table 1.8 Countries with largest drainage area

Country	Arable land (M ha)	Irrigated area (M ha)	Drained area (M ha)	Proportion of drained area (%)
USA	179	22	48	27
China	136	50	20	15
Canada	46	0.7	9	20
Russia	127	5	7	6
Pakistan	22	18	6	27
India	170	55	6	4 (11)*
Mexico	27	6.5	5.2	19
Germany	12	0.4	4.9	41
Great Britain	6	0.1	4.6	77
Poland	14	0.1	4.2	30
Egypt	3.3	3.3	3	91

* Percent of irrigated area; nearly all is surface drainage

It may be seen that in India only 4% of the total cultivated or 11% of the irrigated land has some kind of drainage. This includes mostly surface drainage systems as a flood control measure since field drains even in these areas are lacking. As far as subsurface drainage is concerned, the picture is quite dismal although the first subsurface drainage system was constructed in 1874 in Tamil Nadu in a tank irrigated command. A subsurface drainage system constructed in Maharashtra at Manjri in 1920 is still functioning. The real impetus to this kind of drainage happened during early sixties but could not take up the desired speed. Currently only about 60,000 ha area is covered under subsurface drainage. Vertical drainage has played a key role in lowering the shallow water table in areas underlain with good quality groundwater. To ensure food and nutritional security, probably much more needs to be done through research, extension and policy initiatives so that appropriate drainage measures are adopted by the farmers.

QUESTIONS (THEORY)

1. Define water logging. Write a short note on water logging problem.
2. List the causes of water logging through surface stagnation of water.
3. List the causes that cause rise in water table in irrigation commands.
4. Briefly describe the adverse impacts of water logging on crops.
5. What do you understand by soil aeration? How is it assessed?
6. List the various processes that cause build-up of soil salinity.
7. Describe the drainage problems in various agro-climatic regions of India.
8. Differentiate between prevention, management and reclamation of problem soils.

9. What do you understand by artificial drainage? Name and briefly describe the various types of drainage systems.
10. List the beneficial effects of land drainage.

QUESTIONS (OBJECTIVE)

1. Fill in the blanks

1. Excess water in the root zone due to shallow water table restricts the normal _______ resulting in excessive accumulation of carbon dioxide.
2. Alkali lands are prone to surface stagnation because _________ of these soils is quite less due to their high ESP.
3. Leakage and seepage from the irrigation networks and poor on-farm water management are the major causes of rise in water table in ______________ commands.
4. _________ water table develops on a nearly impermeable layer existing at shallow depth between the soil surface and the real water table.
5. A waterlogged soil is relatively _________ than a non-waterlogged soil because of high specific heat of water.
6. Soil ___________ potential is denoted as Eh.
7. ODR stands for __________diffusion rate.
8.. Salt laden sand blown by sea winds is one of the causes of soil salinization in _______ regions.
9. Land drainage is the removal of excess surface and subsurface water from the land including _________ of soluble salts from the root zone.
10. Interceptor drains are used on sloping lands as well as to intercept ________ from the canal network.

2. Select the correct or most appropriate answer (A, B, C or D) from the following

1. An area having water table at a depth of 2.5 m is characterized as
   (A) Waterlogged (B) Potentially waterlogged
   (C) Safe (D) None of these
2. Soil aeration problem can be assessed be measuring
   (A) Oxygen diffusion rate (B) Redox potential
   (C) Both A or B (D) None of these
3. For roots to conduct their function efficiently, redox potential should be
   (A) Greater than +300 mV (B) +100 to +200 mV
   (C) In the negative range (D) None of these

4. For adequate aeration, the oxygen diffusion rate should be in the range of

   (A) 1-5*10^{-8} g/cm^2/min

   (B) 20-35 *10^{-8} g/cm^2/min

   (C) Less than 10*10^{-8} g/cm^2/min

   (D) None of these

5. Which drainage method is also known as tube well drainage

   (A) Pipe drainage (B) Vertical drainage

   (C) Surface drainage (D) None of these

6. In which method plants are used as drainage material

   (A) Surface drainage (B) Vertical drainage

   (C) Bio-drainage (D) None of these

7. Waterlogged soils have

   (A) Lower temperature than normal soils

   (B) Higher temperature than normal soils

   (C) No difference in temperature than normal soils

   (D) None of these

8. Heat capacity of waterlogged soils is

   (A) Higher than dry soils (B) Lower than dry soils

   (C) Equal to dry soils (D) None of these

9. Under similar agro-climatic conditions, surface stagnation problem is relatively more in

   (A) Saline soils (B) Normal soil

   (C) Alkali soils (D) None of these

10. Improved agronomic practices are generally a part of

    (A) Preventive strategies (B) Management strategies

    (C) Strategies for reclamation (D) None of these

11. The temporary water table developed on an impermeable layer between the soil surface and natural groundwater table of the area is known as

    (A) Static water table (B) Perched water table

    (C) Piezometric head (D) None of these

■■■■

Chapter 2

SURFACE DRAINAGE METHODS

Surface drainage is the oldest drainage practice in the world, being as old as the history of agriculture. It is widely used to remove excess water from the soil surface in rainfed and irrigate lands. Wide variations in climate, topography, soils, crops and farming practices in India impact the need and designs of surface drainage systems. Although a must for sub-humid and humid regions, because of the monsoon climatic conditions, surface drainage is even required in arid and semiarid regions. It is because around 75% of annual average rainfall is received in about 4 months and that too in few heavy storms. It is applied both as a stand alone method as well as in integration with some kind of subsurface drainage. Looking at the importance of this method of land drainage, readers will be acquainted with surface drainage requirement of crops, basic principles of designs, construction and maintenance of open drains in this chapter.

DEFINITION

Surface drainage is defined *'as the removal of excess water stagnating over the ground surface through natural or constructed channels with adequate capacity outlets'*. ICID (1982) defined surface drainage as *'the diversion or orderly removal of excess water from the surface of land by means of improved natural or constructed channels, supplemented when necessary by shaping and grading of the land surface to such channels'*. Flat or nearly flat areas, heavy soils with low permeability and lands in humid or tropical and sub-tropical regions where high intensity storms usually occur, are subject to surface inundation and therefore, require surface drainage. Surface drainage problem may or may not be of local origin. In low lands, it might occur due to the ingress of surface runoff from uplands. Overflow from rivers or natural channels sometimes contributes largely to the drainage problem in an area. The following factors contribute to the problem:

- Uneven land surface causing hindrance in the natural flow of runoff from the area
- Inadequate capacity of the drainage channels being unable to cope with drainage load especially during critical periods
- Inadequate outlet conditions
- Back water flow

Surface drainage is carried out to meet one or more of the following objectives:

- In flat areas for timely removal of water so as to allow optimum growth of plants

- In sloping areas for orderly removal of water such that it does not cause erosion and
- For the collection and disposal of surface irrigation runoff.

The design of a surface drainage system as per definition of ICID (1982) has two components: (a) the shaping of the surface by land forming, which comprises changing the micro-topography of the land to meet the requirements of surface drainage or irrigation; and (b) the construction of open drains (field drains and laterals) to carry the excess water to the main outlet.

Surface drainage has often been considered as a panacea to all drainage problems. However, it is now known that it cannot help to lower the water table effectively although rate of build-up of the water table can be reduced through efficient surface drainage. In the long-term, surface drainage reduces installation and operational costs of subsurface drainage systems.

SURFACE DRAINAGE REQUIREMENT AND CROPS

The adverse effects of temporary water stagnation on the land surface depends on plant species, the variety of the crop, physiological stage of growth at which water stagnation occurs (crop may be more sensitive at one stage compared to another), the time and duration of water logging, the light intensity, temperature and the fertility of the soil. Injury to crops is particularly severe when the crops are flooded on hot days. The damaging effects are generally more when water logging conditions are accompanied by salinity/sodicity (Gupta, 2014). It may be noted that no single factor affects the crop yield and generally, it is a combination of poor aeration, reduced uptake of plant nutrients, imbalance between the uptake of various ions and toxicity effects of various elements/ions.

Not all crops suffer equally from inundation or root submergence. The damage varies from one crop to another as described in Table 1.6 in the previous chapter. In a semi-reclaimed alkali land, one day water stagnation reduced the yield of *berseem* (seed) by only 2% whereas wheat and mustard yields reduced by 8% (Gupta *et al.*, 2004). The yield loss increased and varied from 20% in sorghum to 48% in *berseem* (seed) with water stagnation of six days. For the same duration, yield loss in wheat increased to 39%. Grasses and forages are usually more tolerant to inundation as compared to cereals or other crops. Only limited data for few crops are available to draw elaborate conclusions. But even with limited data, few general observations can be made:

- Crops, in general, are more sensitive to water stagnation at early than later growth stages
- The yield reduction increases with increasing degree (depth of submergence) and duration of submergence.

General guidelines for the periods within which stagnated water should be removed from crop lands are given in Table 2.1.

Table 2.1 Permissible water stagnation period for different crops

Crop	Permissible water stagnation period	Remarks (Pai and Hukeri, 1979)
Vegetables	1	24 h rainfall to be drained in 24 h.
Oil seeds	2	2 days rainfall to be drained in 2 days.
Cotton, maize	3	3 days rainfall to be drained in 3 days.
Rice, sugarcane	7	In the case of rice, panicles should not be allowed to get submerged and depth of ponding for temporary periods should not be more than 0.5 m.

FACTORS AFFECTING DESIGN OF SURFACE DRAINAGE

Surface drainage congestion in monsoon climatic conditions of India is governed by four factors:

1. Climate

Climate, especially the rainfall characteristics of the region, is likely to play the most vital role in determining the extent and degree of surface stagnation. Frequency analysis procedures are employed to determine rainfall events of various return periods, number of storms and the lengths of dry period to plan an effective drainage and reuse strategy. Rainfall probability of 2-10% is used to design drainage ditches and drainage networks.

2. Soil and Topography

Kind of the soils and topography determines the degree of surface stagnation. For similar events, degree is more in a heavy than a light textured soil. Crops are more prone to water congestion in a flat terrain than a terrain with rolling topography.

3. Depth to Water Table

Absorption of water into the soil profile besides other factors is governed by the depth to water table. The degree (depth of submergence) and duration of stagnation is more in a shallow water table region than in a deep water table region.

4. Crop Characteristics

Adverse impacts of surface water stagnation and the need for agricultural land drainage to a great extent such influenced by the kind of crops grown and their tolerance to surface stagnation. A rice crop can withstand surface stagnation much more than maize crop, later being sensitive to surface stagnation. Data in respect of wheat, pigeon peas, cowpeas, pearl millet, maize and groundnut indicate that the yield reduction varies from 5.3 to 23.2% for each day of submergence. It appears that submergence exceeding 1–2 days will result in more than 10% reduction in yield of dry food crops. Even the stage of growth at which stagnation occurs may influence the adverse impacts. As an example, for the maize crop, seedling stage is the most sensitive followed by the silking stage.

DESIGN OF SURFACE DRAINAGE SYSTEMS

Investigations

Various investigations required to plan and design a surface drainage system consists of the following components:

- Topographic surveys
- Geological and soil surveys
- Land use and cropping pattern
- Water sources survey
- Profiles and cross sections of existing streams and ditches
- Stage-frequency analysis of large flows at identified outlets

It may be noted that investigations for surface and subsurface drainage (if required) should be taken together as many of the components and procedures are similar. Since drainage investigations for subsurface drainage are more elaborate, detailed discussions on some of the components of surface drainage investigations are included in the next chapter while some of the components form part of this chapter.

Visual Observations for Problem Identification

A field reconnaissance survey would reveal whether the problem exists. Some simple observations, discussions with local officers of the agricultural and irrigation departments and discussions with farmers will be enough to have a first assessment of the degree and extent of water logging. Additional information obtained through some field measurements will help to firm up the conclusions drawn through these discussions.

- Observe water marks on building walls, wells and other spots is an indicator of the water stagnation depth that can be expected in the area.
- Standing water or wet spots in fields or parts of the fields especially following few days of dry period following a rain event is an indicator of the problem. If crop stand at these spots is poor, then one can be quite sure that the area is prone to water logging. Yellowing of the crop is noticed in areas affected by surface stagnation or shallow water table.
- Lack or absence of surface drains or their poor condition (full of weeds or siltation) is also indicative of water logging.
- Observe wells, gravel pits and deep channels, which show the depth to groundwater. In India water table in open wells is commonly measured and reported. If there are no wells, then one could dig shallow pits or auger shallow observation wells reaching to depths of 1-2 m below the soil surface. If soil horizons are reached which are gluey, wet and may contain black or red mottles, one can assume that soils are poorly drained at that level.

- Vegetative cover is a good index of depth to water table in many areas. Certain grasses/crops and trees such as willows, cottonwood and poplar often thrive in shallow water table areas. Reed grass and sedges are also common. It is a useful practice to look for these features while surveying.
- Salts present in the form of a white crust on the soil surface reveal a problem of subsurface water logging. Since the problem could occur even when the crust is not present, its absence does not mean that there is no problem of water logging.
- During the survey also observe the outfall condition. An improper outlet, its location with respect to the water level in the drain will reveal about the problem.

DRAINAGE COEFFICIENT

The field surface drains are designed on the basis of the crop drainage requirement often given by the drainage coefficient. Crop is the kingpin in the design of this component of the drainage system. The collectors and the main drains of the system are designed on the basis of peak rate of runoff or some multiples of this value. It depends upon the general land use of the area.

Drainage coefficient (DC), the drainage rate for the design of drainage system, is expressed '*as rate of removal of water, in terms of depth or volume of water, from soil surface in a specified period, usually 24 h, so as to provide desired degree of protection to the crops from excess surface water*'. Units of surface drainage are: depth of water to be removed in 24 h, mm/d or flow rate per unit area, cumecs/ha, L/s-ha. For subsurface drainage it is expressed as the rate of removal of water, in terms of depth or volume of water, from the soil profile. For subsurface drainage units are in terms of depth of water to be removed in 24 hours, mm/d.

Since the surface drainage coefficient is the most important factor in the design of the surface drainage system, several methods are used to evaluate the drainage coefficient. Which method to use in a given situation is dictated by the kind of data available with the investigator?

- Statistical analysis of observed discharges
- Calibrated rainfall - discharge relations
- Uncalibrated rainfall - discharge relations
- Empirical methods

Amongst the four listed methods, calibrated rainfall - discharge relations are too complex for use in drainage design and related activities. It calls for the use of Unit hydrograph method (for shallow drainage discharge) and Reaction factor method (for groundwater drainage discharge). Thus, only two methods are discussed in this section while empirical methods are discussed later. For details of some of these methods, readers may refer any textbook on Hydrology and/or Land Drainage (National Institute of Hydrology, 1996; Kumar and Singh, 2005, Ritzema, 1994.

Statistical Analysis of Observed Discharge of Streams

Discharge (Q) is directly measured or indirectly computed using stage level (H) from rating curves showing relationship between Q and H (Fig. 2.1). These relations are developed using Chezy or Manning equations. Long-term series data (>15 years) are needed to generate reliable information on drainage coefficients using this method. Frequency analysis of drainage discharge is performed to determine discharge rates for the selected return periods. Depending upon the type of structure, a design value is chosen. For highway cross road drainage works, 10-50 years return period values are considered. For general farm crops, storm of 48 hours duration having a frequency of occurrence of 2 to 5 years, should suffice. For high value crops sensitive to excess water, protection from the 10-year frequency storm may be desirable.

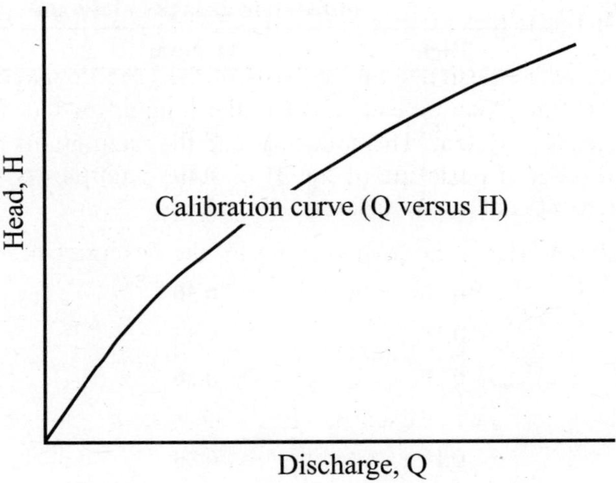

Fig. 2.1 Rating curve for head vs. discharge

Uncalibrated Rainfall - Runoff Relations

Rational formula is commonly used to assess the peak runoff. It is given as:

$$Q_p = CIA/360 \tag{2.1}$$

where, Q_p is the peak discharge, cumecs; I is the uniform intensity of the design storm of duration equal to time of concentration, Tc, mm/h, C is the discharge coefficient and A is the basin area in ha.

Time of concentration, Tc is determined by using the following relation:

$$Tc = L^{1.15}/3080\ H^{0.38} \tag{2.2}$$

Here, Tc is the time of concentration, h; L is the maximum traveling distance in the basin, m and H is the difference in elevation over distance L, m

Discharge coefficient (C) indicates the proportion of the design rainfall that actually discharges rapidly from the basin and contributes to the peak discharge. Values depend on soil infiltration-percolation characteristics of the basin governed by slope and land use (Table 2.2). If the watershed is large and land use varies within the area, weighted average value of C is used in the calculations.

Curve Number Method

For drainage basins where runoff has not been measured previously, the curve number (CN) method is used to estimate the depth of direct runoff or peak runoff from the rainfall depth. The empirical relationship is given as:

$$F/S = Q/(P\text{-}Ia) \tag{2.3}$$

Here, F is the actual retention, mm; S is the potential maximum retention, mm; Q is the accumulated runoff depth, mm; P is the accumulated rainfall depth, mm and Ia is the initial abstraction, mm.

Table 2.2 Guidelines to determine C of the rational formula Eq. (2.1)

Type of land and slope (%)	Infiltration capacity of the soil		
	High	**Medium**	**Low**
Arable land			
5	0.30	0.50	0.60
5-10	0.40	0.60	0.70
10-30	0.50	0.70	0.80
Pasture			
5	0.10	0.30	0.40
5-10	0.15	0.35	0.55
10-30	0.20	0.40	0.60
Forest			
5	0.10	0.30	0.40
5-10	0.25	0.35	0.50
10-30	0.30	0.50	0.60

Source: USDA-SCS (1985b)

After runoff has started, the actual retention F will be equal to (P-Ia-Q), therefore,

$$Q = (P\text{-}Ia)^2/(P\text{-}Ia+S) \tag{2.4}$$

If initial abstraction Ia is 20% of potential maximum retention, then Ia = 0.2S and

$$Q = (P-0.2S)^2/(P+0.8S), \qquad P > 0.2S \tag{2.5}$$

For Indian conditions except for the black soil regions, a change has been suggested in Eq. 2.5 as follows:

$$Q = (P-0.3S)^2/(P+0.7S), \qquad P > 0.3S \tag{2.6}$$

S has been expressed in the form of curve numbers, CN, a dimensionless parameter indicating the runoff response characteristics of a drainage basin. It is expressed as:

$$CN = 25400/\ (254+S) \tag{2.7}$$

$$S = 25400/CN + 254 \tag{2.8}$$

Incorporate S in Eq. (2.6) to calculate the runoff depth.

CN is related to the following features of the drainage basin:

- Land use or cover
- Land treatment or practice in relation to hydrological conditions
- Hydrological conditions
- Hydrological soil group, and
- Antecedent soil moisture condition

Land use represents the surface condition in a drainage basin and is related to the degree of cover. The categories distinguished are: fallow, row crops planted in rows far apart, small grains planted in rows close enough, close seeded legumes, pasture range used for gazing, grassland protected from grazing and woodland.

Land treatment applies mainly to agriculture land use and includes contouring, terracing, management practices such as rotation of crops, gazing control or burning.

Hydrological soil groups

Soil properties greatly influence the amount of runoff. The following hydrological soil groups are defined:

Group **A**: high infiltration rate even when wetted deep well to excessive drained sands and gravels.

Group **B:** moderate infiltration - moderately fine to moderately coarse texture

Group **C:** low infiltration - soils with a layer that impedes downward movement or soils of moderately fine to fine texture

Group **D:** very low infiltration - clay soils with high swelling potential, shallow water table, soils with clay pan at or near surface or shallow soils over nearly impervious material.

Hydrological soil cover complex

To determine the appropriate CN value, tables relating the value of CN to land use, to treatment or practice, to hydrological conditions and the hydrological soil groups are used. Together, these four categories are called Hydrological Soil-Cover Complexes. The values of CN are related to various soil cover complexes in a single table for different AMC groups (Table 2.3). This table is used to estimate the CN for a particular drainage basin. The design runoff depth can then be found using the appropriate equations given earlier.

Antecedent moisture conditions (AMC)

AMC refers to the soil moisture condition in the basin before runoff occurs. For CN method, the AMC classes are categorized as:

AMC I: Soils in the drainage basin are practically dry (Moisture content at or about wilting point)

AMC II: Average conditions

AMC III: Practically saturated from antecedent rainfall.

Table 2.3. Estimation of runoff curve numbers for AMC II (from USDA-SCS, 1985a)

Land use or cover	Treatment or practice	Hydrologic condition	Hydrologic soil group			
			A	B	C	D
Fallow	Straight row	-	77	86	91	94
Row crops	Straight row	Poor	72	81	88	91
	Straight row	Good	67	78	85	89
	Contoured	Poor	70	79	84	88
	Contoured	Good	65	75	82	86
	Terraced	Poor	66	74	80	82
	Terraced	Good	62	71	78	81
Small grain	Straight row	Poor	65	76	84	88
	Straight row	Good	63	75	83	87
	Contoured	Poor	63	74	82	85
	Contoured	Good	61	73	81	84
	Terraced	Poor	61	72	79	82
	Terraced	Good	59	70	78	81
Close seeded legumes or rotation meadow	Straight row	Poor	66	77	85	89
	Straight row	Good	58	72	81	85
	Contoured	Poor	64	75	83	85
	Contoured	Good	55	69	78	83
	Terraced	Poor	63	73	80	83
	Terraced	Good	51	67	76	80
Pasture or range	Contoured	Poor	68	79	86	89
	Contoured	Fair	49	69	79	84
	Contoured	Good	39	61	74	80
		Poor	47	67	81	88
		Fair	25	59	75	83
		Good	6	35	70	79
Meadow (permanent)		Good	30	58	71	78
Woods (farm wood-lots)		Poor	45	66	77	83
		Fair	36	60	73	79
		Good	25	55	70	77
Farmsteads		-	59	74	82	86
Roads		-	74	84	90	92

These classes are based on the 5-day antecedent rainfall (accumulated total rainfall preceding the runoff under consideration). For modeling purposes, one considers watersheds to be AMC II, which is essentially an average moisture condition. For other AMCs, the values determined from Table 2.3 are adjusted by multiplying with the relevant factor given in Table 2.4.

Table 2.4 Multipliers to convert CN of AMC (II) to CN of AMC (I) and AMC (III)

CN of AMC (II)	Multiplier for AMC (I)	Multiplier for AMC (III)
10	0.40	2.22
20	0.45	1.85
30	0.50	1.67
40	0.55	1.50
50	0.62	1.40
60	0.67	1.30
70	0.73	1.21
80	0.79	1.14
90	0.87	1.07
100	1.0	1.0

Example 2.1

How much runoff volume is expected from a rainfall of 10 cm falling on a hectare of land surface if CN is 60? Also calculate the fraction of rainfall that becomes runoff.

Compute the surface storage using Eq. (2.8)

$$25400/60 - 254 = 423.3 - 254 = 169.3$$

Using Eq. (2.5), we get

$$Q = (P-0.2S)^2/(P+0.8S), \qquad P > 0.2S$$

$$Q = (100- 169.3*0.2)^2/(100 +169.3*0.8) = (100 - 33.86)^2/(100+ 135.44)$$

$$Q = (66.14)^2/235.44= 18.58 \text{ mm}$$

Volume of runoff = 10000*18.58/1000 = 185.8 m^3

Fraction of rainfall that becomes runoff = 100 *18.58/100 = 18.58%

Peak runoff from CN method

The peak discharge equation has been derived from a triangular approximation of the hydrograph resulting from a rainfall excess of duration D. The time to peak flow is:

$$T_p = D/2 + T_L = D/2 + 0.6\, T_c \qquad (2.9)$$

Where T_p is the time to peak, hr, D is the duration of excess rainfall, hr, T_L is the time of lag, hr, and T_c is the time of concentration, hr. Time of concentration equal to $T_L/0.6$ is the longest travel time and may be obtained from the equation:

$$T_c = L^{0.8}[(1000/CN) - 9]^{0.7}/[4407\ (S_g)^{0.5}] \qquad (2.10)$$

Here L, is the longest flow length in m, and S_g is the average watershed gradient in m/m.

The peak flow rate is calculated from the equation:

$$q = q_u\, A\, Q \qquad (2.11)$$

Here q is the peak runoff rate (m^3/s), q_u is the unit peak flow rate (m^3/s per ha/mm of runoff), A is the watershed area, ha and Q is the runoff depth, mm as calculated from the CN method.

Unit flow rate is obtained from graphs using the time of concentration and the ratio of initial abstraction to 24-hour rainfall. However, it can be taken as $0.0208/T_p$ such that T_p is in hour (Das, 2012).

Simple Methods to Calculate DC (For field drains)

It is often difficult to determine drainage coefficient with any degree of accuracy although it is the key factor in establishing the needed capacity of the drainage system. Three simple methods to assess surface drainage coefficient for farm drainage are discussed in the following sections.

1. Rainfall-runoff coefficients

If no field data is available, the runoff percentage for different types of soils and for varying intensity of vegetation is adopted to assess DC (Table 2.5). The period of disposal is assessed on the type of crops and their growth stages as summarized in Table 2.1 (Murty and Jha, 2009, Murty and Jha, 2011).

Table 2.5 Recommended runoff percentage for different soils and vegetation intensities

Sl. No.	Type of Soil	Runoff (%)
1	Sandy soils	20
2	Loamy soils (largely cultivated)	30
3	Loamy soils (lightly covered by crops)	40
4	Gardens and cultivated areas	5 to 25

2. Depth-duration-frequency data

Daily rainfall storms series are used to assess the drainage coefficients. The procedure essentially comprises the following steps:

- Using the frequency analysis procedures, evaluate maximum 24 h rainfall depth which may occur with a probability of 10-20%. If storm durations are larger than one day, the analysis may be extended to 2-days or more say *n*-days maximum (R_n).
- Evaluate the basic infiltration rate of the soil and estimate the potential evapotranspiration.

- Estimate the time within which water stagnation is to be removed. It depends upon the tolerance of crop to surface inundation. Let us say it is 'n' days. Multiply the infiltration rate of the soil (mm/d) and expected potential evapotranspiration (mm/d) by the number of days, '*n*'. Select this *n* to assess the rainwater storm to be used in calculations (step 1)
- Subtract the value calculated in preceding step from the value of maximum *n* days storm rainfall. The resulting value divided by number of days, *n* will give the drainage coefficient in depth of water to be removed in 24 h (mm/d).

Mathematically, these steps can be put in the form of an equation as below:

$$q = \frac{1}{n}\left[R_n - n(E+1)\right] \tag{2.12}$$

Here q is the drainage rate in mm/d, *R* is the rainfall in mm, *E* and *I* are potential evapotranspiration and infiltration rate in mm/d and n is number of days. For rice or similar crops, some depth of water is allowed to stand in the fields after the drainage period. To reduce the drainage coefficient, necessary correction for this is incorporated in Eq. (2.7) wherein D_r, the depth of water allowed is subtracted to calculate the drainage coefficient. Eq. (2.7) for such situations is written as:

$$q = \frac{1}{n}\left[R_n - n(E+1)\ - D_r\right] \tag{2.13}$$

Example 2.2

Probability analysis of rainfall data for a gauging station (Karnal) revealed that 3 days storm of 225 mm can occur with a probability of 10%. The basic infiltration rate and average evaporation are 54 mm/d and 6 mm/d, respectively. Calculate the drainage coefficient if it is proposed to remove the water from the crop land within 3 days.

Incorporating the values in Eq. (2.8), we have

$$q = 1/3\ [225\text{-}3(6+54) - 0.0]\ \text{mm/d} = 45/3 = 15\ \text{mm/d}$$

Here D_r is taken as 0.0.

Example 2.3

Calculate the drainage coefficient if in example 2.2, it is proposed to cultivate rice. The infiltration rate is reduced to 30 mm/d as a result of puddling the field. The depth of storage in the rice fields at the end of the drainage period may be assumed as 50 mm.

Incorporating the values in eq. (2.8), we have

$$q = 1/3\ [225\text{-}3(6+30)\ \text{-}50] = 67/3 = 20.3\ \text{mm/d}$$

Many times it is surmised that drainage coefficient can be reduced by cultivating rice crop. But, it may be seen that puddling can make a major difference resulting in even higher values of drainage coefficient in spite of 50 mm of water storage. Such issues must be kept in view while calculating drainage coefficients.

3. Depth-duration-frequency curves

Use of depth-duration-frequency curves is an equally good method to assess the drainage coefficient. In this method, depth-duration-frequency values for 1-5 days as in the previous section are determined using daily rainfall data. The depth-duration curves are plotted on an arithmetic graph paper for the desired frequency say 5-20% probability (Fig. 2.2). Following procedures are then adopted.

Select the relevant curve and the duration within which water is to be removed at this curve. Draw tangents on the line at the selected point to meet the y-axis. The slope of this curve is the drainage coefficient that will remove water stagnating on the land surface above the depth indicated by the point at which tangent meets the y-axis (Fig. 2.2). Conversely, the point on the y-axis can be selected and tangent drawn to work out the slope of the tangent (drainage coefficient) that will remove the water above the pre-determined depth. For rice crop, usually the tangents are drawn from 50 mm depth. For upland crops, tangent is drawn from the origin at a point on the curve (duration) within which drainage water has to be removed. Clearly, the drainage coefficient increases with increasing return periods as revealed by the slopes.

Draw a line parallel to the tangent on the line of given return period passing through the origin. The water drained and water stored at any time can be determined using this line as shown in Fig. (2.2).

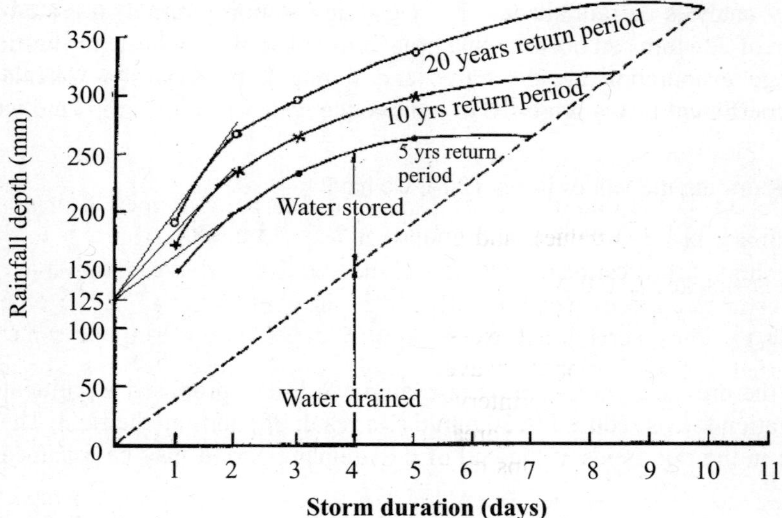

Fig. 2.2 Application of depth-duration-frequency curves to determine drainage coefficient

It may be noted that the values of the drainage coefficients determined in this manner will be higher as infiltration and evaporation have been neglected. Thus, the values so determined need to be corrected by accounting these two abstractions. It can be done by selecting a reasonable point on the y-axis rather than selecting 50 mm for rice and origin for other upland crops.

Drainage coefficients have been assessed by different workers covering different agro-climatic regions of the country using one or the other methods described herein. The most comprehensive analysis has been made by Gupta *et al.* (1971). The data reveal that drainage coefficients in India vary from a low value of 2 mm/d to about 40 mm/d depending upon the crop (rice, grain or vegetable) and the climatic conditions.

EMPIRICAL FORMULAE FOR COLLECTOR/MAIN DRAINS DESIGN

In order to economize the construction of drainage systems, the surface drains are sometimes designed for a discharge ranging from 1/4 to 1/12 of the calculated peak discharge (Dhruvanarayana, 1980) using Rational method, Cooks method or Curve Number Technique. The practical consideration is to remove the excess water due to heavy storms from the cropped fields within the stipulated period as per crop requirement. The drainage systems on Sharda Canal Project in Uttar Pradesh, India have been designed for a capacity of 0.11 m^3/s per km^2 of the catchment. The drainage systems in Punjab have been designed for a capacity of only 0.04 m^3/s per km^2 of the catchment. Two empirical relations namely i) Cypress Creek formula and (ii) Boston Society formula are commonly used to determine drainage coefficient.

Cypress Creek Formula

An empirical relationship relating design drainage rate at the outlet of a watershed and the watershed area is popularly known as the Cypress Creek formula. The formula is valid for mean watershed slopes up to 0.45%. This relationship says that the unit drainage rate (discharge per unit area) progressively reduces with increase in the watershed area as per the following relation:

$$Q = C * A^p \tag{2.14}$$

Here, Q is the runoff rate (m^3/s), A is the area drained (km^2), p = 5/6 (Approximate average value), and C = (0.2098 + 0.0074Y), where Y is the direct runoff volume (mm) estimated by the CN method. Thus, this equation makes use of the CN-estimated runoff which is reflected in the coefficient 'C'. Later on Stephens and Mills (1965) observed that excess rainfall calculated with any other method can also be used in this equation. Few average values of C of Eq. (2.9) for estimating a 5-year return period (recurrence interval) Q are given in Table 2.6. These values have been obtained from the drainage area-design discharge curves for different regions of India and for three major groups of crops (Gupta *et al.*,1971). The design discharge was calculated based on the volume of runoff estimated by the CN method for a constant I_a versus S relation of I_a = 0.2S. To obtain the design runoff for a 2-year return period, the values of Q, calculated using Eq. (2.9) for the C values listed in Table 2.6, are multiplied by a factor varying between 0.79 and 0.84 for different regions of India.

The value of Q is not the same as peak discharge as it is much less than the peak discharge since certain amount of flooding is allowed before the water is drained.

Moreover, the value of p in Eq. (2.14) being less than 1, the rate of increase in the estimated design discharge is not proportional to the increase in the drainage area. Based on the field studies, it has been reported that the Cypress Creek formula over estimates the drainage coefficient for watershed areas up to about 100 km^2.

Table 2.6 Average values of the coefficient (C) in the Cypress Creek formula (Gupta *et al.*, 1971)

Sl. No.	Region	Values of 'C' for different crops		
		Vegetable crops	Grain crops	Rice
1	Punjab and Haryana alluvial soil	1.414	0.354	0.071
2	Indo-Gangetic alluvial soil (North)	3.084	1.050	0.421
3	Indo-Gangetic alluvial soil (South)	2.375	0.668	0.260
4	Indo-Gangetic alluvial soil (Central)	2.820	0.816	0.185
5	Black soil (sub-humid)	4.454	1.113	0.890
6	Black soil (arid and semi-arid)	2.970	0.519	0.372
7	Eastern red soil	3.396	1.050	0.734
8	Southern red soil	2.262	0.565	0.356
9	Assam valley	2.970	0.927	0.446

Boston Society Formula

Peak discharge for the design of surface drains suggested by the Boston Society of Civil Engineers is given as:

$$Q = C\,A^{0.5} \tag{2.15}$$

Here, Q is the peak discharge in cusec (cusecs or ft^3/s), C is a coefficient and A is the catchment area in square miles. Uppal and Sehgal (1965) reported the values of C (Table 2.7) appearing in Eq. (2.15). These proposed values are mainly applicable for the catchments in different parts of Punjab and Haryana states.

Table 2.7 Values of coefficient (C) appearing in the Boston Society formula (Uppal and Sehgal, 1965)

Sl. No.	Catchment Area (Square miles)	Values of 'C' for different rainfall		
		For rainfall less than 20 inches	For rainfall between 20 and 30 inches	For rainfall between 30 and 40 inches
1	0-25	20	50	100
2	25-100	50	100	250
3	100-250	100	200	450
4	Above 250	200	400	600

Note: For rainfall above 40 inches, the value of C is taken as 2500 irrespective of the size of the catchment area

The above formulae have also been modified for various states in India to minimize investments on land drainage. All these formulae to design main and collector drains can be put into the form of an equation:

$$Q = C^* (A/n)^m \qquad (2.16)$$

Here Q is the discharge in cumecs, A is the catchment area in km^2, and C is a coefficient, which depends upon rainfall and n and m are constants. The values of c, n and m are given in Table 2.8.

Table 2.8 Values of C, n and m of Eq. (2.16) for different states to design main/collector drains

State	Values / guideline		
	C	n	m
Punjab and Haryana	3.5 for 500 mm rainfall to 35 for rainfall of 1000 mm	6 up to 260 km^2 4 up to 261-650 km^2 2 >651 km^2	0.5 0.5 0.5
Uttar Pradesh and Andhra Pradesh	0.11	1	1
Madhya Pradesh	0.22 for A>79 km^2 0.44 for A<13 km^2	1	1
Tamil Nadu	0.44 for A= 13 km^2 0.6 for A> 52 km^2	1 1	1 0.67
Odisha	0.3 for discharge of 10-12 cusecs	1	0.83

Note: 1 cumecs = 35.3 cusecs

Computation of Design Flow at Various Sections

Surface drainage network has a hierarchical order beginning with surface field drain at the lowest level followed by collectors, sub-mains if any and main drains - in that order. The field drains intercept excess water and channelize it towards higher order drains. As one or more drains meet, the computation of the total design flow at a particular point on a ditch may involve combining the flow from tributaries or combining the flow from areas in the watershed on which different coefficients were used to compute the drainage flow. Methods used to combine flows from the various parts of any watershed should be directed to the objective of providing the desired protection for each part and for the watershed as a whole. Commonly, the following three methods are used for this purpose.

A method (Method 1) to compute the required capacity of a ditch below a junction is to add the design flows of the two ditches above the junction. This method is used where ditches join at a particular point and drain almost equal areas (about 40 to 50% of the total area). It is assumed that in such situations the time of concentration is likely to be about the same and peak flows will reach the junction at about the same time under normal circumstances.

The second method (Method 2) is to add the tributary areas of the two ditches and compute the design discharge based on the drainage coefficient for the total watershed area. This method is used where a ditch draining a small area, say about 20% of the total area, joins a ditch that drains a much larger (remaining) area. In such a case, the peak discharge from the small ditch will pass before the peak flow of the larger ditch reaches the junction.

Apparently, the first method would give a higher discharge than the second method.

For intermediate conditions, the third method (Method 3) is used to determine the design discharge. It is commonly known as **20-40 rule**. It is used in cases where the watershed area of a lateral is in the range of 20 to 40% of the total watershed. The discharge in this case shall be proportioned to the smaller discharge obtained by use of method 2 to the larger discharge obtained by use of method 1. The method comprises the following steps:

- Check whether the drainage area of a channel is in the range of 20 to 40% of the total area
- Calculate total runoff by method 1
- Calculate total runoff by method 2
- Subtract runoff calculated by method 2 from that calculated by method 1
- Calculate the area drained by the smaller channel as % of the total area drained (say 25%)
- Subtract this value from 20 (25-20 = 5) and calculate its percentage with respect to 20 (100 *5/20 = 25).
- Multiply the discharge calculated in bullet 4 with the percentage calculated in the previous bullet
- Add this value to the discharge calculated by method 2

This final value should be taken as discharge below the junction.

Computing Equivalent Drainage Area for Design Flow at Junctions

When runoff is contributed by different areas at different rates on various parts of the watershed, it is necessary to find either the equivalent areas or the equivalent discharge so that the correct design capacity is carried downstream. Let us say that the discharge from 100 ha land is calculated with equation $Q = C_1 A^{5/6}$. A drain draining 20 ha area whose discharge is given by the equation $Q = C_2 A^{5/6}$ joins the previous drain. In such a case, it is necessary to convert to either $Q = C_1 A^{5/6}$ or $Q = C_2 A^{5/6}$ to determine the discharge at the junction. If it is observed that predominant land use requires runoff removal rate as per $Q = C_1 M^{5/6}$, the discharge from 20 ha is calculated using equation $Q = C_2 A^{5/6}$. Using this Q, equivalent area pertaining to 20 ha is calculated using $Q=C_1 A^{5/6}$. The area thus calculated is added to 100 ha and the new discharge at the junction is calculated using $Q =C_1 A^{5/6}$.

DRAINAGE COMPONENTS AND FIELD PRACTICES

A surface drainage system essentially consists of the following three parts:

- Collection system
- Disposal system
- Outlet

Collection system mainly concerns on-farm drainage with disposal and outlets forming a part of the overall system. On-farm surface drainage of agricultural lands is accomplished by measures such as: (i) land forming (ii) open ditches or terracing and (iii) levees (National Institute of Hydrology, 1996).

(i) Land Forming

Land forming is the process of shaping the surface by modifying the micro-topography of the land to meet the requirements of surface drainage and irrigation. For example, by providing a smooth slope in the direction of drainage, one is able to control the movement of water on to or from the land surface. The land forming can be achieved through land smoothening, land grading or land levelling. Land forming also included construction of shallow field drains, which are easily crossable by farm machinery.

(ii) Land smoothening

Process of land smoothening is sufficient to establish grades in flat lands, where only minor differences in field levels occur. In this process, while general contours of the ground surface are left unaltered, only small ridges are removed and depressions filled up to achieve a smooth surface. In this process, land is tilled under optimum moisture conditions such that soil is broken down to small clods and is moist enough to crumble in the hand. Wooden floats or land planes are used to perform the operation. Being quite cheap, this operation can be repeated every season as a normal practice. This operation is also necessary to finish a land levelling job as the later ends up in an irregular land surface.

(iii) Land grading/levelling

Process of shaping the land surface by cutting, filling and smoothening to form a continuous grade in the direction of land drainage is referred to as land grading. Land levelling, another term for the process is often used synonymously with land grading although in many cases it is misconstrued as table top levelling of the area. It is why the term ''land grading'' is preferred over land levelling. Detailed surveys, planning, and layout are necessary for land grading. Topographic survey is an essential pre-requisite to plan the surface grades and auxiliary drainage system. For efficient drainage, a grade in the range of 0.05% to 0.5% is provided although the preferable value is 0.1%. For large areas, there are four basic methods used to plan land grading operation. These are plane, profile, plan inspection, and the contour adjustment methods. For on-farm land grading such elaborate planning is not necessary.

Currently, laser land levelling has become quite popular and is being widely adopted against the conventional land levelling equipment (Fig. 2.3). It has emerged that about 20% saving in water and 10-15% increase in the yield of wheat can be obtained with laser as compared to conventional levelling. Water standing at the tail ends after an irrigation or rainfall event is a common feature of conventionally leveled fields. Water profiles after irrigation in a laser leveled and conventionally leveled field show a perceptible improvement in the former (Fig. 2.4). The increase in yield and water saving is attributed to improved land levelling index, which is calculated as:

$$LI = \frac{\Sigma |(L_d - L_e)|}{n} \tag{2.17}$$

The parameters L_d and L_e are the desired and the existing elevations at the grid points and n is the number of grid points. It has the units of cm. The lower limit of LI is zero, which indicates perfect levelling. Increasing values of LI reflect poorer quality of land levelling. The data in Table 2.9 show an improvement in the LI from 2.1 to 0.8-0.9 with laser than conventional levelling (Table 2.9).

The laser levelers are now available under custom-hiring mode obviating the need for investment by individual farmers (Ambast *et al.*, 2015). Land levelling operation may need to be repeated once in 3-4 years but during this period it pays for its cost as well as saves water, a precious natural resource (Schwab *et al.*, 1987).

Table 2.9 Performance parameters of laser land levelling

Field	**LI**	
	Before	**After**
Laser (field 1)	1.56	0.80
Laser (field 2)	2.29	0.90
Conventional	3.00	2.10

1: Tripod; 2: Laser beam genearator; 3: Laser beam receiver; 4: Control box; 5: Levelling equipment; 6: Tractor; 7: Laser beam

Fig. 2.3 Equipment for laser land levelling

Fig. 2.4 Water distribution in a laser leveled (left) and a conventionally leveled field (right)

DRAINAGE SYSTEMS

A number of surface drainage systems are in vogue. Some of them are bedding system, parallel field drain system, random drain system and parallel open ditch system.

Bedding System

Bedding system is commonly used to drain flat lands (slopes < 1.5%) and slowly permeable soils where pipe drainage is uneconomical. Bedding is created by ploughing the land to form a series of low beds, separated by parallel field drains. The work can be accomplished either manually or by animal or farm tractors operated equipment. The beds are separated by dead furrows, a kind of shallow open field drains, oriented in the direction of the greatest land slope (Fig. 2.5). The water drains from the beds into the field drains, which discharge into a collector drain constructed at the lower end of the field and at right angles to the field drains. Although the bed width depends upon several factors such as the slope, land use, soil permeability, farm machinery available and nature of the farming operations, the bed width depending upon the hydraulic conductivity, K, varies from 8 to 30 m (Table 2.10).

Table 2.10 Bed widths in relation to hydraulic conductivity of soils

Hydraulic conductivity (m/d)	**Bed width (m)**
Very slow internal drainage ($K \leq 0.05$)	8 - 12
Slow internal drainage ($0.05 < K \leq 0.10$)	15 - 17
Fair internal drainage ($0.10 < K \leq 0.2$)	20- 30

The bed lengths are in the range of 100 to 300 m while the height of the crown from the bottom of the furrow ranges from 20 cm for arable crops to 45 cm for pastures (Fig. 2.5). Current thinking is veering round the view that bedding system is not an acceptable drainage practice especially for row crops. It is because the rows near the field drains are not satisfactorily drained. Moreover, if rows are parallel to the dead furrows, it obstructs the flow to the furrows. Besides the limited capacity of dead furrows, difficulties in farm mechanization and need for regular maintenance of beds are unfavorable factors in practicing bedding as a drainage practice. Moreover, the construction of the system requires significant movement of the top soil from the side of beds to the middle resulting in fertility gradients that impact the crop yield at least in the initial years.

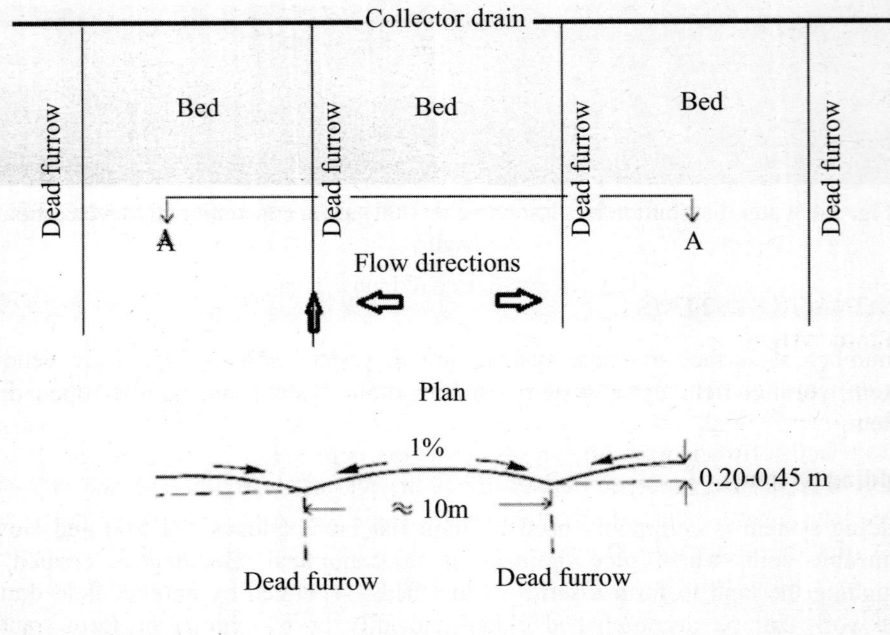

Fig. 2.5 Plan and cross section of a bedding system of land drainage

Local adaptations of the system are emerging in many countries, particularly for heavy textured clay soils. One such alternative known as camber beds is being researched in Ethiopia. Camber beds consist of 6 m wide beds followed by 1 m furrow. Camber beds are constructed using a mould board plough where the soil is turned from the furrow side to form the beds. While the crop is planted on the 6 m wide beds, furrow serves to drain the water. Yields about 50% higher than the conventional farming have been reported.

Parallel Field Drain System

The parallel system is used for flat, poorly drained uniform surface soils, which may have few or numerous small depressions. The channels are spaced farther apart than a bedding system and the channels have greater capacity to carry runoff than the dead furrows (Fig. 2.6). The distance separating the ditches can vary depending upon the soil; maximum lateral flow length to ditch may be 180 m for non-erosive soils but is reduced to 90 m for erosive soils. The depth of the drain should be minimum 20 cm having a minimum cross section area of 0.45 m^2. The side slopes are 8:1 or flatter. The row or field drains are perpendicular to the field lateral. The system is most suitable where farm mechanization is practiced. Land smoothening of field surface between laterals is necessary so that the land remains well graded.

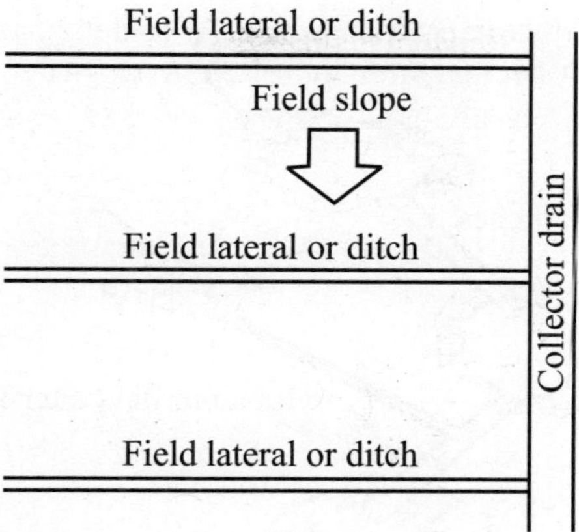

Fig. 2.6 Parallel field drain system

Random system

This drainage system is practiced when a number of depressions are distributed at random over a field. These depressions may be deep or shallow but usually large, making land-forming operation or filling up the depressions uneconomical. In the random field drainage system, field or row drains and field laterals meander from one depression to another to drain the same to the outlet ditches (Fig. 2.7). Location and depth of the lateral ditches are determined by the topography of the area, outlet conditions, and capacity of the channel. While the minimum cross section area should be 0.45 m^2, the grade should neither cause erosion nor sedimentation. The minimum grade is 0.05%. The maximum and minimum grades for sandy and clayey soils are about 0.2% and 0.5%. Minimum velocities range from about 0.3 to 0.6 m/s for depths of flow less than 1 m. The slope can be as flat as 10:1 (drain depth > 60 cm) or 4:1 if fields are farmed parallel to the drains. The capacity of the drain is not a major issue if the drainage areas are less than 2 ha provided the minimum design specifications are met. However, where the area is larger than 2 ha, the capacity should be based on a 10-year return period storm, making allowances for infiltration and interception losses. The channels follow a route that requires minimum cut and ensures minimum interference in the farming operations. The system is less costly and provides more returns on investment. The maintenance cost is relatively more than other systems of drainage. The spoil can be used to fill up low areas further away from the field drain.

Parallel Lateral Open Ditch System

The parallel lateral open ditch system besides providing surface drainage also helps to control the groundwater table. It is why this system is referred to as open subsurface drainage system. The layout of the system is similar to the field drain system but the ditches are deeper. These drains cannot be crossed with farm machinery. The minimum size of open ditches is 0.3 m deep and the side slopes are steeper than 6:1. The design specifications are given in Table 2.11 (Schwab *et al.*, 2005). When financial or other constraints do not allow construction of tile/pipe drainage, this system of drainage is recommended till finances are generated or made available.

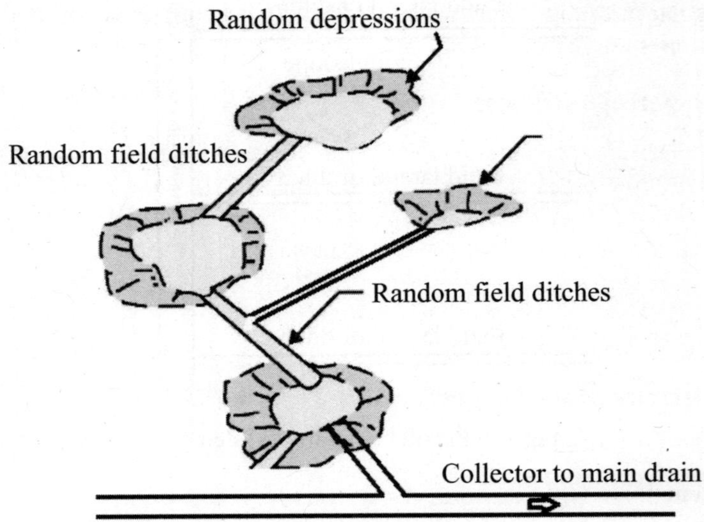

Fig. 2.7 Random ditch system for surface drainage of scattered depressions

Levees

Many times construction of levees is also included in surface drainage yet in most cases this issue finds mention under flood control in flood plains and coastal regions. Levees are embankments or dikes that do not allow the runoff from uplands to enter the affected areas. For agricultural drainage, application of this system is only limited. Nonetheless, it is probably the scale of work that will determine whether to include this activity under surface drainage or flood control, surface drainage being on a smaller scale.

Table 2.11 Ditch specifications for water table control

Sl. No.	Parameter	Sandy soil	Other mineral soils	Organic soils, peat and muck
1	Maximum spacing	200 m	100 m	60 m
2	Minimum side slope	1:1	1:1	Vertical to 1:1[a]
3	Minimum bottom width	1.2 m	0.3 m	0.3 m
4	Minimum depth	1.2 m	0.8 m	0.9 m

Note: [a] Vertical for raw peat to 1:1 for decomposed peat and muck.

Cross Slope Drains

These drains are a channel type graded terrace and function both for surface drainage and for erosion control. The cross-slope ditch system is adapted primarily for soils with poor internal drainage where subsurface drainage is not practical and for land with slopes of 4% or less having numerous shallow depressions (Schwab *et al.*, 2005). These lands are generally too steep for bedding or field drains because farming up and down the slope results in excessive erosion. Cross-slope drains have either a triangular or a trapezoidal shape with a cross section area varying from 0.4-0.7 m^2. Side slopes range from 1:4 to 1:10. Depth may range from 0.15 to 0.25 m.

Diversion or Interceptor Drains

Diversion or interceptor drains are constructed at the downstream of uplands or on sloping lands. A filter strip is usually constructed on the up slope of the ditch to prevent silting of the drains. For areas less than 2.5 ha the diversion drains can be constructed like terraces but for larger areas these should be constructed as grassed waterways.

DESIGN OF FIELD DRAINS

The design of surface drains comprises:

- Hydrologic design
- Hydraulic design

Hydrological design

This refers to selecting drainage coefficient for the design of the drain considering rainfall, watershed characteristics and the land use/land cover. The methods of computation of drainage coefficient have been discussed in the earlier sections of this chapter. The design discharge is obtained by the multiplication of drainage coefficient with drainage area and converting the same in the desired units.

Example 2.4

Calculate the design discharge of the drainage system to drain an area of 2 km^2 if the drainage coefficient is 2 mm/day.

Drainage coefficient = 2/(1000*24) m/hr

Area = 2*100*10000 = 2000000 m^2

Design discharge = 2000000* 2/(1000*24) = 166.7 m^3/hr

Hydraulic Design

Hydraulic design refers to working out the hydraulic dimensions of different components of the cross sections of the drains to carry the designed runoff. Manning's uniform flow formula is commonly used to estimate the capacity of the drains:

$$Q = A\,V \tag{2.18}$$

Such that velocity of flow is given as:

$$V = (1/n)\,(R^{2/3})\,(s^{1/2}) \tag{2.19}$$

Here, V is the mean velocity of flow in the channel, m/s, n is the Manning's roughness coefficient, R is the hydraulic radius of flow, = A/P, m, A is the flow area of channel, m^2, P is the wetted perimeter, m, s is the channel bed slope, m/m and Q are the design discharge, m^3/s. For Eq. (2.19) to be dimensionally homogeneous, n in MKS system has the unit of $T/L^{1/3}$.

Table 2.12 Manning's n for wet cross section well covered with weedy or grassy vegetation

Channel description	n-value
Small channel (water depth <0.75 m)	
Sandy soil	0.050
Clayey soil	0.060
Medium channel (water depth 0.75-1.5m)	
Sandy soil	0.035
Clayey soil	0.050
Large channels (water depth >1.5 m)	0.020-0.025
Concrete lined channel	0.012-0.017

A brief reference to hydraulic grade line may be in order at this stage. The hydraulic grade is the sum of the pressure head (p/ã) and elevation head (z). For open channel flow (in which the pressure head is zero), the hydraulic grade elevation is the same as the water surface elevation. When the hydraulic grade is plotted as a profile along the length of the conveyance section, it is referred to as the hydraulic grade line (HGL). Under the assumption of uniform flow conditions the bottom slope is the same as the slope of the energy grade line and the water surface slope or slope of HGL. It is why the drain slope can be used in the design of channel section. If flow is not uniform, then the use of drain slope is not justified and one should use the slope of HGL instead.

Table 2.13 Permissible velocities (m/s) in channels without and with grass cover

Material	Bare	Medium grass cover	Very good grass cover
Very light silty sand	0.30	0.75	1.50
Light loose sand	0.50	0.90	1.50
Coarse sand	0.75	1.25	1.70
Sandy soil	0.75	1.50	2.00
Firm clayey loam	1.00	1.70	2.30
Stiff clayey or gravely soil	1.50	1.80	2.50
Coarse gravel	1.50	1.80	No good grass cover expected
Shale hard pan and soft rock	1.80	2.10	Good cover unlikely
Hard cemented	2.50	-	-

Manning's roughness coefficient, n, is selected from Table 2.12 or other such tables in the literature. The channels are designed in such a way that the flow velocities remain within a range of acceptable minimum and maximum values so that neither scouring nor sedimentation occurs in the channels (Kumar & Singh, 2005). The permissible velocities in drains for bare soil and soils with grass cover are reported in Table 2.13 (Smedema & Rycroft, 1983). Although bed gradients depend upon local land slope and outfall conditions, the bed slope may not be less than 0.2% to avoid siltation and not more than 0.5% to avoid channel scouring. Maximum permissible side slopes are flatter in sand than heavy clays. For fine sands, the slope may vary from 1:2 to 1:3 (vertical: horizontal) and for heavy clays from 1:1 to 1:2. For loam soils, it may range from 1:1.5 to 1:2 and coarse sand from 1:1.5 to 1:3. Side slopes generally become more stable once they get vegetated but design should normally be based on the initial period after excavation.

Cross Section of Channels

For a given area A, slope S and value of n, a section that carries the maximum discharge with minimum wetted perimeter P is known as the hydraulically efficient section. It is popularly referred to as Best Hydraulic section. The semicircle has the least perimeter among all sections with the same area; hence it is the most hydraulically efficient of all sections. It can be shown as follows:

Wetted area A, and wetted perimeter of the half a circular conduit are given as:

$$A = \pi r^2 /2$$

$$P = \pi r$$

$$R = A/P = \pi r^2/2\pi r = r/2 = y/2 \text{ (Since } r = y\text{, the depth of flow)}$$

Hence, half a circular conduit itself is a best hydraulic cross section.

However, there are practical problems in the construction of such a section. Therefore, sections are made rectangular, parabolic, trapezoidal or triangular sometimes with rounded corners. The dimensions of such channels can be determined from considerations of least wetted perimeter for a given area as per the procedure shown for a rectangular channel:

The area, A for a rectangular channel having the bed width b and depth d is given as:

$$A = b\,d \tag{2.20}$$

$$P = b + 2\,d \tag{2.21}$$

$$P = A/d + 2\,d \quad \text{[Incorporating } b = A/d \text{ from Eq. (2.20)]} \tag{2.22}$$

For P to be minimum for a given A

$$dP/dd = 0$$

$$\text{or} \quad -A/d^2 + 2 = 0$$

$$\text{or} \quad A = 2d^2 \tag{2.23}$$

This gives $b = 2d$

$$\text{or} \quad R = A/P = d/2 \tag{2.24}$$

Here R is the hydraulic radius. It may be noted that channels with the same cross sectional area can have different R. For example, a rectangular section of 2 m bottom width and 1 m depth and 1 m bottom width and 2 m depth have the same area but different values of R, being ½ and 2/5 respectively.

Example 2.5

Calculate the best hydraulic rectangular cross-section to convey Q=10 m^3/s discharge with n= 0.02 and S= 0.0009.

For the best rectangular hydraulic cross section,

$$R = y/2$$

$$A = 2\, y^2$$

$$Q = A\,(1/n)\,(R^{2/3})\,(s^{1/2})$$

$$10 = 2y^2\,(1/0.02)\,(y/2)^{2/3}\,(0.0009)^{1/2}$$

$$y^{8/3} = 5.29$$

$$y = 1.87$$

$$b = 2y = 3.74$$

In a similar manner, it should be possible to show that the bottom width for the most efficient cross section and minimum volume of excavation for a trapezoidal channel is given by

$$b= 2d \tan \theta/2 \qquad (2.25)$$

Where θ is angle formed by side slope.

If the trapezoid is half hexagon then

$$P = 2\sqrt{3}\, d$$

$$R = d/2$$

For a parabolic cross section, the area is given by 2/3 t d and the wetted perimeter $P = t+8\, d^2/3t$. Therefore, $R = t.\ 2d/\ (1.5\, t^2 + 4d^2)$. Here t is the top width. The approximate relation d =1.5 R is commonly used in the design of parabolic channels.

It is not always possible to design for the most efficient cross section. As a compromise between hydraulic efficiency and other considerations, the ratios between bed width (b) and water depth (d) for trapezoidal channels are approximated (Table 2.14; Smedema and Rycroft, 1983). Once b and d are decided, cross sectional area and wetted perimeter of the channel are worked out by using the relations:

$$\text{Cross sectional area } A = bd + zd^2 \qquad (2.26)$$

$$\text{Wetted perimeter } P = b + 2d\,(z^2 + 1)^{1/2} \qquad (2.27)$$

Where z is the side slope z:1 (Horizontal : Vertical)

Depending on the method of construction, minimum sections should have:

- Bed widths of 0.50 -1.0 m
- Side slope of 1:1 or 1: 1.5, and
- Water depth of 0.30 -0.50m.

Table 2.14 Recommended bed width, b and depth, d ratios for open channels

Drain depth	Recommended b/d ratios for trapezoidal channels
Small channel, d=0.75 m	1 for clay to 2 for sand
Medium channel, d=0.75 to 1.5 m	2 to 3
Large channel, d=1.5m	3 to 4

Other Considerations

(i) Freeboard

Freeboard means providing extra depth above the designed depth, to take care of the following:

- Margin of safety when actual discharge exceeds the designed discharge capacity
- To take care of deviations between design and construction
- Non- uniform settlement of land after construction
- Operational flexibility
- Accommodating transient flow conditions (design is based on uniform flow)
- Increase in hydraulic roughness due to weed growth etc.

For small channels having 0.25-0.5 cumecs discharge, the freeboard is taken as 10 to 25% of the designed depth. Lower values are associated with the higher return periods. Other guidelines for freeboard state that a freeboard of 30 cm should be provided for channels having capacity < 2 cumecs, which is increased to 50 cm for large channels having capacity > 5 cumecs.

(ii) Berms

Berm is the horizontal distance left at the ground level between the toe of the bank and the top edge of cutting. Berms are needed to serve the following functions:

- Prevent sloughing of ditch banks caused by heavy soil loads near the edge of the drain
- For movement of machinery during maintenance
- Eliminate the need for moving spoil banks in future operations
- Provide work area during construction of spoil bank
- Prevents excavated material from falling back into the drain.

The size of the berm depends upon the depth of the drain and type of soil and adjacent land use. Minimum recommended berm widths vary from 1.2 m for drain depths in the range of 0.6-1.2 m to about 3 m if drain depths are in the range of 1.8 - 2.4 m.

(iii) Spoil bank

In the construction of drainage channels, if earthwork in excavation is more than filling, the extra earth needs to be disposed of economically. The best way is to construct spoil banks along the edge of the embankment. The dimensions of the spoil bank will be more than the earth work calculated as the loose earth has more volume than the volume excavated. The best use of the spoil and how far it can be spread is determined by the type of excavated soil, the adjacent land use, the need for roads and the method of maintenance to be employed. The spoil should be spread to slope away from the ditch. In all cases a path should be established on the berm or on the spread spoil which is sufficient to take care of the movement and operation of the equipment needed for maintenance of the ditch. In some locations spoil can be shaped and used to good advantage for farm roads. In humid areas, the spoil banks are spread in a manner so that some cultivation is possible.

General Consideration for Field Drains

Surface runoff reaches the field drains by flow through row furrows or by sheet flow. In the transition zone between drain and field, flow velocities should not induce erosion. Field drains for a surface drainage system should be shallow and have flat side slopes to allow farm equipment to cross them and may be easy to maintain with ordinary mowers. They are often constructed with land planes as used in land forming. Simple field drains are V-shaped. According to ASAE (1980), the dimensions of V-shaped field drains are determined by the construction equipment, maintenance needs, and cross ability of farm equipment (Table 2.15). Side slopes should not be steeper than 6 to 1. Nevertheless, long field drains in areas with high rainfall intensities may require a higher capacity than provided by a simple V-shaped channel.

Table 2.15 Recommended dimensions for field laterals

Sl. No.	Type of Drain	Depth (m)	Recommended side slope (H : V)	Maximum side slope (H : V)
1	V-shaped	0.3 to 0.6	6 : 1	3 : 1
2	V-shaped	>0.6	4 : 1	3 : 1
3	Trapezoidal	0.3 to 1.0	4 : 1	2 : 1
4	Trapezoidal	>1.0	1.5 : 1	1 : 1

Without increasing the drain depth too much, the capacity can be enlarged by increasing the bottom width, creating a shallow trapezoidal shape. All field drains should be graded towards the lateral drain with grades between 0.1 and 0.3%. Field laterals less than 1 m deep are usually constructed either manually or by JCB or with motor graders or dozers. The soil is placed near either side of the lateral. Scrapers are needed if the excavated soil is to be transported some distance away. Under wet conditions, excavators are used.

Commonly used cross sections for various components of surface drainage adopted in the Chambal command of Rajasthan, India (RAJAD, 1995) are given in Table 2.16. In this table field drain is the graded channel that collects water from field and is provided along d/s boundary of each field. Carrier drains serve as collector drains for field drain located about 500 m apart to serve as outlet into the main drain. Seepage drain is located along canal (parallel to canal) to intercept surface/ subsurface seepage flow, Main drain or sub-main drain is either natural drain or excavated drain draining 1000 to 10,000 ha. An example to illustrate the design of a collector is given in Example 2.9.

Table 2.16 Cross section of surface drains adopted in Chambal Command, Rajasthan, India (design runoff rate: 1.1 L/s-ha)

Type of drain	Base width (m)	Side slope (H:V)	Depth (m)	Density (m/ha)
Main drain	0.75-1.20	1.5: 1	2.0	7 -10
Seepage drain	0.60-0.75	1 :1	1.0	15 -20
Carrier drain	1.0-1.20	1:1	1.0-1.20	15-20
Field drain	0.60-0.75	1:1	0.9-1.20	20-40

Example 2.6

A trapezoidal ditch with a bottom width of 1.2 m and side slopes of 1V:4H is constructed on a slope of 0.010 m/m and has a depth of flow of 0.5 m. It is lined with a good stand of native grasses. Show that this channel is capable to handle a drainage coefficient of 3 cm/d from an area of 250 ha.

Quantity of water to be removed in 24 h

$$= (3.0/100) * 250 * 10000 \text{ m}^3$$

Discharge rate Q= $0.87 \text{ m}^3/\text{s}$

Now determine the flow area A and hydraulic radius R:

$$A = db+zd^2 = 0.5\,(1.2 + (4 \times 0.5)) = 1.6 \text{ m}^2$$

$$P = b+2d(1+z^2)^{1/2} = 1.2 + (2 \times 0.5)(1 + 16)^{1/2} = 5.32 \text{ m}$$

$$R = A/P = 1.65/5.32 = 0.30 \text{ m}$$

If Manning's n is taken as 0.080 then

$$Q = (1/n)\, A\, R^{2/3}\, S^{1/2}$$

$$Q = (1/0.080)(1.6)(0.30)^{2/3}\,(0.01)^{1/2} = 0.89 \text{ m}^3/\text{s}$$

Since capacity of the channel is more than the designed discharge, the channel is capable to drain the area.

Example 2.7

Design an open ditch to drain 170 ha of land having a drainage coefficient of 1.2 cm/day. The soil is silt loam and maximum permissible slope of the channel bed is 0.1%. Assume *n* as 0.04.

Quantity of water to be removed in 24 h

$$= (1.2/100) * 170 * 10000 \text{ m}^3$$

Dividing by 24 * 60 * 60, we get

Discharge rate Q= 0.236 m^3/s

Side slopes recommended for silt loam = 1.5:1

$\tan\theta = 1/1.5 = 0.666$ or $\theta = 33.66$ or $\theta/2 = 16.83$

Assuming d= 0.5 m, b=2 d tan θ/2

= 2 * 0.5* 0.3025 = 0.30 m

≈ 30 cm

$$\text{Length of side} = \sqrt{(1.5\times0.5)^2+(0.5)^2} = 0.901 \text{ m}$$

P= 2 * 0.90+0.30= 2.10 m

$$A = \frac{(3*0.5+0.30)+0.30}{2}\times 0.5 = 0.53\,\text{m}^2$$

$$R = \frac{.53}{2.1} = 0.252\,\text{m}$$

$$S = \frac{0.1}{100} = 0.001\,\text{m/m}$$

$$V = \frac{(0.252)^{2.3} * (0.001)^{1/2}}{0.04} = 0.315\,\text{m/s}$$

The velocity is within permissible limit.

Q= 0.53*0.315 = 0.167 m^3/s

Since the discharge to be handled is less than the design discharge, assumed dimensions will not suffice. Therefore, one more iteration should be made by increasing d by 0.1 m.

Assuming d = 0.6 m, b = 2 d tan θ/2

=2*0.6*0.3025 = 0.30 m

≈ 36 cm

$$\text{Length of side} = \sqrt{(1.5\times0.6)^2+(0.6)^2} = 1.12\,\text{m}$$

P= 2*1.12+0.36= 2.6 m

$$A = \frac{(3\times0.6+0.36)+0.36}{2}\times 0.6 = 0.756\,\text{m}^2$$

$$R= \frac{.756}{2.6} = 0.29 \text{ m}$$

$$S = \frac{0.1}{100} = 0.001 \text{ m/m}$$

$$V = \frac{(0.29)^{2/3} \times (0.001)^{1/2}}{0.04} = 0.346 \text{ m/s}$$

The velocity is within permissible limit.

$$Q= 0.756*0.346 = 0.261 \text{ m}^3/\text{s}$$

The capacity of the drain is higher than the design discharge and hence the dimensions are OK in this respect.

Example 2.8

Design a surface drain to carry runoff from a 500 ha watershed with a drainage coefficient of 5 L/s-ha. Given that channel bed slope is 0.2% and soil type is clay. Permissible velocity may be taken as 0.6 m/s.

Since the soil type is clayey, we select the following values from the tables

Manning's roughness coefficient = 0.060 (Table 2.12)

Side slopes = 1 : 1

Permissible velocity = 0.6 m/s

$$V = (1/n)\,(R^{2/3})(s^{1/2})$$

$$= (1/0.06)\,(R^{2/3})(0.002^{1/2})$$

$$= 0.745\,(R^{2/3}) = 0.6$$

or $R^{2/3} = 0.805$ m

$$R = 0.722$$

Assuming d = 2R, we have d= 1.45

From continuity equation, Q=AV

So $A = (5*500*1000/10^6)/0.6 = 4.17 \text{ m}^2$

Now $A = bd+zd^2 = bd + d^2 = 4.17$ (2.28)

Incorporating d = 1.45, we get

$$b = (4.17 - 2.10)\,/1.45 = 1.43$$

Such that b = 1.43 and d = 1.45

Using the guidelines illustrated in the previous sections, the design cross section of the drain is given in Fig. 2.8.

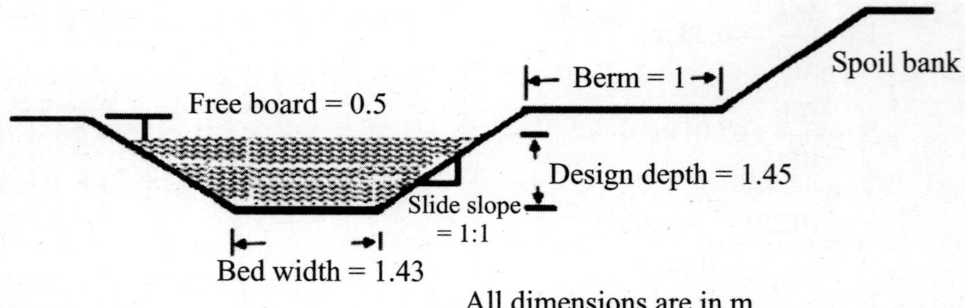

Fig. 2.8 Design cross section of drain for Example 2.8

Example 2.9

A collector drain to drain an area of 1200 ha is to be designed for the state of Haryana. Calculate the dimensions of the channel, if the land slope is 0.1%, average annual rainfall is 500 mm and side slopes are H:V = 1:1. Assume Manning´s n = 0.030. Also assume any other parameter required on the basis of agro-climatic conditions of the state.

For the state of Haryana, incorporating the values of C, n and m in Eq. (2.16) from Table 2.8, we have,

$$Q = 3.5^* (A/6)^{0.5} = 3.5^* (12/6)^{0.5} = 4.95 \text{ cumecs}$$

If we assume b = 2d

$$A = bd+zd^2 = bd + d^2 = 3\ d^2$$

$$P = b+2d\ (z^2+1)^{1/2} = b + 2.83\ d = 4.83\ d$$

$$R = A/P = 3\ d/4.83 = 0.62\ d$$

Applying the Manning´s equation

$$V = 1/n\ (R^{2/3} \times S^{½})$$

For n = 0.030

$$V = 1/0.030 \times (0.62d)^{2/3} \times (0.001)^{½}$$

$$= 33.33 * 0.488\ d^{2/3}\ *0.0316 = 0.514\ d^{2/3}$$

$$Q = A \times V = 3\ d^2 \times 0.514\ d^{2/3}$$

$$= 1.542 \times d^{8/3}$$

$$4.95 = 1.542 \times d^{8/3} \text{ or } d^{8/3} = 3.21 = 1.55 \text{ m}$$

With 20% free board, d = 1.55 + 0.31 = 1.86 m

$$b = 3.72 \text{ m}$$

Top width, T= b+2 * z * d = 3.72 + 2*1.86 = 7.44 m

OUTLET

To ensure adequate disposal of excess water from all parts of the drainage system, there must be uninterrupted flow from the point where the water begins to collect to the disposal system and to an adequate outlet. The stage of a stream at a time when the drainage system is discharging at the design rate determines the adequacy of the stream as an outlet. A study of the frequency of high water stage is needed for large streams to determine their adequacy as an outlet and to establish the elevation of the design hydraulic grade line for the open ditch at the outlet.The outlet must be capable of handling the flow without causing damages above or below the point of discharge. This may require extending the channel improvement programme further downstream. A comparison of alternate locations of the point of outlet may also be needed.

CONSTRUCTION

Construction plans for individual farm ditches usually include:

- Drainage plan map
- Profiles
- Cross sections
- Ditch designs
- Structural details
- Specifications

All items of the plan should be identified with the name of the job(s), landowner(s), and location of farm(s). The scale used should be shown.

The plans should be discussed with the landowner or his representative to make sure he fully understands the proposed work. Cost estimates should be provided on request.

Prior to construction, conduct a site investigation of the general layout, consistent with the work plan. Drainage ditches should be located to provide the most effective drainage of the agricultural land. Topography, existing ditches and drains, bridges, farm boundaries and other physical features all influence ditch location. Transfer the map plan on to the ground with appropriate benchmark pillars.

The hydraulic grade line is established and L-section prepared. The hydraulic grade line is established below the elevations of the principal fields so that all points in the area are adequately drained. As such, natural outlets such as estuaries, rivers, lakes or swamps, or old ditches usually fix the general location and grade line of an open ditch. The control points should also take care of the head loss by surface flows from the field to the surface drains and to the outlet. Control points also need to be established at hydraulic grade lines of lateral ditches that enter the drain. If it is not possible to meet the drainage needs, the land use may be modified to more tolerant crops. Wherever feasible, smooth curves should be used for alignment rather than sharp bends to improve the hydraulic property and stability of the ditches. Open ditches should terminate in an adequate outlet.

Construction is done using manual labour or machines or a combination of both. A machine that can accomplish digging, hauling and placing of the excavated material is most desirable (Fig. 2.9). The construction work should begin from the outlet and progress upstream, drain by drain. Provision of some allowance for initial sedimentation in a ditch after construction allows the drain to function at the designed capacity after the ditch stabilizes. The amount of this allowance depends on the erosion from adjoining lands, the erosiveness of soils exposed in the ditch and the sediment from laterals and tributaries. The best practice for this purpose is to over-excavate the ditch (in depth only) as a construction practice. In some cases bed width is increased but not the top width.

Fig. 2.9 Construction of drain using machines

MAINTENANCE OF OPEN DRAINS

After initial construction, open drains may lose their effectiveness unless properly maintained due to one or more of the following reasons.

- Sedimentation
- Uncontrolled growth of vegetation
- Bank erosion
- Damage to side slopes

A good maintenance program therefore, is just as necessary as proper design and construction. Every effort should be made to pre-decide a maintenance program with the drainage department, group or landowners responsible for the system. Maintenance is divided into two categories i.e. preventive maintenance before failure and corrective maintenance after partial or complete failure. Preventive maintenance is usually cheaper and prevents or delays the failure of the system. Likewise timely maintenance is more economical than delayed maintenance.

Preventive Maintenance

Prevention of sedimentation, control of excessive vegetative growth in the channels, prevention of bank erosion and repair of damaged portions of the drain are some of the works that are needed to prevent drain failure. Sedimentation occurs when the water entering the drain carries sediments. Minimizing the erosion in the watershed through the construction of a simple drop structures, sedimentation basins and diversion ditches can reduce the sediment deposits in the drains. Scouring and deposition of silt may also occur along the bed of the drain, particularly at points where the grade changes sharply and there is change in flow velocities. Settling basins and low check dams can be constructed under these situations. These activities will considerably reduce the cleaning works in the drains.

In a properly designed and constructed drain, the damage due to bank erosion and side slopes will be minimal. Wherever necessary, the spoil banks should be seeded to grow permanent vegetation. Cattle and human trespass should be minimized to maintain side slopes.

Corrective Maintenance

After the first construction, changes in cross section, grade or alignment of the drain may be required for the drain to function well. Due to variation in soil type, side slopes at some places may cave in and fill the drain. The side slope at these vulnerable points needs to be reshaped to a more stable slope. Sharp curves causing meandering of the drains may necessitate modifications in the ditch alignment. Either the drain may be straightened or the curve may be flattened.

Weed growth or vegetation reduces the capacity of the drains. As such, drains fail to carry the design discharge. Excessive weed growth in channels has been found to reduce the capacity of the channels as much as 75% or even more. Control of excessive tree, brush, grass and weed growth along the sides of the channels as well as in the channels is an important maintenance activity. Manual removal of weeds is cumbersome and expensive. Therefore, mechanical and chemical means should be employed to control weeds. Mowing, grubbing, clearing, grazing, burning and spraying are used to control weed growth. Burning of weeds or spraying of weedicides may be required more than once in the growing season. The practice of grazing should be used with caution as it might damage the side slopes. Depending upon the method of maintenance, side slopes of the channels should be decided right in the construction plan (Table 2.17). These guidelines override the guidelines on the basis of soil type discussed earlier.

Stakeholders Participation in Maintenance

Participatory irrigation management rules stipulate that drains falling in the jurisdiction of the water users society shall be maintained by the society. Nonetheless, maintenance program should be discussed and methods of maintenance should be agreed upon with the stakeholders. The maintenance plan for group jobs should show an annual maintenance schedule and the practices to be followed should be listed. It should include details for carrying out the maintenance operations and for periodic inspection. The method for payment of maintenance costs needs to be agreed upon and included in the group agreement.

FIELD ASSESSMENT OF BENEFITS OF SURFACE DRAINAGE

Several organizations have established that surface drainage helps in alleviating the adverse effects of water logging resulting in higher productivity. Benefits of improved surface drainage through well maintained drains are illustrated through the data given in Table 2.18. Similar results in Ukai-Kakrapar command in Gujarat have shown that yield benefits range from 20% or more in paddy and sugarcane to about 50% in the case of Indian bean. Good economic returns are achieved in heavy textured Vertisols, as is apparent from the data presented in Table 2.19 (Bhandarkar and Singh, 2000 and Singh and Ramana Rao, 2014). Even more encouraging results have been reported by Uttar Pradesh *Bhumi Sudhar Nigam* indicating that reclamation of alkali soils is almost impractical without appropriate provisions of surface drainage. As such, it is an important component of reclamation projects in Uttar Pradesh. It is now established that improved yields will more than compensate the area lost in surface drainage.

Table 2.17 Ditch side slopes for use with various maintenance methods

Type of maintenance	Recommended minimum side slopes	Remarks
Mowing	3 : 1	Flatter slopes desirable for ordinary farm wheeled tractors. Special equipment is used on steeper slopes
Grazing	2 : 1 or flatter	For ditches greater than 1.25 m deep, use ramps.
	1/2 : 1 or flatter	For ditches greater than 1.25 m deep, use ramps
Dragline	1 : 1	Suitable for use in stable soils on ditches greater than 1.25 m deep
Blade equipment	3 : 1	Flatter slopes desirable

Table 2.18 Effect of surface drain maintenance on water table, soil condition and crop yield in Tawa Command (MP)

Treatment	Evaluation parameters			
	No. of days water table < 100 cm in a year	Time (days) for field to become workable	Soybean yield (t/ha)	Wheat yield (t/ha)
Ill- maintained drains (no modification)	120	20	1.12	2.54
Well maintained drains	101	13	1.55	3.08

Source: IARI (2014)

NON-CONVENTIONAL DRAINAGE SYSTEMS

Conventional surface drainage systems comprising mains, sub-mains or collector drains implemented by the states, mostly as flood control measure, have failed to improve agricultural productivity. It is attributed to the following factors:

- Drains fail to function when these are required the most
- Leaving the construction of field drains to the farmers who fail to construct the field drains
- No link to the systems for land locked areas
- Poor upkeep and maintenance resulting in dilapidated state of drainage systems

Table 2.19 Cost : Benefit analysis of drainage technology for maize and pigeon pea

Cultivation practice	Maize		Pigeon pea	
	B:C ratio	Net benefits (Rs./ha)	B:C ratio	Net benefit (Rs./ha)
Surface drainage at 20 m spacing (SD)	1.35	7300	1.53	12500
Conventional (No drainage)	1.16	3330	1.27	6500
Crop cultivation under low lying waterlogged condition	0.41	(-)11700	0.47	(-) 12000

Therefore, these systems must be supplemented by some kind of field drainage systems, which should be under the direct control and supervision of the farmers. Since farmers are usually disinterested in surface drainage, surface drainage needs to be integrated with one or more of other utilities such as crop diversification, groundwater recharge, multi-enterprise agriculture etc. It should encourage the farmers to construct/utilize drainage structures.

On-farm Ponds

On-farm ponds, being under the direct control of the farmers, are able to provide effective drainage, conservation of rainfall and possibilities of its reuse during dry spells. A three-tier rainwater management technology has been proposed and tested at CSSRI, Karnal (Gupta and Dhruvanarayana, 1974). In a rice-based cropping system, the three tiers of the system are (i) rainwater is stored on the cropland till such time and extent that it will not adversely affect the crop (Fig. 2.10), (ii) excess water from the fields is stored in on-farm ponds and (iii) the stored water is subsequently utilized for irrigation during dry spells. These ponds can also be used for aquaculture. If used for aquaculture, the value added water can be used for irrigation. If not, the excess water can be used for artificial groundwater recharge (Fig. 2.11). The results reveal that the technology helps to conserve rainwater within the farm. More than 80% or even 100% rainwater can be stored within the farm. Besides ensuring land drainage, it helps to reduce the irrigation water requirement of rice crop. The net benefit to groundwater recharge ranges from 8- 40 cm depending upon the total rainfall in the season (Gupta and Pandey, 1979). The best results emerge when the pond is used as a multi-enterprise agriculture hub with aquaculture. The technology is now becoming popular with demonstration farms coming up in large number that practice aquaculture and vegetable and horticultural crops on the dykes (Singh *et al.*, 2007).

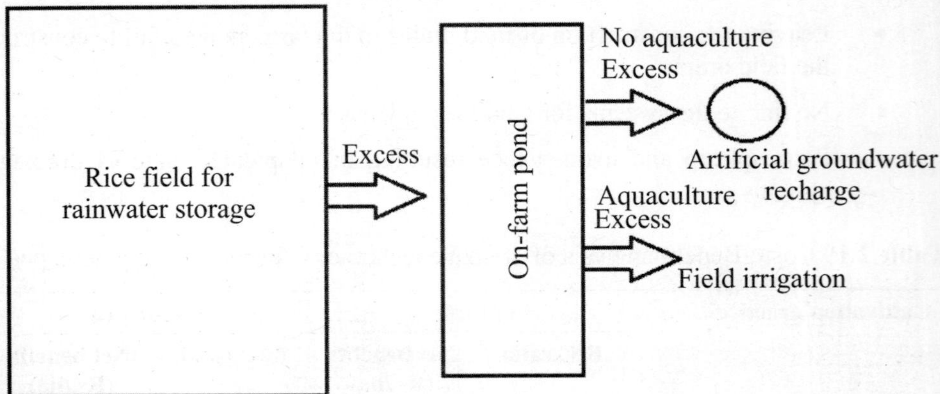

Fig. 2.10 A schematic view of rice fields (not to scale) showing drainage of surplus rainfall

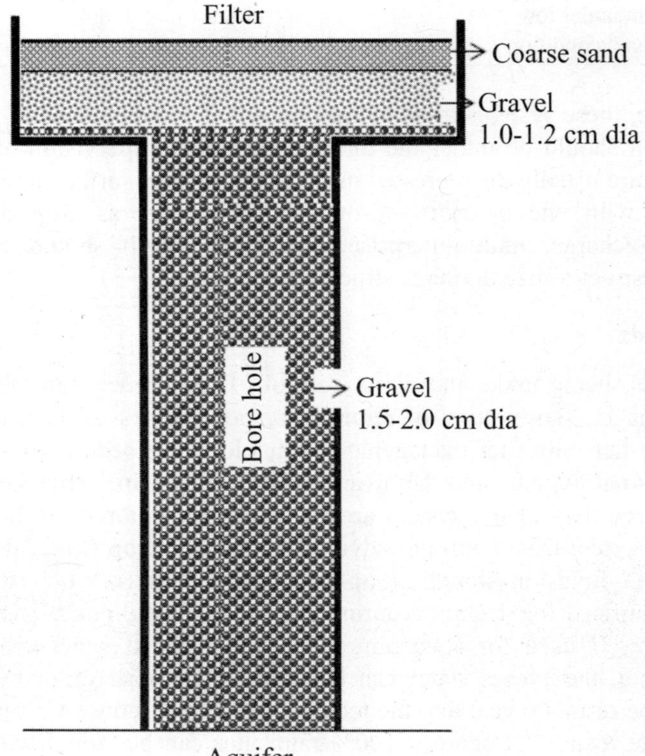

Fig. 2.11 A low cost filter shaft for drainage and artificial groundwater recharge

Integrated Drainage and Groundwater Recharge

Farmers often become complacent as field drainage is not required in every season or even every year. Discussions with communities reveal that they are also not willing to invest in the recharge structures, so necessary to reverse the falling trends of groundwater in many regions. Therefore, attempt has been made to combine artificial recharge structures that may also serve the purpose of drainage (Kaledhonkar *et al.*, 2008). It will obviate the need to convince the stakeholders to operate and maintain the recharge structures. Special low cost shafts have been used to provide drainage and artificial groundwater recharge (Fig. 2.11). Use of cavity wells for irrigation, drainage and groundwater recharge has been demonstrated by Goel and Gupta (1986). The defunct cavity pits can be put to use for drainage and groundwater recharge (Fig. 2.12).

Fig. 2.12 Cavity pits for drainage and artificial groundwater recharge

QUESTIONS (THEORY)

1. Define surface drainage. What are the objectives of a surface drainage system?
2. Define drainage coefficient. Which methods are used to measure drainage coefficient.
3. Describe the Rational method to determine peak runoff.
4. Describe curve number method to determine the total runoff.

5. How will you compute the drainage discharge at the junction of two drains when the area drained by one drain is 30% of the total area drained by the two drains.
6. Describe the construction method for surface drains.
7. Write a note on maintenance of open drains and describe preventive and corrective maintenance activities.
8. Write short notes on the following:
   Cypress creek formula, Boston Society formula, Bedding system, Random system of drainage, Laser land leveling, Characteristics of a good outlet

QUESTIONS (OBJECTIVE)

1. Fill in the blanks

1. Vegetables are more __________ to surface stagnation than arable crops such as wheat.
2. Water __________on building walls, wells and other spots are the indicator of the water stagnation depth that can be expected in the area.
3. Surface drainage coefficient can be expressed in units of __________ or flow rate per unit area, cumecs/ha, L/s-ha.
4. In rational formula peak rate of runoff Q_p is given as CIA/360. Here, I is the uniform intensity of the design storm of duration equal to time of __________.
5. Hydrological soil group D is characterized by soils that have very __________infiltration rate.
6. In Antecedent moisture condition I, the moisture content of the soils is at about ______ point.
7. For similar conditions, curve number of AMC II will be __________ as compared to AMC III.
8. In Boston Society formula peak runoff rate is directly proportional to the square root of the _____.
9. The 20-40 rule is used to design __________ of the drains at junctions.
10. Parallel lateral open ditch system besides surface drainage also helps to control ______________.
11. ___________ uniform flow formula is commonly used for hydraulic design of channels.
12. For the best hydraulic cross-section of a rectangular channel, the hydraulic radius is half the _________ of flow.
13. For a parabolic cross section, the cross sectional area is given by 2/3 t d such that *t* is the _______ and *d* is the depth of flow.

14. Channels with the same cross sectional area can have different __________radius.

15. Curve number technique can also be used to assess __________ runoff rate.

2. Tick mark correct or most appropriate answer (A, B, C or D) from the following

1. Which is most sensitive to water stagnation?
   (A) Tomato (B) Cotton
   (C) Rice (D) Sugarcane ( )
2. Drainage coefficient can be described in units of
   (A) mm/day (B) Cumecs/h
   (C) Liters/h-ha (D) Any of these ( )
3. Which formula uses the direct runoff calculated by curve number method
   (A) Rational formula (B) Boston Society formula
   (C) Cypress creek formula (D) None of these ( )
4. When used in FPS system, peak discharge in Boston Society formula, is proportional to
   (A) Area (B) Square root of the area
   (C) Square of the area (D) None of these ( )
5. In bedding systems, bed width is a function of
   (A) Hydraulic conductivity (B) Drainable pore space
   (C) Soil texture (D) None of these ( )
6. Parallel open ditch systems helps in
   (A) Surface drainage (B) Control water table
   (C) Both A and B (D) None of these ( )
7. For a hydraulically best cross section of a rectangular channel the hydraulic radius, R is given as (d is the depth of flow)
   (A) $R = d$ (B) $R = 2d$
   (C) $R = d/2$ (D) None of these ( )
8. For a hydraulically best cross section of a trapezoidal channel the bed width b is given as (d is the depth of flow)
   (A) $b= 2d \tan \theta/2$ (B) $b= 2d \tan \theta$
   (C) $b= d \tan \theta/2$ (D) $b= d \tan \theta$ ( )

9. If the cross sectional area, A is given as $A = bd + zd^2$, such that b is bottom width and d is depth of flow, the z is

   (A) Side slope z : 1 (Horizontal : Vertical)

   (B) Side slope z : 1 (Vertical : Horizontal)

   (C) None of A and B

   (D) Any of A and B ( )

10. The drainage coefficient of a watershed of 1000 hectares is 1.44 mm/day. How much average rate of discharge is expected at the outlet

    (A) 6 m^3/min (B) 10 m^3/min

    (C) 100 m^3/min (D) None of these ( )

11. Which one of the methods can be used to calculate peak discharge?

    (A) Rational formula (B) Curve Number method

    (C) Any of A and B (D) None of A and B ( )

12. The open ditches for drainage should satisfy the condition(s)

    (A) Adequate capacity to carry the design discharge

    (B) Flow neither causes scouring nor silting

    (C) The side slopes are stable

    (D) All of these ( )

13. The average rate of discharge from a watershed of 1000 hectares is 10 m^3/min. What is the drainage coefficient for the watershed?

    (A) 6.0 mm (B) 14.4 mm

    (C) 1.44 mm (D) None of these ( )

14. Curve number method is used to calculate

    (A) Runoff depth (B) Peak flow rate

    (C) Both A and B (D) None of these ( )

15. In Antecedent moisture content I, the moisture content of the soils is

    (A) About saturation (B) About wilting point

    (C) Average conditions (D) None of these ( )

■■■■

Chapter 3

DRAINAGE INVESTIGATIONS

INTRODUCTION

Improved drainage through artificial means is an essential pre-requisite to reclaim waterlogged salt affected lands. Surface drainage discussed in the previous chapter is the key element to improve drainage of agricultural lands in monsoon climatic conditions. On the other hand, it needs to be augmented by some kind of subsurface drainage to reclaim salt affected lands in the presence of a shallow water table. Traditionally, removal of excess groundwater to control the water table or to remove excess salts or to prevent build-up of salts in the root zone using water as the medium of salt transport, is known as subsurface drainage. A system that is designed to achieve these objectives is known as subsurface drainage system (Fig. 3.1). The figure illustrates lateral drain spacing and depth that in turn controls the post drainage groundwater level in the treated area. Subsurface drainage has been found to be quite effective and eco-friendly technology in areas with poor quality groundwater. A large number of subsurface drainage studies conducted in India aimed at the development of appropriate location-specific drainage and reclamation technologies for solving the problems of water logging and salinity in canal commands. This chapter deals with the basic principles of drainage investigations required to assemble design information and provides guidelines based on the current experiences and performances of subsurface drainage systems at selected locations.

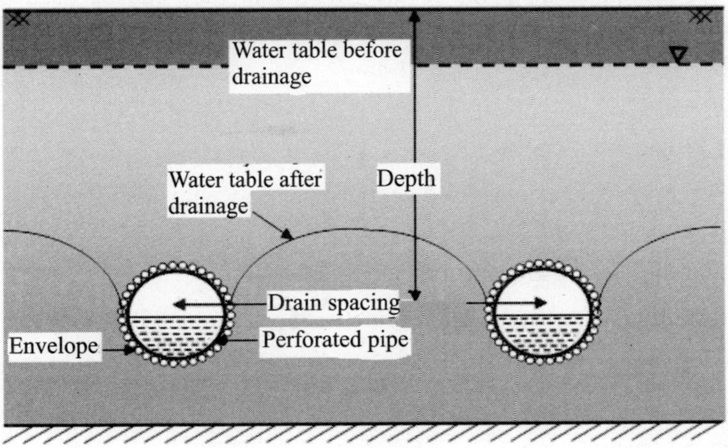

Fig. 3.1 Line drawing of a subsurface drainage system for water table control and leaching of salt

STAGES IN DRAINAGE INVESTIGATIONS

Drainage investigations are essentially required to assess the technical feasibility of a subsurface drainage project and to design and construct an efficient and cost effective system. A prerequisite for the sound planning of a survey program for this purpose is knowledge of the basic survey needs of a particular project. The engineer-manager must therefore, be knowledgeable about engineering surveying, drainage design and drainage construction so that a sound survey program is formulated. The surveys and investigations for a drainage project should reveal the following information:

- Inventory of the project site conditions
- Causes of the excess water
- Identification of appropriate solution
- Data needed to establish drainage system design parameters
- Construction problems likely to be encountered, if any

In general, the following four stages of project preparation are taken into account. Depending on the project, certain stages may be omitted or combined.

- **Identification stage**: Mostly on the basis of available information with hardly any analysis or appraisal.
- **Reconnaissance/Pre-feasibility stage**: Information collected through reconnaissance
- **Feasibility stage**: Information collected through semi-detailed type of field investigation (mapping scale 1:10,000/50,000) for presentation of the proposed plan in sufficient details to demonstrate technical soundness and to enable cost estimation within ~10% accuracy
- **Final stage**: Information collected through detailed type of investigations, elaboration of all plans to the extent that they can be used as working documents. These include detailed plan and design, construction drawings and specifications.

For the design of the system, following information is required (Ritzema, 1994).

- Topographic details
- Soil profile description and depth to impermeable layer, etc.
- Position and fluctuations of the groundwater table, depth to water table, artisan pressure and groundwater quality
- Infiltration rate, hydraulic conductivity and drainable pore space
- Climatological and meteorological data; drainage coefficient
- Data on soil salinity/alkalinity
- Crops and their tolerance to water logging
- Irrigation practices and water requirement
- Surface drainage network and the outlet conditions

The data on these aspects is obtained through systematic surveys and investigations including laboratory and field investigations. The extent and degree of the investigations depends upon the investigator's knowledge of the area, the manpower and the funds available. The amount of data already in hand or available with different agencies operating in the area, size of the project and experience of executing similar projects in the vicinity can also play a major role in deciding the intensity of investigations. For example; the data on the following aspects may be available with different agencies operating in the area:

- Basic topographical maps
- Soil survey reports
- Water table data
- Well logs
- Irrigation and drainage maps
- Crops, crop yields and cropping patterns
- Land ownership

The drainage investigators must look for drainage works carried out by any other agency in the area adjoining to the present one. If the geologic, soil and topographic conditions are alike, it will help in getting a great deal of insight into the drainage problems and their solutions. Even realistic drainage coefficients can be evaluated from the drain discharge of an operating drainage system in the adjoining area. Drainage wings of the state can provide information on surface conditions, out falls and sources of excess water. At present, these departments may not be well equipped with providing information on sub-soil drainage but for surface features, advantage of their experience must be exploited. It may be noted that large projects require more intensive investigations. It is advisable to foresee the minimum data requirement and plan the best means of obtaining the same.

Following kinds of surveys are undertaken:

- Reconnaissance surveys
- Topographic surveys
- Soil and Geologic Surveys
- Photogrammetric surveys
- Groundwater surveys
- Construction surveys

Besides, special investigations are needed to assess the water transmission characteristics such as

- Infiltration rate
- Hydraulic conductivity
- Drainable pore space

RECONNAISSANCE SURVEY

To begin with, reconnaissance or an exploratory survey based on available information and some limited fieldwork is conducted. Secondary information from available maps (geological, topographical, soil and land use), reports related with surface and groundwater, climate, cropping pattern and yields and the logs of shallow/deep wells are collected to have preliminary information of the project site. The topographic map of area/state is also used to get some relevant information of the problem. An isohyetal map of the state is used to assess the annual average rainfall of the area. The secondary data also reveal the trend of rise and fall of the groundwater and its quality. The soil maps are used to get information on soil texture and problem of soil salinity/alkalinity. During the reconnaissance survey, first hand observations are made on water table in open wells that indicates the depth to water table. Any white incrustation on the soil surface confirms the problem of salts. The existing grades, side slopes of existing channels, drain flows; high water marks on buildings or bridges can give insight into the problems.

During the reconnaissance survey, enquiries with local people can reveal much needed information. Seasonal variations of the drainage problem, response of the crops to shallow water table and their viewpoints on reclamation and management practices in the area can prove to be quite helpful. Such visits are also used to locate on the base map information on physical and surface water boundaries, source of irrigation water and natural drainage outlets. The reconnaissance survey helps to draw some inferences on:

- Preliminary diagnosis of the drainage problem
- Rough outline of possible solution(s)
- Delineation of project area and its subdivisions
- Evidence that the proposed project is promising and desirable.

TOPOGRAPHIC SURVEY

Topographic survey is the basis for further investigations and to plan a drainage project. Therefore, data collection should begin with the preparation of a detailed topographic map of the area. For this purpose, a map on 1:2400 to 1:6000 scale is appropriate. In RAJAD, National Topography maps of Survey of India scale 1:50,000 were used as base maps for identifying land features and locating investigational points at sites. Other maps at a scale of 1:25,000 showing canal and surface drains were also used to support these maps. Suitable grid system is developed to create common reference points for all information. The grid may comprise a block of 1' longitude and 1' latitude. Each block is further divided into 100 tetragons with an area of about 77.22 ha each. Each tetragon can further be subdivided into 9 sub-grids representing 8.5 ha area. Such a grid helps in conducting semi-detailed soil surveys and avoids bias of the investigators.

Topographic map of the area should be prepared well in advance and should show all physical features, natural or man-made, which can influence drainage design. Great judgment is required to prepare a topographic map and to represent relevant information on it. The scale and contour interval should adequately describe the area. Whereas, for land drainage, minor field elevation differences are unimportant, they become essential when designing a field drainage system. A 15 cm contour interval may usually suffice. Apart from elevations of the fields, elevations of all critical locations such as water elevations in existing surface channel network and potential outlets etc. should be indicated. Boundary conditions are constraints that govern the pattern of flow from source to sink and are finally related to installation of a drainage system. Such details must appear on the map of the area. In addition, locations of existing drains, roads, railways or pipelines, farm boundaries, land use pattern should be marked, as it will add to the value of the map. The map prepared in this way also reflects on the possible cause(s) of the problem. The topographic map provides information on the general slope of the area both in the longitudinal and transverse directions (Fig. 3.2; Rao *et al.*, 1986). Based on the topographic map of the Sampla site and of the project area within the site, collector was planned in the longitudinal direction with a slope of about 0.2%, while the laterals were laid in the transverse direction with a slope of 01 to 0.15%.

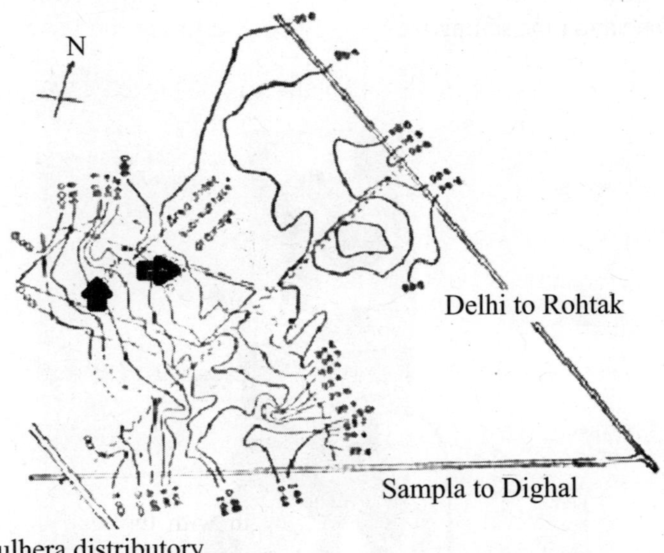

Fig. 3.2 Topographic map of the Sampla (Haryana) drainage project

Equipment Required

Topographic surveys can be accomplished in a number of ways. It depends upon the equipment available and the details required. The plain table with a telescopic alidade provides good to excellent information on topography. In present day drainage investigations, more sophisticated equipment such as total station is used (Fig. 3.3). The total station is essentially an electronic survey instrument. The data in this equipment are stored in data logger/REC module attached with the instrument. Although costly, it is more efficient and precise. Work can be accomplished much faster than with the traditional equipment. Raw data can be uploaded to the computer for further processing and preparation of base/topo maps. Summary characteristics of a typical total station and its capabilities are described as follows:

Input parameters

- Angle unit: degree or gon
- Distance units: feet or meter
- Pressure unit: Expressed in height of mercury column (inches or mm)
- Temperature: ºF or ºC
- Prism constant (Usually 0.03 m)
- Offset distance
- Face 1 or Face 2 selection
- Automatic point number incrimination
- Height of instrument
- Height of reflector
- Point numbers and code numbers for occupied and sighted stations
- Date and time settings

Fig. 3.3 A total station used for topographic surveys in large projects

Capabilities

- Monitor: battery status, signal attenuation, horizontal and vertical axes status, collimation factors
- Compute coordinates: northing, easting, elevation
- Traverse closure and adjustments
- Topographic reductions
- Remote object elevation
- Distance between remote points

- Inversing
- Resection
- Horizontal and vertical collimation corrections
- Vertical circle indexing
- Records search and review
- Load external programmes
- Downloading to computer
- Transfer of computer files to the data collector for layout purpose

SOIL AND GEOLOGICAL SURVEY

In this kind of survey, information on soil texture, depth to impermeable layer, cross section of the profiles, particle size distribution, structure and macro pores, engineering properties and soil salinity and alkalinity problems are collected. Reports of soil surveys conducted by the Department of Agriculture or Soils Departments of the State Agricultural Universities and pre-irrigation surveys of irrigation projects can be used to get much information. Soil maps of the state are usually available with many organizations. State Atlas is also a handy source of information. For primary data collection, the soil survey is conducted on a pre-decided grid, where soil and water samples are collected to determine the required information.

Digging Soil Profile(s)

Ideally, one should dig a few soil test pits around the site to capture any differences in the soil profile. Consider digging a soil test pit in a wet and dry area of the field for comparison sake. A soil profile exposed to a depth of 1.5-2.5 m should suffice. The following observations should be made.

- Note the depth and thickness of the layers.
- For each layer, consider the soil texture.
- The presence of visible cracking in the soil profile.
- Depth of rooting, shallow root systems indicate poor drainage.
- Colour of different depths.

These pits are also used to take soil samples from appropriate depths. It is always better to consult a specialist for examination of the profiles to formulate recommendations.

Profile studies are supported by the boring tests, which are carried out at enough locations necessary to yield information that is quite representative of the area. The U.S. Bureau of Reclamation recommends boring up to 7 m depth. Usually borings are made on a grid pattern at pre-determined locations but additional bores are made whenever significant changes are noticed from one bore to another. A change or break in topography can also mean a change in sub-soil horizons. It reflects the need for additional bores. During this operation color, structure and consistency of different layers are also recorded along with mottling and mineral deposition etc.

Cross Section of the Profile

The profile cross section is drawn by selecting a line through the area and plotting ground surface elevations and the depths of the textural changes as determined from borings along the line indicating textural classes between the plotted points of change. Such a cross-section helps in correlating the borings. This will also show continuity or breaks in the subsoil stratigraphy.

Soil Samples

A great deal of economy in the cost of drainage investigations can be achieved by proper planning and by making a proper grid. For this purpose usually three procedures are used.

A temporary grid is usually made in small projects. Any grid, say 100 × 100 m, can be made by staking the lines. All investigations are done on this grid as per procedure described towards the end of this section. Revenue grid is used in line with the grid system of the revenue maps, upon which every plot of 1 acre (0.4 ha) can be systematically numbered. The revenue grids have successfully been used in Haryana Operational Pilot Projects (Fig. 3.4). Another way to construct a grid is to use Survey of India maps as proposed in the RAJAD project and explained previously under topographic survey (Garg and Bhargava, 2001).

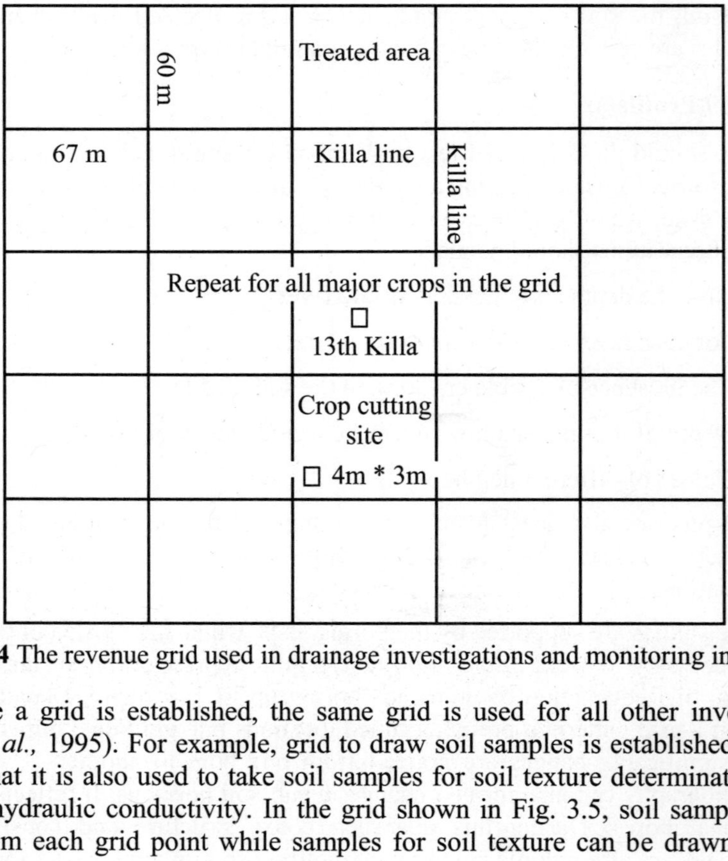

Fig. 3.4 The revenue grid used in drainage investigations and monitoring in HOPP

Once a grid is established, the same grid is used for all other investigations (Gupta *et al.*, 1995). For example, grid to draw soil samples is established in such a manner that it is also used to take soil samples for soil texture determination and to measure hydraulic conductivity. In the grid shown in Fig. 3.5, soil samples can be drawn from each grid point while samples for soil texture can be drawn on every third grid point while hydraulic conductivity can be measured on every alternate point of soil texture grid. That means the grid chosen for soil texture measurement is 3 times the grid for soil sampling while for hydraulic conductivity evaluations it is double the grid size for soil texture analysis (Fig. 3.5). The grid dimensions are decided depending upon the funds and manpower resources available in the project.

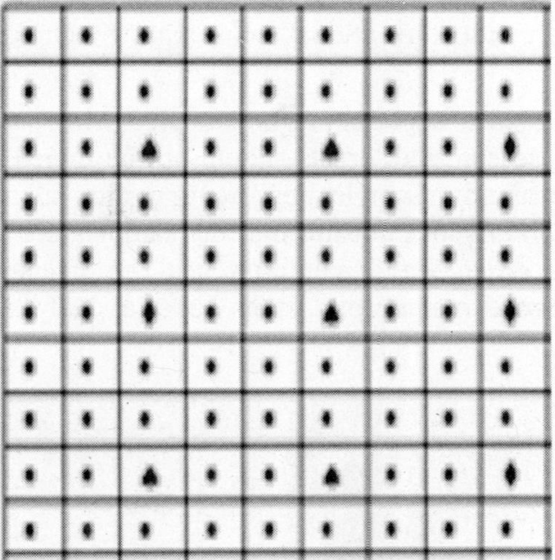

• Soil samples for salinity; ♠Soil samples for salinity and soil texture; ♦Soil samples for salinity, texture and determination of hydraulic conductivity

Fig. 3.5 Grid set-up to draw samples for soil salinity, texture and hydraulic conductivity

The method for soil sampling depends on the purpose for which a sample is collected. In a land reclamation program, the surface (0-15 cm) and subsurface (15-30 cm) sample are taken for laboratory analysis. Thus, for each grid point, two composite samples should be sufficient. When it is intended to improve a patchy field, model soil samples should be composited from a few representative samples taken from within these spots.

The depth to which samples should be drawn depends on the land use. If the intention is to grow shallow to medium rooted crops like rice, wheat etc., it is sufficient to obtain samples up to a depth of 30 cm. A major fraction of the roots of these crops is confined to this zone. If, on the other hand, it is proposed to grow deep-rooted crops like cotton, it is necessary to take samples up to a depth of about 60 cm. In case the intended land use is to plant trees (horticultural or forest tree species), it is essential to carry out a detailed study of the soil profile up to a depth of 150 to 180 cm. The performance of tree species on long-term basis is strongly influenced by properties of the lower soil layers including the presence of a hard layer due to $CaCO_3$ concretions etc. Since soil texture and hydraulic conductivity are measured at different depths, a part of these samples can be used to investigate soil salinity variations with depth.

For samples up to 30 cm depth, a cut in the soil can be made with a spade and a thin slice of soil taken to a desired depth (0-15 and/or 15-30 cm) with the help of trowel (Mason's spatula). An auger is used, if sample from deeper soil layers are needed (Fig. 3.6). For conventional diagnostic analysis for drainage purposes, it is sufficient to have about 250-500 g of soil sample. Remote sensing equipment such as Electromagnetic Induction Meter (EM 38) can be used, which measures the bulk salinity in the upper 0.6 and 1.2 m depth of soil profile. It helps to reduce the time required for salinity surveys. It also offers an easy way to monitor soil salinity in an intensive manner than with conventional soil sampling techniques and at much lesser cost.

Important: The slick spot should be sampled when spots are best identified during the cropping season. Salt crusts should not be mixed with the soil samples. Collect salt crust separately. Samples representing slick spots should not be mixed with samples from good spots. Ensure that the samples are not taken immediately after rains/irrigation. Also avoid sampling near an irrigation channel, drainage channel, dykes, farmyard manure pits, spots under a shady tree and from threshing floors or roads.

Fig. 3.6. Soil sampling with a screw auger

Water Samples

Proper diagnosis of the nature and the amount of salts present in water is essential to assess the possible hazards it may pose to the soil or the crop during and after the reclamation. Collection of representative sample is, therefore, basic for correct diagnosis and for making sound recommendations for the management of poor quality water. As far as possible, plastic bottles are used for collection of water samples. Glass bottles can also be used if plastic bottles are unavailable. The bottles should be thoroughly cleaned and rinsed 2-3 times with water before a sample is drawn. Avoid metal containers for collection of water samples.

In case of gravity canal, take the sample of the water from the channel after irrigation has been in progress for about 15-20 minutes. For groundwater, the sample should be taken only after the tube well has run for about 30 minutes. The samples from the observation wells and piezometers should be drawn as per procedure mentioned in section on measurement of water level. Most routine salinity appraisal tests require about 100 ml of water. However, to be on the safer side and to allow specialized tests, if need be, as much as 500 ml of sample should be sufficient.

The water containers should be properly screwed and if possible wax sealed. Containers should be properly labeled for identification. For microbiological investigations, if needed, the containers should be appropriately sterilized before taking the sample.

Soil Texture

Particle-size distribution and the soil texture may give an indication of the permeability, water retention, plasticity, aeration, base-exchange capacity and adaptability of the soil for construction purposes. Many of the design parameters are also related to the soil texture. Thus, first guess of these parameters can be made with this information. It also helps to decide/justify the drain spacing. Normally, light textured soils are easily drained than the heavy textured soils. Information on soil texture also helps to decide the side slopes of the channels and to assess the roughness coefficients. Particle size distribution curve of the soil at the drain depth also determines the need of filter/envelope and helps in designing these materials.

Feel and Touch Method under Field Conditions

In the field, soil texture is assessed by feel of the soil with the fingers. The general definitions of soil textural classes in terms of field experience and feel included in the USDA Soil Survey Manual (1951) are as follow:

Sand: Sand is loose and single grained. The individual grains can be seen or felt readily. Squeezed in the hand when dry, sand falls apart when the pressure is released. Squeezed when moist, it forms a cast but crumbles when touched.

Sandy Loam: A sandy loam is a soil containing a high percentage of sand but having enough silt and clay to make it somewhat coherent. The individual sand grains can be readily seen and felt. Squeezed when dry, a sandy loam forms a cast that falls apart readily. If squeezed when moist, a cast can be formed that bears careful handling without breaking.

Loam: It is a soil having a relatively even mixture of different grades of sand, silt and clay. It is mellow with a somewhat gritty feel but is fairly smooth and slightly plastic. Squeezed when dry, it forms a cast that bears careful handling, and the cast formed by squeezing the moist soil can be handled freely without breaking.

Silt Loam: A silt loam is having a moderate amount of fine sand and only a small amount of clay; over half of the particles are of the size called silt. When dry, a silt loam appears cloddy but the lumps can be broken readily; when pulverized, it feels soft and floury. When wet, the soil runs together readily and puddles. Either dry or moist, it forms a cast that can be handled freely without breaking; when moistened and squeezed between thumb and fingers, it does not ribbon but has a broken appearance.

Clay Loam: A clay loam is fine-textured soil that usually breaks into clods or lumps, which are hard in a dry state. When the moist soil is pinched between the thumb and finger, it forms a thin ribbon that breaks readily, barely sustaining its own weight. The moist soil is plastic and forms a cast that bears much handling. When kneaded in the hand, it does not crumble readily but works into a heavy compact mass.

Clay: Clay is fine-textured soil that usually forms very hard lumps or clods when dry and is very plastic and usually sticky when wet. When the moist soil is pinched between the thumb and finger, it forms a long, flexible ribbon. Some clay soils are very high in colloids, are friable and lack plasticity at all conditions of moisture.

Organic soils: The soils in which organic matter content ranges from 20% to as high as 95% are referred as organic soils. They are generally classified on the basis of degree of decomposition of the organic deposits. If deposits are only slightly decayed or non-decayed are called peat. In peat soils, the kinds of plants that make up the deposits are easily identifiable. Well decomposed deposits in which the original plants cannot be identified are called muck. The word mucky as mucky loam and mucky silty loam is used in the textural class name for horizons of mineral soils that contain 15% or more partially decomposed organic matter.

A recheck is made through mechanical analysis in the laboratory. Sand, silt and clay contents are determined in the laboratory using pipette or hydrometer methods. The fractions of these particles so determined are used to evaluate soil texture using soil textural diagram (Richards, 1954).

Sieve Analysis

For this purpose soil samples are drawn from the crop root zone, intermediate layers and at the drain depth. The particle sizes larger than 0.074 mm are usually analyzed by means of sieving. Method of sedimentation is used to analyze soil material finer than 0.074 mm. Sometimes it is recommended that sedimentation analysis be carried out with particle sizes that pass through a 2 mm sieve. In that event, the pan could be placed below a 10 mesh sieve.

First the soil is oven dried and then all the lumps are broken into small particles before they are passed through the sieves. US standard mesh sieves are commonly used. A dry sample of known weight is placed in a set of precision-screened sieves with successively finer opening size. A sieve set consists of a cover, 4, 10, 20, 40, 60, 100, 140, 200 mesh sieves and a pan. The sieve number (Mesh) and the opening sizes for the US standard sieves are given in Table 3.1. The sieves are placed in a standard shaking machine. The machine is operated for 5 min at amplitude of 0.75 mm with a frequency of 50-60 Hz. When the material finer than the screen size has passed through the individual screen, material retained on each sieve is weighed and expressed in the form as reported in Table 3.2. The percent fraction (%) retained on a certain sieve, P is calculated with the help of Eq. (3.1). If a sizable portion of soil is retained on the No. 200 sieve, it should be washed. It is accomplished by placing the sieve and retained soil in a pan and pouring clean water on the screen. A spoon or glass rod is used to stir the slurry. Recover the soil which is washed through; dry and weigh it. The weight of soil recovered should be subtracted from the weight retained on the No. 200 sieve and added to the weight retained in the pan.

Table 3.1 US standard mesh and aperture openings of the sieves

US standard mesh	Sieve aperture (mm)	US standard mesh	Sieve aperture (mm)	US standard mesh	Sieve aperture (mm)
3.5	5.66	18	1.00	**80**	**0.177**
4	4.76	20	0.841	**100**	**0.149**
5	4.00	25	0.707	**120**	**0.125**
6	3.36	30	0.595	**140**	**0.105**
7	2.83	35	0.500	**170**	**0.088**
8	2.38	40	0.420	**200**	**0.074**
10	**2.00**	**45**	0.354	230	0.063
12	**1.68**	**50**	0.297	270	0.053
14	**1.41**	**60**	0.250	325	0.044
16	**1.19**	**70**	0.210	400	0.037

Bold numbers are commonly used sizes

$$P = 100\ gr/G \tag{3.1}$$

Here, gr is the dry weight of the soil in g retained on a certain sieve and G is the dry weight of the total soil sample used in the particle size analysis. Cumulative soil mass retained is calculated from the cumulative mass retained on each sieve (Table 3.2). Cumulative soil mass retained on the sieve is plotted as a curve against size of the sieve and is designated as particle size distribution curve (Fig. 3.7). Sometimes, the graph is also plotted between the cumulative passing and grain size, grain size being on the log scale.

The shape of the cumulative particle size distribution curve is an indication of the uniformity of the soil. A steep curve indicates a uniform soil while a flat curve reveals a well-graded soil. Humps in the curve indicate a mixture of two or more uniform soils. These are only few specific instances but additional conclusions can be drawn on the basis of the shape of the curve. The curve is used to calculate the effective diameter of the grains, d_{10}, such that 10% of the grains are finer and 90% are coarser (retained). It is a measure of the fineness of the aquifer material. Uniformity coefficient is determined by taking the ratio of the d_{60} and d_{10} where d_{60} is 40% retained and d_{10} is the 90% retained on the curve. If CU is more than 4, it is highly non-uniform material and if the CU is less than 2, it indicates a uniform material. Another parameter that is calculated with the curve is C_c, the coefficient of curvature or gradation given as:

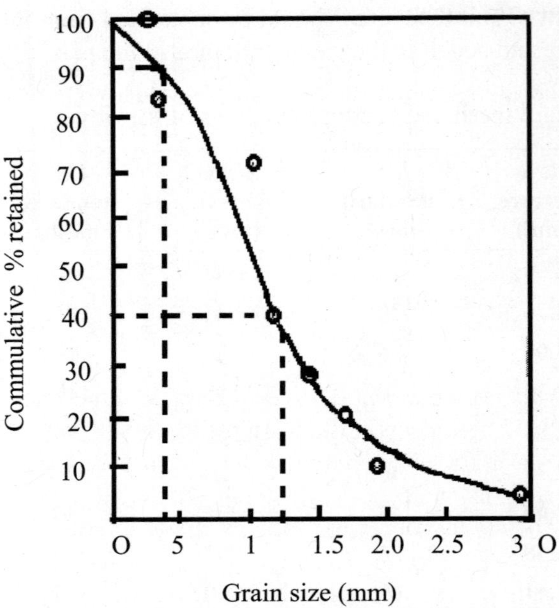

Fig. 3.7 Grain size distribution curve

$$Cc = \frac{(D_{30})^2}{(D_{60} \times D_{10})} \tag{3.2}$$

Here D_{60} is the grain diameter for 60% passing, D_{30} is the grain diameter for 30% passing and D_{10} is the grain diameter for 10% passing. A well graded soil will have a uniformity coefficient greater than about 4 for gravels and 6 for sands and a coefficient of gradation between 1 and 3 (for gravels and sands).

Table 3.2 Presentation of sieve analysis data

Sieve opening (mm)	Retained on sieve		Cumulative retained		Cumulative passing	
	Weight (g)	%	Weight (g)	%	Weight (g)	%
2.80	30	5.1	30	5.1	560	94.9
2.00	40	6.8	70	11.9	520	88.1
1.70	50	8.5	120	20.4	470	79.7
1.40	60	10.2	180	30.6	410	69.5
1.18	60	10.2	240	40.8	350	59.3
1.00	170	28.8	410	69.6	180	30.5
0.31	80	13.6	490	83.2	100	16.9
0.20	60	10.2	550	93.4	40	6.8
0.15	40	6.8	590	100.2 or 100	0	0

Depth to Impermeable Layer

An impermeable layer is not completely impermeable to water. Rather it is a relative term. Practically speaking, when the hydraulic conductivity of the layer is around 10% of the weighted mean hydraulic conductivity of the upper soil layers; the layer is designated as impermeable layer. Sometimes, a layer having hydraulic conductivity less than 20% of the weighted permeability of the zones above is also treated as an impermeable layer. Generally, an extensive stratum having a texture of clay, silty clay, silty clay loam or heavy clay loam acts as a barrier.

Most drainage design formulae require information on this aspect. Location of drains with respect to ground surface and the impermeable layer affect the drainage design (lateral drain spacing, Table 3.3). It is important to locate the impermeable layer as accurately as possible. Many times it may be quite undulating, particularly on sloping lands. Sufficient deep borings are made, logged and plotted to locate this surface. If the barrier is not encountered up to 7 m, it may be worthwhile to try the existence of any such layer up to 13 m depth. For shallow subsurface drains, the impermeable layer below this depth has little influence on drainage design.

The drains are always located above the barrier. The occurrence of the first impervious layer will fix the lower boundary of the flow region. In case depth to impermeable layer varies widely, it is useful to divide the fields into blocks for the purpose of system design. It may also be noted that:

- The deeper depths of impervious layer will be a case for vertical drainage
- Confining of high-pressure artesian aquifer shall be a case for relief well drainage
- Occurrence of hardpan within 50-70 cm of soil depth will prohibit use of horizontal subsurface (pipe) drainage
- Soil stratification and drain depth might restrict the drain discharge.

Table 3.3 Effect of depth to impermeable layer on drain spacing for drain pipes of 100 mm diameter

Depth to impermeable layer below drains (meters)	Spacing (m)	Percentage change over the previous depth
0	9.5	-
0.5	15.9	67.4
1	19.4	22.0
2	23.4	20.6
3	26.9	15.0

Particle Size Distribution, Structure and Macro-pores

Particle size distribution, aggregates and soil structure generally determine the hydraulic conductivity of the soil. In sandy soils, even in a completely structure less state, the hydraulic conductivity may not fall below 0.6 m/d. In most clayey sub-soils, hydraulic conductivity is 0.3 m/d except in well aggregated soils. Shape of the aggregates, their size, degree of aggregation and structural stability also influence the water transmission characteristics to a large extent. If intake characteristics are to be considered favorable, the majority of soil aggregates at least 75% by weight must remain intact. Though it is known that the texture affects the feasibility of mole drainage, its stability is also affected by water stable aggregates. Rycroft and Alcock (1974) suggested that a test of stability to water is the most reliable test of finding out the stability of mole drains in a given soil. Another useful application of particle size distribution of soil at the drain depth is in designing drainage envelopes.

Soil Salinity and Alkalinity

The soil samples taken during field investigations are processed for laboratory analysis to determine, EC, pH, SAR and ESP. Cations and anions are also determined for few samples. Based on these parameters, the soil is characterized as saline, alkali or saline-alkali as per guidelines given in Chapter 8. Although pH is sufficient to assess the problem of alkali and its degree in alluvial soils of the Indo-Gangetic plains, for other soils, determination of ESP will be essential to assess the problem and to determine gypsum requirement (GR). The data are used to calculate SAR of the soils. This information is used in assessing the need of envelops.

PHOTOGRAMMETRIC SURVEY

Engineering survey methods used in the planning, design, and construction of land drainage have changed rapidly in recent years. With photogrammetric methods one can save lot of time and funds as much of the field work is eliminated. Photogrammetric methods are capable of providing contours and some of the design data required in land drainage. Aerial photographs, remotely sensed data and Geographical Information System (GIS) can be used to assess the extent and even degree of salinity and water logging. The use of photogrammetric mapping must be spot checked and should be augmented by field surveys. Depending upon the type of land drainage, field surveys may be required to establish horizontal and vertical survey control; to check topographic maps prepared by photogrammetric methods, as well as to supplement topographic details in areas of heavy tree cover. Since the methodologies are not very well documented, one must exercise due care and even consult experienced photogrammetrist to plan a sound survey programme.

GROUNDWATER SURVEYS

Any remedy through drainage emerges only if it is possible to pinpoint the source of excess water in the area. Both surface and groundwater conditions need to be considered since both are interconnected. The shallow water can be traced to one or more of the following factors:

1. Periodic flooding of the land due to overflow of rivers, drainage streams or breaches in the canals causing water table to rise
2. Submergence of low lying areas from run-off or return flow and seepage from irrigated uplands
3. Accumulation of rainwater due to restricted drainage

4. Rise in water table due to seepage from canals network and deep percolation from over irrigation
5. Presence of artesian pressure.

Depending upon the factors responsible, the nature of solution varies. For example, for problems listed at serial 1, diversion bunds and surface drainage will be appropriate. For the problem listed at serial 2, diversion drain will be suitable. For the other kinds of problems, surface, subsurface or an integration of these two systems will be appropriate. In all cases, a survey of the history of hydrology, climatic trends and irrigation status of the area are necessary along with the groundwater data.

Precipitation

Precipitation records of the daily rainfall at the rain gauging station(s) in and around the project site must be collected for at least last 20 years or so. The data are analyzed to determine the distribution of rainfall, its trend and the frequency of occurrence. A correlation between the water table fluctuation and the rainfall will reveal whether precipitation is the source of excess water or not. A hydrograph of a well in the area with superimposed monthly precipitation can give a good insight into this issue. In the absence of a good correlation, other sources of water need to be investigated. Frequency analysis of the rainfall data is utilized to evaluate surface drainage requirement of the area. To illustrate this point, let us consider the analysis of rainfall data of the Sampla (Haryana) gauging station. The frequency analysis revealed that as much as 171 mm of rainfall can occur in 3 days with a return period of 5 years (Table 3.4). Such data sets represent the real situation for most stations in semiarid regions of India. Since subsurface drainage alone cannot evacuate so much water, it is necessary to provide surface along with subsurface drainage to control the water table.

Table 3.4 Rainfall depth of different durations and frequencies at Rohtak gauging site

Duration (days)	**Rainfall (mm) for return period (years)**		
	5	**10**	**20**
1	105	1220	132
2	140	159	174
3	171	195	216
4	180	207	225
5	189	216	228

GROUNDWATER INVESTIGATIONS

The main purpose of groundwater investigations is to get information on the position of the water table and seasonal/annual fluctuations of the groundwater, which will determine the severity and extent of the problem. For these investigations, a network of observation wells on a grid is established. To investigate artesian pressure, piezometer clusters at appropriate locations are also installed. Usually both the investigations are undertaken simultaneously although intensity of piezometers is much less unless the limited tests reveal the existence of artesian pressure as a cause of the drainage problem.

Open Wells

Open wells, a very common feature of the landscape in rural India, have been used to measure the groundwater table behaviour. Central Ground Water Board and State Directorate of Groundwater have fully utilized the potential of these wells in monitoring groundwater table behavior during pre-monsoon (June) and post-monsoon (October) months. Although multiple agencies are engaged in this task, it may be useful to obtain data from all of them to prepare a unified map. Recently, few ideally located wells have also been selected to undertake monthly reading and for water quality analysis. These data provide wealth of information to a designer. The secondary data of this nature helps to reduce the intensity of monitoring. A major limitation of these wells is their high density at few points while quite sparse in other regions. Thus, there may be a need to install more observation wells to have a better insight into the regional problem.

Observation Wells

Observation wells are installed at strategic locations usually on a grid throughout the area or are appropriately meshed with the existing open wells. These wells may be cased or uncased. The simplest form of an observation well (though of temporary nature) is an auger hole left uncased in the field, which may be employed to measure water table depth during an exploratory survey.

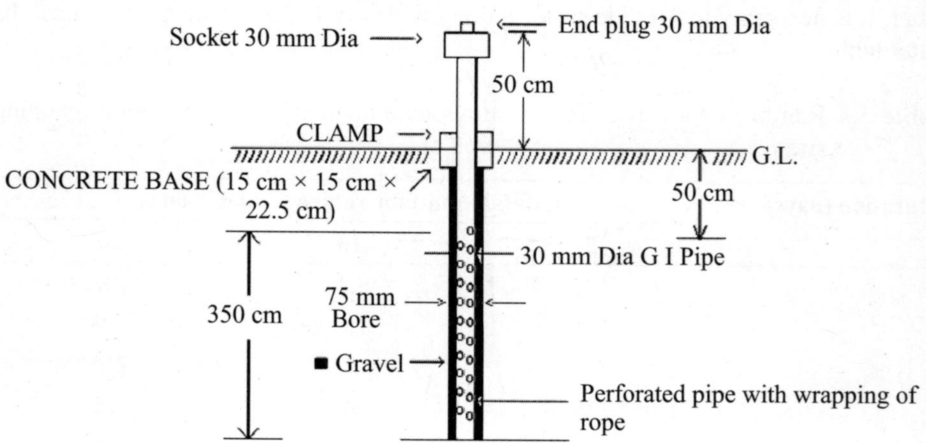

Fig. 3.8 A schematic view of an observation well (not to scale)

A cased well is usually 2.5-3.75 cm diameter plastic or galvanized iron pipe. Perforations are made in these pipes throughout except for the top 0.5 m (Fig. 3.8). The perforations in the observation well are drilled in five to six rows and are covered with jute or coir rope to prevent fine particles from entering the pipe or clogging of the tiny holes. The pipe is placed in the center of a bore hole made with a hand or power auger. The annular space between the pipe and wall of the hole over the entire length of the perforated portion is filled with gravel or coarse sand. The gravel pack facilitates the flow of water from the soil into the pipe and vice versa. The rest of the annular space is filled with fine materials (clay, clayey fine sand) whose permeability is low enough to prevent leakage along the pipe. To keep the pipe in position and to prevent any direct contact between the surface water and the backfill material, a concrete seal is provided around the pipe, extending from the ground surface to a depth of about 15 cm. The lower end is sealed while the upper end is provided with a cap, which is opened to record the observation.

A rough guide to decide the number of wells/observation wells is given in Table 3.5.

Table 3.5 Wells/observation wells required in relation to the size of the drainage area

Project area (ha)	Number of observation wells	Number per 100 ha
100	20	20
1000	40	4
10000	100	1
100000	300	0.3

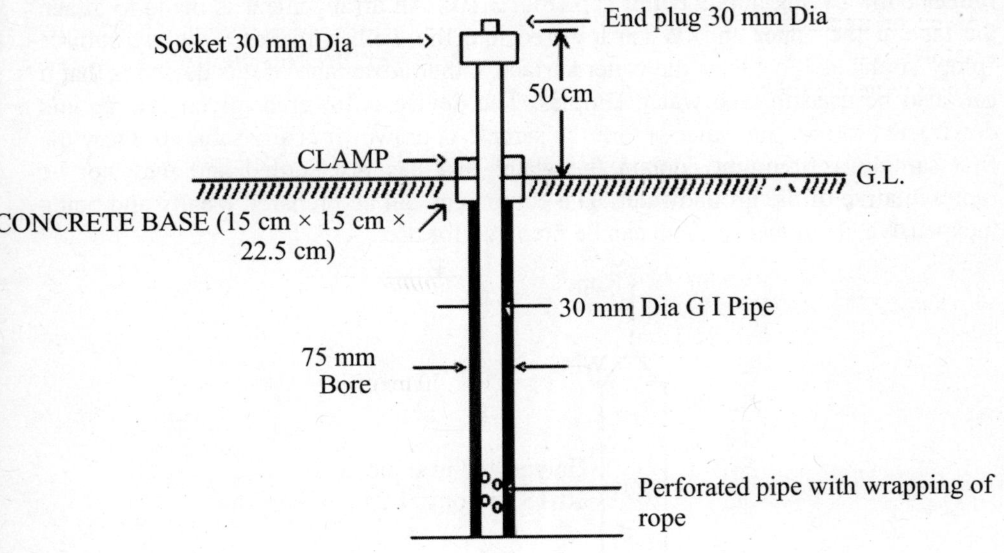

Fig. 3.9 Schematic cross-section through a piezometer (not to scale)

Piezometers

Piezometers are reliable tools for determining hydraulic head at a particular depth while the observation wells are used to measure fluctuations of the groundwater levels. A piezometer is an open-ended non-perforated galvanized iron (or sometimes plastic) pipe of 4 to 5 cm diameter, driven or jetted into the ground to the depth at which the hydraulic head is to be measured. Sometimes, perforations are made only at the bottom end for about 10-15 cm length (Fig. 3.9). Initially, a battery of piezometers is installed at one location with the lower end of each piezometer terminating at different depths. The number is increased if the problem of artisan pressure is indicated. The water level in the pipe corresponds with the hydraulic head at the lower end where the pipe terminates.

Measurement of Water Level

There are many ways to measure the water elevations in observation wells and piezometers. Electric water level indicators are now available for smaller diameter wells. Pressure loggers or electronic water level loggers are also used for accurate and continuous measurements. Float recorders are used at one or two points to record continuous variation of water levels at the drainage site. The other simple methods are:

- Chain and bell sounding device.
- Below pipe method.

The methods are described in the following paragraphs.

The most commonly time tested method is the chain and bell sounding device fitted at the bottom end of a thin steel or metallic tape. The device is a 1-1.5 cm galvanized iron pipe of about 15-20 cm length. The bottom end is sealed with a rubber cork leaving about 1 mm gap (Fig. 3.10). An arrangement is made to attach the tape at the upper end. When lowered into the well, it makes a clearly audible "plop" sound as it contacts the water surface. Other advantage of this device is that it can also be used to take water samples. The device is lowered, given few up and down jerks to mix the water before the sample is drawn. It is advisable to throw the first sample as it might contain the water that has just settled and may not be representative of the groundwater. The equipment can be prepared locally and being inexpensive, a number of them can be prepared for use.

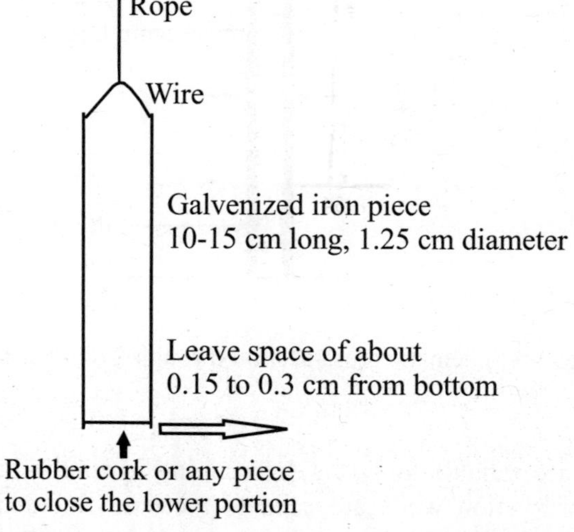

Fig. 3.10 Chain and bell sounding device to measure the groundwater levels

Another way is the blow pipe method that uses a hollow aluminum pipe of 10 mm diameter, which is inserted in a plastic tube marked in mm scale. It is lowered in the hole gradually and air is blown through the tube. As soon as the pipe reaches the water level, there is a particular sound indicating the water table. The depth of water table with reference to this point can be found.

Observation Well Hydrograph

Once the water table data are available, these data are used to prepare several kinds of map to draw information related to source of water and/or design of the system. Well hydrograph are prepared by plotting well data against time on a graph paper. A hydrographs provides insight into the causes of the rise and fall in the water level at a given site. In irrigated areas, groundwater table tends to buildup through the irrigation season, reaching a peak at the end of the season. In monsoon climatic regions, the water level is at its maximum at the end of the season (generally October) and minimum before the onset of monsoon (May or June). Sometimes hydrographs depicting water table in the month of June or October are also prepared to assess rise in water table during the monsoon season.

Groundwater hydrographs illustrating the historical record of aquifer water levels measured within a well clearly reveal the trend of the groundwater table. The hydrograph that is shown in Fig. 3.11 reveals a gradual rise in water table upon introduction of canal irrigation. Average values of all the observation wells are also plotted as a hydrograph to get a regional trend of the water table. However, in some cases, it might not give sound conclusions.

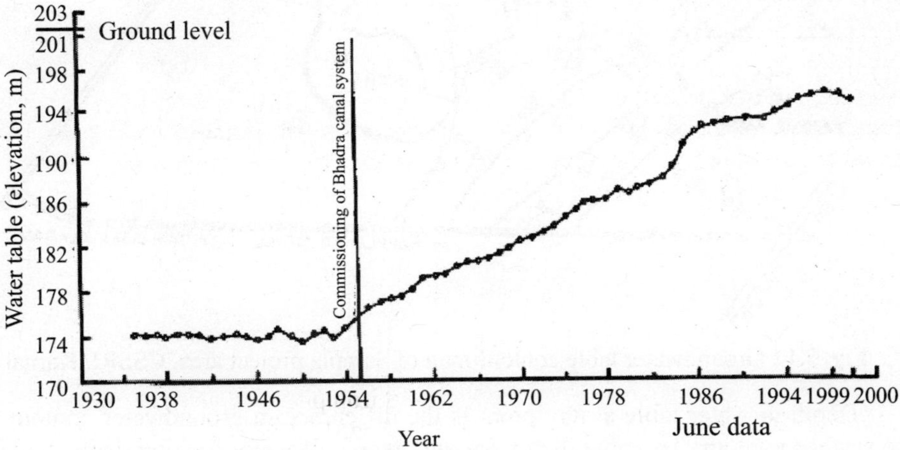

Fig. 3.11 Hydrograph depicting the rising water trend upon introduction of canal irrigation system

Groundwater Contour/Depth Maps

Water table contour maps show the phreatic surface (water table) and are prepared by plotting absolute water levels or elevations of the water table for all observation points on a topographical map. When superimposed on to a topographic map, it shows the relationship between ground surface and the water table. A water table contour map provides direct visual observation of the slope of the water table. In general, there is horizontal movement of groundwater in the direction of the slope of the water table. Radical changes in direction or slope may indicate unusual barrier formation, perched water table, artesian conditions or high and low permeability areas leading to clues for further investigations. Arrows are drawn on to the map to indicate the direction of groundwater flow (Fig. 3.12). In this figure, it is apparent that the groundwater flow is from Dulhera distributary side to the Rohtak-Delhi road marked in the map. To intercept groundwater, drain lines are laid perpendicular to the direction of flow or parallel to the groundwater contours.

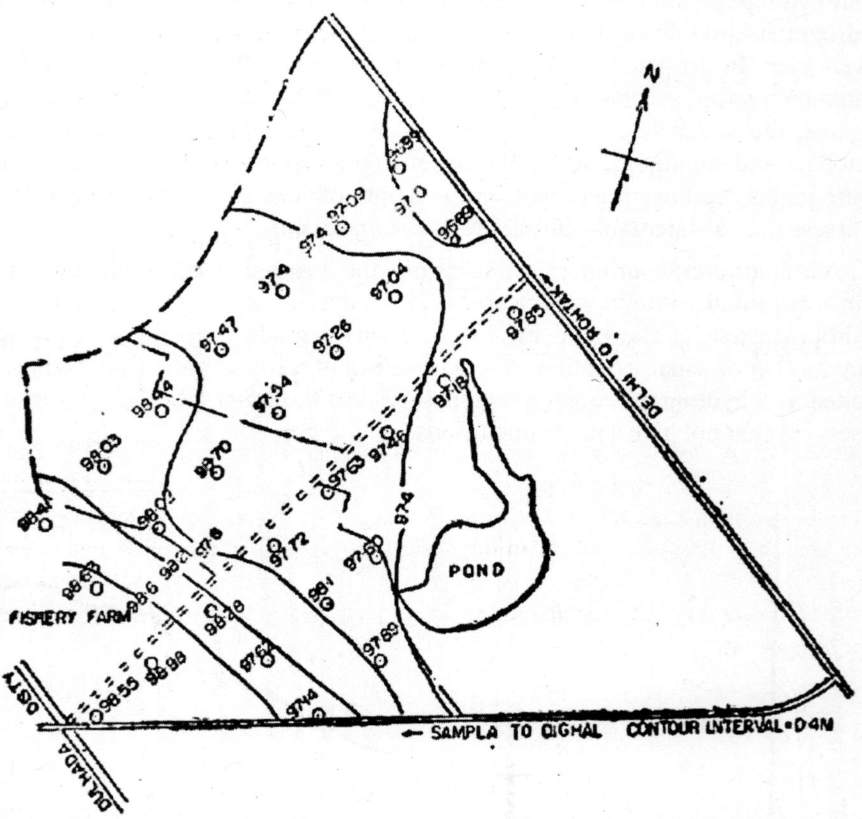

Fig. 3.12 Groundwater table contour map of Sampla project area, CSSRI, Karnal

Depth to water table at any point is the difference in groundwater contour and the surface contour. To know the extent and degree of water logging, isobath map is prepared. An isobath map shows lines of equal depth to water table (Fig. 3.13). This map is generally used to justify a project. It can also be used to assess the yield losses due to shallow water table when the crop yield- depth to water table relation is known. Commonly, a drainage project is recommended for an area when 80% of the area has a water table within critical limits. Water table fluctuation maps are also prepared to understand the variations in water table. A shallow water table with little fluctuation will mean perpetually waterlogged areas.

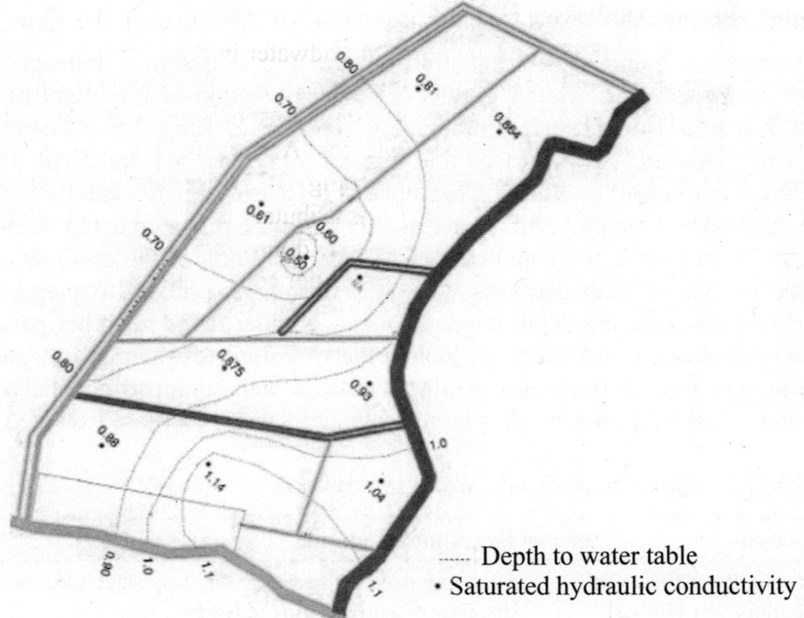

Fig. 3.13 Depth to water table map of Hatnapur site in RAJAD (Source: RAJAD, 1995)

Artesian Conditions

An observation well indicates the true water table (Fig. 3.14 left). The two observation wells in Fig. 3.14 (center) indicate the presence of a perched water table. Observations from piezometers reveal artesian conditions, if any. Artesian pressure is indicated if water level elevation in the piezometer is above the true water table (Fig. 3.14 right). If the piezometric level lies above the drain depth, such areas require relatively more intensive drainage because such areas with artesian pressure are difficult to drain.

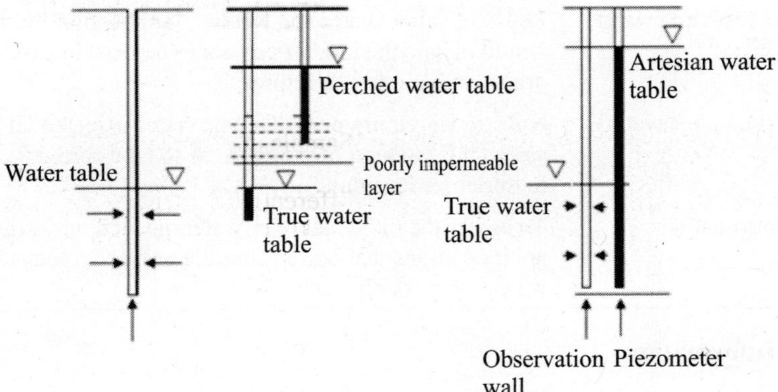

Fig. 3.14 An observation well (left), and two observation wells showing presence of a perched water table (center) and a piezometer showing the presence of artesian pressure (right)

Water Quality Analysis

A number of chemical determinations are needed to design a drainage system and later on manage salt affected lands. A brief description of important investigations and their utility in subsurface drainage is reported in Table 3.6. Although, all these investigations are not a part of this manual, only EC, pH and SAR and ESP are included in relevant section of Chapters 8 and 10 as these are essentially required for the diagnosis of the problem. Water quality maps are prepared using the information. An EC map reveals the magnitude and spatial variation of the groundwater salinity. Since EC varies with depth as well, it is therefore, prudent to prepare the water quality maps showing depth wise variations. Similar maps for other parameters can also be prepared to have a quick look at the variations of water quality parameters in the project area. It is because of this reason; a well-equipped chemistry laboratory should be established whenever large-scale drainage projects are executed.

Table 3.6 Kind of analysis and use of information

Analysis	Use of information
Total salts or Electrical Conductivity (EC) of soil/water	Leaching need for reclamation, crop selection, reuse and disposal of saline drainage water
Exchangeable Sodium Percentage (ESP), pH (soil) and pH (water)	pH of the soil in many cases being an indicator of ESP is used to have quick estimates of alkali problem. Need for amendment application, pH (water) is also a general indicator of water quality
Sodium Adsorption Ratio (SAR) of soil/water	Need to assess leaching, envelope requirement in combination with clay % age. SAR of water helps to decide reuse of saline drainage effluent and need for amendment application
Residual Sodium Carbonate (RSC)	Reuse of saline drainage effluent and need for amendment application
Iron, Manganese	Soil: Ochre formation; Water: Ochre formation and need to take corrective steps
Sulphates (Soil and water)	Reuse of saline water and fertilizer application. In drainage, it would dictate the kind of cement to be used in case of tile drains and in other structures
Nitrates (Drainage water)	Nutritional requirement of saline water irrigated lands, to assess the leaching of nitrate so as to take corrective measure to minimize leaching.
Specialized analysis	To assess the toxic nature of water, needed only when there are indications that boron, fluoride or heavy metals in the water can possibly occur.

Related Information

(i) Irrigation supplies

The source of irrigation water and its distribution network in the area need to be investigated. In a canal irrigated area, information on the delta of the command, and system of supply, irrigation intensity and irrigation method is useful to estimate seepage and field percolation losses to the drainage system.

(ii) Crop data

Crops grown, crop calendar, yield and water table depth relationship, yield and soil salinity relationship and drainage requirement of crops (optimum average water table depth during crop period and water table drawdown rate for root zone aeration in a specified period) are collected. Besides, crops tolerant to water logging and soil salinity or both needs to be selected for cultivation at least in the initial years of reclamation (Maas and Hoffman, 1977). Advantage of the salt tolerant crop varieties should be taken while selecting the crops.

(iii) Construction Surveys

After the design survey, project implementation will next require a construction survey. Construction surveys provide the horizontal and vertical control for construction, information required for calculation of final quantities and data for development of as-built plans.

WATER TRANSMISSION CHARACTERISTICS

Infiltration Rate

Infiltration rate of water through the soil surface is extremely important in soils because it determines the total water storage in the profile. From the standpoint of soil salinity, it determines how much and how fast the leaching of a given soil can be accomplished. In a saline environment, it will also determine whether salts will be leached (pushed downward), accumulate or flushed with runoff. From the hydrological point of view, it is probably the most important hydrological process in a watershed and reveals the relative contribution to runoff and groundwater recharge

Infiltration rate is a dynamic parameter. When excess water is applied continuously to the soil surface, not only infiltration rate decreases with time, the rate of the decrease also reduces with increasing time (Table 3.7). The asymptotically decreasing curve has a tendency to approach a final minimum value designated as the basic infiltration rate. Initial moisture content has a great bearing on the infiltration rate curve of a soil. The infiltration rate curve of a wet soil always lies below that of an initially dry soil. The two curves may meet at a time when the infiltration rate approaches the basic infiltration rate. Whereas, total infiltration will determine the surface drainage requirement, basic infiltration rate is important to design a subsurface drainage system and to evaluate leaching capability of soils and the fastness with which a soil can be leached.

The infiltration rate of a soil is usually expressed in terms of an equation given by Kostiakov (1932) which is expressed as:

$$I = at^{-b} \tag{3.3}$$

in which I is the infiltration rate, t is the time and a and b are constants. The value of b is always less than one. While the constant a describes the instant infiltration rate at $t = 1$, exponent b describes the decay rate of the infiltration rate.

The cumulative infiltration, I_c is obtained by integrating Eq. (3.3) and is written in the same form as Eq. (3.3):

$$I_c = ct^d \tag{3.4}$$

here c and d are constants and related to a and b such that c=a/(1-b) and d= 1-b. On a plain arithmetic graph, the plots of the data (Table 3.7) as per Eq. (3.3) or Eq. (3.4) appear as asymptotically decreasing or increasing curves (Fig. 3.15). One can assess the basic infiltration rate of the soil from the infiltration rate curve. For this case basic infiltration rate is 1.2 cm/h. Both Eqs. (3.3 and 3.4) will plot as straight lines on a log-log paper.

The average infiltration rate at a particular elapsed time can be given as

$$I_{av} = I_c/t = c.t^{d-1} \tag{3.5}$$

Two basic drawbacks of this equation are that it does not represent the real physical phenomena when t tends to 0 and when t tends to ∞. In the first case, infiltration rate is infinity while in the second event it tends to zero. Both the conditions are not bound to happen in reality. As t tends to ∞, the infiltration rate should be equal to the basic infiltration rate. In spite of these discrepancies, the equation is widely used in irrigation studies where time spans are small. Besides, a modification to this equation is also suggested. Eq. (3.3) with this modification is written as:

Table 3.7 Instantaneous and cumulative infiltration rate of the soils at Sampla

Elapsed time (min)	**Volume of * water used (ml)**		**Accumu-lated intake (cm)****	**Intake during elapsed time (cm)**	**Infiltration rate (cm/h) (col 5 /col 1) *60**
	Instant-aneous	**Cumulative**			
5	636	636	0.333	0.333	4.0
10	556.4	1192.4	0.625	0.29	3.5
15	381.6	1574	0.824	0.2	2.4
20	333.9	1907.9	1.000	0.175	2.1
25	238.5	2146.4	1.125	0.125	1.5
35	413	2559.4	1.342	0.216	1.3
45	381.6	2941	1.542	0.2	1.2
55	381.6	3322.6	1.742	0.2	1.2
85	1144.4	4467	2.342	0.6	1.2
115	1144.4	5611.4	2.942	0.6	1.2
145	1144.4	5725.8	3.542	0.6	1.2
205	2480.1	8205.9	4.842	1.3	1.3

* Observed, ** Column 3/area of the inner cylinder

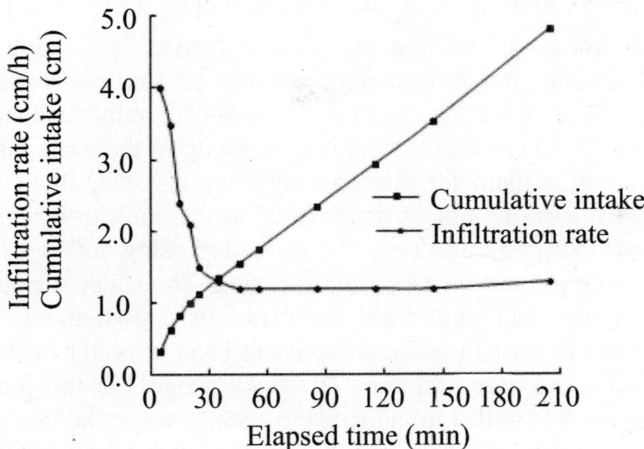

Fig. 3.15 Instantaneous and cumulative infiltration rate curves on a normal paper. Note the basic infiltration rate of around 1.2 cm/h

$$I = at^{-b} + e \tag{3.6}$$

Here *e* is equal to the basic infiltration rate of the soil.

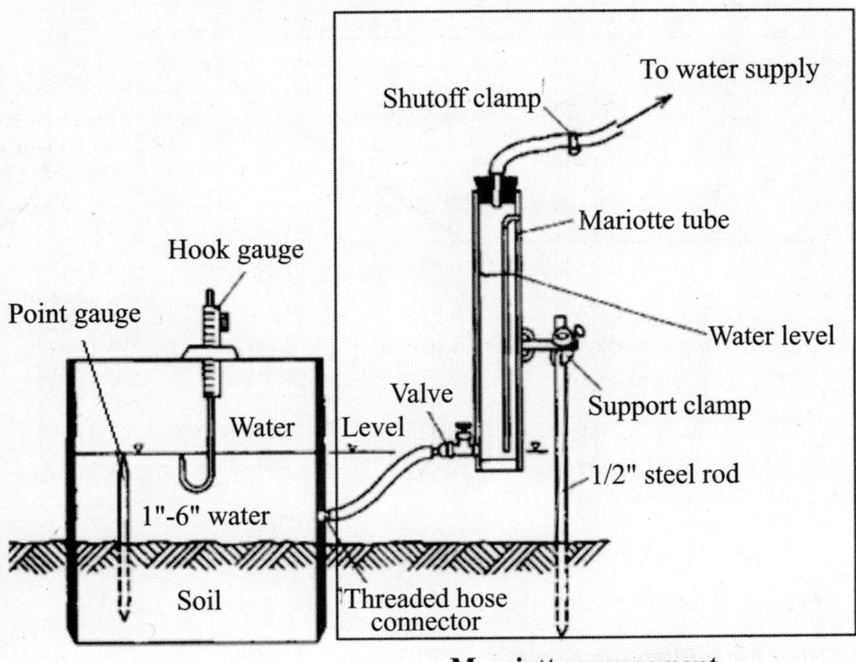

Fig. 3.16 A view of the set-up to measure infiltration rate
(Note: Only one out of the three Marriott, hook or point gauge arrangement is used)

Equipment and procedures

The field measurement of infiltration rate is carried out using double ring infiltrometers. A double ring infiltrometer consists of two concentric rings (only inner ring is shown; Fig. 3.16). The inner ring, made of 2 mm rolled steel, is 22.5-30 cm in diameter and 25-30 cm deep. Outer ring made up of the same material is 1.5-2 times of the inner ring in diameter. Both the rings are installed about 10 cm deep in the soil so that there is minimum of disturbance to the soil during installation. If a single ring is used, then the outer ring is constructed using soil dyke all-round the inner ring in the same proportion as is the outer ring. The water is poured into these cylinder in such a way that minimum disturbance is caused to the soil. Either a Marriott arrangement is used to maintain constant head of water or it is maintained by adding water up to the edge of a point or a hook gauge. For this purpose, either a hook or a point gauge is installed to indicate the water level in the inner cylinder. The quantity added is used to find out the infiltration rate in the time interval between the first addition and the subsequent additions. Usually, replicated tests are conducted at suitable locations in the field. Number of tests needed to arrive at a realistic average value is governed by the expected heterogeneity of the soils. Besides heterogeneity, time and manpower at one's disposal also determine the number of tests. The objective should be to make sure that average values represent the infiltration rate of the soils.

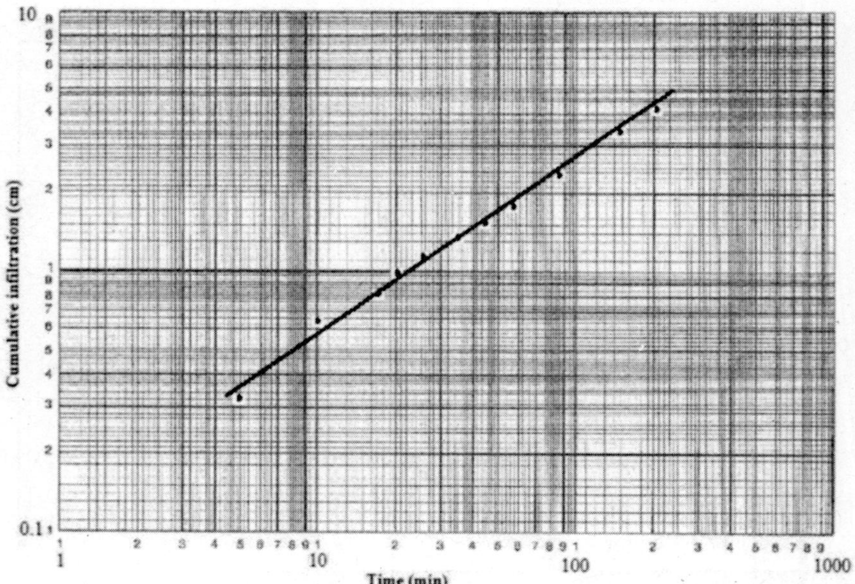

Fig. 3.17 Plot of cumulative infiltration on log-log scale to fit Eq. (3.4)

Curve fitting for infiltration curves

Kostiakov relation given by Eq. (3.3) and Eq. (3.4) are usually fitted to assess the infiltration rate or the cumulative infiltration. If cumulative infiltration is plotted, the resulting straight line is used to evaluate the parameters c and d. The plot of data given in column 1 and column 4 of Table 3.7 on a log-log scale results in a straight line

(Fig. 3.17). Value of c is determined from the intercepts of the straight lines on y-axis at t = 1. It may be noted that the straight line meets the y-axis at 0.13 such that the value of c =0.13. The value of d is given by slope of the straight line; in this case slope of the line being 0.66 gives the value of d as 0.66. The value of a and b can be determined either by plotting the infiltration rate curve on a similar paper or by the relations b= 1-d and a =c (1-b).

Infiltration Rate and Vertical Hydraulic Conductivity

The infiltration rate of water, I in the soil is governed by Darcy's law such that,

$$I = K_\theta (h+z-P)/z \tag{3.7}$$

Here h is water depth at the soil surface, m, z is the depth to the wetting front below the soil surface; and P is the soil water pressure at the wetting front inside the transmission zone. With elapse of time moisture content in the transmission zone becomes saturated such that K_θ tends to K_{sat} and P tends to zero in medium to heavy textured soils and h becomes quite small in comparison to z such that

$$I = K_{sat} (z)/z = K_{sat} \tag{3.8}$$

The method is not very accurate (especially for coarse textured soils) but the basic infiltration rate can be considered as the vertical hydraulic conductivity (Eq. 3.8). It is then possible to assess the anisotropy of the soil and radial hydraulic conductivity. From practical point of view, infiltration rate measured at different layers will give the vertical hydraulic conductivity of the soil layer (Fig. 3.18).

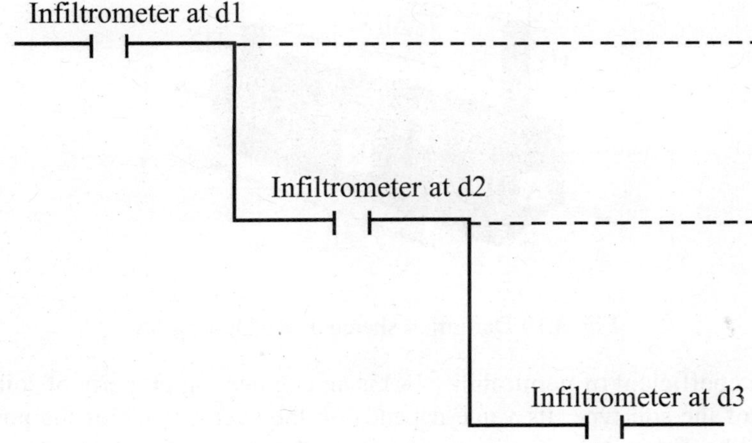

Fig. 3.18 A schematic view of measuring vertical hydraulic conductivity at different depths

HYDRAULIC CONDUCTIVITY

The hydraulic conductivity of a soil is *'a measure of the soil's ability to transmit water when submitted to a hydraulic gradient'*. Hydraulic conductivity is defined by Darcy's law, the proportionality coefficient, K, of Darcy's law being designated as the hydraulic conductivity. It specifies the quality of a soil horizon that enables water or air to move through it. It expresses the interaction between the fluid and the media and has the units of velocity, cm/s. Accurate values of hydraulic conductivity are vital to the design of a subsurface drainage system.

Darcy's law states that there is a linear relationship between flow velocity (v) and hydraulic gradient (i) for any given saturated soil under steady laminar flow conditions. If the rate of flow is q (volume/time) through cross-sectional area (A) of the soil mass, then Darcy's law can be expressed as

$$q = v*A = -K*i*A \tag{3.9}$$

Here K is the coefficient of permeability or hydraulic conductivity of the soil, i is the $\Delta h/L$ such that Δh is the difference in total heads and L is the length of the soil mass. The flow velocity (v) is also called the Darcian velocity. For i is equal to 1, the flow velocity is equal to K. The negative sign indicates that the flow is in the direction of decreasing head. For the set-up shown as Fig. 3.19, q can be given as:

$$q = K* A* (H1 - H2) / L \tag{3.10}$$

Note that i in Eq. (3.7) is equal to (H1 – H2) / L and negative sign is taken care of by writing H1-H2 instead of H2-H1.

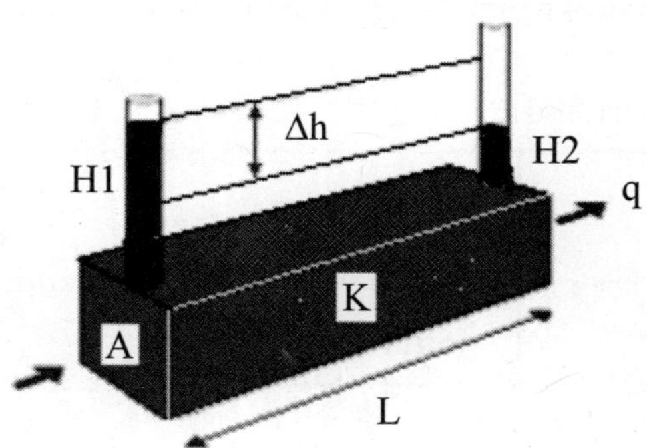

Fig. 3.19 Definition sketch of the Darcy´s law

The coefficient of permeability (K) is an engineering property of soils and is a function of the soil type. Its value depends on the average size of the pores and is related to the distribution of particle sizes, particle shape and soil structure. The ratio of permeability of typical sands/gravels to those of typical clays is of the order of 10^6. A small proportion of fine material in a coarse-grained soil can lead to a significant reduction in permeability.

Hydraulic conductivity of the saturated soil is independent on the pressure gradient as water conducting porosity is not influenced by the change in pressure gradient. However, under unsaturated conditions, hydraulic conductivity varies with soil moisture suction, since suction, soil moisture content and water conducting porosity are closely interlinked. Some of the commonly used terms to describe the soils with respect to hydraulic conductivity or for that matter other properties are as follows:

Homogeneous soils: *'A soil is described as homogeneous, when hydraulic properties such as hydraulic conductivity, density or Young's modulus etc. are same in all directions'.*

Heterogeneous soils: *'A soil is heterogeneous, if hydraulic properties vary spatially'.*

Isotropic: *'The condition in which hydraulic properties of the aquifer are equal in all directions'.*

Anisotropic soils: Since most soils show some kind of stratification, hydraulic properties vary being more in one than the other direction. Thus, *'the conditions under which one or more of the hydraulic properties of an aquifer vary according to the direction of flow are called anisotropic soils'.*

The hydraulic conductivity of the soils can be assessed or determined by the following methods:

- Correlation between hydraulic conductivity and other soil characters
- Laboratory methods
- Field methods

Few of the methods used under the three categories are described in the following sections.

Soil Texture and Hydraulic Conductivity

To estimate a representative value of the hydraulic conductivity is rather difficult because of wide spatial variability. In view of the difficulties encountered, it has been common in drainage design to determine the hydraulic conductivities as related to other soil properties such as soil texture. Although, soil texture and hydraulic conductivity relationship may appear arbitrary; such tables have been used for drainage design in the USA and the Netherlands quite successfully (Table 3.8). In place of testing however, it is still the most desirable since in many cases the conductivity values vary by a factor of 10 or more for soils of the same texture group (Table 3.8).

LABORATORY METHODS

In the laboratory, the value of K is determined by several different instruments and methods such as the permeameter, pressure chamber and consolidometer. A common feature of all these methods is that a soil sample is placed in a small cylindrical receptacle representing a one-dimensional soil configuration through which the circulating liquid is forced to flow. Depending on the flow pattern imposed through the soil sample, the laboratory methods are classified as either a constant-head test with a steady-state flow regimen or a falling-head test with an unsteady-state flow regimen.

Table 3.8 Relationship between hydraulic conductivity and soil texture

Soil texture	Hydraulic conductivity (cm/h)
Clay	.0025 - .0380
Silty clay	.0381 - .0635
Silty clay loam	.0635 - .1016
Clay loam	.1016 - .1778
Sandy clay loam	.1778 - .3810
Silt loam	.3810 - .8890
Loam	.8890 -1.7780
Very fine sandy loam	1.778 - 3.048
Fine sandy loam	3.048 - 4.318
Sandy loam	4.318 - 6.350
Loamy fine sand	6.350 - 8.890
Loamy sand	8.890 -11.430
Fine sand	11.430 -15.240
Sand	15.240 -20.320
Coarse sand	20.320 -30.480

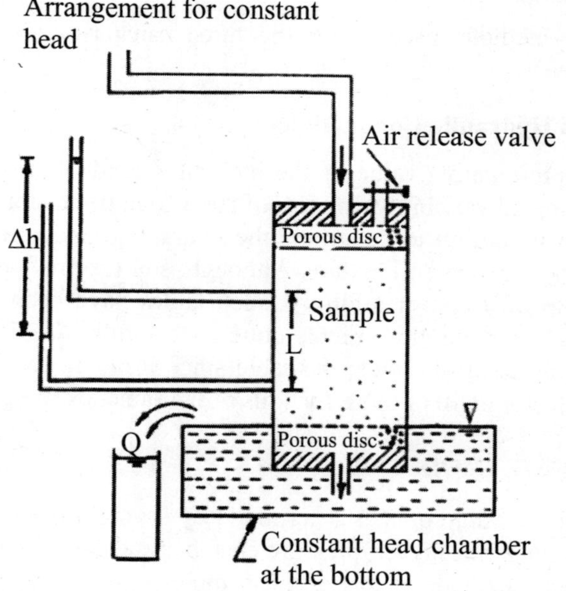

Fig. 3.20 A schematic diagram of constant head permeameter

Constant Head Method

The constant-head test with the permeameter is one of the most commonly used methods to determine the saturated hydraulic conductivity of coarse-grained soils. The test is based on the direct application of Darcy's law to a soil liquid configuration. It represents a one-dimensional steady flow of a percolating liquid through a saturated column of soil from a uniform cross-sectional area. In this method, a cylindrical soil sample of cross-sectional area *A* and length *L* is placed between two porous plates that do not provide any extra hydraulic resistance to the flow. A constant head difference, $\Delta H = H_2 - H_1$, is then applied across the test sample (Fig. 3.20). By measuring the volume Q of the water that flows through the system during time *t*, the saturated hydraulic conductivity *K* of the soil is determined as follows:

$$K = QL/\Delta h.t.A \tag{3.11}$$

Here, K is coefficient of permeability, Q is quantity of water discharged, L is difference between water level in the two manometers, A is cross-sectional area of specimen, t is total time of discharge and Δh is difference in head on manometers.

Falling-Head Method

The falling head test with the permeameter is primarily used to determine K value of fine-grained soils. Like the constant-head method, the falling-head test also operates in accordance with direct application of Darcy's law to a one-dimensional, saturated column of soil with a uniform cross-sectional area. The falling-head method differs from the constant-head method in that the liquid that percolates through the saturated column is kept at an unsteady-state flow regimen in which both the head and the discharged volume vary during the test. In the falling head test method, a cylindrical soil sample of cross-sectional area *A* and length *L* is placed between two highly conductive plates. The soil sample column is connected to a standpipe of cross-sectional area *a,* in which the percolating fluid is introduced into the system (Fig. 3.21). By measuring the change in head in the standpipe from h_0 to h_t during a specified interval of time *t*, the saturated hydraulic conductivity is determined as follows (Klute and Dirksen, 1986):

$$K = aL/At \{Ln\ h_0/h_t) \tag{3.12}$$

Here a is cross sectional area of stand pipe, L is length of soil column, A is cross section area of soil column, t is time interval to head drop, h_0 is total head before test and h_t is total head after test.

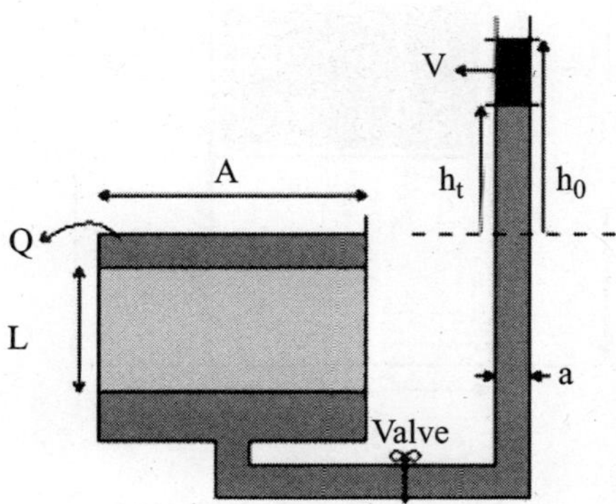

Fig. 3.21 A schematic diagram of falling head permeameter to measure hydraulic conductivity

An important issue in using the laboratory methods to measure K is related to the soil sampling procedure and preparation of the test specimen. The sampling process, if not properly adopted, usually disturbs the matrix structure of the soil and results in a misrepresentation of the actual field conditions. Undisturbed sampling of soils is possible, but it requires the use of specially designed techniques and instruments.

FIELD METHODS

For saturated zone, auger hole method and piezometer method are used while for unsaturated zone inverse auger hole, double ring method and Guelph/disk permeameter are used. This section describes auger hole method and inverse auger hole method commonly used in drainage studies.

Auger Hole Method in the Presence of Water Table

The method to determine hydraulic conductivity in the presence of water table is given by Hooghoudt commonly known as 'auger hole method' (USBR, 1978). The method consists of making a borehole of 10-20 cm in diameter and 1.0-1.5 m deep (Fig. 3.22). After the water in the bore comes in equilibrium with the groundwater level, water is pumped out of the borehole so that the water level in the bore is lowered by 20-50 cm. Measurements of the rise in water table with time are observed (initially at an interval of 10-20 seconds increasing it to 1-2 minutes depending upon the rate of rise). These values are used to calculate the hydraulic conductivity. Other information required for use in the equation is the depth of the borehole, the groundwater level in the bore or initial water level and borehole diameter.

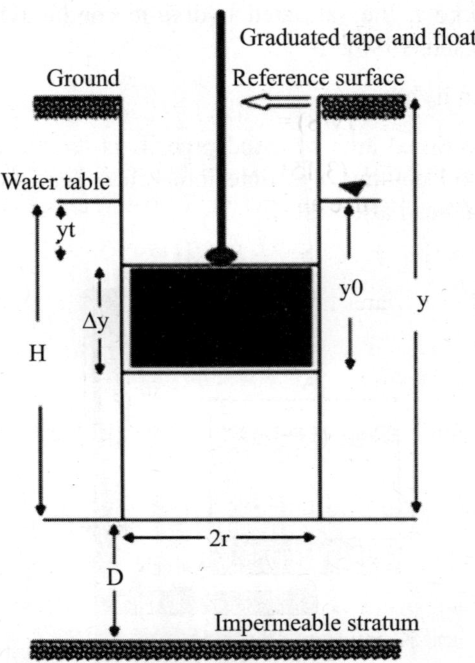

Fig. 3.22 A schematic view of the *in-situ* auger hole test

Hooghoudt's Equation to Measure Hydraulic Conductivity

Hooghoudt perfected the method to determine hydraulic conductivity of the soils using an auger hole below the water table (Fig. 3.22). While deriving this equation, following assumptions were made:

- The water table around the auger hole is not lowered when water is pumped out of the auger hole. This condition is reasonably met if the pumping is done quickly within a short period of time.
- Water flows horizontally into the sides of the auger hole and vertically up through the bottom of the hole.

The water flowing into the whole fills up the hole with water. The rate at which hole is filled up is directly proportional to the circumference of the hole and inversely proportional to the cross-sectional area of the hole. The rate of rise in water table due to water flowing horizontally from the circumference of the hole is given as:

$$dy/dt = -K\,(2\,\pi\, r\, H/\,\pi\, r^2)\,(y/S) \tag{3.13}$$

Here, K is the hydraulic conductivity, m/d, y is the rise in water table, cm, r is the radius of the auger hole, cm, t is time, s, H is the depth of the auger hole below groundwater table, cm and S is a constant that depends upon r, H and D. Here D is the depth of the impermeable layer below the bottom of the auger hole, cm. S has the dimensions of length and is given as:

$$S = rH/0.19 \tag{3.14}$$

The coefficient 0.19 is valid only if S is in m and for other units the value has to be changed accordingly.

The rate of rise in water table due to water entering the hole from the bottom of the hole can be written as:

$$dy/dt = -K\,(\pi\, r^2/\,\pi\, r^2)\,(y/S) = -\,Ky/S \tag{3.15}$$

Adding Eq. (3.13) and Eq. (3.15), to get rate of rise in water table both due to water entering from circumference and bottom of the hole, we have

$$dy/dt = -\,(2\,KH/r - K)\,(y/S) = -\,K\,(2H{+}r)y/rS \tag{3.16}$$

If in time t =0 to t =t, water table rises from y_0 to y_1, we integrate Eq. (3.16) to get

$$\mathrm{Ln}(\,y_0/y_t) = K\,(2H{+}r)\,(\Delta t/rS) \tag{3.17}$$

Δt is the time difference between the two measurements being equal to t if the initial time is 0.

Thus K can be written as

$$K = [2.3\ rS\ \log_{10}(\,y_0/y_t)]/(2H{+}r)\Delta t \tag{3.18}$$

If the hole terminates on the impermeable layer, there would be no flow from the bottom of the hole and K can then be written as:

$$K = [2.3\ rS\ \log_{10}(y_0/y_t)]/(2H\Delta t) \tag{3.19}$$

Now, we can incorporate the value of S and also convert seconds into day by multiplying with 86400 such that

$$K = [86400 * 2.3\, r^2 \log_{10}(y_0/y_t)]/0.19 * (2 + r/H)\Delta t \tag{3.20}$$

$$K = [1046,000\, r^2 \log_{10}(y_0/y_t)]/(2 + r/H)\Delta t \tag{3.21}$$

Or, when the hole rests on impermeable floor, then

$$K = [86400 * 2.3\, r^2 \log_{10}(y_0/y_t)]/(2 * 0.19 * \Delta t) \tag{3.22}$$

$$= 523,000\, r^2 \log_{10}(y_0/y_t)]/\Delta t) \tag{3.23}$$

However, for the similar set-up, Ernst (1950) derived the formulae which are now commonly used to determine the hydraulic conductivity using auger hole method.

For D > H/2, hydraulic conductivity is calculated by the relation (van Schilfgaarde, 1974);

$$K = \frac{4000\, r^2}{(H + 20\, r)[2 - (y/H)]\, y} - \frac{\Delta y}{\Delta t} \tag{3.24}$$

Here K is in m/d, r, H, y and Δy in cm and Δt in seconds. The parameter y is average depth of the water table in the auger hole given by $y_0 - ½\,\Delta y$.

When D = 0, the auger hole rests on the impermeable layer and equation used for estimation of the hydraulic conductivity is:

$$K = \frac{3600\, r^2}{(H + 10\, r)[2 - (y/H)]\, y} - \frac{\Delta y}{\Delta t} \tag{3.25}$$

The parameters are the same as described for Eq. (3.24). For intermediate values, still there is a lack of a suitable formula. In such events values from these two equations are worked out to select an intermediate value or Eq. (3.25) is used.

To obtain acceptable accuracy the following conditions should be met with.

$2r > 6.25$ cm and < 14 cm

$H > 20$ cm and < 200 cm

$y > 0.2\, H$

$\Delta y < 1/4\, y_0$

$S > H$

The calculation procedure to determine hydraulic conductivity is given in Table 3.9 wherein all the related data are also given.

Example 3.1

Calculate the hydraulic conductivity using the data given in Table 3.9.

For solution of the problem refer Table 3.9.

Table 3.9 Calculation of hydraulic conductivity using auger hole method

Location : Beri Study Area	**Field No. : 41/13**	**Date : October, 2004**

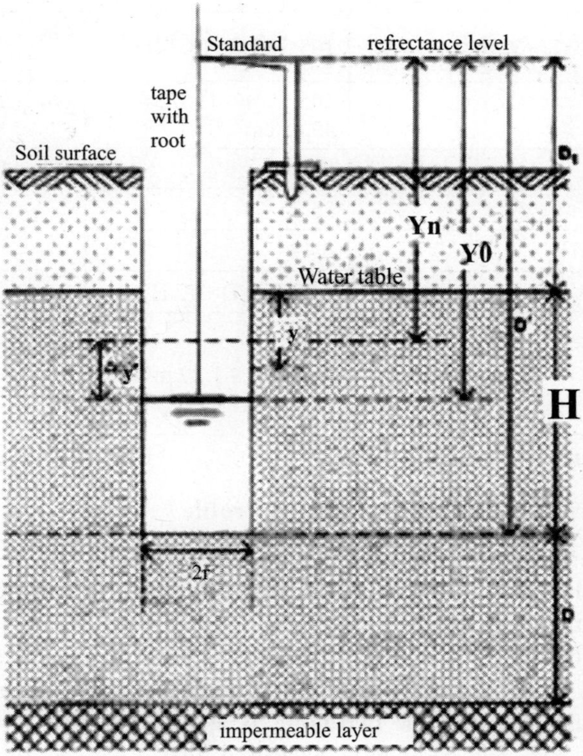

Auger hole Measurement

t (sec)	Y_n	$Y_n - Y_{n-1}$
0	105.50	- -
10	102.50	3.00
20	99.10	3.40
30	96.20	2.90
40	93.30	2.90
50	91.80	1.50
60	89.70	2.10
70	87.50	2.20
80	85.60	1.90
90	83.90	1.70
100	82.30	1.60
110	80.80	1.50
120	79.50	1.30
130	78.20	1.30
140	76.80	1.40
150	75.50	1.30
160	74.30	1.20
170	73.00	1.30

Measurements for Auger hole Method

Where,	**Input**	**Check**	**Formula for check**
Radius of the auger hole (cm), r =	4	OK	IF(r >7,"NO OK" AND IF(r <3,"NO OK")
Depth of auger hole below level of the standard (cm), D'	150	------	
Depth from standard to ground water (cm), D_1	49	------	
Depth of impermeable layer (cm)	300	------	
Depth of the auger hole below water table (D' - D1) in cm, H D_2 =	101	OK	IF(H>200,"NO OK" and IF(H<20,"NO OK"))
Depth impermeable layer -depth of the auger hole (cm); D	199		

Calculate		Check	Formula for check
t=	50		
Y_n =	91.8 (cm)		
$\Delta y = Y_0 - Y_n$	105.5-91.8 =13.7 cm	OK	IF($\Delta y <(Y_n - D_1)/4$, "OK",)
y	105.5 – 49 -13.7/2 = 49.65 cm	OK	(y > H/5,"OK")
r/y	0.08	When D>1/2 H	
H/r =	25.25	$C = \dfrac{4000 \frac{r}{y}}{\left(20 + \frac{H}{r}\right)\left(2 - \frac{y}{H}\right)}$	
y'/ H=	0.49		
C =	4.72 (for D>1/2 H)	When $0 < D \leq 1/2$ H	
K =	C Δy /dt = 1.29 m/d	$C = \dfrac{3600 \frac{r}{y}}{\left(10 + \frac{H}{r}\right)\left(2 - \frac{y}{H}\right)}$	

Source: Mohan (2011)

Measurement in Two-Layered Profile

The method to determine K in a homogeneous profile can be easily extended to a two-layered profile. In a layered profile, initially a shallow auger hole is made which terminates at least 20 cm above the interface of two layers but contains at least a minimum of 40 cm thick water body. A deep auger hole is also made going up to at least 50 cm below the interface into the second layer (Fig. 3.23). The minimum distance between the two bores is about 3 m. The value of hydraulic conductivity from the deeper hole is measured first followed by the shallower hole. If the hydraulic conductivity of the shallower hole is K1 and the hydraulic conductivity as measured from the deeper hole is K_{12} then K2, the hydraulic conductivity of the second layer is:

$$K2 = [K_{12} (H1 + H2) - K1\ H1]/H2 \tag{3.26}$$

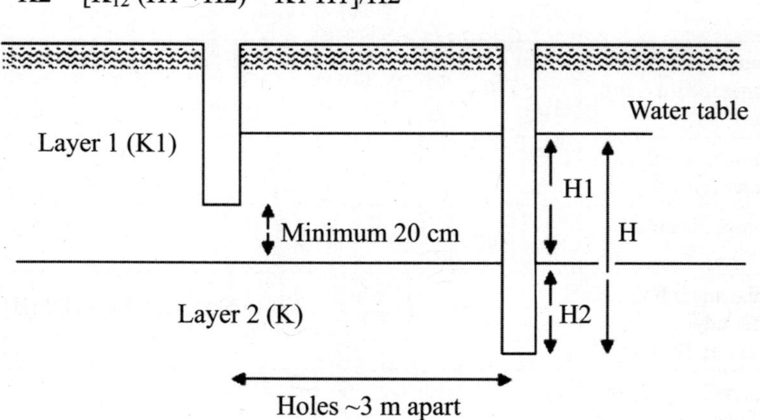

Fig. 3.23 Set-up to measure hydraulic conductivity in a two-layered profile

Auger Hole Method in the Absence of Water Table

Two methods used in such situations are referred as inverse auger hole methods. In the first procedure a borehole similar to the one made for hydraulic conductivity tests in the presence of water table is made. To minimize the disturbance of the soil in the hole while filling the water, the auger hole is filled with gravel or sand. The hole is then filled with water to a constant level say H cm from the bottom and the water level is maintained at the constant level by suitable arrangement (Fig. 3.24 Top). For good results, value of H should exceed 10 times the value of the radius of the auger hole. The constant head is maintained till the flow into the soil is constant. Zangar (1953) equations are than used to calculate the value of K as:

For $S > 2H$

$$K = \frac{Q}{2\pi H^2}\left[\ln\left(H/r + \sqrt{(H^2/r^2 - 1)}\right) - 1\right] \tag{3.27}$$

For $S < 2H$

$$K = \frac{3Q}{\pi H (3H + 2S)} \ln H/r \tag{3.28}$$

In these equations, Q is the flow rate at equilibrium, H is the hydraulic head, r is the radius of the auger hole (Fig. 3.24 Top) and S is the depth of the impermeable layer from the bottom of the auger hole.

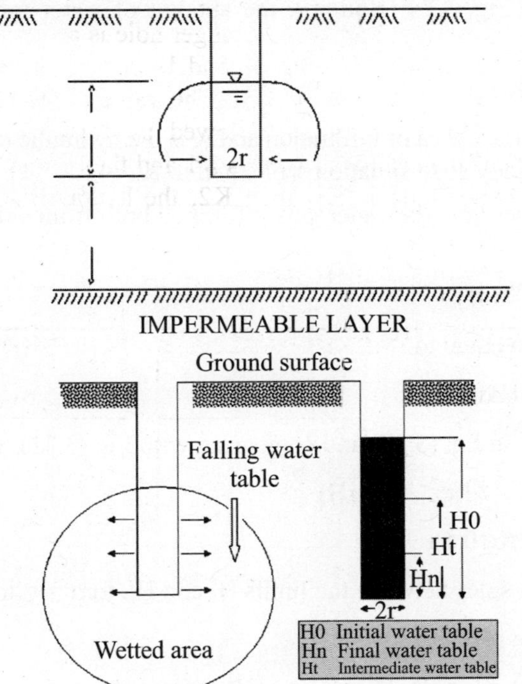

Fig. 3.24 Inverse auger hole methods to determine hydraulic conductivity, constant head (left) and falling water table (right)

Another variation of this test measures the fall of the water level in the auger hole. This modification dispenses with the arrangements needed to keep a constant water level in the hole. In this method, the borehole is filled with water and water is left to drain away freely. The process is repeated several times so as to saturate a considerable area until infiltration has attained more or less a constant value. After the last filling, the rate of drop of the water level in the hole is recorded. The Performa for data collection for this method is shown as Table 3.10. The hydraulic conductivity can then by calculated by the equation given in ILRI (1994) and is derived as follows:

Table 3.10 The Performa to arrive at the values of (h_t + r/2)

Time (s)	h_{t1} (cm)	$H_t = D - h_{t1}$	$H_t + r/2$
	Water level in the bore from top of the auger hole or land surface	D is the depth of the auger hole from top or ground surface, subtract values in column 2	Add r/2 in values of column 3

By referring to Fig. 3.24 (Bottom), the amount of water infiltrated, Q can be given by:

$$Q = vA = K\,A \tag{3.29}$$

Here A is the surface area of infiltration and K is the hydraulic conductivity which upon reaching the steady state situation will equal to K.

Since in the auger hole the water gets infiltrated both from sides and bottom, the area is given as

$$A = \pi\, r^2 + 2\pi\, rH \tag{3.30}$$

Further Q is also equal to

$$Q = -\pi\, r^2\, dH/dt \tag{3.31}$$

Incorporating A in Eq. (3.29) and comparing with Eq. (3.31), we get

$$-\pi\, r^2\, dH/dt = K(\pi\, r^2 + 2\pi\, rH) \tag{3.32}$$

$$r\, dH/dt = 2K((0.5)r + H) \tag{3.33}$$

Integrating both sides between the limits H_0 and H_t, for t = t_0 to t we have

$$\int 2K/r\,(dt) = \int dH\,[((0.5)r + H)] \tag{3.34}$$

$$(2K/r)\,(t - t_0) = \log\,(H_0 + 0.5r) - \log\,(H_t + 0.5r)$$

$$K = r\,[\log\,(H_0 + 0.5r) - \log\,(H_t + 0.5r)]/\,(t - t_0)$$

$$K = 1.15\,r\,\frac{\log(H_o + r/2) - \log(H_t + r/2)}{(t - t_o)} \tag{3.35}$$

$$= 1.15\,r\tan\alpha \tag{3.36}$$

Here H_o is the initial height of the water column in the hole (cm) at time t_o, H_t is the water level (cm) in the auger at any time t, r is the radius of auger hole (cm), t and t_o are the times (s) since the beginning of the test and K is the hydraulic conductivity (cm/s).

The most appropriate way to use Eq. (3.35) is to plot (H_t + 0.5 r) against t on a semi-logarithmic paper, the former being on the log scale. It will result in a straight line as shown in Fig. (3.25). The slope of this line is equal to tan α, which is used in Eq. (3.36) to obtain the value of K in cm/s. Multiplication of this K value by 864 gives K in m/d.

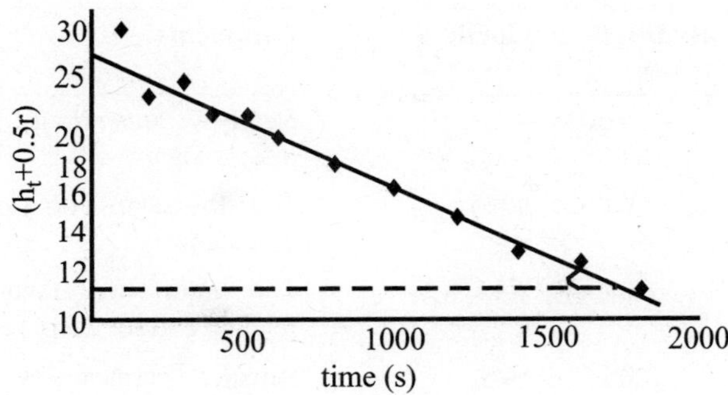

Fig. 3.25 Plot of data against time for inverse auger hole method

Number of Tests

No set criteria are fixed to suggest as to how many number of observations should be taken to assess the average hydraulic conductivity? At each site, a minimum of two tests should be conducted, keeping the holes few meters apart. If the results of two tests are significantly different, then a third test should be conducted. A common recommendation is that the number of hydraulic conductivity measurements should be around 35 when the sample size of an auger hole is in the range of 1/2 - 1 m^3. The required density of test sites is related to soil variability and the expected drain spacing (Table 3.11). Generally, the number of measurements should be large enough to cover the whole area.

Table 3.11 Maximum number of hectares represented by one test site

Soil conditions	Probable drain spacing		
	<30m	**≈ 50m**	**>75m**
Heterogeneous	<5	5-10	10-15
Less heterogeneous	5-10	10-25	20-40
Homogeneous	10-25	25-50	40-75

Characteristics of the Hydraulic Conductivity

Hydraulic conductivity depends on the properties of both the soil and the water. A high value is often associated with coarse textured soils or soils with highly developed structure. High porosity is usually associated with high hydraulic conductivity but the geometry of soil pores and pore-size distribution influence the hydraulic conductivity more than the total pore space. A few large pores are more effective in contributing to high permeability than many small pores as is the case in sandy and clay soils respectively. Smith & Browning (1946) grouped the soils on the basis of hydraulic conductivity (Table 3.12). Pertinent comments are also listed in respect of leaching and drainage of these soil groups to serve as a useful guide for characterizing the soils. Hydraulic conductivity of 0.1 cm/h is considered reasonable below which leaching and irrigation may present serious difficulties.

Table 3.12 Classes of hydraulic conductivity for saturated sub-soils

Class	Hydraulic conductivity (cm/h)	Comments
Extremely slow	0.0025	Nearly impervious, leaching is insignificant
Very slow	0.0025-0.025	Poor drainage. Too slow for artificial drainage
Slow	0.025- .254	Too slow for favorable air-water relation and for deep root development
Moderate	0.254-2.540	Adequate permeability
Rapid	2.54 -25.4	Excellent water holding relations as well as excellent permeability
Very rapid	>25.4	Associated with poor water holding

Spatial Variation and Average Hydraulic Conductivity

It is always difficult to assign an average value to water transport parameters since large spatial variations are encountered in field soils. Hydraulic conductivity, infiltration rate, pore water velocity, diffusivity and dispersion coefficient all show a high variability with coefficient of variation going as high as 200% in case of saturated hydraulic conductivity and may be much higher in other cases. In case of hydraulic conductivity, both lateral and vertical variability have been observed. Factors such as climate, topography, parent material, presence of organisms and human intervention can accentuate the variability. The variation of hydraulic conductivity in the 0-110 cm layer at Sampla revealed that it varied from a low value of 0.06 m/d to as high as 10.00 m/d. Similar variations were observed in other layers.

In order to assess an effective average value of the measured hydraulic conductivity, one needs to increase either the sample size or the number of sampling spots. With limited size and data, some statistical means of studying sample variation have been evolved to arrive at a realistic average value. Since the values of hydraulic conductivity are log-normally distributed, it is recommended that geometric rather than arithmetic mean of the values better represents the mean hydraulic conductivity.

Example 3.2

For a test site 14 observation of hydraulic conductivity are: 1.0228, 0.0838, 0.9881, 2.3863, 1.0140, 0.1570, 0.0717, 0.4919, 1.3612, 0.9825, 0.8215, 1.0851, 0.3032 and 0.3627. What is the average hydraulic conductivity for drainage design?

$$\text{Geometric mean} = (n_1{*}n_2{*}\ldots.\ n_n)^{1/n} = (1.0228{*}0.0838{*}\ 0.9881{*}\ldots.\ 0.3627)^{1/14}$$
$$= 0.6281 \text{ m/d}$$

If sufficient numbers of value are available, frequency analysis procedure can also be used to calculate effective hydraulic conductivity. In this procedure the following relation assesses the probability of occurrence:

$$P = R/\,(M{+}1) \tag{3.37}$$

Where P is the probability of occurrence, R is the rank of the value after the values have been ranked in the descending order and M is the total number of values of the hydraulic conductivity (Table 3.13). The values of the hydraulic conductivity are plotted against P on a log-normal probability paper since the values of hydraulic properties are assumed to be log-normally distributed. The plot will show as a straight line in the graphs (Fig. 3. 26). The value corresponding to 50 % probability is used as the representative value of K in the design.

Table 3.13 Ranking of rainfall to assess probability using Weibull′s formula

R	Value (cm/d)	Probability
1	2.3863	0.067
2	1.3612	0.133
3	1.0851	0.2
4	1.0228	0.267
5	1.0140	0.333
6	0.9881	0.4
7	0.9825	0.467
8	0.8215	0.533
9	0.4919	0.6
10	0.3627	0.667
11	0.3032	0.733
12	0.1570	0.8
13	0.0838	0.867
14	0.0717	0.933

Vertical Hydraulic Conductivity and Anisotropy

The value of K in horizontal and vertical directions can be different due to layering of the profile or differences in the pore geometry or any other feature of the pores in the horizontal and vertical directions. This variation can play an important role in land drainage, because the groundwater flow direction can be vertical or horizontal or radial depending upon the relative values of the hydraulic conductivity in the two directions. The vertical hydraulic conductivity is generally given by basic infiltration rate at specific depths (Fig. 3.18). If hydraulic conductivity in horizontal direction is indicated by K_h and in vertical direction by K_v, then the effective hydraulic conductivity (also the hydraulic conductivity in the radial direction), K_r, is computed from the geometric or logarithmic mean of K_h and K_v:

$$K_r = \sqrt{(K_h . K_v)} \tag{3.38}$$

Standard formulae are available to calculate drain spacing for anisotropic soils.

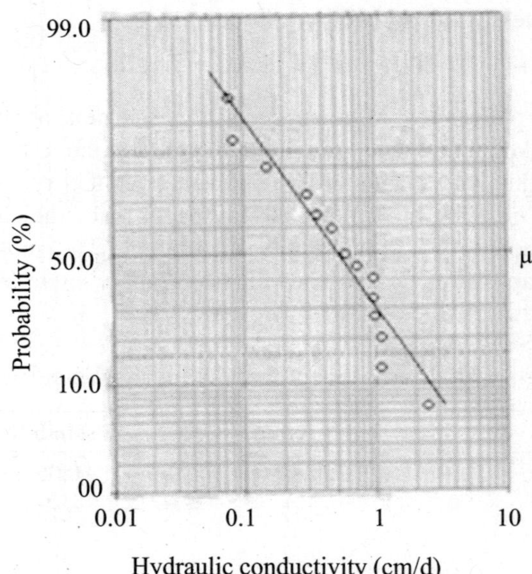

Fig. 3.26 Probability plot of hydraulic conductivity on a log-normal probability paper

DRAINABLE PORE SPACE

Drainable pore space is used to estimate the lateral drain spacing using transient (unsteady) theory of subsurface drainage design. The drainable pore space is defined *'as the volume of water released or taken into storage in an unconfined aquifer per unit cross sectional area per unit decline or rise of water table as a fraction of total volume of soil'*. It is also known as effective porosity or specific yield. The drainable porosity is a function of the soil moisture suction. If this function is indicated by f(h), the quantity of water draining out of unit column when the water table is lowered from h_1 to h_2 is given by

$$Q = \int_{h_2}^{h_1} f(h)\, dh \tag{3.39}$$

The nature of the function f(h) depends upon the soil and is generally complex. If a straight line function is assumed (f(h) = ah) with slope a, the quantity of water drained is given by

$$q = \int_{h_2}^{h_1} ahdh = a\,(h_1^2 - h_2^2)/2 \tag{3.40}$$

Drainable pore volume is usually a dynamic quantity yet this dynamics is difficult to evaluate. Moreover, complexity of design formulae does not permit the use of dynamic values for field drainage designs. Therefore, a constant value is assigned to drainable pore space. Even then, considerable difficulties are encountered in evaluating drainable pore space. Like hydraulic conductivity drainable pore space is also assessed or determined using the following methods:

- Correlation between other soil characters and drainable pore space
- Laboratory methods
- Field methods

Both the laboratory and field procedures are expensive and time consuming.

Correlation Methods

Drainable pore space is related to a few soil properties that are easier to evaluate. There is a definite correlation between specific yield and the hydraulic conductivity. As the drain spacing is directly related with hydraulic conductivity and inversely with drainable pore space, it is argued that such a relationship is also practically justified. A curve prepared by the USDA is commonly used to determine the drainable pore space once the hydraulic conductivity is known (Fig. 3.27). Since the fluctuation of the water table in a drained area takes place in the zone between the drains and the maximum allowable water table height, the drainable porosity corresponding to the weighted average hydraulic conductivity of this zone is used to evaluate drainable pore space.

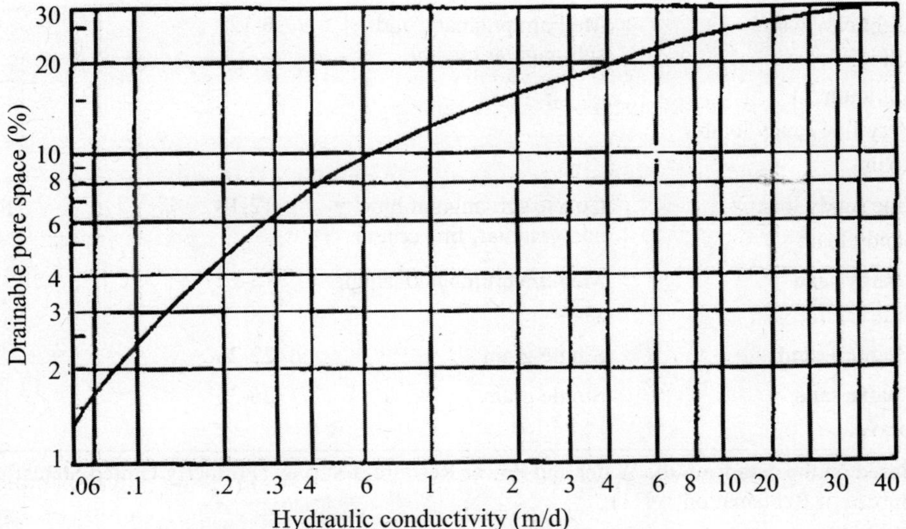

Fig. 3.27 Relationship between hydraulic conductivity and drainable pore space

Mathematical relations have also been developed. For example; $f = K^2$ (f is in percent and K in cm/d, Smedma & Rycroft, 1983), $f = 0.025 + 0.006$ K (f is in fraction and K in m/d, Chossat, 1987) and $f = 3.86\ K^{0.37}$ (f is in fraction and K in m/d, RAJAD, 1995) have been used to determine drainable pore space, f.

Another approach utilizes the relation between drainable pore space and soil texture and soil structure both of which have considerable influence on the drainable pore space. The representative values of the volumetric drainable porosity for different textures of soil can be taken as 3-5% for heavy clays, 6-9% for medium textured soils and 10-15% for sandy soils. Exact estimates can be made from the data presented in Table 3.14.

Table 3.14 Influence of soil texture and structure on drainable pore space

Texture	**Structure**	**Drainable pore space (%)**
Clay Heavy clay loam	Massive, very fine or fine columnar	1-2
Clay Clay loam Silty clay Sand clay loam	Very fine or fine prismatic, angular, blocky or platy	1-3
Clay Silty clay Sandy clay Silty clay loam Clay loam Silty loam Silt Sand clay loam	Fine and medium prismatic, angular, blocky and platy	3-8
Light clay loam Silt Silt loam Very fine sandy loam Loam	Medium prismatic and sub-angular blocky	6-12
Fine sandy loam Sandy loam	Coarse sub-angular blocky and granular, fine crumb	12-18
Loamy sand Fine sand	Medium crumb and single grain	15-22
Medium sand	Single grain	22-26
Coarse sand Gravel	Single grain	26-35

Based on the data from the Water and Power Resources Survey (Formerly United States Bureau of Reclamation, 1971).

Laboratory Methods: Soil Characteristic Curves

To have independent estimates of the drainable pore space is a difficult task as no standard procedures have been developed as yet. One of the procedures utilizes soil water characteristic curves of a soil. A soil water characteristic (SWC) curve describes the relationship between the amount of water held by the soil (soil water content expressed on weight or volume basis) and the force by which it is held (capillary pressure or suction or tension referred to as soil matric potential/tension expressed in bars or kPa/MPa). These curves are also known as soil moisture release curve or soil moisture retention curve or simply the pF curve. Since SWC is a hydraulic property related to size and connectedness of pore spaces; hence strongly affected by soil texture and structure and by other constituents including organic matter. In the laboratory, a pressure membrane apparatus is used to subject soils to specific matric potentials and their soil water contents are determined gravimetrically to prepare the SWC curve. Several regions of this curve are of particular interest depending upon the purpose. In relation to water management the field capacity (the force with which moisture is held at this point varies from 1/10 to 1/3 of an atmosphere or -10.1 to -33.77 KP_a) and the permanent wilting point (the force with which moisture is held by the soil at this point corresponds to -1500 KP_a) are important as they represent the upper and lower limit of water available to plants. From drainage point of view, the volume of water in percent of the total volume between saturation and 100 cm suction or -9.806 Kp_a is taken as the representative drainable pore space. Some changes in the upper value of the suction have been suggested and can be considered while taking a decision depending upon the soil type.

Example 3.3

A soil (bulk density 1.40 g/cc) contains 30% moisture by dry weight at saturation. When the soil moisture suction is increased to 100 cm, the moisture content is 20 %. What is its drainable pore space?

At saturation 1 cc of soil will contain = 1.4* 0.30 = 0.42 cc of water

At 100 cm head it will have = 1.4 * 0.20 = 0.28 cc of water.

Assuming the density of water to be 1.00, the drainable porosity at 100 cm head is 0.42-0.28 = 0.14 cc of water per cc of soil.

It is the amount of water that will drain out of 1 cc of soil at the soil surface when the water table is lowered to 100 cm below the soil surface.

Therefore, drainable pore space = 14%

Hanging Water Column Method

The hanging water column method is commonly used to determine water retention curves, being easiest requiring no external vacuum source and pressure regulators. An advantage of the hanging water column method is that the air pressure in the soil sample is always at atmospheric pressure and has therefore, no effect on the shape of the air–water interface because of the presence of trapped air as it has with the pressure cell method. Hanging water column apparatus essentially consists of a glass funnel with a porous plate, a burette, and a flexible transparent tube connecting the glass funnel and the burette (Fig. 3.28). A number of set-ups have been proposed and any one that is convenient can be used.

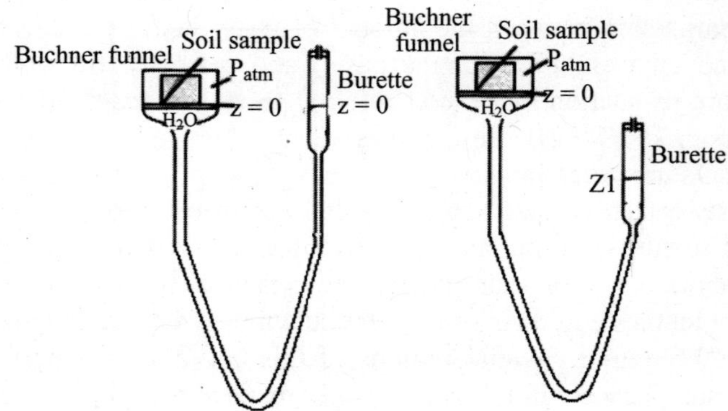

Fig. 3.28 Hanging-water column apparatus: (left) Initial saturated soil column; (right) lowered burette

This apparatus is used to determine drainable porosity in the laboratory. The procedure is described in the following paragraphs:

Undisturbed soil sample is taken from the field in a metal ring of known volume and is placed in the apparatus. The soil sample is saturated in the laboratory by maintaining the water level in the burette at the same level as the top of the porous plate. (Fig. 3.28, left). The saturation may take 24 hours or more depending on the soil type.

To impart suction to the saturated soil sample, the burette is lowered by a certain distance (usually in steps of 10 cm) resulting in water draining slowly (Fig. 3.28 right). The drained water enters the burette and raises water level in it. At a given suction, if rise in water level is more, the burette is readjusted such that the originally applied suction is closely maintained. When there is no further rise in water level in the burette, the elevation difference between the top of the porous plate and the water level in the burette are noted down, which gives the value of average suction applied to the soil sample. The difference between the initial and the final burette readings gives the volume of water drained from the soil sample due to the applied suction. This process is repeated by lowering the burette in steps of 10 cm initially and more afterwards until the desired suction (corresponding to the maximum possible depth of the subsurface drain or any other criteria) is obtained. Since the drain depths are within 2 m, the maximum suction may be around this value although the method is most suitable in the tension range of 0-150 cm.

Based on the cumulative volume of water drained as given in Table 3.15, drainable porosity is calculated by dividing the volume of water drained with the volume of the soil sample. For 100 cm suction, drainable pore space is given as: 100 (6.7/77) = 8.7%.

Table 3.15 The cumulative water drained from a hanging water column apparatus (Sample volume 77 cm^3 with 35 cm^3 of water at saturation)

Capillary pressure (cm)	Water drained (cc)
0	0.0
10	0.1
20	0.3
30	0.7
40	1.8
50	2.9
60	4.0
70	5.0
80	5.9
90	6.4
100	6.7

Field Methods

A field method to measure drainable pore space uses water balance approach. Water balance over a short period of time is calculated say for few days of a continuous heavy rainfall. Knowing runoff, evaporation and field soil moisture storage, the total rainfall that percolates to the groundwater is estimated. The total rise in water table during this period yields the drainable pore space, f by the following relation;

$$f = \frac{\text{Change in groundwater storage}}{\text{Change in groundwater level}} \tag{3.41}$$

The comparative values of drainable pore space estimated by the field method (water balance approach), graphical technique (Fig. 3.27) and soil texture pore space data (Table 3.14) are presented for a soil of the IARI farm, New Delhi (Table 3.16). The results show that there is reasonable similarity in the values from the three approaches.

Table 3.16 Drainable pore space for sandy loam soil (IARI New Delhi)

Method	Drainable pore	Remarks
Field method (Eq. 3.41)	12	-
Graphical method	11	Hydraulic conductivity 3.0 cm/h
Soil texture- pore space rating table	12-18	Sandy loam

Head- Discharge Data from Operational Drainage Systems

The most useful field method is based on using drain discharge and water table drawdown from an operating drainage system or pilot drains. Depending upon the data in hand simple method or theoretically elegant method can be used (Sanjay *et al.* 1994; See Chapter 6). In the simple method, the drain flow during a given duration is used to calculate drainable pore space as follows:

The length of the drain = L

Number of drains contributing to flow = n

Drain spacing = S

Depth to water table initial = h_1 (Observed)

Depth to water table after the drain flow = h_2 (Observed)

Total area drained = L*n*S

Total volume of soil drained = L*n*S* (h_1 –h_2)

Volume of water drained during the period (Observed) = Q

Drainable pore space = Q/ L*n*S* (h_1 –h_2)

The above calculated value of drainable porosity is the average of the soil layer given by (h_1 –h_2) and also an average over the time period. The computed value of drainable porosity is approximate as the entire depth of the soil has been assumed to have experienced a uniform water table decline of (h_1 –h_2) cm, which may not be true in subsurface drained lands, water table being curved between the adjacent parallel drains. Provisional values of the drainable pore space should be improved upon as and when more data are available.

Example 3.4

The drainable pore space of a soil is 5%. What will be fall in water table if 10 mm of water is extracted from the groundwater?

Using Eq. (3.41)

f = (Change in groundwater storage)/(Change in groundwater level)

Change in groundwater level = Change in groundwater storage/(f/100)

The fall is given as (10/5)100 = 200 mm or 20 cm.

Drainable pore space, like hydraulic conductivity, varies widely in space and is log-normally distributed. On this basis, it is recommended that geometric mean of the values should give a good estimate of the mean drainable pore space. In drainage works, a value of specific yield within 10% of its actual value would be reasonable keeping in view the uncertainty attached with other drainage parameters.

SUBSURFACE DRAINAGE COEFFICIENT

Drainage coefficient (DC) is the most important parameter that decides the lateral drain spacing and hence the cost of the system. Therefore, the need to select an appropriate value for this parameter has always been emphasized. The drainage coefficient is *'the rate of water removal and is usually expressed in terms of depth of water over the drainage area to be removed in 24 h'*. For subsurface drainage systems, drainage coefficient is assessed both for steady and unsteady state situations. Factors governing DC are: geographical location- rainfall (a most critical source of excess water), topography and land use, drainage requirement of the crops (design water table level) and salinity (salt tolerance of prevalent crops). Drainage coefficient is usually determined for the following five situations:

- Irrigation period
- Leaching for land reclamation and leaching requirement
- Average water balance during monsoon period
- Yearly water balance
- Yearly salt balance

Irrigation Period: Soil Conservation Service, USBR (1971) Approach

The approach is based on the water balance in an irrigated area. Following equation is used:

$$Q = (P+LR+C)\,Di/100F \tag{3.42}$$

Here

Q is the DC, mm/d; P is the deep percolation; LR is the leaching requirement (LR) of the soil both expressed as % of depth of irrigation; C is the seepage losses from canal/water courses as % of depth of irrigation; Di is depth of irrigation water in an application, mm; and F is the irrigation frequency (irrigation interval), days. It may be useful to consider the contribution of rainfall, R_f to on-farm groundwater recharge although in the above equation this factor has not been accounted.

The estimates for P and C in the form used in Eq. (3.42) might not be handy and therefore have to be assessed based on long-term data on rainfall, evaporation, cropping intensity, irrigation application, losses from water bodies and the experience of the field staff involved in irrigation and agriculture in a particular command (Ram, 2008). The leaching requirement of the soil is computed from the irrigation water salinity and the crop tolerance levels (desired level of salinity) (See next section).

Example 3.4

Determine the drainage coefficient with the following data:

P = deep percolation: 20% of irrigation depth, Di

C = average seepage losses from canal system: 14% of Di

Di = average irrigation depth: 70 mm per application

LR = 5% of Di

F = irrigation interval: 20 d

Incorporating these values in Eq. (3.42), we have

$$Q = [(20+5+14)*70]/(100*20) = 1.36 \text{ mm/d}$$

Leaching Requirement Approach

Leaching is a pre-requisite after drainage to leach down the accumulated salts, commonly referred as one time leaching. It is a short period operation. Although quantities of water passed may be quite high, but considering the short duration, this aspect is not taken into account to calculate drainage coefficient. Leaching requirement (LR) needed to maintain salt balance in the root zone on the other hand is a regular operation especially when saline water is used for irrigation in drained lands. Therefore, it is considered in working out drainage coefficient in the following manner:

$$LR = EC_{iw}/(5EC_e - EC_{iw}) \quad (3.43)$$

Here

EC_{iw} is the EC of irrigation water, dS/m and EC_e is the desired soil EC_e in the root zone, dS/m.

In the computation of LR, an EC_e should be used that reflects the need of the most sensitive crop grown in the area. Some of the crops such as paddy, sugarcane, pulses are quite sensitive to salts. In order to realize full yield potential of most crops, the saturated soil extract EC_e should be maintained at 2 dS/m within the root zone.

Example 3.5

Calculate the LR if salinity of irrigation water (EC_{iw}) is 0.3 dS/m and EC_e for the most sensitive crop of the area is 2.5 dS/m.

$$LR = 0.3/(5*2.5-0.3) = 0.0245 \text{ or } 2.45\% \text{ of irrigation water}$$

In the earlier example LR was taken as 5%. This LR now can be used in eq. (3.42) instead of 5% to improve over the calculated drainage coefficient in example 3.4.

Average Water Balance during Monsoon Season

In most parts of the country, 75-80% of rainfall is received during monsoon period (June-September). In waterlogged and saline areas, this period is critical from aeration point of view and leaching for reclamation (Kalendhonkar, 2013). Therefore, DC may be calculated for this critical period and compared with other methods to select an appropriately value.

- Compile long-term weather data to assess annual rainfall, % of yearly rainfall and normal value of rainfall during monsoon and number of rainy days during monsoon
- Find out the rainfall contribution to groundwater recharge usually in the range of 10-30%.
- If some irrigation water is applied, deep percolation losses as in the case of irrigated areas should also be considered. Leaching requirement need not be considered in this calculation.

$$DC = [(\text{Rainfall* Recharge factor}) + \text{contribution due to irrigation}] / \text{no. of days in the monsoon season.} \quad (3.44)$$

This gives DC in mm/d where rainfall is in mm and recharge factor is in fraction.

If investigations reveal that some period other than the monsoon season is critical, that season should also be included and compared.

Example 3.6

For the Hisar drainage project site, the average annual rainfall during July, August and September is about 336 mm. Assume that 20% of the total rainfall during the period contributes to groundwater. Calculate the drainage coefficient assuming that irrigation water adds another 30 cm to the groundwater during this period.

Using Eq. (3.44), we have

DC = [(336*0.2) + 30]/92 = 1.06 mm/d

Yearly Water Balance

Simple or complex equations are used to arrive at the excess water that contributes to groundwater during the year. Since the method of water balance has been discussed in details in one of the chapters of this book, detailed discussions are avoided herein. It may suffice to say that excess water so calculated divided by 365 will give an average drainage coefficient.

Yearly Salt Balance

To avoid resalinization of reclaimed lands, the mass of salt input should not exceed the mass of salt output. In monsoon climatic conditions, it is desirable to go in for annual rather than seasonal or short-term salt balance. A detailed discussion on this issue has been made under the monitoring and evaluation section of the Chapter 6. The readers may refer to that section to assess the drainage coefficient on the basis of yearly salt balance.

Once all calculations are made, the most optimum value may be chosen on the basis of the experience of the designer and guidelines (Table 3.17). The drainage coefficient for some of the sites in India is observed to be in the range of 1-5 mm. The guidelines emerging for this parameter can straightway be used to design subsurface drainage systems (Table 3.18)

Table 3.17 Guidelines to decide the drainage coefficient

Range of DC (mm/d)	Conditions
<1.5	Soils having a low infiltration rate
1.5-3.0	Moist soils with higher rate for more permeable soils and where cropping intensity is high
3-4.5	Extreme conditions of climate, crop and salinity management and under poor irrigation practice
>4.5	Special conditions, e.g. rice irrigation on light textured soils

Table 3.18 Guidelines for selection of drainage coefficient

Climatic conditions	Range (mm/d)	Optimum value (mm/d)
Arid	1-2	1
Semi-arid	1-3	2
Sub-humid	2-5	3

Drainage Coefficient for Non-Steady State Design

Under the conditions of periodic irrigations or high intensity rainfall conditions, steady state conditions will not be applicable. In such cases, water table needs to be lowered within a short period to ensure proper aeration. Usually, it is assumed that water table must be lowered by 30 cm below the ground surface once the water table has reached the soil surface either due to excess rainfall or irrigation. The time within which such lowering of the water table should be achieved is taken as 2.0 days. Recent experiences in Indian monsoon climatic conditions have shown that the crops yield are affected more by the soil salinity rather than aeration. As such, time within which such lowering of the water table is achieved can be increased from 2.0 days to 2.5-3.0 days. It helps to increase the drain spacing and as a result reduced investments.

COLLATION AND APPLICATION OF DATA

All available survey data for the project area needs to be assembled in readily usable forms of maps and Tables. Once it is done, one can proceed to design and prepare the layout maps.

- Decide depth of drain
- Compute drain spacing using selected approach-steady or unsteady state, design size of laterals and collectors (Chapter 4)
- Decide about drain envelope requirement, select the envelope material and design (Chapter 5)
- Decide layout pattern, fix positions of effluent disposal outlet ditches and go for construction (Chapter 6)
- Design specifications for outlet and other structures and show their location on the layout map (Chapter 6)
- Draw a management plan for land reclamation (Chapter 8), acid soils (Chapter 9) and reuse of saline water (Chapter 10)
- Conduct economic analysis (Chapter 11)

QUESTIONS (THEORY)

1. How will you proceed to undertake drainage investigations? Name various kinds of survey that are undertaken. Briefly describe at least two of them.
2. Discuss various methods used to monitor water table. What is the role of observation well and piezometers in the measurement?

3. Discuss the role of hydrographs, contour maps of the groundwater and isobaths maps in assessing the drainage problems. Illustrate each of these with neat sketches.
4. Describe the double ring infiltrometer method to determine infiltration rate.
5. Describe the Kostiakov equation of infiltration. Discuss the procedure to assess the parameters of this equation. What are its limitations?
6. Describe the constant and falling head method of finding hydraulic conductivity in the laboratory.
7. Calculate the hydraulic conductivity if the auger hole data are as under:

   Depth of hole below ground surface: 154 cm

   Depth of water table below ground surface: 74 cm

   Radius of hole: 4 cm

   Depth of impermeable layer below hole bottom: 40 cm

   Assume any other value, if required

   Recorded values

Time (sec)	Water level below Ground surface (cm)	Head (cm)	Change in head (cm)	Time interval (sec)
0	116.8	42.8	-	-
20	115.6	41.6	1.2	20
40	114.4	40.4	1.2	20
60	113.3	39.3	1.1	20
80	112.2	38.2	1.1	20
100	111.2	37.2	1.0	20

8. Derive the Hooghoudt′s equation to assess hydraulic conductivity using auger hole method above the water table.
9. Derive the equation to find hydraulic conductivity using falling water table inverse auger hole method in the absence of water table.
10. Differentiate between the methods to measure hydraulic conductivity using auger hole method in the presence and absence of water table
11. Describe the method to calculate the hydraulic conductivity in a two layered profile using auger hole method.
12. Name various methods to determine drainable pore space. Discuss the role of hydraulic conductivity in determining the drainable pore space.
13. Discuss the field methods of determining drainable pore space.
14. How will you proceed to determine subsurface drainage coefficient under various situations?

QUESTIONS (OBJECTIVE)

1. Fill in the blanks

1. An exploratory survey based on available information and some limited fieldwork is known as _____________ survey.
2. For topographic surveys in large projects, a _______ station is used.
3. In the field, soil texture is assessed by ________ of the soil with the fingers.
4. For a highly non-uniform material coefficient of uniformity is more than ______.
5. For a uniform material coefficient of uniformity is less than ______.
6. Aerial photographs, remotely sensed data and ___________ Information system have eased the works related to drainage investigations.
7. Observation wells can be either cased or __________.
8.. Observation wells are used to measure the static _____________.
9. Piezometers are reliable tools to determine ________ head at a particular depth.
10. A water table contour map provides direct visual observation of the _______ of the water table.
11. A cluster of __________ is provided around a single point to measure hydrostatic pressure, each of them __________at different depth.
12. A _______ gauge is used to measure the water in an infiltrometer.
13. The basic infiltration rate gives an approximate estimate of the ________ hydraulic conductivity.
14. A soil is said to be _________, if hydraulic properties vary spatially.
15. A soil is said to be ___________, if hydraulic properties of the aquifer are equal in all directions.
16. An __________ auger hole method is used to measure the hydraulic conductivity if water table is deep enough.
17. The drainable pore space is defined as the volume of water released or taken into storage in an unconfined aquifer per unit cross sectional area per unit decline or rise of water table as a fraction of total _________ of soil.
18. Drainable pore space is well correlated with _____________.
19. _______ water column method is used to measure drainable pore space in the laboratory.
20 infiltration rate data are fitted/analysed with the help of ________ equation.

2. Tick mark correct or most appropriate answer (A, B, C or D) from the following:

1. Reconnaissance survey is
   (A) Groundwater investigations
   (B) Detailed hydrological survey
   (C) Preliminary survey
   (D) None of these
2. An equipment to conduct topographical survey
   (A) Total station (B) EM 38
   (C) Augers (D) None of these
3. Soil texture can be assessed/determined from
   (A) Feel and touch method (B) Hydrometer method
   (C) Pipette method (D) All of these
4. Soil material is highly non-uniform if uniformity coefficient is
   (A) Less than 4 (B) More than 4
   (C) Between 2 and 4 (D) None of these
5. An impermeable layer is defined as the one having hydraulic conductivity ____________ of the hydraulic conductivity of the overlying layers
   (A) 10-20% (B) 50%
   (C) 40% (D) None of these
6. Water table is monitored using
   (A) Cased wells (B) Uncased wells
   (C) Open wells (D) All of these
7. Piezometer observations reveal
   (A) Hydrostatic pressure (B) Water table
   (C) Both A and B (D) None of these
8. Piezometers are usually installed as
   (A) A battery around a point terminating at the same depth
   (B) A battery around a point terminating at different depths
   (C) Single piezometers as in the case of observation wells
   (D) None of these
9. Which instruments are used to measure water level
   (A) Chain and bell sounding device
   (B) Electric water level recorder
   (C) Pressure logger
   (D) All of these

10. In the modified Kostiakov equation of infiltration rate $I = at^{-b} + e$, e stands for (I is the infiltration rate, t is time and a, b and e are constants)

    (A) Basic infiltration rate (B) Initial infiltration rate

    (C) None of A and B (D) None of these

11. The cumulative infiltration, I_c is given as $I_c = ct^d$ such that t is the time and c and d are constants, then d is

    (A) Always positive but less than 1

    (B) Always negative

    (C) Always positive but can be less or more than 1

    (D) None of these

12. The Darcy′s equation is given as V = -Ki, such that V is the velocity, K is the coefficient of soil permeability and I is the gradient. The negative sign indicates

    (A) flow is in the direction of increasing head

    (B) flow is in the direction of decreasing head

    (C) None of A and B

    (D) None of these

13. In a function of the type $I = kt^n$, the plot of I and t will be a straight line on

    (A) Semi-log paper (B) Normal arithmetic paper

    (C) Log-log paper (D) None of these

14. The average hydraulic conductivity is determine by

    (A) Arithmetic mean of the values

    (B) Harmonic mean of values

    (C) Geometric mean of the values

    (D) None of these

15. The hydraulic conductivity at three places in an area is 1.6, 2.0, 2.4 cm/d, the average hydraulic conductivity is

    (A) 2.0 (B) 1.973

    (C) 2.89 (D) None of these

16. Drainable pore space of a soil is 10%. What will be fall in water table if 10 mm of water is extracted from the groundwater?

    (A) 100 mm (B) 10 mm

    (C) 1 mm (D) None of these

17. The drainable pore space is determined by measuring the soil water content at saturation and soil moisture content at a suction of

    (A) 100 cm of water (B) 200 cm of water

    (C) 300 cm of water (D) None of these

18. Which of the following method is used to determine hydraulic conductivity above the water table?

    (A) Auger hole method

    (B) Inverse auger hole method

    (C) Both A and B depending upon situation

    (D) None of A and B

19. If Kh and Kv are the horizontal and vertical hydraulic conductivities, then the radial hydraulic conductivity is given as

    (A) Kh * Kv (B) Kh * Kv

    (C) √ (Kh * Kv) (D) None of these

20. All conditions being the same, subsurface drainage coefficient is least in

    (A) Arid regions (B) Semiarid regions

    (C) Sub-humid regions (D) Humid regions

■■■■

Chapter 4

DESIGN OF SUBSURFACE DRAINAGE SYSTEMS

INTRODUCTION

Subsurface drainage is accomplished by a series of perforated drains laid out at a pre-determined distance that take the water out of the area through collectors and main drain (Fig. 4.1). While planning and designing a drainage system, one has to decide upon the criteria to be used, type and layout of the system, depth and spacing of drains and the discharge capacity or the diameters of the drains (Fig. 4.2). Amongst all the factors, depth, spacing of the lateral drains and capacity of the laterals and collectors are the key elements of drainage design. In this endeavor, it should be possible to vary one or more of the key parameters to achieve an equally effective design. The cost also varies from one layout to another and thus several layouts need to be prepared to arrive at the most economic plan. In the present chapter layout of the drainage systems, drainage criteria and determination of three parameters namely the depth, spacing and diameter of drains are included. Other issues will form the part of the forthcoming Chapters.

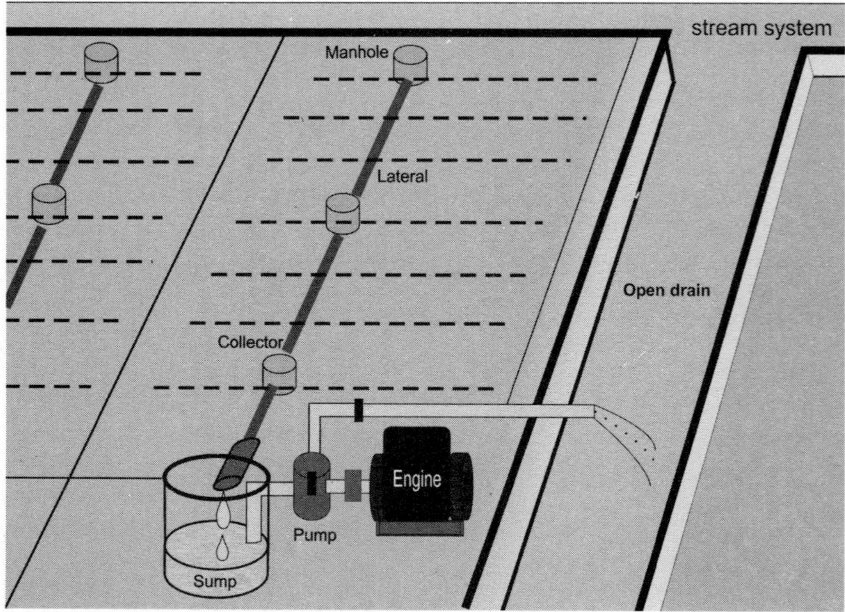

Fig. 4.1 Schematic diagram showing various components of the subsurface drainage systems

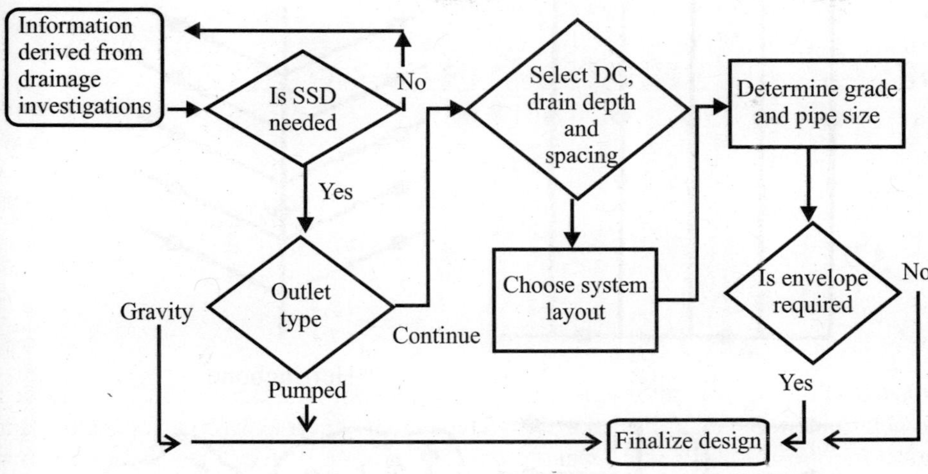

Fig. 4.2 Process flowchart to design a SSD system

LAYOUT

The layout of a subsurface drainage system is designed in a manner so as to utilize the natural slope. The layout should be such that it minimizes the lengths of main and collector drains and also the earthwork. The idea is to minimize the cost of the system. Although, layout usually depends upon the designer yet it has been classified into four general types namely: parallel, herringbone, double main and random (Fig. 4.3).

Parallel System

The parallel system also called the grid iron pattern is the most preferred system of land drainage. This system is mostly used on flat, regularly shaped fields having soils of uniform permeability. Here, parallel lateral drains are aligned perpendicular to the collector/main drain (Fig. 4.3). Laterals can enter either from one or from both the sides of the collectors. In sloping lands, field laterals are aligned across the direction of flow while the collectors are laid down the slope.

Herringbone System

In the herringbone system, parallel laterals drains enter the main drain at an angle from one or both sides (Fig. 4.3). Varying the angle of confluence with the main drain helps in achieving the desired grade of the lateral drains. The main/collector drain is aligned along the main slope.

Double Main System

This layout is used when a depression runs through the field to be drained (Fig. 4.3). Thus, as the name suggests, a main drain is placed on each side of the depression. Laterals enter the main drains as in the case of herringbone or parallel systems.

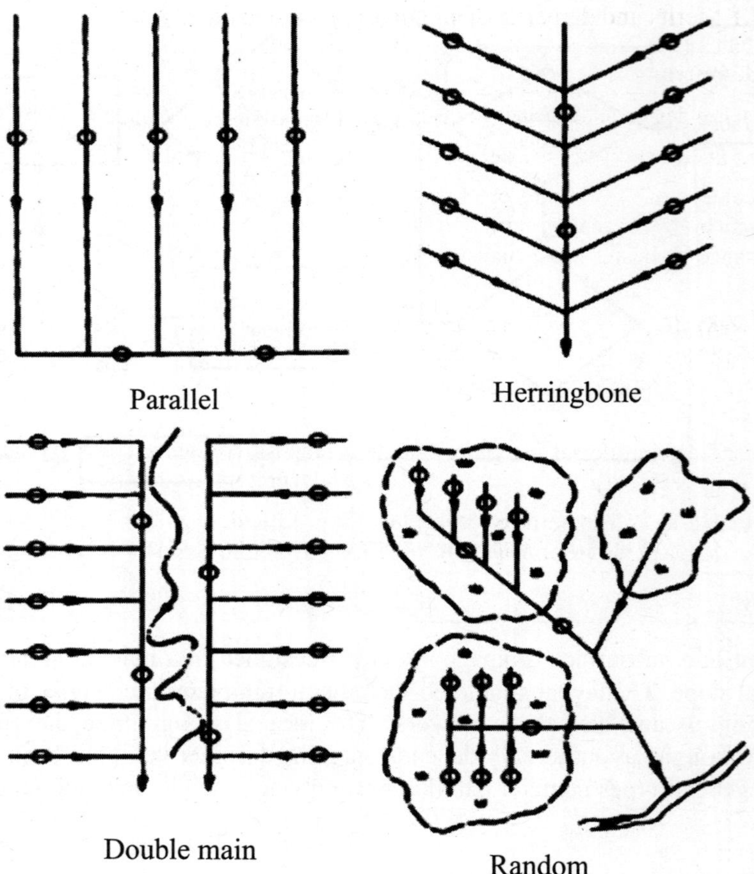

Fig. 4.3 Commonly used layouts of subsurface drainage systems

Random System

A random system of drains is used when the topography is undulating/rolling and numbers of scattered isolated wet areas are encountered. In this system drains meander from one depression to another connecting them to an outfall (Fig. 4.3). Within the random system, different wet spots can have parallel or herringbone systems. Often this system is cheaper than the other systems but operation and maintenance is difficult and costly. The system loses its advantage once the size and/or number of depressions is large.

Singular or Composite System of Drainage

In a singular pipe drainage system, each field pipe (lateral drain) discharges into an open main or collector drain (Fig. 4.4, Left). On the other hand, the field drains discharge into a pipe collector in a composite system. The collectors discharge into an open drain (Fig. 4.4, Right). In this case, collector system itself may be composite with sub-collectors and main collector. The merits and demerits of both the systems are described in Table 4.1.

Table 4.1 Merits and demerits of singular and composite layouts

Singular system	Composite system
It is easier to design and cheaper to install	It is complex to design and costly to install
Cleaning of lateral drains is easier	Cleaning of lateral drains is difficult
Large number of outlets into open ditches in a singular system means more points for maintenance and care. Maintenance cost is high	Maintenance cost is less
Open collectors break up the land in small parcels causing inconvenience in farm operations and movement of animals and machinery	No such problems are encountered in buried collectors
Grade can be controlled as per general land slopes in the area	Composite systems cause greater head loss and therefore, require steeper gradients
It is possible to collect surface runoff and reuse the drainage water for irrigation.	It is difficult to collect and dispose of surface runoff through composite drainage systems

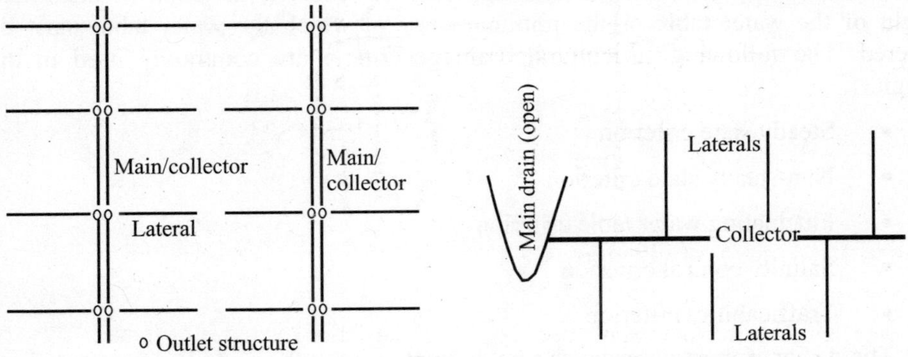

Fig. 4.4 A singular (Left) and a composite (right) subsurface drainage system

Guidelines for Layout

Layout is an art and as many layouts can be prepared as the number of persons preparing the layout. The art of it lies in maintaining the effectiveness of the system and at the same time minimizing the cost. In general

- Full advantage of the land slope should be taken while maintaining the depth of the drains as uniform as the land slope permits.
- Lateral drains should be oriented as far as possible with the field's contours to intercept water as it flows down-slope. Collectors are positioned on steeper grades so as to facilitate the placement of laterals (Fig. 4.5).

In general, a composite system of layout is preferred. An integration of singular and composite systems can also be used.

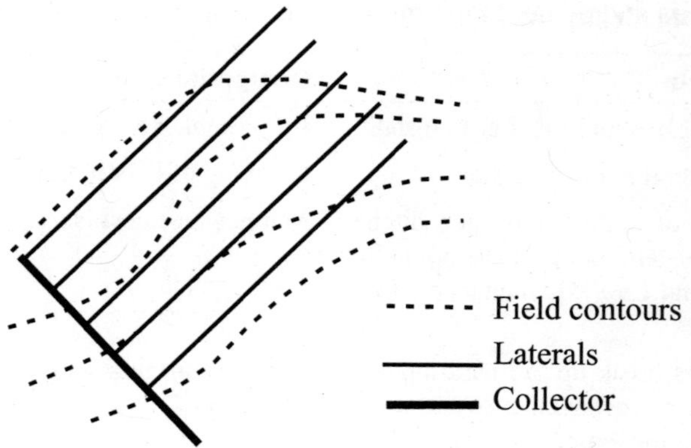

Fig. 4.5 Alignment of field laterals with contours

DRAINAGE CRITERIA

Once it is decided to drain an area using subsurface drainage, the drainage engineer must know how much excess water needs to be removed in terms of maximum height of the water table or the minimum rate at which the water table must be lowered. The following agricultural drainage criteria are commonly used in the design.

- Steady state criterion
- Non-steady state criterion
- Fluctuating water table criterion
- Salinity control criterion
- Trafficability criterion

The designer must compare the field situation with the underlying assumptions on which the drainage theories are based. He should then apply one of them which as per his judgment are most appropriate to solve the local problem.

Steady State Criterion

The steady state concept is based on the criterion that a time constant rate of rainfall or replenishment is removed equally at all distances from the drains. The steady state situation thus ensures equilibrium between recharge and discharge at a given hydraulic head, which remains constant (at a steady state condition). To apply steady state design formulae, the design criterion describes the quantity of water to be removed in a given period of time under conditions of a given water table height. Steady state conditions can occur in flat areas of humid regions where drainage need is determined by long duration, low intensity rainfall. In irrigated fields, steady state may be approached where water is applied continuously or in small amounts at frequent intervals. Although both steady and non-steady state criteria have been applied in India yet the former criterion of Hooghoudt (1940) is more frequently employed than the non-steady state criterion.

Non-steady State Criterion

Although steady state conditions are commonly employed, these are seldom encountered under field situations. In real field situations, the recharge to the water table varies with time and consequently the flow towards the drains is unsteady. In such cases, unsteady state approach describes the fluctuation of the water table as a function of time. Usual criterion is to assume that the drains are capable to lower the water table by a given amount in some pre-decided duration. It is now possible to convert steady state criterion into an unsteady state criterion or vice-versa. The unsteady state situation is encountered where high intensity rainfall is common and where field is irrigated, particularly by surface irrigation method. This criterion notably of Glover and Dumm as reported by Dumm (1954) has been used in the RAJAD project in India.

Fluctuating Water Table Criterion

Although fluctuating water table criterion also forms a part of unsteady state design and may be the most realistic, yet the steady and unsteady (falling water table) state criteria are used to design the subsurface drainage systems. It is because of their simplicity in applying to field conditions. The best design practice should be to prevent the undesirable patterns of water table fluctuations. This presumes that it should be possible to predict the effect of a given water table hydrograph on crop yield. Currently, equations are available to predict the water table hydrograph and assess its impact on crop yield (Sieben, 1964; Hiler, 1977). Equations given by Hiler and Clark (1971) as extended to land drainage (Hiler, 1977) are commonly used in the designs based on this criterion.

Salinity Control

Sustainable irrigated agriculture requires water application in excess of evapotranspiration to maintain salt balance in the root zone. The drainage systems in such situations must ensure that

- Water tables are deep enough when leaching is practiced
- Water table is deep enough during periods when soil salinization might occur due to capillary rise mostly between the two growing seasons.

Trafficability Criterion

Trafficability is the main concern when crop yields are adversely affected due to delay in crop planting as the machines cannot enter the fields. It might also happen at the time of harvesting, which might affect the quality of the produce or cause excessive losses. In this case, design is based on predicting water content or tension in the top soil using different drain spacing and a certain sequence of rainfall and evapotranspiration. Once it is known, it is correlated with traction. Aldabagh (1971) concluded that decreasing drain spacing from 97.5 m to 24.4 m provided 16 more working days during critical periods. However, this criterion is rarely used and has not been used so far in India.

Drainage engineers, besides agricultural drainage criteria, should also employ technical drainage criteria (relating to minimization of the cost of installation and operation of the system while maintaining the agricultural criteria), environmental criteria (relating to the minimization of the environmental damage) and economic drainage criteria (relating to the maximization of the net benefits).

DRAIN DEPTH

Drain depth controls the level of the water table when drain flow has ceased. When the drains are flowing, drain depth interacts with drain spacing and controls the water table height at mid-point between the two lateral drains. Some of the factors that influence the drain depth are discussed in the following sections.

Crop Tolerance to Shallow Water Table

The data on water table depth and crop yield are essential to arrive at a suitable depth at which water table needs to be controlled for a given cropping pattern. Most critical crop in the cropping pattern should be chosen while deciding the drain depth. For example, in rice-wheat rotation, wheat crop (*rabi*) might govern the drain depth whereas in pearl millet-wheat rotation, pearl millet crop (*kharif*) may ultimately decide the drain depth.

Tolerance of crop to water logging varies greatly. There are intra and inter-genic differences in crop tolerance to water logging. Whereas some crops and varieties can tolerate prolonged water logging others might fail to grow under similar conditions. On the other hand, shallow water table may not always be a curse as the crops can draw a part or even their entire consumptive use requirement from shallow groundwater. *Per se*, a shallow groundwater is not harmful to the plants but indirectly, it controls the prevailing moisture conditions and therefore, has an influe-nce on water supply, aeration, and heat properties of soils. Any imbalance results in adverse impacts on crop yield. Many studies have been conducted to assess crop yields against static water levels (Table 4.2, Source: Willamnson and Kritz, 1970). It might not represent the real field situations where water table fluctuates widely. Relative tolerance of crops on the basis of internet search is given in Table 4.3.

Table 4.2 Tolerance of crops to depths to water table for optimum yield

Water table depth (cm) for optimum yield (100%)		
50	**50-100**	**100**
Ladine clover	Alfalfa	Wheat
Orchard grass	Corn	Barley
Fescue	Peas	Oats
Horse beans	Snap beans	Beans
Cabbage	String beans	Sugar beet
Tomato	Soybean	Grain sorghum
Potato	Colza	
Millets		
Squash		
Mustard		
Pepper		

Table 4.3 Relative crop tolerance to shallow groundwater tables (Groundwater at 50 cm)

Tolerance level	Crops
High tolerance	Sugarcane, potatoes, rice, willow, plum, broad beans strawberries, some grasses
Medium tolerance	Sugar beet, wheat, oats, citrus, bananas, apple, barley, peas, cotton pears, blackberries, onion
Sensitive	Maize, tobacco, peaches, cherries, olives, peas, beans, date palm

Sieben (1964) analyzed the Dutch data on water logging and its effect on crop yield on the basis of an index named as Stress due to Excess Water (SEW). To calculate this index, all values during the crop growth period are summed up whenever the groundwater has risen above the 30cm depth (SEW_{30}). On the basis of fluctuating water table, tolerance of four crops is in the order of winter wheat > peas > oats > winter barley. The yield of wheat remains unaffected until this index crosses 200 mark whereas barley yield reduces at an index of about 120.

Gupta (1984) reviewed the available information on water table depth and crop yield from various experiments in India (Table 4.4). It indicates the relative tolerance of the crops with first crop being the most tolerant. It may be pointed out that the optimum depth to water table listed in this table is only qualitative in nature. Quality of groundwater, amount of rainfall and its distribution and irrigation scheduling can greatly modify the yield under similar depths to water table.

Table 4.4 Optimum depth to water table for various crops

Crop	Depth to water table (m)
Rice	≤ .30
Sugarcane	0.60
Wheat	0.60
Cowpea	0.75
Pea	0.75
Barley	0.90
Gram	0.90
Safflower	1.00
Maize	1.20
Cotton	1.22
Pearl millet	1.25
Soybean	1.25

Soil Salinization – Groundwater Table and Drain Depth

In land reclamation projects, it is imperative to minimize soil resalinization following a land reclamation programme. As such, the water table must be lowered to such depths so as to cutoff evaporation through capillary rise of water from the soil surface. Although, upward water flux decreases with increasing depth to water table, generalized relationships between water table depth and upward water flux suggests that water can certainly move at appreciable rate from the water tables as deep as 5 m in heavy textured soils. Talsma (1963) reported that water evaporation from the water table can take place at similar rates as from an open pan evaporation at some depths, being a function of the soil type and rate of evaporation (Table 4.5). Several empirical equations have been proposed to assess the maximal limiting upward flux (cm/d), one proposed by Gardener and Fireman (1958) is given as:

$$q_{max} = \frac{A\,a}{d^n} \tag{4.1}$$

where d is the depth to water table (cm); a (cm^{n+1}/ d) and n are constants; and A is also a constant which depends on n (Table 4.6).

Table 4.5 Water table depths at which upward water movement from the water table equals daily class A pan evaporation rate for two soils

Soil type	Depth to water table (m) for daily evaporation rate (mm/d)					
	1	2	4	5	8	10
Medium clay	1.3	0.9	0.6	0.5	0.4	0.4
Loam	2.0	1.6	1.3	1.2	1.1	1.0

Table 4.6 Experimental values of constants of Eq. (4.1)

Soil texture	n	a (cm^{n+1}/d)	A
Sand	4	1.7×10^8	1.52
Fine sandy loam	3	3.2×10^5	1.76
Loam	2	0.7×10^3	2.46
Clay	2	1.1×10^3	2.46

Within these limits, critical depth to water table is defined as the depth at which evaporation from the soil surface is nearly 10% of its value when the water table is at or near to the ground surface (Talsma, 1963). For good quality groundwater, a range of 0.5 - 1 mm/d is also suggested. It is common to take 1 mm/d as the limiting value. The depth to water table at which such flux values are achieved is referred as critical depth to water table. Critical depth to water table is shallower in coarse than the fine textured soils. Several studies in India established that critical depth to water table is nearly 1.0 m in many cases. The drain depth for salinity control then can be determined by adding to it the depth of the root zone. Khosla *et al.* (1979) reported that flux rate in a sandy loam soil of the Gohana region is reduced to less than 0.1 cm/d, when the water table is at or around 80 cm. Therefore, he suggested that it will be desirable to lower the water table to 1.4-1.7 m below the ground surface to minimize root zone salinization. These values are obtained by adding the crop root zone for common crops of the region being 60 to 90 cm. However, current view does not support the concept of critical depth to water table. According to this view, the evaporation from the soil surface can be minimized or even avoided under shallow

drains provided the water flux in the drained lands is kept downward. Under the Indian monsoon climate, this situation occurs during the monsoon season. Most salts accumulated during the *rabi* season are leached during this period.

Based on the above discussions and other results emerging from various studies, following criteria are adopted to arrive at an optimum drain depth:

- Drains should be laid at sufficient depth so that the system is able to lower the water table to the required depth within the stipulated time. The water table in a drained field is highest at mid-way between the drains. Factors such as precipitation or other sources of recharge, evaporation and other resources of discharge, depth and spacing of drains, soil properties, cross-sectional area of drains and water level in the drains influence the height of the water table. These factors are interrelated by drainage equations based on Dupuit-Forchheimer theory.
- Depth to the impermeable barrier governs the depth of drainage. Drains are not installed below the impermeable barrier.
- The maximum depth of the drains is limited by the capacity of drainage machines being used. Although, in manual laying, it is possible to go to any depth, the cost often becomes the deciding factor then. Increasing top width and difficulties in laying pipes/tiles at deeper depths increases the cost of drainage system which may not be proportional to the benefits achieved through deep drainage.
- Under gravity outflow conditions, depth of drains is governed by the outlet conditions.
- The depth of drains is also controlled by economics. As spacing and depth to a greater extent are inter-related, the cost of digging up to a certain depth is compensated by the increased lateral spacing. The digging beyond this depth becomes uneconomical. The increasing cost beyond a particular depth is attributed to the fact that the depth of water conveying network is also related to the depth of the lateral drains (Fig. 4.6). A general diagram showing the relationship between depth, spacing and cost of construction is given in Fig. 4.7. The vertical line passing through the trough or near about should help to select the drain depth. In general, this depth has been found to vary between 1.5 m to 2.0 m.
- Salinity of the groundwater is another factor that controls the drain depth. It is desirable to lower the water table to deeper depths with increasing salinity of the groundwater to reduce secondary salinization (Table 4.7, Gupta & Gupta, 1997). However, this statement is sometimes criticized as it results in deeper drainage systems. A more detailed discussion follows in a forthcoming sub-section of this chapter. Quantity of drainage effluent is also governed by the depth of the lateral drains. Deeper the drains, more is the discharge. Thus, if salinity of groundwater is more, disposal problems will increase with depth of the drainage system.
- Depth of drainage in certain cases is modified to account for some specific problem/difficulty. For example in the case of earthen ware pipes, the

depth is always kept deeper than the frost penetration depth of the soil. In areas where heavy machinery is used, the maximum depth should ensure that the drains do not fail by repeated passes of the heavy machinery. The minimum depth for mineral soils is 0.6 m while for organic and deep peat and muck soils is 0.75 m and 1.2 m respectively. It is one of the major deciding factors in case of mole and plastic drains. On peat soils, the depth of drains should also account for the settlement of peats subsequent to land drainage.

Table 4.7 Recommended minimum depths to water table as influenced by groundwater salinity

Groundwater salinity		Groundwater table depth (m)
g/L	**dS/m**	
0.1	0.16	1.0
1.5	2.34	1.5
2.2-3.0	3.1-4.7	2.5
3.0 -20.0	4.7-15.6	3.0
20.0-30.0	15.6 -31.2	3.5

- In many cases, soil texture and strata permeability may play an important role in deciding drain depth. If a soil layer of high hydraulic conductivity is available at a reasonable depth, it is advisable to lay pipe/tile lines in the more permeable layer.
- Besides what has been stated in this section, drain depth is also governed by the kind of material available. The static and dynamic load on the conduit should not exceed the load bearing capacity of the conduit. This issue has been taken up in Chapter 5.

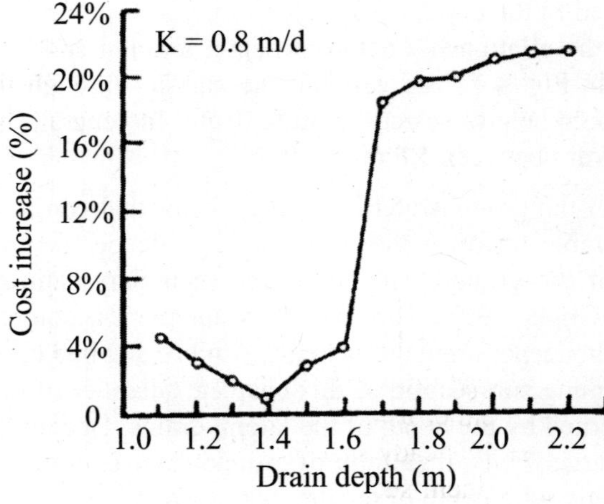

Fig. 4.6 Variation in cost of SSD system with increasing drain depth for sandy loam soils

In practice, drain depths vary greatly. Drain depths used in some of the successful drainage experiment in India vary from 0.75 m to 2.0 m. From installation point of view a depth of 1.2-1.8 m is optimum for salinity control.

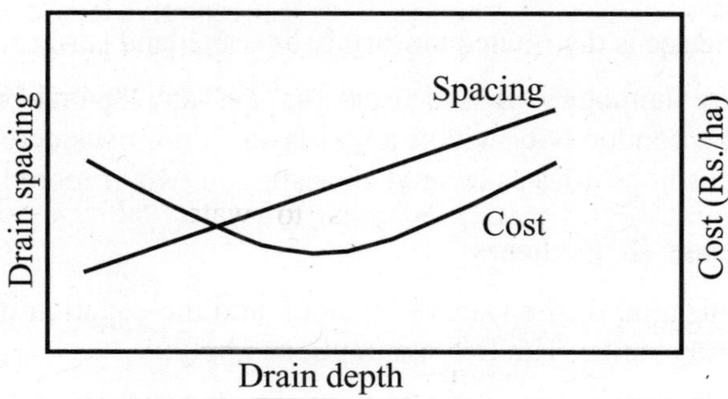

Fig. 4.7 Drain depth, spacing and cost

Guidelines on Drain Depth

On the basis of field experiences and salt and water balance modeling, it emerges that shallow drains can be quite effective in maintaining salt and water balance. As such following guidelines have been framed and can be adopted under the Indian conditions (Table 4.8, Gupta, 2011).

Table 4.8 Guidelines on drain depth

Outlet conditions	Depth of the drains	Optimum depth (m)
Gravity	0.9-1.2	1.1
Pumped	1.2-1.8	1.5

SPACING OF LATERAL DRAINS

Determination of lateral drain spacing has been a matter of intensive and extensive investigations throughout the world. To derive a theoretically sound exact solution of a drainage problem is really complex and time consuming. Therefore, often assumptions are made to derive simpler solutions. Steady state situations can be handled analytically better than unsteady state. Many equations have been derived to assess ditches or tiles spacing employing steady (Inflow = Outflow) state theory.

Steady State

Steady state formulae for agricultural land drainage are commonly employed because their usefulness under field conditions has been extensively tested. Steady state equations have been derived either using Dupuit-Forchheimer theory or the potential theory. Although more than 40 steady state drain spacing formulae have been listed in the literature, only few of them such as Hooghoudt (1940), Dagan (1964), Ernst (1956) and Toksoz and Kirkham (1961) are in common use. This chapter will deal with the first three equations.

In steady state, to describe the groundwater flow to the drains, following assumptions are made:

- The flow is two-dimensional i.e. flow is considered identical in any cross section perpendicular to the drains
- The recharge is distributed uniformly over the land surface
- Soils are homogenous and isotropic i.e. any spatial variation in the hydraulic conductivity within a soil layer is not considered, although one can treat soil profiles as layered consisting of two or more layers.

Dupuit-Forchheimer (D-F) theory

Simultaneous application of the Darcy's equation and the equation of continuity has yielded many useful results. The D-F assumptions imply:

- For small inclinations of the free water surface, all streamlines in a gravity flow system are horizontal
- The velocity along these streamlines is proportional to the slope of the free water surface but is independent of the depth.

These assumptions allow one to reduce the two-dimensional flow to a one-dimensional flow by assuming parallel and horizontal streamlines. Such a flow pattern will occur as long as the impervious soil layer is close to the drain. Moreover, for small inclinations of the free surface, the streamlines can be taken as horizontal because all the equipotential lines are vertical planes. Such an assumption usually leads to an erroneous picture of the flow path near the drain line. If we strictly analyze the two assumptions, it may be realized that there will be no flow through the tile drains. It is an indication of the approximate nature of the equations derived on the basis of this theory.

Potential theory

In this theory, the Laplace's partial differential equation for steady state is solved by identifying the boundary values of a given problem. The analysis on this basis is mathematically more exact but complex. The analysis using this theory provides the drain spacing equation as well as gives a fairly good idea of the flow system. Wesseling (1964) compared the solutions of Kirkham (1958) derived on the basis of potential theory and of Hooghoudt (1940) derived on the basis of D-F assumptions. He observed that the two equations gave almost similar results at least in respect of drain spacing.

DESIGN EQUATIONS (STEADY STATE)

Amongst the three equations namely the Hooghoudt (1940), the Dagan (1964) and the Ernst (1962), the Hooghoudt (1940) equation is most popular and commonly used in drainage design in India. It is also referred as Ellipse equation in the literature.

Hooghoudt Equation

Considering a steady-state flow to vertically walled open drains reaching an impervious layer and linearly distributed recharge, the flow of groundwater (q_x)

through a vertical plane (y) at a distance (x) from the ditch through P(x,y) (using Darcy's equation and D-F assumptions) is given as follows (Fig. 4.8):

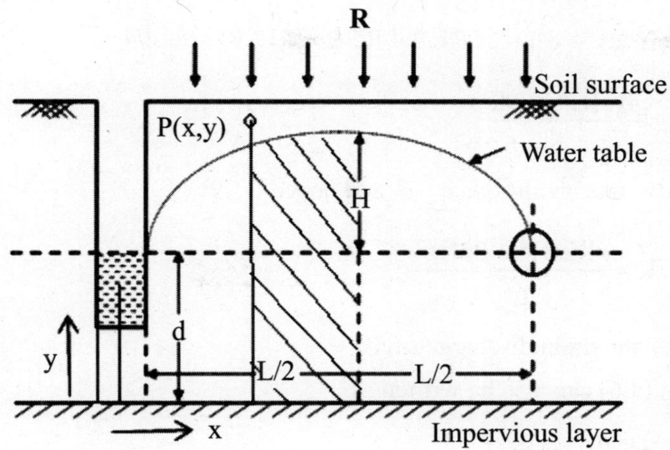

Fig. 4.8 Definition sketch of subsurface drainage under steady state conditions

$$q_x = K.y\ \frac{dy}{dx} \tag{4.2}$$

Here q_x is the unit flow rate in x direction, (m^2/d), K is the hydraulic conductivity of the soil, (m/d), y is the hydraulic head above the base of the impermeable layer at any x and dy/dx is hydraulic gradient at x (-).

Using the principle of continuity, all the water entering the soil in the surface area mid-way between the drains and the vertical plane (y) at distance (x) will pass through this plane on its way to the drains. If R is the rate of recharge per unit area, then the flow per unit time through the plane (y) from the shaded area is given as:

$$q_x = R(1/2L\text{-}x) \tag{4.3}$$

Here R is the rate of recharge per unit surface area (m/d) and L is the drain spacing (m)

Equating Eq. (4.2) and Eq. (4.3)

$$K.y\ dy/dx = R(1/2L - x) \tag{4.4}$$

The boundary conditions for an ideal drain are written as:

$$x = 0 \rightarrow y = d;\ \text{and}\ x = 1/2L \rightarrow y = H + d$$

Such that d is the elevation of the water level in the drain (m) above the impermeable barrier and H is the elevation of water level midway between the drains (m) above the center of the drain.

Integrating Eq. (4.4) $\int K.y\ dy/dx = \int R(1/2L - x)$;

$$K\,y^2/2\Big|_{d}^{H+d} = RL/2 - Rx^2/2\Big|_{0}^{L/2}$$

$$K[(H+d)^2 - d^2] = 2(RL/2 - RL/2 + RL^2/8) = RL^2/4$$

$$L^2 = \frac{4K(H^2 + 2dH)}{R} \tag{4.5}$$

For steady state conditions q = R and hence

$$q = R = \frac{4K(H^2 + 2dH)}{L^2} \tag{4.6}$$

where q is the drain discharge (m/d)

Equation (4.6) can also be written as

$$q = \frac{8KdH}{L^2} + \frac{4KH^2}{L^2} \tag{4.7}$$

If drains are located on the impervious layer (d=0), the first term of Eq. (4.7) becomes negligible. In that case q is given as flow from above the impermeable layer

$$q = \frac{4KH^2}{L^2} \tag{4.8}$$

In case d >>H, the second term of Eq. (4.7) becomes negligible and the flow can be given as:

$$q = \frac{8KdH}{L^2} \tag{4.9}$$

It means the flow is from below the drains.

Considering that in a two layered profile drains are placed at the interface of the two layers having hydraulic conductivities K_t, of the layer above the drains and K_b, of layer below the drains, then Eq. (4.7) can be written as:

$$q = \frac{8K_b dH}{L^2} + \frac{4K_t H^2}{L^2} \tag{4.10}$$

Equation (4.10) therefore, can be applied to a two layered profile provided the drains are placed at the interface and $K_b > K_t$.

Ellipse Equation

Equation (4.6) derived by Hooghoudt in 1936 is also known as Donnan Equation (Donnan, 1946). It can be derived by integrating Eq. (4.4) $\int K.y\, dy/dx = \int R(1/2L - x)$; using boundary conditions

For an ideal drain $x = 0 \rightarrow y = d$; and $x = 1/2L \rightarrow y = b$, such that $b = d + H$

$$L^2 = \frac{4K\,(b^2 - d^2)}{R}$$

It can be seen that this formula is the same as the simplest form of Hooghoudt's formula.

Concept of Equivalent Depth

Hooghoudt equation has been derived using Dupuit-Forcheimer assumptions, which in reality means that the flow is idealized (Fig. 4.9, left). On the other hand, if d>>H, the flow lines become radial (Fig. 4.9, right). To account for the extra resistance caused by radial flow, Hooghoudt introduced the concept of equivalent depth that practically reduces the depth d to a smaller equivalent depth d_e, which is a function of L, d and r_o (radius of drain). The concept implies that *'A permeable layer overlying a fictitious impermeable layer of equivalent thickness such that this value if incorporated in the approximate solution, the design spacing will be equal to that of an exact solution derived assuming radial flow'* (Fig. 4.10). The values of d_e corresponding to different values of L, d and r_o in tables and nomographic forms can be found in standard books on drainage (Kumar & Singh, 2005; Ritzema, 1994). As an example, tabulated values of equivalent depth for a drain radius of 0.1 m for different drain spacing and actual depth to impermeable layer are given in Table 4.9. Since calculators/ computers are handy now, equations have been derived to compute d_e to dispense with the use of tables and graphs.

A closed-form expression for the equivalent depth (d_e) has been developed that can replace the Hooghoudt's tables (Van Der Moolen & Wesseling, 1991)

$$d_e = \pi L / 8 \; \{ \ln(L/\pi r) + F(x) \} \tag{4.11}$$

Here $x = 2\pi\,(d - d_2) / L$ such that d_2 is the depth of drains below ground surface.

$$F(x) = \Sigma\, 4e^{-2nx} / n\,(1 - e^{-2nx}), \text{ with } n = 1, 3, 5, \ldots \tag{4.12}$$

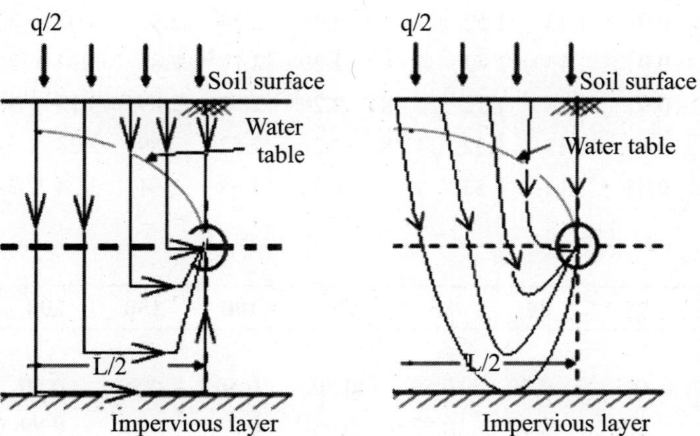

Fig. 4.9 The idealized flow with Dupuit-Forcheimer assumptions (left) and actual radial pattern (right)

Table 4.9 Values of equivalent depth for r_o = 0.1 m [d and L in m (Hooghoudt, 1940)]

Part I

L	**5 m**	**7.5**	**10**	**15**	**20**	**25**	**30**	**35**	**40**	**45**	**50**
d											
0.5	0.47	0.48	0.49	0.49	0.49	0.50	0.50	0.50	0.50	0.50	0.50
0.75	0.60	0.65	0.69	0.71	0.73	0.74	0.75	0.75	0.75	0.76	0.76
1.00	0.67	0.75	0.80	0.86	0.89	0.91	0.93	0.94	0.96	0.96	0.96
1.25	0.70	0.82	0.89	1.00	1.05	1.09	1.12	1.13	1.14	1.14	1.15
1.50	0.70	0.88	0.97	1.11	1.19	1.25	1.28	1.31	1.34	1.35	1.36
1.75	0.70	0.91	1.02	1.20	1.30	1.39	1.45	1.49	1.52	1.55	1.57
2.00	0.70	0.91	1.08	1.28	1.41	1.5	1.57	1.62	1.66	1.70	1.72
2.25	0.70	0.91	1.13	1.34	1.50	1.69	1.69	1.76	1.81	1.84	1.86
2.50	0.70	0.91	1.13	1.38	1.57	1.69	1.79	1.87	1.94	1.99	2.02
2.75	0.70	0.91	1.13	1.42	1.63	1.76	1.88	1.98	2.05	2.12	2.18
3.00	0.70	0.91	1.13	1.45	1.67	1.83	1.97	2.08	2.16	2.23	2.29
3.25	0.70	0.91	1.13	1.48	1.71	1.88	2.04	2.16	2.26	2.35	2.42
3.50	0.70	0.91	1.13	1.50	1.75	1.93	2.11	2.24	2.35	2.45	2.54
3.75	0.70	0.91	1.13	1.52	1.78	1.97	2.17	2.31	2.44	2.54	2.64
4.00	0.70	0.91	1.13	1.52	1.81	2.02	2.22	2.37	2.51	2.62	2.71
4.50	0.70	0.91	1.13	1.52	1.85	2.08	2.31	2.50	2.63	2.76	2.87
5.00	0.70	0.91	1.13	1.52	1.88	2.15	2.38	2.58	2.75	2.89	3.02
5.50	0.70	0.91	1.13	1.52	1.88	2.20	2.43	2.65	2.84	3.00	3.15
6.00	0.70	0.91	1.13	1.52	1.88	2.20	2.48	2.70	2.92	3.09	3.26
7.00	0.70	0.91	1.13	1.52	1.88	2.20	2.54	2.81	3.03	3.24	3.43
8.00	0.70	0.91	1.13	1.52	1.88	2.20	2.57	2.85	3.13	3.35	3.56
9.00	0.70	0.91	1.13	1.52	1.88	2.20	2.57	2.89	3.18	3.43	3.66
10.0	0.70	0.91	1.13	1.52	1.88	2.20	2.57	2.89	3.23	3.48	3.74
∞	0.71	0.93	1.14	1.53	1.89	2.24	2.58	2.91	3.24	3.56	3.88

Part 2

L	**50**	**75**	**80**	**85**	**90**	**100**	**150**	**200**	**250**
d									
0.5	0.50	0.50	0.50	0.50	0.50	0.50	0.50	0.50	0.50
1	0.96	0.97	0.97	0.97	0.98	0.98	0.99	0.99	0.99
2	1.72	1.80	1.82	1.82	1.83	1.85	1.00	1.92	1.94
3	2.29	2.49	2.52	2.54	2.56	2.60	2.72	2.70	2.83
4	2.71	3.04	3.08	3.12	3.16	3.24	3.46	3.58	3.66

L	50	75	80	85	90	100	150	200	250
5	3.02	3.49	3.55	3.61	3.67	3.78	4.12	4.31	4.43
6	3.23	3.85	3.93	4.00	4.08	4.23	4.70	4.97	5.15
7	3.43	4.14	4.23	4.33	4.42	4.62	5.22	5.57	5.81
8	3.56	4.38	4.49	4.61	4.72	4.95	5.68	6.13	6.43
9	3.66	4.57	4.70	4.82	4.95	5.23	6.09	6.63	7.00
10	3.74	4.74	4.89	5.04	5.18	5.47	6.45	7.09	7.53
12.5	3.74	5.02	5.20	5.38	5.56	5.92	7.20	8.06	8.68
15	3.74	5.20	5.40	5.60	5.80	6.25	7.77	8.84	9.64
17.5	3.74	5.30	5.53	5.76	5.99	6.44	8.20	9.47	10.4
20	3.74	5.30	5.62	5.87	6.12	6.60	8.54	9.97	11.1
25	3.74	5.30	5.74	5.96	6.20	6.79	8.99	10.7	12.1
30	3.74	5.30	5.74	5.96	6.20	6.79	9.27	11.3	12.9
35	3.74	5.30	5.74	5.96	6.20	6.79	9.44	11.6	13.4
40	3.74	5.30	5.74	5.96	6.20	6.79	9.44	11.8	13.8
45	3.74	5.30	5.74	5.96	6.20	6.79	9.44	12.0	13.8
50	3.74	5.30	5.74	5.96	6.20	6.79	9.44	12.1	14.3
60	3.74	5.30	5.74	5.96	6.20	6.79	9.44	12.1	14.6
∞	3.88	5.38	5.76	6.00	6.26	6.82	9.55	12.2	14.7

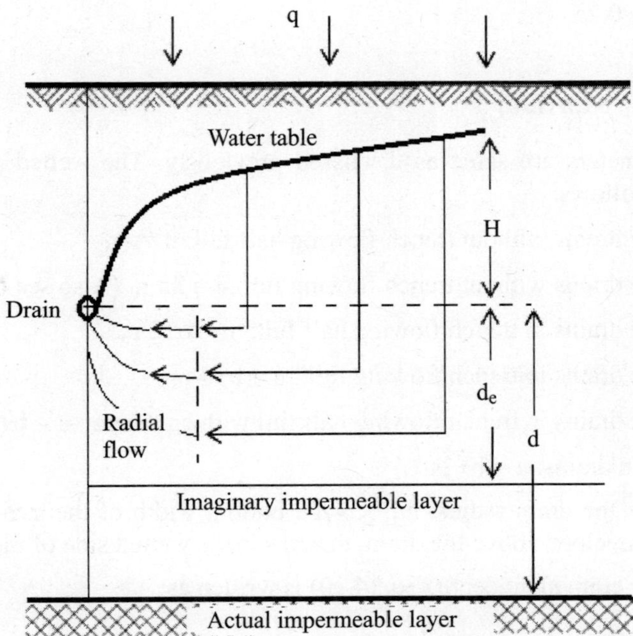

Fig. 4.10 Concept of impermeable layer
(Note the actual depth d and equivalent depth d_e)

Simplified versions are given as follows:

For $0 < d/L < 0.3$

$$d_e = \frac{d}{1 + d/L\,[(8/\pi)\,Ln\,(d/r_e) - \alpha} \tag{4.13}$$

Here effective drain radius, r_e is calculated as the sum of the drain radius, r_0 and thickness of the gravel envelope, fthick, as follows:

$$r_e = r_0 + \text{fthick}$$

α is calculated with the help of the following equation:

$$\alpha = 3.55 - 1.6d/L + 2(d/L)^2 \tag{4.14}$$

For $d/L > 0.3$,

$$d_e = \frac{\pi L}{8\,[Ln\,(L/r_e) - 1.15)]} \tag{4.15}$$

Slightly different equations have also been given in the literature that can be used to find out equivalent depth to impermeable layer:

For $0 < d/L < 0.25$

$$d_e = \frac{d}{1 + \,[(8d/\pi L)\,Ln\,(d/u)]} \tag{4.16}$$

For $d/L > 0.25$,

$$d_e = \frac{\pi L}{8\,Ln\,(L/u)} \tag{4.17}$$

All parameters are same as discussed previously. The wetted perimeter, u is calculated as follows:

- Pipe drains without trench flowing half full: $u = \pi r_0$
- Pipe drains without trench flowing full: $u = 2\pi r_0$ (Also see Fig. 4.11)
- Pipe drains in trench flowing half full: $u = b + 2r_0$
- Pipe drains in trench flowing full: $u = b + 4r_0$
- Pipe drains in trench flowing half full with envelope: $u = b + 2(2r_0 + p)$
- Open drains: $u = b + 2s$

Here r_0 is the drain radius, m, b is the bottom width of the trench, m, p is the height of the envelope above the drain, m and s is the wetted side of the open drain.

Using the equivalent depth, Eq. (4.10) is written as:

$$q = \frac{8K_b d_e H}{L^2} + \frac{4K_t H^2}{L^2} \tag{4.18}$$

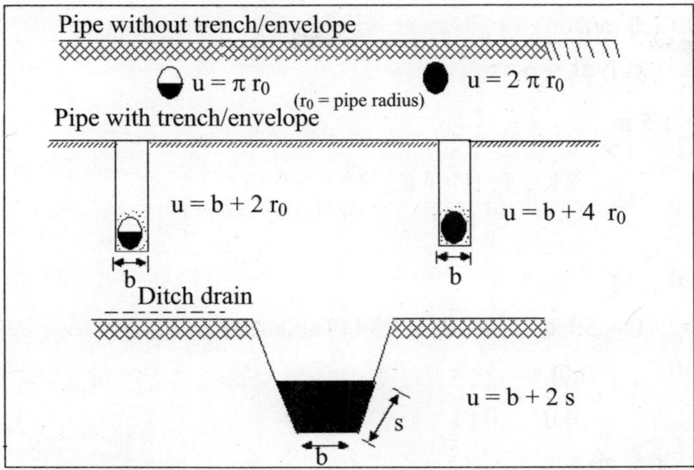

Fig. 4.11 The calculation of u values for various situations

Limitations

The main limitations of Hooghoudt equation arise not because of the assumptions made in deriving this equation but in its practical use:

- The calculations require trial and error to arrive at the design spacing. It is due to the nature of equivalent depth to impermeable layer which is not given explicitly but as a function of the drain spacing.
- The equation is inapplicable for layered soils where hydraulic conductivity of layer below the drains K_b is less than the hydraulic conductivity of the layer above the drains K_t.
- The Hooghoudt's equation is not applicable when there is more than one soil layer below the drains.
- Later studies have also shown that this equation is not applicable to sloppy lands where slopes exceed 4 to 5%.
- The equation is limited to situations where the horizontal hydraulic conductivity equals or exceeds the vertical hydraulic conductivity.

Example 4.1

In an irrigated area, drain pipes of 0.1 m radius are to be installed at a depth of 1.5 m below the ground surface. While the hydraulic conductivity of this layer is 1.0 m/d, a relatively impermeable soil layer exists at a depth of 3.0 m below the soil surface. The hydraulic conductivity above this layer but below 1.5 m depth is estimated at 7.5 m/d. Calculate the lateral drain spacing. Assume that drainage coefficient is 10 mm/d and mid-way water table depth below the soil surface is 0.75 m.

Knowing

$q = 0.01$ m/d

$K_b = 7.5$ m/d

$K_t = 1.0$ m/d

h = 1.5- 0.75 = 0.75 m

d = 1.5 m

$$L^2 = \frac{8 K_b d_e H + 4 K_t H^2}{q}$$

First trial

For L = 50, d= 1.5, d_e = 1.34 (Table 4.9)

$$L^2 = \frac{60.3}{0.01} + \frac{2.25}{0.01}$$

$\therefore L = 79.1$ m

For L = 75, d = 1.5, d_e = 1.39 (Table 4.9)

Solving the equation with the new values, one gets

L = 80.4 m

Since the selected value and the calculated values are nearly the same, the drain spacing for this case can be taken as 80 m.

Following example illustrates the application of the Eqs. (4.13) and (4.15) for calculating the equivalent depth to impermeable layer.

Example 4.2

For the drainage of an irrigated area, drain pipes with a diameter of 0.1 m are to be installed at a depth of 1.8 m below the ground surface. The hydraulic conductivity of this layer is 50 cm/d. A relatively impermeable soil layer exists at a depth of 5.0 m below the drains. From auger hole tests the hydraulic conductivity above this layer but below 1.8 m depth was estimated at 75 cm/d. If the drainage coefficient is 4 mm/d, what drain spacing must be applied if the mid-way water table depth below the soil surface is 1.05 m? The thickness of the envelope may be taken as 10.0 cm.

Assume drain spacing = 50 m

r_o = 0.1/2 = 0.05 m

fthick = 0.05

d/L = 5/50 = 0.1

As the value of d/L is < 0.3, Eq. (4.14) and (4.13) will be used to calculate the equivalent depth to impermeable layer.

$\alpha = 3.55 - 1.6\,(0.1) + 2\,(0.1)^2$

= 3.41

$r_e = r_o +$ fthick = (0.1/2) + 0.1 = 0.15 m

$$d_e = \frac{5}{1+\{8d/\pi L * \text{Ln}\,(5/0.15-3.41)\}} = 2.68$$

H= 1.8-1.05 = 0.75

Using Hooghoudt equation for layered profile Eq. (4.18):

$$L = \left[\frac{8\times 0.75\times 2.68\times 0.75 + 4\times 0.5\,(0.75)^2}{0.004}\right]^{1/2} = 57.4$$

As the value calculated is higher than the assumed value, a second trial is recommended.

Second trial:

Assume L = 60 m

D/L = 5/60 = 0.083 (< 0.3)

$\alpha = 3.43$

$d_e = 2.91$ m

$$L = \left[\frac{8\times 0.75\times 2.91\times 0.75 + 4\times 0.5\,(0.75)^2}{0.004}\right]^{1/2} = 59.6$$

We may end up here since the calculated spacing is nearly equal to the assumed spacing.

Dagan Equation

Dagan equation for homogenous soil profiles is commonly used in drainage design because the drain spacing can be explicitly determined without trial and error (Dagan, 1964). The equation has been developed with the assumption that the flow to the drain is composed of a radial flow in the area between the drain and a distance (1/2 D√2) away from the drain. In the area between ½ D√2 plane and mid plane between the drains, flow is mainly horizontal. The relationship between various parameters is given by the following equations (Fig. 4.8):

$$H = \frac{q\,L}{K} F_{HD} \tag{4.19}$$

Here $$F_{HD} = \frac{1}{4}\left(\frac{L}{2d} - E\right) \tag{4.20}$$

$$E = \frac{2}{\pi}\text{Ln}\left(2\,\text{Cosh}\,\frac{\pi\, r_o}{d} - 2\right) \tag{4.21}$$

Such that Cosh x is the hyperbolic function of $x = (e^x + e^{-x})/2$, H is the hydraulic head midway between the drains above the drain level, m, d is the depth to impermeable layer from the drain level, m, L is the drain spacing, m, q is the drain discharge corresponding to hydraulic head, H, m/d, K is the average hydraulic conductivity of the soil profile above the impervious layer, m/d and r_o is the radius of the drain, m.

On putting the value of E (-ve values) and then putting the value of F_{HD}, we get a quadratic equation. By solving this equation and discarding the negative value, we get the drain spacing as shown in the following example:

Example 4.3

For the drainage of an irrigated area, drain pipes with a radius of 0.1 m are to be installed at a depth of 1.8 m below the ground surface. The hydraulic conductivity of this layer is 1.0 m/d. A relatively impermeable soil layer exists at a depth of 3.0 m below the soil surface. From auger hole tests the hydraulic conductivity above this layer but below 1.8 m depth was estimated at 7.5 m/d. If the drainage coefficient is 10 mm/d, what drain spacing must be applied when the mid-way water table depth below the soil surface is 1.05 m?

$q = 0.01$ m/d

$H = 0.75$ m

$K = 4.84$ m/d (Average of 1.0 m to 3.0 m layer)

$r_0 = 0.1$ m

$d = 1.2$ m

$$E = \frac{2}{\pi} \ln\left(2 \operatorname{Cosh} \frac{\pi r_o}{d} - 2\right) = -1.70 \tag{4.22}$$

The equation can be written as

$$H = \frac{qL}{k}\left(\frac{1}{4}\left(\frac{L}{2d} - E\right)\right) = \frac{qL_2}{8kd} - \frac{qEL}{4k} \tag{4.23}$$

On putting the values of the parameters in the above equation, we get the quadratic equation as

$$0.0002152\, L^2 + 0.000878\, L - 0.75 = 0 \tag{4.24}$$

Solving for L and rejecting –ve value, we get

$$L = \frac{-b \pm (b^2 - 4ac)^{1/2}}{2a} \tag{4.25}$$

$L = \{-0.000878 \pm [(0.000878)^2 + 4\,(0.0002152*0.75)]^{1/2}\}/\ 2*0.0002152$

$L = \{-0.000878 \pm 0.02541\}/\ 0.0004304$

$L = 57$ m

Ernst Equation

The Ernst equation also based on steady state principals is applicable to any kind of two-layered soil profile, being especially useful when the top layer has considerably low hydraulic conductivity than the bottom layer. Moreover, this equation can also be used even if the drains are located either above or below the interface of the two soil layers. Rao *et al.* (1986) used this equation to design subsurface drainage system at Sampla (Haryana), India.

To obtain a generally applicable solution for soil profiles consisting of layers with different hydraulic conductivities, Ernst (1956, 1962) divided the flow into drains in three components namely the vertical, the horizontal and the radial. Consequently, the total available head (h) is divided into a head loss caused by the vertical flow (h_v), the horizontal flow (h_h) and the radial flow (h_r):

$$h = h_v + h_h + h_r \quad (4.26)$$

Vertical flow is assumed to take place in the layers between the water table and the drain level and can be obtained by applying Darcy's law:

$$q = K_v \frac{h_v}{D_v} \text{ or } h_v = q \frac{D_v}{K_v} \quad (4.27)$$

Where D_v is the thickness of the layer in which vertical flow is considered (m), and K_v is the vertical hydraulic conductivity (m/d). In view of the difficulty in measuring K_v, it is often replaced with K_h ignoring the resulting error.

The horizontal flow is assumed to take place below the drain level. Quite analogous to Hooghoudt solution (eq. 4.9), we get:

$$h_h = q\, L^2 / 8\, \Sigma\, (Kd)_h \quad (4.28)$$

Here $\Sigma(Kd)_h$ is the transmissivity of the soil layers through which the water flows horizontally (m^2/d).

Radial flow is also assumed to take place below drain level. The head loss caused by radial flow can be expressed as:

$$h_r = q \frac{L}{\pi K_r} \text{Ln} \frac{a\, D_r}{u} \quad (4.29)$$

Where K_r is the radial hydraulic conductivity (m/d); a is geometry factor of the radial resistance; D_r is the thickness of the layer in which the radial flow is considered (m) and u is the wetted perimeter of the drain (m). The values of u need to be decided carefully as discussed in Fig. 4.11.

The equation is applicable to situations where $D_r < L/4$. The geometry factor, a, depends on soil profile and position of the drain. In a homogenous soil profile a=1. In a layered soil, it depends on whether the drains are in the top or bottom soil layer. If the drains are in the bottom layer, the radial flow is assumed to be restricted to this layer, and again a=1. If the drains are in the top layer, the value of 'a' depends on the ratio of the hydraulic conductivity of the bottom (K_b) and top layer (K_t).

Using the relaxation method, Ernst (1962) distinguished the following situations:

$K_b/K_t < 0.1$, the bottom layer can be considered homogenous and the case is reduced to a homogenous soil profile such that a=1.

$0.1 < K_b/K_t < 50$, a depends on the ratios K_b/K_t and D_b/D_t (Table 4.9)

$K_b/K_t > 50$, a = 4.

Here D_t is the thickness of top layer below the drain, D_b is the thickness of the layer between top layer and the impervious layer.

The expressions for vertical flow (Eq. 4.23), the horizontal flow (Eq. 4.24) and radial flow (Eq. 4.25) can now be substituted in the Eq. (4.21) such that:

$$h = \frac{qD_v}{K_v} + \frac{qL^2}{8\Sigma\,(Kd)_h} + \frac{qL}{\pi K_r}\ln\frac{a\,D_r}{u} \tag{4.30}$$

Table 4.10 Ernst method for steady state conditions in layered soils– the geometry factor (a) obtained by relaxation method

K_b/K_t	D_b/D_t					
	1	**2**	**4**	**8**	**16**	**32**
1	2.0	3.0	5.0	9.0	15	30
2	2.4	3.2	4.6	6.2	8.0	10
3	2.6	3.3	4.5	5.5	6.8	8.0
5	2.8	3.5	4.4	4.8	5.6	6.2
10	3.2	3.6	4.2	4.5	4.8	5.0
20	3.6	3.7	4.0	4.2	4.4	4.6
50	3.8	4.0	4.0	4.0	4.2	4.6

or

$$h = q\left\{\frac{D_v}{K_v} + \frac{L^2}{8\Sigma\,(Kd)_h} + \frac{L}{\pi K_r}\ln\frac{aD_r}{u}\right\} \tag{4.31}$$

If the design drainage rate (q) and the available total head (h) are known, the quadratic equation for spacing (L) can be directly solved to get the value of L.

Let us now consider a situation of drain placement in a two layered profile with $K_t < K_b$. If the drain is placed in the bottom layer as shown in Fig. 4.12 (Top), then

- Vertical resistance in the bottom layer is neglected as the hydraulic conductivity of the bottom layer is more than the top layer.
- Transmissivity of the top layer can also be neglected since the hydraulic conductivity as well as the depth of the layer is less.
- Radial flow is restricted to the layer below the drain level and a =1.

Therefore, Eq. (4.31) reduces to:

$$h = q\left\{\frac{D_v}{K_t} + \frac{L^2}{8\,K_b D_b} + \frac{L}{\pi K_b}\ln\frac{D_r}{u}\right\} \tag{4.32}$$

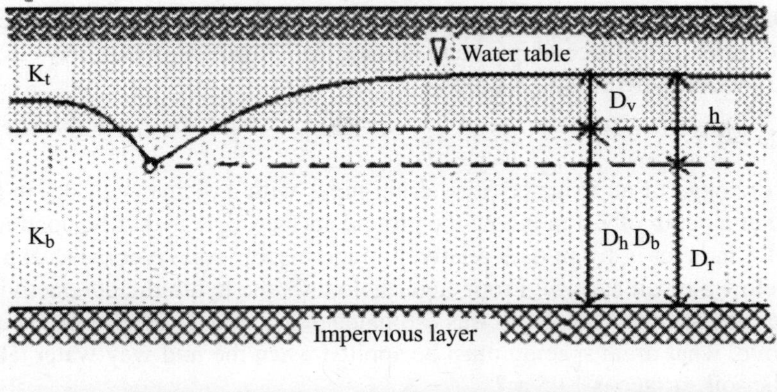

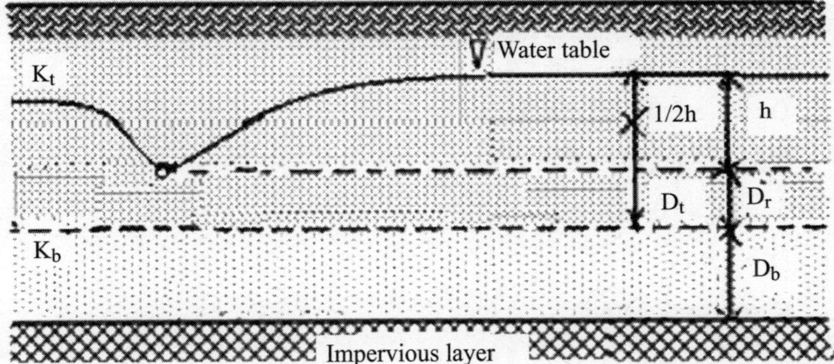

Fig. 4.12 Geometry of the Ernst equation for $K_t < K_b$

If the drains are placed in the top layer of a two-layered profile with $K_t < K_b$ as shown in Fig. 4.12 (bottom), then

- There is no vertical flow in the bottom layer such that $D_v = h$.
- Radial flow is restricted to the region in the top soil layer below the drain level.
- Transmissivity $\Sigma\,(Kd)_h = K_b D_b + K_t\, D_t$ in which $D_t = D_r + 1/2\ h$.
- The value of a is calculated using Table 4.10.

For such a case, Eq. (4.31) reduces to:

$$h = q\left\{\frac{D_v}{K_t} + \frac{L^2}{8\,(K_b D_b + K_t D_t} + \frac{L}{\pi K_t}\ln\frac{aD_r}{u}\right\} \tag{4.33}$$

Pandey & Gupta (1983) compared the equations of the Hooghoudt, the Ernst, the Dagan and the van Beers for similar design problems. The comparison revealed that any of the four equations can be used to design lateral drain spacing. Equation of van Beers showed the maximum deviations but in the practical range even the errors due to this equation were less than 10%. The other three equations gave equivalent results within practical limits of accuracy and within the most practical range of parameters. One of the limitations of the Hooghoudt equation can be overcome by using Dagan equation or Ernst equation as these equations eliminate the necessity of trial and error. The second problem arising in situations where $K_b < K_t$ can be resolved by using the Ernst's equation.

Example 4.4

For the drainage of an irrigated area, drain pipes with a diameter of 0.1 m are to be installed at a depth of 2.0 m below the ground surface. The hydraulic conductivity of the top 1.8 m layer is 1.0 m/d. A relatively impermeable soil layer exists at a depth of 3.0 m below the soil surface. From auger hole tests the hydraulic conductivity above this layer but below 1.8 m depth was estimated at 7.5 m/d. If the drainage coefficient is 10 mm/d, what drain spacing must be applied when the mid-way water table depth below the soil surface is 1.05 m?

The drains are installed in the second layer below the top layer. We can use the Ernst equation. The parameters are:

$D_v = h = 2.0\text{-}1.05 = 0.95$ m

$D_h = 1.2 - 0.2 = 1.0$ m

$u = \pi r_o = 0.157$

$D_r = 1.0$ m

$K_v = K_t = 1.0$ m/d

$K_h = 7.5$ m/d

$K_r = 7.5$ m

a=1.0

Equation (4.31) with a = 1.0 is used to calculate the drain spacing as:

$$h = q\left\{\frac{D_v}{K_t} + \frac{L^2}{8\,K_b d_b} + \frac{L}{\pi K_b}\,\text{Ln}\,\frac{D_r}{u}\right\}$$

On putting the values of parameters in the above equation, we get the quadratic equation as

$$\frac{0.95}{0.01} = \frac{0.95}{1.0} + \frac{L^2}{8*1*7.5} + \frac{L}{3.142*7.5}\,\text{Ln}\,\frac{1.0}{0.157} \tag{4.34}$$

$$0.016666\,L^2 + 0.07857\,L - 94.05 = 0$$

Solving as in the Dagan equation and neglecting the -ve value, we get the value of L as

L = 73.2 m

SELECTION OF STEADY-STATE EQUATION

Two important factors in selecting the most appropriate steady-state equation are the soil profile and the relative position of the drains in the profile. Table 4.11

summarizes some of the more common field situations and appropriate equation for each of them.

Table 4.11 Summary of the steady-state equations

Serial No.	Soil profile	Location of drain	Equation
1	Homogeneous	On the top of the impervious layer	Hooghoudt or Donnan equation: $q = \frac{4K(H^2 - D^2)}{L^2}$
2	Homogeneous	Above the impervious layer	Hooghoudt equation with equivalent depth: $q = \frac{8Kdh + 4Kh^2}{L^2}$
3	Two layers ($K_t > K_b$)	Drains at interface of the two soil layers	Hooghoudt equation with equivalent depth: $q = \frac{8K_b dh + 4K_t h^2}{L^2}$
4	Two layers ($K_t < K_b$)	In the bottom layer	Ernst equation: $h = q\left(\frac{D_v}{K_t} + \frac{L^2}{8K_b D_b} + \frac{L}{\pi K_b} \ln \frac{aD_r}{u}\right)$
5	Two layers ($K_t < K_b$)	In the top layer	Ernst equation: $h = q\left(\frac{D_v}{K_t} + \frac{L^2}{8(K_b D_b + K_t D_t)} + \frac{L}{\pi K_t} \ln \frac{aD_r}{u}\right)$ where, $D_t = D_r + ½\, h$, and $D_v = h$.

UNSTEADY STATE EQUATIONS

In areas with periodic irrigation or high intensity rainfall, the assumption of steady state is not justified because the water is stored during irrigation or rainfall and released from the soil on cessation of irrigation/rainfall. Since such unsteady state situations are more commonly encountered, unsteady state theories are used to derive drain spacing equations. The main difficulty in the use of unsteady state equations has been the lack of information on drainable pore space essentially required in the use of these equations. In spite of all the difficulties and complexities, these equations simulate a more realistic situation of the drainage problem as it allows the evaluation of the drainage system based on either water table height or drain outflow rate. The unsteady state drainage theory is described under the two heads namely:

- The falling water table
- The fluctuating water table

The unsteady state approach, alike the steady state, is based on the Dupuit-Forchheimer (D-F) assumptions. Based on the D-F assumptions, Boussinesq (1903) conceived a continuity equation describing unsteady state water table fluctuation above a horizontal impervious layer in an unconfined aquifer. The Boussinesq´s equation is used to study the rate of fall of water table from some initial conditions after the cessation of irrigation or to study the water table fluctuation as recharge input varies with time. The Boussinesq´s equation is non-linear and difficult to solve. Various linearization attempts have been made to solve this equation such as Glover (Dumm, 1954); Werner, 1957; Kraijenhoff, van de Leur, 1958; Massland, 1959; Tezidis,1968. Glover solution also known as USBR solution or Glover-Dumm solution is widely accepted and is included in this section.

The differential equation for unsteady state flow based on Dupuit's Forchherimer assumptions, can be written as:

$$KD \frac{\partial^2 h}{\partial x^2} = \mu \frac{\partial h}{\partial t} - R \tag{4.35}$$

or when recharge rate R equals zero

$$KD \frac{\partial^2 h}{\partial x^2} = \mu \frac{\partial h}{\partial t} \tag{4.36}$$

Here KD is the transmissivity of the aquifer (m^2/d), R is the recharge rate per unit surface area (m/d), h is the hydraulic head as a function of x and t (m), t is time (days) and μ is the drainable pore space, dimensionless (m/m).

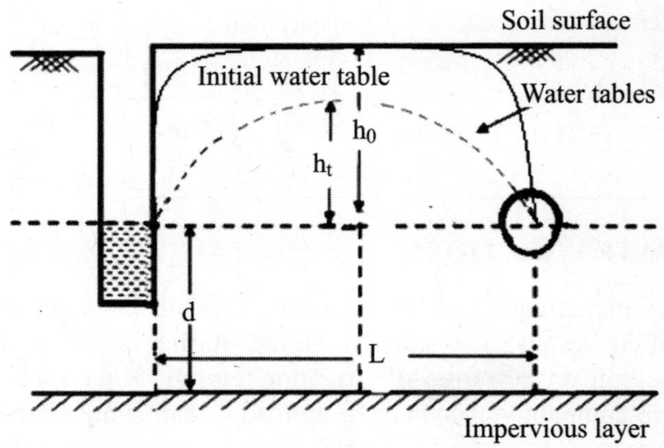

Fig. 4.13 Geometry of a transient state drainage system

Glover-Dumm Equation

Glover (reported by Dumm, 1954), solved the Boussinesq Eq. (4.36) for irrigated areas. It is assumed that the excess water to be disposed of is the difference between net input of water and available moisture storage in that portion of the profile. For example, excess water between irrigations must be removed to maintain the water table below the root zone. The geometry of the transient state water table shown in Fig. 4.13 describes a falling water table after a sudden rise due to instantaneous recharge. Here the initial water table is nearly horizontal to the soil surface. This situation is encountered in irrigated areas, where the water table rises sharply during the application of irrigation water and then recedes. For a homogeneous profile, the change in storage in soil profile with no recharge to groundwater is given by (Fig. 4.14) and is written as:

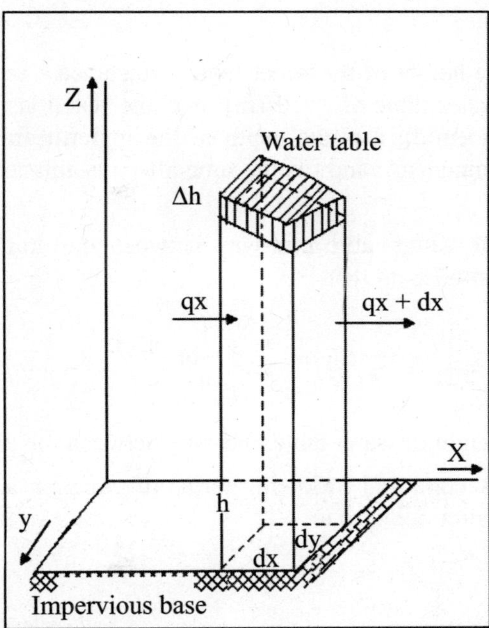

Fig. 4.14 Change in storage under a falling water table

$$\Delta W = \mu \, \Delta h \; dx.dy \tag{4.37}$$

Where ΔW is the change in water storage per unit surface area over the time considered (m), Δh is the change in the level of the water table over the time considered (m).

The change in storage considered over an infinitely small period of time, dt, is

$$dw = \mu \, \frac{\partial h}{\partial t} dx \, . dy \tag{4.38}$$

The continuity principle now requires that the total difference of outgoing minus incoming flow in x and y directions equal the change in storage. Hence, the continuity equation is written as

$$-\left[\frac{\partial}{\partial x}\left(h\frac{\partial h}{\partial x}\right)+\frac{\partial}{\partial y}\left(h\frac{\partial h}{\partial y}\right)\right] dx \, . dy = \mu \frac{\partial h}{\partial t} dx \, . dy \tag{4.39}$$

Considering flow in one direction and h =d (h >>Δh), the differential equation is used for unsteady state condition (Dumm, 1954). This solution is based on Glover formula describing the lowering of initially horizontal water table as a function of time, place, drain spacing and soil properties.

$$h\ (x,t) = \frac{4\,h_0}{\pi} + \sum_{n=1,3,5}^{\infty} \frac{1}{n}\, e^{-n2\alpha t}\, \sin \frac{n\pi x}{L} \tag{4.40}$$

$$\alpha = \frac{\pi^2\, Kd}{\mu L^2} \tag{4.41}$$

Here, h(x,t) is the height of the water table at distance x at time t (m); h_0 is the initial height of the water table at t = 0 (m); α is the reaction factor (d^{-1}); K is the hydraulic conductivity (m/d); d is the depth of the impermeable layer below drain (m), L is the drain spacing (m); and t is the time after instantaneous rise of the water table (d).

The height of the water table mid-way between the drains can be found by substituting x =1/2 L into Eq. (4.40)

$$h_t = \ h\left(x\ =\ \frac{1}{2L}\right) = \frac{4}{\pi} h_0 + \sum_{n=1,3,5}^{\infty} \frac{1}{n} e^{-n2\alpha t} \tag{4.42}$$

where h_1 is the height of water table mid-way between the drain at t > 0 (m)

If αt > 0.2, the second and following terms of equation are small and can be neglected, so this Equation reduces to

$$h_t = 4/\pi\ h_0\ e^{-\alpha t} = 1.27\ h_0\ e^{-\alpha t} \tag{4.43}$$

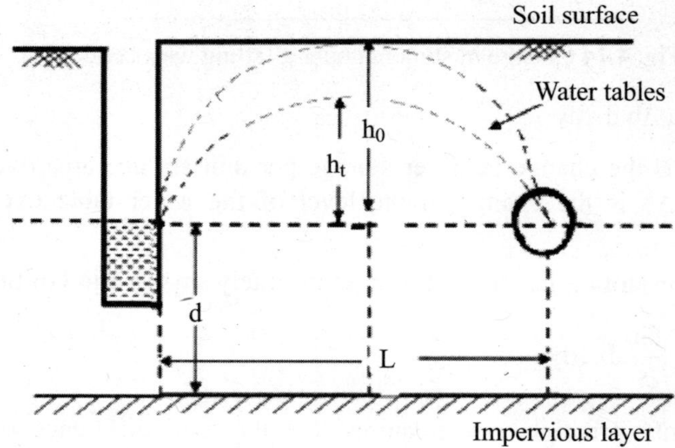

Fig. 4.15 Geometry of a transient state drainage system

If instead of horizontal, the initial water table has the shape of a fourth degree parabola as shown in Fig. 4.15, Eq. (4.43) becomes (Dumm, 1964).

$$h_t = 1.16\ h_0\, e^{-\alpha t} \tag{4.44}$$

By substituting Eq. (4.41) with Eq. (4.44), the expression for the drain spacing is given as.

$$L^2 = \left[\frac{\pi^2 Kdt}{f \ln\left(\frac{1.16h_0}{ht}\right)} \right] \tag{4.45}$$

or

$$L = \pi \left[\frac{Kdt}{f \ln\left(\frac{1.16h_0}{ht}\right)} \right]^{1/2} \tag{4.46}$$

Here t is the time period during which the water table is lowered from h_0 to h_t (days).

The drain discharge at any time ′t′ expressed per unit surface area is given as (Using Darcy′s law)

$$q_t = - 2Kd/L\ [dh_t / dx]\ x=0 \tag{4.47}$$

Differentiating Eq. (4.40) with respect to x, neglecting all terms where n>1 and after substituting x=0, Eq. (4.47) is written as follows:

$$q_t = 8Kd/L^2\ [ho\ e^{-\alpha t}] \tag{4.48}$$

Substituting eq. (4.43) in Eq. (4.48), we get

$$q_t = 2\pi Kd\ h_t / L^2 \tag{4.49}$$

Similarly the relationship of drain discharge using Eq. (4.44) is given as:

$$q_t = 2.2\pi Kd\ h_t / L^2 \tag{4.50}$$

It may be noted that it is similar to part of the Hooghoudt equation that describes the flow from below the drain level except the factor 2π or 2.2π instead of 8 in the Hooghoudt equation.

The head and discharge equations can also be put in a different form. For example Eq. (4.44) can be written as:

$$h_{t+1} = 1.16\ h_t\, e^{-\alpha\Delta t} \tag{4.51}$$

It states that head at any time t+1 is related to head at any time t as per Eq. (4.51) such that $\Delta t = (t+1) - t$.

Since q and h are linearly related, the statement is also true for discharge such that

$$q_{t+1} = 1.16\ q_t\, e^{-\alpha\Delta t} \tag{4.52}$$

Comparing Eq. (4.51) and Eq. (4.52), it can be shown that

$$q_{t+1}/q_t = h_{t+1}/h_t = e^{-\alpha\Delta t} \tag{4.53}$$

Hooghoudt′s Equivalent Depth

In using Eq. (4.45 or 4.46), computation of Hooghoudt's equivalent depth corresponding to the actual depth of impermeable layer below the drains should be made as described in the steady state Hooghoudt equation. If d >>h_0, then d_e so calculated may be used in Eq. (4.46). However, if d is small in comparison to h_o or contribution to drain flow from above the drains is anticipated, a correction to the d_e value needs to be made. USBR suggested that to account for flow above the drains, a value equal to $h_o/2$ should be added to the calculated value of d_e. Later on it emerged that the following equation is a better proposition to assess the d_{ec} to be used in Eq. (4.45 or 4.26):

$$d_{ec} = d_e + \frac{(h_o + h_t)}{4} \tag{4.54}$$

Example 4.5

For the drainage of an irrigated area, drain pipes with a radius of 0.1 m are to be installed at a depth of 1.8 m below the ground surface. The hydraulic conductivity of soil profile up to a depth of 3.0 m is 3.62 m/d. A relatively impermeable soil layer exists at a depth of 3.0 m below the soil surface. If the drainage pore space is 0.14, what drain spacing must be applied if the water table is to be lowered from the ground surface to 1.5 m above the drains in 2 days?

K = 3.62 m/d

t = 2 days

f = 0.14

h_0 = 1.8 m

h_t = 1.5 m

$$L = \pi \left[\frac{K\, d_e t}{\mu}\right]^{1/2} \left[\text{Ln}\, 1.16 \frac{h_0}{h_t}\right]^{-1/2} \tag{4.55}$$

On putting the values of parameters in the above equation (taking square of both sides), we get the equation as

$$L^2 = \pi^2\, (3.62 \text{ x } d_e \text{ x } 2 / 0.14)\ (\text{Ln } 1.16 \text{ x } 1.80/ 1.50)^{-1}$$

$$= 510.4\, d_e \text{ x } 3.0235$$

$$= 1543.2\, d_e$$

Now by hit and trial method, we get the value of L as below.

Without correction

For L = 40, for D=1.2 d_e = 1.066 (From Table 4.9)

$L = (1543.2 \text{ x } 1.1)^{-0.5} = 41.2$ m

For L = 42, for D=1.2 d_e = 1.1

L = 42

The spacing assumed is OK.

With correction

For $L = 40, D=1.2\ d_e = 1.1$

$d_e = d_e + (h_0 + h_t)/4$

$= 1.925$

$\therefore\ L = 54.5$

For $L = 55, D=1.2\ d_e = 1.12$

$d_e = d_e + (h_0 + h_t)/4$

$= 1.945$

$L = 54.8$

or $L = 55$ m

Apparently, in some cases correction applied can make significant change in the calculated drain spacing.

Example 4.6

As per agricultural drainage criteria the maximum permissible height of the WT is 1 m below the soil surface. Irrigation water is applied every 10 days, and field application losses-percolating to water table are 25 mm per irrigation. What would be the drain spacing, if drains of a radius of 0.10 m are installed at a depth of 1.8 m and the depth of impermeable layer is 9.5 m below the soil surface? Take the average K as 1.0 m/d and the drainable porosity as 0.05.

For examples of this nature also, Glover-Dumm equation is applied. Field application losses in this case can be regarded as an instantaneous recharge i.e. R =0.025 m, the rise of the WT is

$R/f = 0.5$ m

Since the WT must not rise above the desired depth i.e. 0.8 m above the drains, the initial water table before the irrigation begins should be at 0.3 m above the drains. Thus we have,

$K=1.0$ m/d

$f= 0.05$

$D=9.5-1.8=7.7$ m

$r =0.10$ m

$h_{10} =0.3$ m

$h_0 =0.8$ m

$T=10$ d

Substituting the above data in Eq. (4.46) and simplifying, we get:

$$L^2 = \pi^2 (1.0 \text{ x } d_e \text{ x } 10 / 0.05) \ (\text{Ln } 1.16 \text{ x } 0.80/ 0.30)^{-1}$$

$$L = 41.8 \ d^{1/2} \ \ \text{m}$$

First estimate; assume L=80 m, so for D=7.7 m and r = 0.10 m, d_e =4.41 m (Table 4.9) and L = 88 m

This is more than the assumed spacing.

Second trial: Assume L= 90 m, corresponding to D = 7.7 m and r= 0.10 m, d_e = 4.63 m and we get

L= 90 m

Therefore, the designed spacing is 90 m.

Equivalence between Steady and Unsteady State

The water table behavior in an irrigated land under steady and non-steady state condition is described in Fig. 4.16. Usually the peak water table is higher in non-steady than under steady state conditions. If an estimate of q is made independently, the drain spacing assessed from steady and unsteady state equations will differ. An equivalence of results is expected, if q_s, i.e. q for steady state condition is derived using q_{ns}, q for the unsteady state. For example, the total flow under unsteady state case can be calculated from the following equation:

$$q_{ns} = \frac{(h_o - h_t) \text{ x f}}{t} \tag{4.56}$$

It should be then adjusted for the heads under which this flow is received under the two conditions. However, this approach of equivalence is rarely followed in drainage designs. In most cases, a more appropriate approach is to assess q_s through field water balance. The lateral drain spacing is then calculated independently using steady and non-steady approaches. A decision on drain spacing is then taken besides costs, on the surface drainage conditions, crops to be grown and the subsurface drainage conditions.

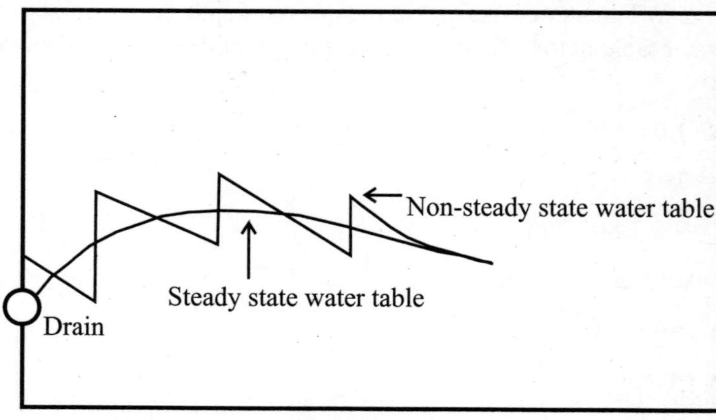

Fig. 4.16 Variations in water table depth in steady and non-steady state drainage

To show the equivalence between steady and non-steady state conditions, few attempts have been made to derive non-steady state equations using steady state equation. Most commonly used is the integrated Hooghoudt equation.

Integrated Hooghoudt Equation

Bouwer & van Schilfgaarde (1963) adopted a steady state equation to predict the rate of fall of groundwater in drained lands. They assumed that the instantaneous drainage rate can be taken equal to the steady state drainage rate corresponding to the water table height midway between the drain. A steady state relationship is used to describe the rate of fall of the water table midway between the drains by introducing a correction factor C in the equation as below.

$$q = f\,C\,(dh/dt) \tag{4.57}$$

Here C is a correction factor to take care of the steady state assumption. For $0.02 < h_0/L < 0.05$, $C = 0.80$ and for $h_0/L > 0.15$, $C = 1.0$. By substituting for q of Hooghoudt's equation in Eq. (4.59) and integrating from t = 0 to t, they obtained the following equation for the transient conditions.

$$L^2 = \frac{8\,K\,d_e\,t}{C\,f\,Ln\left[\dfrac{h_t\,(2d_e + h_o)}{h_0\,(2d_e + h_t)}\right]} \tag{4.58}$$

All the terms have already been defined. The drain spacing can be obtained by hit and trial method.

Example 3.7

For the drainage of an irrigated area, drain pipes with a radius of 0.1 m are to be installed at a depth of 1.8 m below the ground surface. The hydraulic conductivity of soil profile up to a depth of 3.0 m is 3.62 m/d. A relatively impermeable soil layer exists at a depth of 3.0 m below the soil surface. If the drainage pore space is 0.14, what drain spacing must be applied if the water table is to be lowered from 1.8 m above the drains to 1.5 m above the drains in 2 days? Take C as 0.8.

$K = 3.62$ m/d

$D = 1.2$ m

$t = 2$ days

$f = 0.14$

$H_0 = 1.8$ m

$H_t = 1.5$ m

$C = 0.8$

On putting the values of different parameters in Eq. (4.58), we get

$$L^2 = \frac{57.92\,d_e}{0.112\,Ln[1.8\,(2\,d_e + 1.5)\,/\,1.5\,(2\,d_e + 1.8)]}$$

For $L = 60, D = 1.2 \quad d_e = 1.11$ (Table 4.9)

$L = 74.0$ m

For $L = 74, D = 1.2 \quad d_e = 1.12$

$L = 74$

So $L = 74$ m

FLUCTUATING WATER TABLE

De Zeeuw-Hellinga Equation

To determine the hydraulic heads and the drain discharge over a period of time intervals of equal length and constant recharge during each period of time, De Zeeuw-Hellinga (1958) proposed two equations given as under.

$$q_t = q_{t-1}\, e^{-\alpha \Delta t} + R\,(1 - e^{-\alpha \Delta t}) \tag{4.59}$$

and

$$h_t = h_{t-1}\, e^{-\alpha \Delta t} + \frac{R\,\Delta t\,(1 - e^{-\alpha \Delta D})}{0.8\, f\, \alpha} \tag{4.60}$$

From these two equations, the discharge and hydraulic heads are calculated for use in the fluctuating water table theory for drainage design. Once the hydraulic heads are calculated in this manner, a plot of the depth to water table (Depth of drains – hydraulic heads) is prepared or directly, hydraulic heads are plotted, as shown in Example 4.8. If there are periods during which water table is within the root zone and may prove to be detrimental to the crop, a new design of the drainage system is used to calculate the hydraulic heads. The exercise is repeated until a design is found which does not result in shallow water table at critical periods.

Example 4.8

Calculate the daily water table fluctuations and corresponding discharge from a given drainage system whose characteristics are: lateral drain spacing = 40 m, drain depth = 1.2 m, the hydraulic conductivity times the equivalent depth to impermeable layer = 2.5 m^2/d, drainable pore space = 0.05. If the overlying profile above the drains is 20 mm below field capacity and the initial hydraulic head is 0.1 m, calculate future hydraulic heads and drain discharges for consecutive rainfall of 0.026, 0.021, 0.010, 0.025 and 0.005 m/d. Assume ET as 0.002 m/d.

$Kd = 2.5\ m^2/d$; $L = 40$ m; $H_0 = 0.1$; Storage = 20/1000 = 0.020 m;

ET = 0.002 and f = 0.05

Using the first term of the Hooghoudt Eq. (4.7), a first guess of drain discharge is made as follows.

$q_0 = 8Kdh_0/L^2 = 8*2.5*0.1/40*40 = 0.001$ m/d

Based on the data given, a table is constructed to assess the daily recharge given as

$$\text{Recharge} = \text{Rainfall} - \text{ET} - \text{Storage} \quad (4.61)$$

For t =1, Recharge = 0.026 – 0.002 – 0.020 =0.004

$$\alpha = \pi^2 \, Kd/f\, L^2 = 0.31$$

Then h_1 from Eq. (4.60)

$$h_1 = (0.1*0.73) + (0.004*.27)/0.012 = 0.16$$

Similarly, knowing q_{t-1}, new q is also calculated using Eq. (4.59). All the calculations except q values are reported in tabular form below.

Day	**Rainfall (m)**	**ET (m)**	**Storage (m)**	**Recharge (m)**	**h (m)**	**q (m)**
0	-	-	-	-	0.10	0.001
1	0.026	0.002	0.020	0.004	0.16	0.002
2	0.021	0.002	0.0	0.019	0.54	
3	0.010	0.002	0.0	0.008	0.57	
4	0.025	0.002	0.0	0.023	0.96	
5	0.005	0.002	0.0	0.003	0.77	
6	0.0	0.002	0.0	0.0*	0.56	
7	0.0	0.002	0.0	0.0	0.41	
8	0.0	0.002	0.0	0.0	0.30	
9	0.0	0.002	0.0	0.0	0.22	

*The negative values at and below this value are treated as zero

The hydraulic heads, h are plotted against time (t =0 is plotted as 1) as shown in Fig. 4.17. The curve shows that the water table with the designed drainage system would come within 30 cm for a day. Similar analysis for other storms in the season should reveal the total number of days for which such situations will occur. Simulated values for two consecutive years for the Sampla site in Haryana are shown in Fig. 4.18 (Hirekhan *et al.*, 2007). If the water table remains within 30 cm or the known detrimental depth below the soil surface, then design of the system should be changed and all calculations should be repeated until a perfect design is achieved.

This concept has been taken forward wherein values quantifying the excessive soil water conditions (SEW_{30}) are calculated using the following equation (Sieben, 1964).

$$SEW_{30} = \sum_{i=1}^{n} (30 - X_i) \quad (4.62)$$

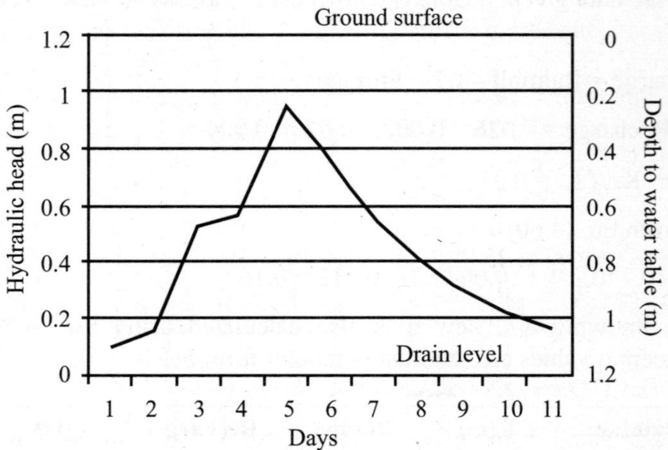

Fig. 4.17 Variation of hydraulic heads or depth to water table below the ground surface

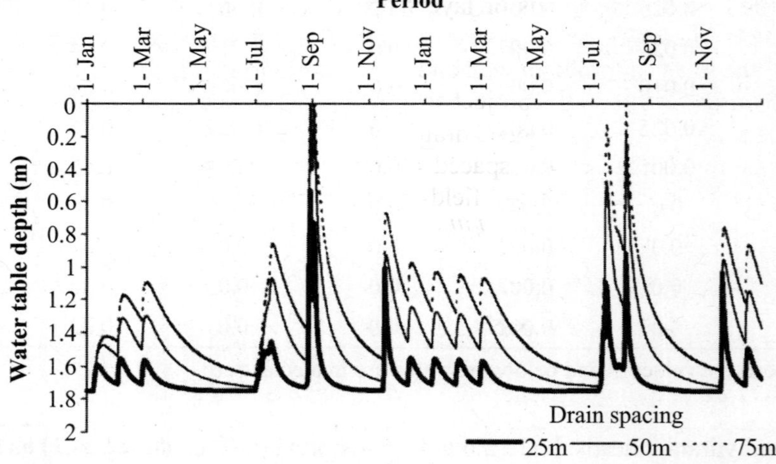

Fig. 4.18 Simulated water table levels in subsurface drained land with continuous normal rainfall for two consecutive years

Where X_i is the water table depth below the ground surface in cm on i^{th} day and n is the number of days in the growing season. The SEW_{30} units are depth day and therefore, sometimes it is also called depth day index. All values less than zero (negative) are neglected. The sum is only for positive values smaller than 30 so that the sum is a measure of depth above the 30 cm level (Wesseling, 1974). Larger the SEW_{30} values, poorer the drainage conditions, indicating need for improvements. The impact on crop yield can also be determined with the help of the following equation. The magnitude of stress day index, SDI helps to determine the anticipated crop yield losses (Ravelo *et al.*, 1982).

$$SDI = \sum_{j=1}^{M} (CSj * SDj) \tag{4.63}$$

Here M is the number of growth stages, CSj is the crop susceptibility factor for stage j and SDj is a stress day factor for stage j. The SEW_{30} can be used as the SD factor (Hardjoamidjojo *et al.*, 1982). It may be noted that only few studies have been made so far on this issue (Kanwar *et al.,* 1988). The application of SWAP to predict excess water stress utilizing SEW_{30} index for different spacing under three rainfall situations revealed that SEW_{30} exceeding 100 cm/day was encountered with drains spaced at 75 m, while it remained within 50 cm/day with drains spaced at 50 m during normal and/or above normal rainfall years. No excess water stress problems occurred in drains spaced at 25 m or in years with below normal rainfall in any of the test spacing (Verma *et al.*, 2010). Fortunately, this concept is now a part of the most popular model DRAINMOD to calculate yield penalty on crops.

RECOMMENDATIONS ON DRAIN SPACING

Drain spacing calculated from various equations for a similar physical problem may vary unless meticulous calculations are made to simulate the physical problem. It is attributed to the assumptions made in the derivation of the equations, from treating the profile as a homogeneous or layered profile and inherent errors in investigated parameters. Therefore, it is always better to select an appropriate equation that simulate the physical problem well and chose appropriate parameters. In some cases, physical infrastructure in the project area may govern the drain spacing. For example, in Haryana, although the designed drain spacing using Ernst equation is around 50 m, field test proved that drains spaced 75 m apart are equally effective. After the consolidation of land holding, fields in Haryana are 60*67. Since farmers are interested to put the drains on *killa* lines, the drain spacing for large projects in Haryana is either 67 m or 60 m, depending upon the orientation of the system (Mohan *et al.*, 2012). Based upon experiences at different locations in the country, general guidelines for drain spacing are given in Table 4.12.

Table 4.12 Lateral drain spacing for different textured soils

Soil texture	Drain spacing (m)
Light textured	100 – 150
Medium textured	50 – 100
Heavy textured including Vertisols	30 –50

INTERCEPTOR DRAINAGE FOR SLOPING LANDS

Most drainage installations constructed on sloping lands require drains perpendicular to the water table gradient. A major question remains to design an efficient system that is capable to intercept the maximum flow. Donnan (1959) gave a solution to design an interceptor drain resting on a sloping impervious layer with a slope of tan α. Since the slope of the water table is parallel to the slope of the impervious layer, the amount of seepage water flowing downhill is given as (Fig. 4.19):

$$Q_1 = K\, H_1 \tan \alpha \tag{4.64}$$

Where Q_1 is the flow rate per unit width (m^2/d); K is the hydraulic conductivity of top layer (m/d); H_1 is the height of the water table above the impervious layer before installation of the drain and tan α is the slope of the impervious layer. If the water table below the drain following construction of drain is H_2, then the amount of water going beyond the downhill is given as:

$$Q_2 = K\, H_2 \tan \alpha \tag{4.65}$$

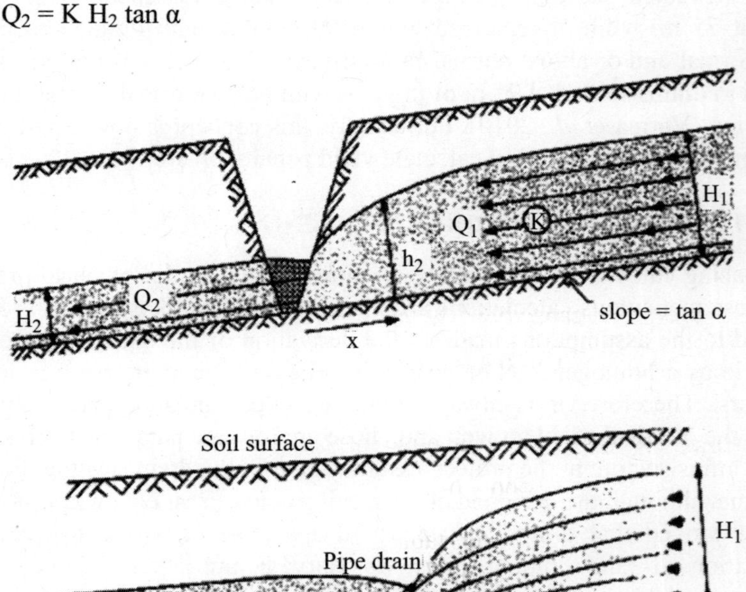

Fig. 4.19 Drainage of sloping lands (Top) an open interceptor and (Bottom) pipe interceptor

The intercepted flow $(Q_1\text{-}Q_2)/Q_1$ then is calculated by Eq. (4.64) and Eq. (4.65) as:

$$\text{Intercepted flow} = (H_1 - H_2)/H_1. \tag{4.66}$$

The elevations of the water table hx, at any distance X uphill of the drain are calculated by the equation:

$$X = (1/\tan \alpha)\, H_1 \,\text{Ln}\, [(H_1\text{-}H_2)/(H_1\text{-}h_x)] + (H_2 - h_x) \tag{4.67}$$

Example 4.9

An irrigation scheme (500 x 1000 m) is located in a sloping area. The deep percolation losses are 1 mm/d. The soil consists of a permeable layer, 6 m thick having hydraulic conductivity of 2.5 m/d, on top of an impervious layer with a slope of tan α = 0.04. To control the water table in the area downhill from the irrigated area at a level of 2 m below the soil surface, an interceptor drain is proposed. Calculate the required depth and capacity of the interceptor drain and the uphill elevation of the water table after the construction of the interceptor drain.

To control the water table at 2 m below the soil surface, the height of the interceptor drain above the impervious layer has to be

$H_2 = 6.0 - 2.0 = 4.0$ m

The percolation losses result in a seepage flow per meter width,

$Q_1 = 500 \times 0.001 = 0.5\ m^2/d$

Elevation of the water table above the impervious layer before the construction of the drain using Eq. (4.64) is

$H_1 = 0.5/2.5 * 0.04 = 5.0$ m

After the construction of the interceptor drain, the seepage flow downhill from the drain is given by Eq. (4.65) as

$Q_2 = 2.5 \times 4.0 \times 0.04 = 0.4\ m^2/d$

Thus the required capacity of the interceptor drain per m length is

$q_i = Q_1 - Q_2 = 0.5 - 0.4 = 0.1\ m^2/d$

In other words, 20% of the percolating water will be intercepted. As the scheme is 1000 m long, the discharge at the outlet of the interceptor drain will be

$Q = 1000 \times q_i = 1000 \times 0.1 = 100\ m^3/d$

The elevation of the water table uphill from the interceptor drain can be determined with the help of Eq. (4.67)

$X = (1/\tan\alpha)\ H_1\ \text{Ln}\ [(H_1 - H_2)/(H_1 - h_x)] + (H_2 - h_x)$

$X = 125\ \text{Ln}\ (5 - 4)/(5 - h_x) + 25(4 - h_x)$

Thus:

$h_x = 4.0$ m at $X = 0$ m

$h_x = 4.2$ m at $X = 27.7$ m

$h_x = 4.4$ m at $X = 63.4$ m

$h_x = 4.6$ m at $X = 99$ m

$h_x = 4.8$ m at $X = 181$ m

$h_x = 4.9$ m at $X = 265$ m

CAPACITY/SIZE OF DRAINS

Laterals

The Manning equation is used to determine the diameter of the laterals and collectors. It is also used to determine the length of the laterals and the maximum area to be drained with a given size of the lateral. The Manning equation in its most familiar form is written as:

$$Q = \frac{1}{n} A\ R^{2/3}\ S^{1/2} \qquad (4.68)$$

Such that Q is the flow rate (m^3/s); A is the cross-sectional area (m^2); S is the hydraulic gradient; R is the hydraulic radius, A/P (m); P is the wetted perimeter (m) and n is the Manning's roughness coefficient.

Manning´s roughness coefficients are given in Table 4.13. Besides, values of n for corrugated pipes can also be derived from the following relation:

$$n = 0.015 + 0.02\ d - 0.013\ d^2 \tag{4.69}$$

Here d is the inside diameter of the corrugated pipe (m).

Table 4.13 Values of Manning coefficient 'n'

Pipe and flow characteristics	Minimum	Normal	Maximum
Pipes flowing full or partly full			
Concrete or asbestos-cement pipe	0.011	0.013	0.015
Clay tile	0.011	0.013	0.017
Pipes flowing full			
Corrugated plastic pipes d = 40, 50 mm[1]		0.0140	
Corrugated PE			
d = 50, 75 mm [2]		0.016	

d is inside diameter of the drain pipe (mm); Values from [1] Wesseling & Homma (1967), [2] Broughton *et al.* (1992), Ritzema (1994).

Size

Subsurface drains are designed to carry the required capacity without pressure flow. For a full-flowing circular pipe (hydraulic radius = d/4), Eq. (4.68) is written as

$$Q = \frac{0.312\ d^{2.67}\ S^{1/2}}{n} \tag{4.70}$$

If Q is expressed in m^3/d, then

$$Q = \frac{2.7 \times 10^4\ d^{2.67}\ S^{1/2}}{n} \tag{4.71}$$

Normal practice is to 'over design' the pipe to allow for partial silting and for misalignment during construction. It is possible by increasing the discharge Q by a factor varying from 1.33 for continuous plastic pipes installed in stable soils to 2.0 for the tile drains where siltation is likely to occur. Ritzema (1994), on the other hand mentions that use of Manning equation to design the drainage system has a in-built factor of safety of about 58%. It is because the Manning equation has been derived for uniform flows (The flow remains constant over the entire length of the pipe) whereas in the perforated drain pipes flow is non-uniform (The flow is variable being zero at the far end to highest at the point where it joins the collector). Therefore, Manning equation can be applied without applying any additional factor of safety.

Drain Diameter and Maximum Area Drained

From Eq. (4.70), drain diameter can explicitly be determined.

$$d = 1.548\ (n\ Q)^{0.375}\ S^{-0.188} \tag{4.72}$$

$$\text{Also}\quad Q = q\ A_{max} = q\ L\ B_{max} \tag{4.73}$$

Although Eq. (4.73) is good for all practical purposes, many times a more accurate discharge equation is also used i.e. $Q = q\ L\ (B_{max} + L/2)$. Here, q is the drain discharge per unit area, $m^3/d\text{-}m^2$, A_{max} is the maximum area that can be drained by a lateral of given size, m^2, L is the drain spacing, m and B_{max} is the maximum allowable length of a lateral for a given size of drain pipe, m.

From Eq. (4.70) and Eq. (4.73). we get

$$B_{max} = \frac{0.312\, d^{2.67}\ S^{1/2}}{n\, q\, L} \times 24 \times 3600 \tag{4.74}$$

It may be noted that the maximum length of the lateral increases with internal diameter of the pipe and slope of the lateral. For clay and cement concrete tiles, the minimum diameter prescribed is 10 cm, while for PVC and PE pipes, it is 75 mm. The computed pipe diameters are to be rounded off to the next higher commercially available size. The new selected value of the drain diameter is used in Eqs. (4.72 and 4.74) to calculate the maximum length of the lateral. Length of lateral should be less than the B_{max}, otherwise one need to change the diameter of the drain.

Although the above procedure has worked well, the following equations are also used directly to calculate the pipe diameters or the maximum length of the corrugated laterals for a given spacing without bothering about the factors of safety due to siltation etc. (Ritzema, 1994).

For corrugated PVC pipes

$$Q = q^*A = 2.0 * 10^6 * d^{2.67} * S^{0.5} \qquad \text{if} \quad d \le 100 \text{ mm} \tag{4.75}$$

$$Q = q^*A = 2.5 * 10^6 * d^{2.67} * S^{0.5} \qquad \text{if} \quad d > 100 \text{ mm} \tag{4.76}$$

The corresponding equations for rigid smooth PVC pipes derived on the basis of Darcy-Weisbach equations are:

$$Q = 4.6 * 10^6 .\ d^{2.714}\ S^{0.572} \qquad \text{for} \quad d \le 100 \text{ mm} \tag{4.77}$$

$$Q = 5.79 * 10^6 .\ d^{2.714}\ S^{0.572} \qquad \text{for} \quad d > 100 \text{ mm} \tag{4.78}$$

Here, A is the drainable area (m^2) equal to length of the pipe multiplied by drain spacing and q is the drainage coefficient (m/d).

Collector Size

Flow rate in collectors also increases in the direction of flow but the increase is not gradual. The increase occurs in sudden large increments at places where laterals join the collectors. So the collector sizes for different segments between two adjacent junctions need to be calculated. Therefore, before one is able to design the size of the collectors, a tentative layout plan needs to be developed. The tentative plan for the site for which calculations have been made in this section is shown in Fig. 4.20. Any change in this plan will require recalculations of the collector sizes. On the basis of the sizes obtained, one or two sizes of the collector are selected.

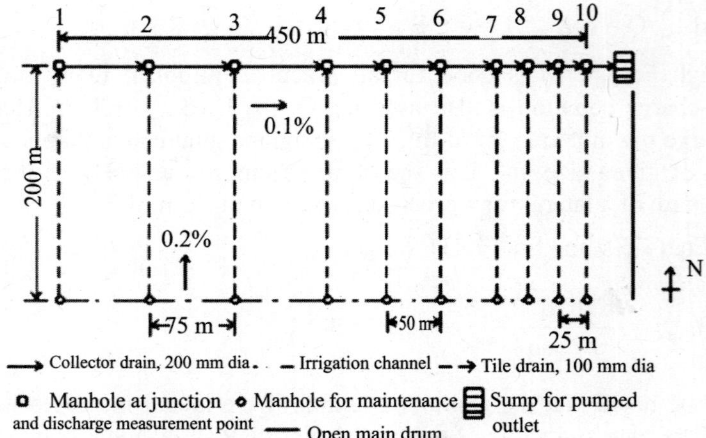

Fig. 4.20 The tentative plan of the system to calculate collector size at numbered sections

In irrigated areas, all the laterals will not contribute to flow at the same time. The collectors receive water at different rates from the laterals on account of variable pattern of irrigation. Although no precise data are available, but for irrigated areas certain correction factors to discharge have been proposed depending upon the

Table 4.14 Correction factors to discharge in collectors for irrigated land

Area drained (ha)	Correction factor to discharge
0-40	1.00
40-80	1.00-0.92
80-120	0.92-0.87
120-160	0.87-0.82
160-200	0.82-0.79
200-240	0.79-0.76
240-280	0.76-0.74
280-320	0.74-0.72
320-480	0.72-0.65
480—640	0.65-0.60
640-800	0.60-0.56
800-960	0.56-0.54
960-1120	0.54-0.52
1120-1280	0.52-0.50
1290-5000	0.50

drainage area from which discharge is being received at a particular junction (Table 4.14). These factors are recommended for use only for irrigated lands. In monsoon climatic conditions, where discharge during the monsoon is expected equally from all the laterals, one may not use these coefficients. The procedure described in Example 4.9 is valid for the monsoon climatic conditions.

Example 4.9

Calculate the lateral and collector diameters to be used in a drainage system laid at a gradient of 0.002 m/m and spaced at 25, 50 and 75 m. If drainage coefficient is 10 mm/d, calculate the maximum length and maximum area drained by a lateral. It is proposed to use cement concrete tiles.

Size of the lateral pipe required to carry the designated flow rate is given by Eq. (4.72) as:

$$d = 1.548\ (n\ Q)^{0.375}\ S^{-0.188} \tag{4.79}$$

$q = 10$ mm/d $= 0.01$ m/d

$S = 0.002$

$n = 0.014$ (assessed from values given in Table 4.14)

$L = 25$ m, 50 m, 75 m and

$Q = q \times L \times B_{max} / (24 \times 3600)$ m^3/s

Ss per layout plan, B_{max} is 200 m

$Q = 0.0005787$, 0.0011574 and 0.0017361 for $L = 25$, 50 and 75 m, respectively

Incorporating these values in Eq. (4.72), we get

For L= 50 m

$$d = 1.548\ (0.014 \times 0.0011574)^{0.375}\ (0.002)^{-0.188}$$

$$= 1.548 \times 0.015981 \times 3.2166$$

$$= 0.08 \text{ m} = 8.0 \text{ cm}$$

Similarly, for L= 25 and 75

$d =$ 6.2 cm and 9.2 cm

Since for clay and cement concrete tiles, the minimum diameter prescribed is 10 cm, the computed diameter being less than 10 cm, it is uniformly taken as 10 cm for all the drain spacing.

Maximum Length of Lateral

Maximum length of lateral that can be employed is given by Eq. (4.74)

$$B_{max} = \frac{0.312\, d^{2.67}\, S^{1/2}}{n\, q\, L} \times 24 \times 3600$$

$q = 10$ mm/d $= 0.01$ m/d

$S = 0.002$

$L = 25$ m, 50 m, 75 m

$n = 0.014$ and

$d = 0.01$ m

For $L = 50$

$$B_{max} = 0.312 \times 0.1^{2.67} \times 0.002^{.5} \times 24 \times 3600/ (0.014 \times 0.01 \times 50)$$

$$= 368 \text{ m}$$

Similarly, for L = 75 m and 25 m, B_{max} = 245 m and 736 m, respectively.

Since the maximum drain length is 200 m, the drain diameter used serves the purpose.

Collector Size

The collector size is given by

$$d = 1.548 (n\, Q_c)^{0.375}\ S^{-0.188}$$

where

$Q = q \times A$

Q_c = Cumulative discharge passing through the pipe at the section under consideration

$q = 0.01$ m/d

$S = 0.001$

For the third row of Table 4.15, the area drained by all the laterals spaced at 75 m is included. So

$$Q_1 = q \times 75 \times 200 / (24 \times 3600)\ \ m^3/s = 0.001736\ m^3/s$$

$$Q_2 = q \times 75 \times 200 / (24 \times 3600)\ \ m^3/s = 0.001736\ m^3/s$$

$$Q_3 = q \times 75 \times 200 / (24 \times 3600)\ \ m^3/s = 0.001736\ m^3/s$$

$$Q_c = 3 \times 0.001736 = 0.005208$$

Putting these values in the above equation, we get

$$d = 1.548 \times (0.014 \times 0.005208)^{0.375}\ 0.001^{-0.188}$$

$$= 0.139 \text{ m} = 13.9 \text{ cm}$$

If the design drain discharge is increased by 1.33, we get

$$d = 0.156 \text{ m} = 15.6 \text{ cm}$$

Computing each step of the Table 4.15, we finally arrive at the last section of the collector for which collector diameter, d = 18.8 cm without correction and 21.0 cm with correction. The total line can now be divided into 2 or maximum 3 segments and accordingly, commercially available pipes can be chosen.

Table 4.15 Calculation of segment wise collector size

Section	Drainage Area (m^2)	Q (m^3/s)	Cumulative Q (m^3/s)	d (cm)	d* (cm)
1	75 x 200	0.001736	0.001736	9.25	10.3
2	75 x 200	0.001736	0.003472	12.0	13.4
3	75 x 200	0.001736	0.005208	13.9	15.6
4	62.5 x 200	0.001446	0.006654	15.3	17.1
5	50 x 200	0.001157	0.007811	16.2	18.1
6	50 x 200	0.001157	0.008968	17.1	19.1
7	37.5 x 200	0.000868	0.009836	17.7	19.7
8	25 x 200	0.000578	0.0104147	18.1	20.1
9	25 x 200	0.000578	0.01099	18.5	20.5
10	25 x 200	0.000578	0.1157	18.8	21.0

*Factor of safety of 1.33 against siltation incorporated

Example 4.10

Determine the diameter of the laterals if a drainage system has: q = 2 mm/d; spacing (L) = 80 m; length of laterals (B) = 250 m; drains are to be installed at a slope = 0.001. Corrugated PVC pipes are used in construction of the system. Design the size of the smooth PVC collectors, if 22 laterals discharge into a collector along with equal number of laterals each 150 m long entering the collector on the opposite side. It is proposed to use only one size of pipe for the collector.

We can approach such calculations in a backward manner as well. Prepare a table of maximum drainable areas for the different commercially available pipe diameters (Table 4.16, upper part) using Eqs. (4.75), (4.76) and (4.73).

Area to be drained by each lateral = 250 m x 80 m = 2 ha

The area per lateral falls in between the maximum drainable areas for pipes 65/57 mm and 80/72 mm. In such a case one should go for a larger size of the lateral. As such, the lateral pipes of 80/72 mm diameter would be appropriate.

Slope of collector = 0.001; length of collector = 1760 m; Smooth PVC pipes are available with inside diameter of 100 mm, 150 mm, 200 mm, 250 mm, 300 mm, respectively:

Prepare a table of maximum drainable areas using Eq. (4.77), Eq. (4.78) and Eq. (4.73) (see Table 4.16, lower part);

Determine area drained by the collector:

A = BL = (250 +150) m x 80*22 m = 70.4 ha

Data in Table 4.16 show that one needs to use 250 mm pipes to take care of the discharge.

Table 4.16 Maximum discharges and drainable areas with pipe drains of different diameters

Type of pipe	Diameter (mm)		Q (m^3/d)		Drainable area (ha)
	Outside	Inside	Practical		
			d ≥ 100 mm	d ≤ 100 mm	
	60	52	-	23.5	1.2
Corrugated	65	57	-	30	1.5
	80	72	-	56	2.8
PVC	100	91	-	104	5.2
	125	115	244	-	12.2
	160	148	477	-	24.0
	-	100	-	134	6.7
	-	150	495	-	25
PVC collectors	-	200	1065	-	53
	-	250	1930	-	97
	-	300	3141	-	157

DRAINAGE OF PADDY FIELDS

Lowland paddy is grown in many Asian countries including India, where water is kept ponded within certain limits during the crop growing season. Drainage for this crop might be required at one or more of the following stages:

- To remove the water used in puddling
- Drainage to prevent diseases/pest proliferation
- Drainage at the time when lodging is anticipated
- To provide suitable trafficability before the harvest
- To minimize methane production from paddy fields

In some countries, notably in China and Japan, one mid-season drainage is provided at late tillering stage to enhance crop productivity. Although it may appear that surface drainage can take care of the drainage requirement, yet in many cases surface drainage alone may not be sufficient to provide good environment, especially for mid-season drainage or drainage at the time of harvest. Intensive field experiments revealed that methane (CH_4) emission from two rice paddy fields in Niigata, Japans were 71% lower in subsurface drained than non-drained paddy field (Shiratori *et al.*, 2007). The subsurface drainage system lowered the groundwater level and top of the grey soil layer to the drainage pipe level, enhanced soil permeability, and resulted in more oxidized soil conditions in the fallow season. The lower total and hot water extractable carbon in the plowed soil layer of the drained field versus the non-drained field strongly suggests that the organic substrate that gives rise to CH_4 decomposed more quickly in the drained field.

Delauox equation is used in Japan for the designing of drainage systems for paddy fields. The equation is given as:

$$S = (H-h_1)/\tan b \tag{4.80}$$

Here S is the drain spacing, m; H is the water table height at mid-spacing, m; h_1 is the height of water table right above the drain; and tan b is the average gradient of groundwater surface (0.028 to 0.038 for humic clay and clay sub-soils)

In China, among others following equation is used for drainage of paddy fields.

$$S = K\,(H + h_0)/q\,a \tag{4.81}$$

Here K is the saturated hydraulic conductivity, m/d; h_0 is the depth of the water ponded on the soil surface, m, q is the percolation rate, m/d; and a = 0.5 + 0.174 $(H/D)^{1.5}$, such that D is the depth of the impermeable layer below the drain.

Besides what is stated, commonly used equations of Hooghoudt and Glover-Dumm are also used to design drainage systems for the paddy fields. Such designs have been found quite effective in waterlogged saline and waterlogged acid sulphate soils. In the latter case, pipe drains with 15 and 30 m spacing improved soil conditions within two years and significantly improved crop yield. Data collected over a period of 14 years showed an yield increase of 1.1 t/ha (43%) compared to non-drained areas. A useful proposition to provide flow control devices in paddy fields or selected rice growing blocks is discussed in Chapter 6.

QUESTIONS (THEORY)

1. Name and describe the various layout patterns of subsurface drainage.
2. Describe singular and composite systems of subsurface drainage listing their advantages and limitations.
3. Name and describe the various drainage design criteria used in subsurface drainage design.
4. Discuss the various criteria that are used to decide depth of subsurface drains.
5. Derive the Hooghoudt equation for the design of lateral drain spacing in subsurface drainage.
6. Describe the limitations of Hooghoudt equation.
7. A drainage system having drain radius of 10 cm is to be installed at a depth of 1.8 m below the soil surface. If depth of impermeable layer is 6.8 m below the soil surface and hydraulic conductivity of the soil as 0.8 m/day, what drain spacing must be applied if the water table is to be kept 1.2 m below the soil surface? Take recharge rate as 2 mm/d. Assume any other parameter required, if necessary.
8. A drainage system having the drain radius as 10 cm is to be installed at a depth of 1.5 m below the soil surface. If depth of impermeable layer is 3 m below the soil surface and hydraulic conductivity of the soil as 1.0 and 1.2 m/day, what is the design drain spacing using Hooghoudt equation if the water table is to be kept 1.0 m below the soil surface? Take recharge rate as 5 mm/d. Assume any other parameter required, if necessary. Draw a neat sectional sketch showing the system.

9. Using neat diagrams explain the concept of vertical, horizontal and radial flows used in the deriving the Ernst equation.
10. Derive the Ernst equation for the steady state conditions.
11. Using the Boussinesq equation, derive the Glover Dumm equation for the non-steady state conditions.
12. Design the drain spacing using Glover-Dumm equation with the data set: hydraulic conductivity, 1.2 m/day, depth to impermeable layer as 7.5 m, drainable pore space as 0.07 and radius of the drain as 10 mm. It may be assumed that water table at mid way between the drains is to be brought down from 0.5 m to 0.8 m in 2 days.
13. Explain the use of De Zeeuw-Hellinga equation for use in the design of subsurface drainage systems based on fluctuating water table.
14. Explain the concept of stress due to excess water (SEW).
15. Write a short note on drainage of sloping lands
16. Write down the procedure to calculate the drain diameter for laterals.
17. Estimate the discharge to be carried by a lateral of a subsurface drainage system if the spacing of the laterals is 50 m and the length of the lateral is 200 m. Assume that the drainage rate is 9 mm/day.
18. Determine the drain diameter if the length of the lateral is 300 m and lateral drain spacing is 60 m. The drains are laid at a slope of 1%. Assume that the drainage rate is 6 mm/day. Assume suitable values of other parameters required during calculations.
19. Write down the procedure to calculate the drain diameters for collectors.
20. Write a short note on the need and design of subsurface drainage for paddy fields

QUESTIONS (OBJECTIVE)

1. Fill in the blanks

1. In the ___________ system, parallel lateral drains enter the main drain at an angle from one or both the sides
2. In a ___________ pipe drainage system, each field lateral drain discharges into an open main or collector drain.
3. Under gravity outflow conditions, ___________ of drains is governed by the outlet conditions.
4. As per Dupuit-Forchheimer (D-F) assumption, for small inclinations of the free water surface, all streamlines in a gravity flow system are _________.
5. Equivalent depth to impermeable layer is ________ than the actual depth impermeable layer.
6. The wetted perimeter, u for pipe drains in trench and flowing half full is given as $u = b+2\ r_0$. Here r_0 is the radius of the drain and b is _________ width of the trench.

7. Hooghoudt´s equation is not applicable for layered soils where hydraulic conductivity of layer below the drains K_b is ________ than the hydraulic conductivity of the layer above the drains K_t.
8. The concept of ________ depth to impermeable layer is used in unsteady state equation given by Glover and Dumm.
9. In the perforated drain laterals, flow is ________ non-uniform.
10 De Zeeuw-Hellinga equations are used for drainage designs based on ________ Fluctuating water table theory.

2. Tick mark the correct or most appropriate answer for each question

1. The layout in which parallel laterals enter the main drain at an angle from one or both sides

   (A) Herringbone (B) Random

   (C) Double main (D) Parallel

2. Used to drain scattered isolated wet areas

   (A) Herringbone system (B) Double main system

   (C) Parallel system (D) Random system

3. The drainage system in which each field pipe (lateral drain) discharges into an open main/collector drain

   (A) Composite (B) Singular

   (C) Both A and (B) (D) None of these

4. Which of the four is the more commonly used drainage design criterion?

   (A) Trafficability (B) Fluctuating water table

   (C) Steady state (D) Salinity control

5. Hooghoudt equation is derived on the basis of

   (A) Trafficability (B) Fluctuating water table

   (C) Salinity control (D) Steady state

6. Which of the drainage criterion is rarely used

   (A) Steady state (B) Non-steady state

   (C) Trafficability (D) Fluctuating water table

7. The drain depth is decided on the basis of

   (A) Crop characteristics

   (B) Capability of the available machinery

   (C) Cost of drainage

   (D) All of these

8. Drainage depth in certain cases is decided on the basis of specific problem/difficulty such as
   (A) The depth to be kept deeper than the frost penetration depth of the soil
   (B) The minimum depth should ensure that the drains do not fail by repeated passes of the heavy machinery
   (C) The depth of drains should take into account the settlement of peats subsequent to land drainage.
   (D) All of these depending upon the specific issue
9. Cost of land drainage
   (A) Increases with depth of drains
   (B) Decreases with depth of drains
   (C) Initially decreases and later increases
   (D) None of these
10. Which is not Dupuit-Forchheimer (D-F) assumption in the design of subsurface drainage
    (A) For small inclinations of the free water surface, all streamlines in a gravity flow system are horizontal
    (B) The velocity along these streamlines is proportional to the slope of the free water surface but is independent of the depth.
    (C) The recharge is distributed uniformly over the land surface
    (D) All of these
11. The major limitation of Hooghoudt´s equation for a two-layered profile is
    (A) Drains have to be established in layers which has high hydraulic conductivity
    (B) Drains have to be established in layers which has low hydraulic conductivity
    (C) Drains have to be established at the interface of the two layers
    (D) None of these
12. Equivalent depth of impermeable layer is
    (A) More than the actual depth of the impermeable layer
    (B) Less than the actual depth of the impermeable layer
    (C) Equal to the actual depth of the impermeable layer
    (D) None of these
13. Equivalent depth to impermeable layer is a function of
    (A) Actual depth to impermeable layer
    (B) Drain spacing
    (C) Drain radius
    (D) All of these

14. The wetted perimeter of a lateral pipe flowing half full without trench is given as

    (A) π r (B) 2 π r

    (C) πr^2 (D) None of these

15. Equivalent depth to impermeable layer is used to account for

    (A) Extra resistance caused by radial flow

    (B) To reduce the depth of impermeable layer

    (C) To take care of the low hydraulic conductivity of the impermeable layer

    (D) None of these

16. Wetted perimeter of open drain of bottom width b and one of the wetted side s both in m is given as

    (A) b+s

    (B) π r (r is the equivalent radius)

    (C) b+2s

    (D) None of these

17. The Dagan equation is useful as a steady state design equation because

    (A) No hit and trial process is needed to arrive at lateral drain spacing

    (B) No need to determine equivalent depth to impermeable layer

    (C) Both A and B

    (D) None of these

18. To use the Ernst equation for a two-layered soil profile, the hydraulic conductivity of the two layers K1 (top) and K2 (bottom layer)

    (A) K1>K2

    (B) K1 =K2

    (C) K1<K2

    (D) No condition is attached as it can be used under all the situations

19. The basic equation to derive the unsteady state drainage design equation is

    (A) Boussinesq equation (B) Dumm equation

    (C) Glover equation (D) Hooghoudt equation

20. Using the Glover Dumm equation it can be shown that (q and h are discharge and hydraulic heads such that q and h at t+1 is a one-step higher than q and h at t, the previous time step)

    (A) $q_{t+1}/q_t \neq h_{t+1}/h_t$ (B) $q_{t+1}/q_t = h_{t+1}/h_t$

    (C) $q_{t+1}/q_t > h_{t+1}/h_t$ (D) $q_{t+1}/q_t < h_{t+1}/h_t$

21. In unsteady state situations, the lateral drain spacing is

    (A) Directly proportional to drainable pore space

    (B) Inversely proportional to cube root of the drainable pore space

    (C) Inversely proportional to squre root of the drainable pore space

    (D) None of these

22. Hooghoudt´s equivalent depth should also be used in unsteady state equation derived by Glover-Dumm

    (A) No

    (B) Yes

    (C) Depending upon one´s choice

    (D) None of these

23. Fluctuating water table theory is used in design for which equations have been given by

    (A) Hooghoudt

    (B) Glover Dumm

    (C) De Zeeuw-Hellinga

    (D) Boussinesq

24. In a subsurface drainage system, SEW_{30} values are 120 in design A while the same is 150 in the design B for the same crop growing season. From drainage point of view, which of two is a better design of the system?

    (A) Design (B)

    (B) Design (A)

    (C) No difference between the two designs

    (D) None of these

25. SEW_{30} is an index based on the values of water table whenever it is in between the soil surface and 30 cm below the ground surface during the crop growing season such that index is given by (X_i is the depth to water table on i_{th} day of the crop growing season of n days)

    (A) $SEW_{30} = \sum_{i=1}^{n}(30 - X_i)$ (B) $SEW_{30} = \sum_{i=1}^{n}(X_i - 30)$

    (C) $SEW_{30} = \sum_{i=1}^{n}(X_i)$ (D) None of these

26. An equation similar to a non-steady state equation can be derived by integrating a steady state equation given by

    (A) Hooghoudt (B) Ernst

    (C) Dagan (D) All of these

27. All other conditions remaining the same, lateral drain spacing is

    (A) Light textured soils> Medium textured soils> Heavy textured soils

    (B) Medium textured soils = Light textured soils = heavy textured soils

    (C) Heavy textured soils> Medium textured soils > Light textured soils

    (D) None of these

28. Drainage installations constructed on sloping lands require drains

    (A) Parallel to the water table gradient

    (B) Perpendicular to the water table gradient

    (C) Any of (A) and (B)

    (D) None of these

29. Capacity of the drains is determined with the help of

    (A) Darcy′s equation (B) Bernoulli′s equation

    (C) Manning′s equation (D) None of these

30. Manning′s roughness coefficient is

    (A) Higher for smooth plastic pipes than corrugated pipes

    (B) Higher for corrugated plastic pipes than smooth PVC pipes

    (C) Equal for smooth and corrugated plastic pipes

    (D) None of these

31. If q is the drainage rate in m^3/day and L is the lateral drain spacing, m then the B_{max}, the maximum length of the drain, m that can be drained is given by

    (A) $Q/qL = B_{max}$ (B) $Q*q*L = B_{max}$

    (C) $Q*L/q = B_{max}$ (D) None of these

32. Drainage for paddy fields is required to

    (A) Drain the fields when lodging of the crop is anticipated

    (B) To provide suitable trafficability before the harvest

    (C) To minimize methane production from paddy fields

    (D) All of these

33. In Manning's equation, hydraulic radius has the units of

    (A) L (B) L^2

    (C) L/T (D) None of these

34. Interceptor drains are used in areas

    (A) Along the canals to arrest seepage

    (B) In hilly lands to intercept groundwater before it resurfaces

    (C) Both (A) and (B)

    (D) None of (A) and (B)

35. Flow in lateral drain pipes is

    (A) Uniform (B) Non-uniform

    (C) Turbulent (D) None of these

36. Hooghoudt equation in a two-layered profile can be used to determine drain spacing, if

    (A) Drain is laid in the upper layer

    (B) Drain is laid at the interface of the two layers

    (C) Drain is laid in the bottom layer

    (D) None of these

37. Minimum grade requirement is higher for

    (A) Higher diameter pipes than lower diameter pipes

    (B) Lower diameter pipes than higher diameter pipes

    (C) Requirement is same for all diameter pipes

    (D) None of these

■■■■

Chapter 5

DRAINAGE MATERIALS

INTRODUCTION

Nearly 9000 years ago, subsurface drainage systems were assembled in Mesopotamia using gravel and stones and/or permeable voluminous substances, e.g. bundles of small trees and shrubs tied together at the bottom of a trench (van Schilfgaarde, 1971). Ridge tiles were introduced as drains in Scotland and Europe. Later, these were replaced by simple horseshoe drains, horseshoe drains on sole plates, flat-bottomed D-shaped drains and finally round pipes. Drain pipes made from wood boards or box drains, bricks, and horseshoe-shaped ceramic tiles, circular clay tiles, concrete tiles, bituminized fiber perforated pipes, perforated smooth plastic pipes and corrugated plastic pipes have been used (Stuyt *et al.*, 2000). Even in recent times, drainage materials have gone through an important evolution especially during the last 50 years or so. Clay/concrete tiles with gravel envelopes have been replaced by light weight flexible corrugated perforated polyvinyl chloride (PVC) and high density polyethylene (HDPE) corrugated plastic drain pipes. Simultaneously, a switchover occurred from bulky gravel filter to cheaper and labor saving synthetic envelope materials. Drain industry has now come up with light weight drain pipes pre-wrapped with drain envelopes of different kinds. Since the performance of a subsurface drainage system is affected by the kind and quality of materials used in assembling the system, proper selection of drain pipes and envelope materials assumes a key role. Some of the possible causes of poor performance of the drainage systems as related to drainage materials are due to: the use of poor quality drainage materials and blocking of the drain tubes due to silting up phenomenon in unstable soils. Apparently, soil texture and the chemical composition of the salt present in the soil will influence the choice of the drainage materials. This chapter presents the kinds of materials available and used, tests required and guidelines to select appropriate drainage materials. Although many locally available and fabricated materials (bamboo, bricks, vitreous sewerage pipes and sand drains etc.) have been used in drainage construction, current installations mostly use plastic pipes. As such, greater emphasis is placed on the selection and tests related to these pipes. Need of drain envelopes and their design procedures are included.

MATERIALS FOR DRAIN PIPES

Several kinds of tiles/pipe materials are used in subsurface drainage. Based on the flexibility of pipes, these have been categorized into two groups:

Rigid conduits: The materials include clay, concrete, asbestos and locally available materials such as bricks, stoneware and sewage pipes/tiles. Metal pipes are not used except some pieces at emergent locations such as outlets or road crossings etc.

Flexible conduits: These include plain or corrugated plastic pipes made of polyvinyl chloride (PVC) and high density polyethylene (HDPE).

Besides, several accessories namely, end caps, couplers (Fig. 5.1), reducers (Fig. 5.1), pipe fittings such as cross piece, T, Y piece, elbow and clip on junction, saddle bend combined and saddle T combined are used to assemble the drainage system. Some of the accessories are also shown in Fig. 5.2.

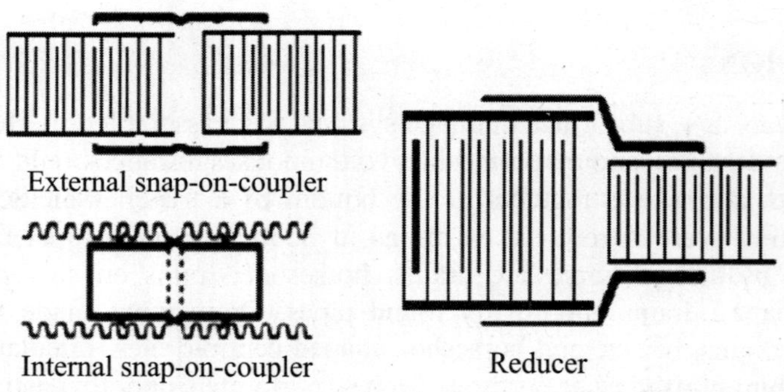

Fig. 5.1 A view of an external and internal snap-on-couplers and reducer

Fig. 5.2 Few accessories used in assembling the drainage systems

Clay Tiles

Clay tiles, porous or glazed, 300 mm long and having internal diameters (ID) of 50, 65, 75, 80, 100, 130, 160, and 200 mm have been used in Europe when large-scale drainage works were undertaken in and around 1970. Sometimes lengths of 333 mm are used so that 3 tiles make 1 m length. As per standards, 3% variations in diameter and length are permissible for all sizes and lengths of pipes. Visual tests are performed to see the shape of the tiles and for the presence of fissures and cracks.

The tiles should be tested in the laboratory for its mechanical strength. Baked clay tiles made of clay (Fig. 5.3), shale, fire clay or a mixture of these materials, should meet the following general criteria:

- Tiles uniformly burnt should be regular, dense, strong and sound, and free from irregularities such as cracks and laminations. Tiles with visible shortcomings such as badly formed pipes should not be used.
- The ends should be straight cut without any inside fillets
- These should be made from homogeneous raw material and should be free from impurities like stones, lime or any other foreign material
- Tiles should be free from minerals or chemicals that cause slaking or disintegration of tiles
- The dry tiles when tapped should give a clear high sound
- If clay tiles with sockets are used then depth of socket should very from 150 mm for 100 mm diameter to 565 mm for a 450 mm diameter tile.

Fig. 5.3 Clay tiles manufactured locally in Assam (Source: Singh, 2007) and industry manufactured tiles (Inset) (Inset with permission from RSPB: Droy, 2010)

Clay tiles are classified as light duty, medium duty, heavy duty and extra heavy duty. Light duty tiles have the lowest strength and are used under general conditions. Heavy duty and extra heavy duty are much stronger and are used in more severe and extreme conditions, respectively. The wall thickness in United Kingdom varies from 12 to 24 mm, increasing with increasing diameter of the pipe usually given by 0.08d + 8 mm, where *d* is the inside pipe diameter in mm. Wall thickness should not vary more than 5% plus or minus of the nominal diameter. Sometimes weight of 1000 tiles is also specified and should exceed the minimum values, being 1400 kg for 60 mm and 2000 kg for 80 mm diameter pipes. The wall thickness of clay tiles should provide the minimum crushing strength (individual tile) of 9.9, 14.4, 18.4 and 26.3

KN/m for the four kinds of tiles respectively for tile sizes up to 150 mm (BIS, 2009). These values for an average of 5 tiles should be 11.7, 16.0, 20.4 and 26.3 respectively. Five hour boiling test is conducted to determine the water absorption by the tiles. It should not exceed 16% by weight of the tile and 13% as an average of 5 tiles.

Pipe sections are abutted against each other and water enters through the joints that exist due to imperfect fit between the ends of the pipe section. The maximum gap between individual pipes may not exceed 3 mm. Pipes with perforations in the form of punched holes can also be used. Holes of 6.0 ± 1.5 mm in diameter are arranged in rows parallel to the axis of the tile. The holes are approximately 75 mm center to center along rows. Minimum number of recommended rows and perforations are given in Table 5.1. Clay tiles with holes are also being manufactured or can be specially ordered.

Table 5.1 BIS recommendations for perforations in clay tiles

Nominal diameter (mm)	Minimum rows of perforations	Minimum number of perforations per row for nominal length (m)				
		0.30	0.45	0.50	0.70	0.90
100,125,150	4	3	5	7	9	11
100,125,150	2	5	8	11	13	15
200,250,300	4	3	5	7	9	11
380 to 450	6		5	7	9	11

Clay tiles are very durable and highly resistant to weathering and deterioration under varying soil conditions e.g. saline soils, alkali soils, soils containing sulphates and corrosive chemicals etc. Although clay tiles are lighter than concrete tiles, these have excellent bearing strength. Clay tiles are fragile and require careful handling. Clay tile should be used with caution in cold climates where extreme freezing and thawing reversals occur.

Concrete Tiles

Concrete pipes/tiles have virtually become obsolete upon introduction of plastic pipes. Concrete pipes of various diameters, going up to 500 mm, are manufactured. The quality of concrete tiles depends largely on the use of well graded, sound aggregates, sufficient good quality cement and optimum mix of water. Large diameter pipes exceeding 300 mm are made from reinforced cement concrete. Concrete tiles are classified as standard, extra quality, heavy duty-extra quality, or special quality (BIS, 1979). Standard is for general use and extra quality and heavy duty-extra quality are for situations requiring stronger tiles. Special quality tiles are used in acidic soils (pH < 6.0). Some general quality guidelines for good concrete tiles are:

- Tiles/pipes should be well formed, have good finish, should be free from cracks and chips. The tiles should be free from local dents or bulges, greater than 3 mm in depth and twice the thickness of the tiles in length in any direction
- The tiles must be properly cured

- Pipe ends are either straight or have a collar or a spigot and groove. In straight tiles, the ends should be square. If two tiles are placed together in a straight line, the gap between the two adjacent tiles should not exceed 3 mm. The water entry is through the joints between pipe sections.
- The deviation in straightness in any tile shall not exceed 3 mm for every meter length
- Special treatment is needed when used in acid soils (pH < 5.5) and/or soils containing sulphates. If the sulphates of sodium or magnesium exceed 300 ppm, use sulphate-resistant cement in preparing the tiles. Steam curing at 100°C or higher may also be needed.

The crushing strength for light and medium duty 80-300 mm diameter tiles should be minimum 9.9 and 14.4 kN/m for individual tiles and 11.7 and 16.9 KN/m for the average of 5 tiles. For heavy duty tiles, minimum crushing strength for individual tiles is 18.4, 19.7, 20.4 and 22.3 for tiles up to 150 mm diameter and 200, 250 and 300 mm diameter respectively.

For water absorption test, three pieces shall be taken from each tile that has withstood the crushing strength test. One of the pieces shall be taken from one end of the tile, second piece from the opposite end and the third from the center or near center of the tiles. Each piece shall be free from visible cracks having a minimum area of 128 cm^2 as measured on one surface of tile. Thickness of the piece shall be equal to the thickness of tiles. Water absorption with boiling test should not exceed 9% by weight in any class of tiles.

The main advantage of these tiles is that these can be fabricated at site. Even when plastic pipes are used to assemble the system, concrete pipes are used as outlets for tile and plastic drains. These are also used as substitute when other pipes are unable to withstand surface load where minimum cover cannot be ensured or under road crossings. These are also used to construct auxiliary structures.

Other Pipes

The asbestos pipes made from a mixture of clean asbestos fiber and ordinary or rapid hardening Portland cement and water are also used in land drainage although not so commonly as clay and concrete pipes. The pipe used in drainage shall be seamless, compact and homogeneous. Crushing strength of the asbestos pipes should meet the standards as laid down for concrete tiles.

Locally available materials are used only at experimental scale or for small plots. Use of stoneware pipes as drainage conduit is shown in Fig. 5.4.

Metal pipes are not ordinarily used in subsurface drainage. However, metal pipes are used to serve specific purposes as follows:

- At outlets for tile and plastic drains and under road crossings
- As a part substitute when the other pipes are unable to withstand surface load where minimum cover cannot be ensured
- For auxiliary structures
- For installations through pockets of quicksand or unstable soils where a single piece of pipes needs to be used.

Fig. 5.4 Use of stoneware pipes in subsurface drainage in Andhra Pradesh

Plastic Pipes

Straight walled smooth plastic pipes are being used in land drainage since 1960. Corrugated pipes, introduced around 1970, soon replaced the smooth pipes (Fig. 5.5). The plastic pipes are made of polyvinyl chloride (PVC), polyethylene (PE) and polypropylene (PP). The main advantage and limitations of plastic pipes over clay and concrete pipes is listed as under:

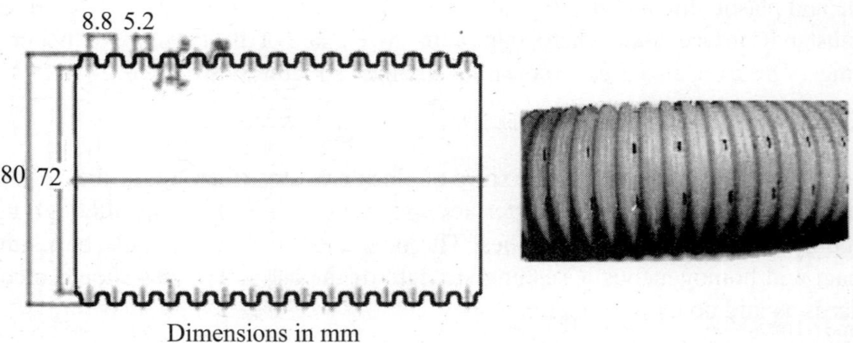

Fig. 5.5 Corrugations on 80 mm diameter corrugated plastic pipe (left) and perforations in the valley of corrugated pipes (right: with permission from RSPB: Droy, 2010)

Positive features

- Lightweight makes handling easier, gives less alignment problems
- More uniformly distributed perforation area, facilitating access of water
- Easy wrapping with envelope materials
- Safer implementation without misalignment.

- Less labour - intensive leading to lower labour costs
- Inert to all common soil chemicals.

Limitations

- Vulnerability to deterioration from UV-radiation when exposed to sunlight for long periods, especially if made of PVC.
- Increased brittleness at low temperatures.
- Increased deflection risk at high temperatures and excessive stretch during installation.
- Risk of collapse under sudden load, e.g. by trench wall caving or stones.
- Unable to resist fire
- The disposal of PVC waste has some environmental issues as PVC pipes upon burning produce hydrochloric acid.

Plastic pipes are classified as standard or heavy duty according to pipe stiffness and elongation percentage (Fig. 5.6). Standard corrugated drain pipes usually have outside diameters (OD) of 40, 50, 65, 80, 100 and 125 mm such that internal diameter (ID) is given as $ID \sim 0.9*OD$. The height of the corrugations varies from 8-13% of the pipe diameter for small to larger diameter pipes. The water entry is through the punched holes/slots in the valley of the corrugations. The price of plastic pipes increases sharply with increasing diameter (Mohan Lal *et al.*, 2006). The price differential between a 4 cm and a 10 cm diameter pipe is roughly four times. The advantages and limitation of corrugated pipes over smooth bore pipes are as follows.

Fig. 5.6 Corrugated plastic pipes for collectors (left) and for laterals (right)

Advantages

- Corrugated pipes have higher stiffness on per unit weight basis than smooth pipes.
- Corrugated pipes are more flexible and thus can have a longer working length in the form of coils. The length of pipes may vary from 5 to 6 m for smooth bore semi-rigid pipes to specified coils of length 30, 50, 70 or 100 m up to 160 mm nominal outside diameter.
- Corrugated pipes are easily amenable to mechanical laying.

Limitations

- For a given perforation pattern corrugated drains have higher entrance resistance than smooth pipes (Dierickx, 1980)
- Due to their flexibility, corrugated pipes pose problem while laying in trenches in straight line.
- The hydraulic roughness coefficient of corrugated pipes is higher than smooth pipes. Therefore, for the same flow about 20% higher diameter pipes are required than smooth pipe.

Perforations in the corrugated pipes are 2-4 mm long and 0.6-2.0 mm wide made in the valley of the corrugations. The longer dimensions of the perforations shall be along the circumference of the pipe. Numbers of slots per meter length should be 100 distributed evenly around the circumference of the pipe. The open area should be minimal 800 mm^2 although a range of 960-2400 mm^2 is suggested. Smooth pipes generally come without holes. To have an inflow area of about 1200-1800 mm^2 or 2-3% of the surface area of the pipe, 60 slots/m lengths having 0.8 to 1.0 mm width and 25-30 mm length are made. The slots should be uniformly distributed along the circumference of the pipes in four rows.

PVC Versus PE Pipes

Any of these materials can be selected on the basis of economic considerations since a good product can be made from any of these materials. Of all, PVC is the most common because of higher bearing capacity and low cost. The properties of PVC and PE pipes are summarized below:

PVC pipe	PE pipe
• High strength for a given pipe design and weight and thus, for the same strength cheaper than PE pipes	• To achieve the same strength of pipe, more material is needed, making it costlier than PVC pipes
• Softens only at very high temperature and therefore, it poses no problems in this regard	• Softens at temperature between 50° and 100°C • Virtually impact proof
• Being brittle at low temperature *i.e.* 0°C to –10°C and susceptible to impact damage, the pipes are not recommended if outside temperatures are likely to fall below 3°C. The risk that cracks will be formed is too high.	• Degradation due to UV rays can be avoided by simply adding carbon black at the manufacturing stage
• Gets degraded due to exposure to ultra-violet (UV) rays, needs greater care during storage to prevent exposure	

TESTING FLEXIBLE PIPES

Indian manufacturers follow one of the three test standards ASTM D2152-80 (1986), DIN (1987) and BIS (1980). The same standard is used to test the quality of pipes normally at the manufacturer's premises. Pipes are selected randomly for visual and dimensional tests. Depending upon the lot size 5 pipes/coils from a lot of 300, 8 from a lot of 300-1000 and 13 for a lot of >1000 pipes/coils are selected (BIS, 2004). If 2 or more than 2 samples fail, second sampling is done and tests repeated.

Visual and Physical Tests

The pipes must meet the following properties that are visible to the naked eye and physical characters that are measured in the manufacturer premises:

- The pipes should be homogenous
- The pipes should be uniform in colour, appearance and free of chalking, sticky, or glossy material
- The pipe walls shall be free of cracks, holes, blisters, voids, foreign inclusion or other defects
- The length shall not be less than 99% of the stated quantity
- The actual inside diameter shall not be less than the nominal diameters minus 3%.
- The minimum weight of corrugated PVC pipe depends upon the corrugations. The manufacturer must specify the weight in kg/m. The density of the pipe shall be between 1.40 and 1.46 g/cm^3.
- Water inlet area specifications given in the relevant standard being followed by the manufacturer must be met. For example, ASTM F800 (1989) specifies a minimum inlet area of 1800 mm^2 for 80/72 mm diameter pipes and 2100 mm^2 for other sizes up to 294/258 mm diameter pipes (Mohan *et al.*, 2012)

The samples, which have passed the visual and dimensional tests, are further tested for the following properties (Table 5.2).

1. Acetone test etc.
2. Elongation
3. Bending
4. Impact test
5. Deformation or stiffness

Ami's (1987) manual on drain pipe tests provides detailed procedures. Besides, it is always desirable to consult the relevant standard when conducting the tests.

TESTS ON FITTINGS

The fittings shall not reduce or impair the overall integrity of the function of the tubing line and shall not significantly reduce the nominal diameter of the tubing being joined. The maximum allowable gap between fitting and tubing shall not exceed the perforation width. The fittings are tested for joint integrity, fitting strength test and alignment (Table 5.3).

Table 5.2 Test procedure for plastic pipes

Test name	Purpose	Procedure in brief	Pass/fail criteria
Acetone test	Quality of extruded PVC	A specimen > 13 mm in length is conditioned at 23±2^{0}C is immersed in acetone of 0.7857 g/mL density in a sealed container	Failure occurs if 50% attack as described by lifting, raising or removing on outside, inside or middle wall
		After 20 minutes, it is taken out and inspected	Failure also occurs if 10% attack on more than one surface
Elongation test	As pipes are stretched and elongated during mechanical laying, this test assesses the elongation characteristics	A 1270 mm long sample conditioned at 23±2°C for >4 hr should be used. Apply initial tare loads 1.3 kg for 80 mm going up to 7.03 kg for 455 mm pipe. Mark the gauge length at centre 762 mm. Apply additional test loads of 6.43 kg for 80 mm going up to 35.2 kg for 455 mm pipes. Measure the length after 3 min. Elongation, E is given as:	Pipes shall not elongate more than 7.5%
		E = 100*mm of stretch/ 762(%)	
Bending test	To measure the bending characteristics of the pipes	Tests are conducted on 3 specimens conditioned at 0°c ± 1°C for 24 h.	No cracks or splits should be visible
		Within 30 s after removal from the chamber, the piece is bent by hand on the mandrel with a radius of 3 times the pipe outside diameter and kept as such for 10 minutes	
		Samples are inspected for splits and cracks	
Impact test	To test the strength of the pipes against impacts	10 specimens conditioned at 10°C for 24 h are used and tested within 60 s after removal from the cold chamber	The pipes shall not show signs of fracture, cracking, rupture, or splitting. One failure is acceptable
		A weight of 2.5 kg is dropped from 0.55 m height and inspected for	

Test name	Purpose	Procedure in brief	Pass/fail criteria
		cracks, rupture, fracture and splits (ASTM D2444-93, 1984)	
Deformation test or stiffness test (The ratio of the deflection in pipe inside diameter to the initial inside diameter expressed as the percentage of the initial inside diameter).	The force per unit length of specimen by the resulting deflection in the same units at the prescribed percentage deflection.	Plot the graph deflection versus load and measure loads corresponding to 5% and 10% deflection. Calculate the pipe stiffness (PS) for any given deflection as follows: $PS= F/\Delta Y$ Where PS pipe stiffness. in kN/m^2, Here F load at 5% and 10% deflection, ΔY at 5% and 10% deflection.	Minimum stiffness at 5% for pipes of 160 mm dia shall be 210 KN/m^2 while for higher diameter it should be 250 KN/m^2. For 10% deflection these values should be 175 and 210 KN/m^2 respectively. When used under roads and other drainage applications, the stiffness should be 1.6 times these above values.

Source: Summarized from ASTM D2152-80, DIN-1987 and BIS-2004

LOAD ON CONDUITS

Drain pipes must be durable and have adequate strength against the load likely to develop on the pipes. For example, clay and concrete tiles may fail due to breaking if soil load on the tiles is excessive. Two kinds of loads are encountered on drain pipes:

- Static load or weight of overburden soil
- Concentrated loads due to passage of equipment or vehicles

While soil loads increase with depth, concentrated or wheel loads is critical only if tiles are laid at shallow depths.

Types of Soil Loading

The width of the trench at the top of the tile determines the type of soil loading, therefore, the shape of the trench above the drain is unimportant. Usually two situations namely ′ditch conduit conditions′ and ′projecting conduit conditions′ are encountered. The former applies to narrow trenches. The latter applies to trenches wider than 2 to 3 times the outside diameter of the pipe. In the first case (ditch conduit situation), weight of soil settling above the drain is partially carried by neighboring soil. The soil load under this situation is computed by:

$$Wc = C_d \, w \, B_d^2 \tag{5.1}$$

Table 5.3 Test procedure for pipe fittings

Test name	Quantity of sample/fitting required	Procedure in brief	Pass/fail criteria
Joint integrity	3 fitting of each type with tubing samples of at least 150 mm in length	Assemble joint fittings to the appropriate tubing in accordance with manufacturer recommendation on both sides of the fitting.	The maximum allowable gap between fitting and tubing shall not exceed the associated tubing perforation width.
		Use hanging weight procedure as described in elongation test Maintain the tare and test load for 3 min.	Failure occurs if there is separation
Fitting strength test	3 fitting of each type with tubing samples of at least 150 mm in length	Attach tubing to both sides of the fitting except for the end cap. Load the connected tubing and fitting between parallel plates at the rate of 12.7 mm/min. The vertical inside diameter is reduced by at least 20% of the nominal diameter of the fitting. Remove and inspect for damage	There shall be no cracking, splitting or creasing of fitting
Alignment	3 fitting of each type with tubing samples of at least 150 mm in length	Assure that the assembly or joint is correct and complete. If tubing is bent, it should be hand straightened prior to the test. Lay the assembly or joint on a flat surface.	It should accommodate straight line flow.

Here Wc is the total load on conduit per unit length, kg/m, C_d is the load coefficient for ditch conduits (Fig. 5.7), w is the unit weight of fill material, kg/m^3, B_d is the width of ditch at the bottom of the conduit, m and H is the depth to top of tile, m (Fig. 5.7). The values of C_d can be calculated either from the following equation and Table 5.4 or directly from Fig. 5.7.

$$C_d = [1 - e^{(-2\,k\,f.\,H/Bd)}]/(2.kf) \tag{5.2}$$

Here k is the ratio between internal pressure and external vertical pressure and f is the friction coefficient between the backfill material and trench vertical sides. Values of $k.f$ are selected from the Table 5.4.

In the projecting conduit situation, settling of soil in trench at sides is more than directly above the tile. Therefore, some of the weight of soil in the trench at the sides of the tile is transferred to the soil directly above the tile. The load on the tile is computed by:

$$Wc = C_c\, w\, B_c^2 \tag{5.3}$$

Here C_c is the load coefficient for projecting conduits (Fig. 5.7) and B_c is the outside diameter of tile. C_d and C_c are functions of the soil's frictional coefficient, height of the fill and width of the trench (B_d) or tile (B_c).

Values of C_c can be calculated from Fig. 5.7.

Table 5.4 Values of k.f for various backfill materials

Backfill type	**k.f**
Cohesion less granular material	0.192
Sand and gravel	0.165
Saturated top soil (other than clay)	0.15
Clay	0.13
Saturated clay	0.11

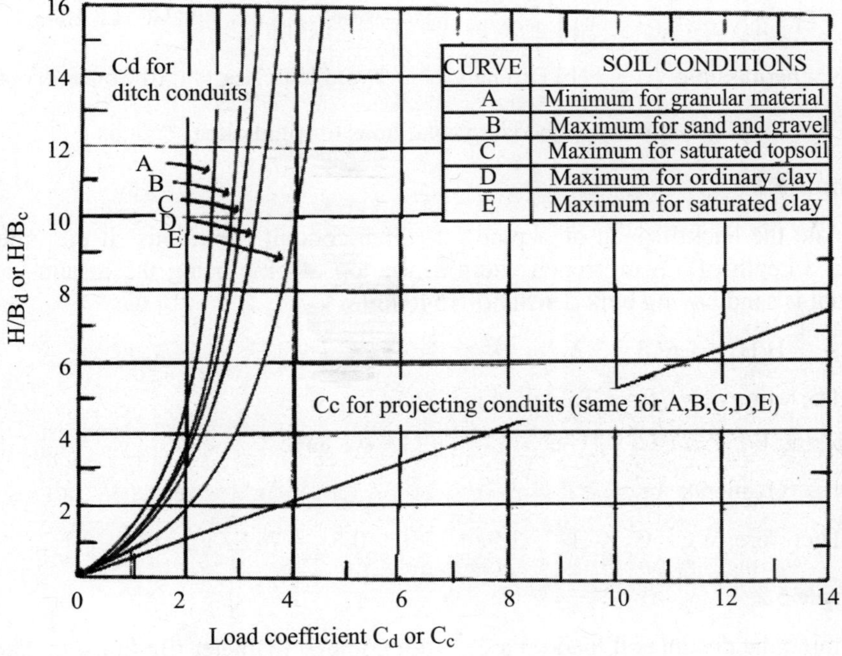

Fig. 5.7 Load coefficients for rigid pipe drains

Ditch conduit condition is preferable to projecting conduit condition. In instances where the actual loading condition is unknown, the soil load should be calculated for both the conditions. The smaller soil load is the actual condition.

While assessing safety against loads on the conduits, one must allow for variation in tile strength and bedding conditions (Spangler, 1957). In Fig. 5.8a, since no attempt is made to shape the foundation, such a situation is impermissible in tile drainage. The load factor is 1.5 for ordinary bedding conditions where bottom of the trench is shaped for at least half of the conduit width. The pipe should be surrounded with fine granular soil extending at least 15 cm above the top of the pipe (Fig. 5.8b). For first class bedding, the bottom of the trench must be shaped to 0.6 times the width of the pipe and should be placed in fine granular material that should extend to about 30 cm or more above the top (Fig. 5.8c). The load factor for such a case is 1.9. The load factor is up to 3.4 for concrete cradle, which is seldom provided in farm drainage (Fig. 5.8d). The tile strength should be multiplied by these load factors to arrive at the actual load it can support.

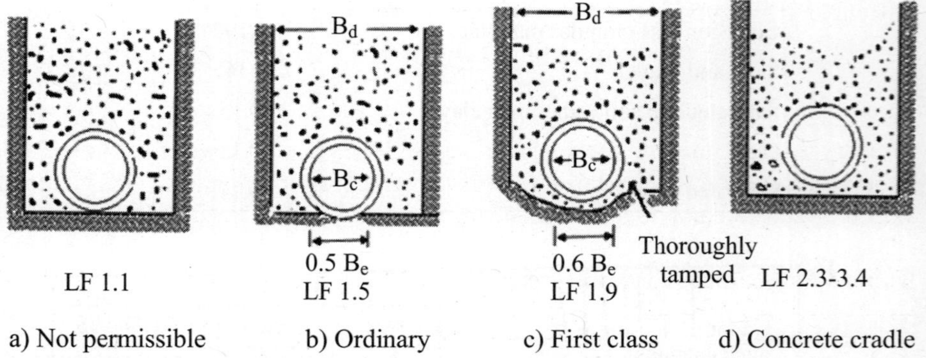

Fig. 5.8 Example bedding conditions for rigid ditch conduits

Example 5.1

Calculate the backfill load on a pipe for 'ditch conduit conditions' if the pipes are laid at a depth of 1.6 m, trench width at the top of pipe being 0.5 m and backfill material is sand having bulk density of 1540 kg/m^3.

$$H/B_d = 1.6/.5 = 3.2$$

The value of k.f from Table 5.4 is 0.165.

Using Eq. (5.2), $C_d = [1 - e^{(-2\,k\,f.\,H/Bd)}]/2.kf$, we have

$$C_d = 1.98$$

Therefore, $Wc = C_d\,w\,B_d^2 = 1.98 * 1540 * (0.5)^2 = 762.3$ kg/m

Example 5.2

Determine the design soil load on a 250 mm nominal diameter tile (outside diameter = 305 mm) installed 2.13 m deep in a trench 46 cm wide if the backfill is ordinary clay weighing 1922 kg/m^3. The safety factor is 1.5. What quality clay tile is required, assuming ordinary bedding conditions?

Depth to the top of the tile, H = 2.13 – 0.30 = 1.83 m. For $H/B_d = 1.83/0.46 = 4.0$, Read from Fig. 5.7, $C_d = 2.4$. Substitute in Eq. (5.1) to obtain the load for ditch conditions.

$$W(c) = 2.4 \times 1922 \times 0.46^2 = 976 \text{ kg/m}$$

Read from Fig. 5.7 for $H/B_c = 1.83/0.305 = 6.0$ m, $C_c = 10.8$. Substitute in Eq. (5.3) to obtain the load for projecting conditions.

$$W(c) = 10.8 * 1922 * 0.305^2 = 1930 \text{ kg/m}$$

The lower value 976 kg/m is the correct load; hence ditch conditions apply. To allow for variation in tile strength and bedding conditions, a safety factor of 1.5 results in a design load of 976 x 1.5 = 1464 kg/m.

From Fig. 5.8, the load factor for ordinary bedding is 1.5 and the supporting strength of light-duty tile is 1194 kg/m, based on the three-edge bearing method. Strength for ordinary bedding conditions is 1194 x 1.5 = 1791 kg/m, which is sufficient to support a design load of 1464 kg/m.

Knowing the three-edge strength of the rigid conduits, one can go backwards and calculate the maximum allowable trench depths for various trench widths. For ditch or projecting conditions, the maximum allowable trench depths for light ordinary straight tile are shown in Table 5.5. For ease of tabulation, trench widths have been rounded to the nearest 0.1 m. These depths apply only for loads due to the weight of the soil and do not include wheel loads.

Table 5.5 Maximum allowable trench depths for rigid clay tile

3-edge bearing strength (kg/m)	Nominal diameter (mm)	Trench width (m)					
		0.3	0.4	0.5	0.6	0.8	1.0
1194	100-150	*	2.8	2.2	2.2	2.2	2.2
1194	200	*	2.9	2.2	1.8	1.8	1.8
1194	250		3.0	2.1	1.8	1.5	1.5
1194	300			2.1	1.8	1.5	1.5

*Any depth is permissible; Soil density for wet clay 1922 kg/m^3; bedding load factor 1.5; safety factor 1.5.

Concentrated load

The portion of a concentrated load at the soil surface transferred by the soil to a tile is computed by

$$Wt = 1/L(I_c\, C_t\, T) \tag{5.4}$$

Here Wt is the average load; L is the length of tile; I_c is the impact coefficient (1 for static concentrated loads, 1.5 – 2.0 for wheel loads moving at 32.2 km/h or 20 mi/h); C_t is the load coefficient (Table 5.6); and *T* is the concentrated load

Example 5.3

Determine the total live and soil load transmitted to 0.3 m length of tile for the conditions given in example 5.1, if the concentrated wheel load from a truck is 2000 kg.

From Example 5.1, the soil load was computed as 976 kg/m or for 0.3 m length (976 x 0.3) = 293 kg (assuming uniform distribution along the 1 m length). From Table 5.6, read for *H* =1.83 m approximately 1.8 m) and trench width of 0.46 m, the percent load transmitted (by interpolation) as 1.6 %.

Total load on 0.3 m length = 293 + (0.016 x 2000) = 325 kg.

At this depth the live load is only 11% (0.016 x 2000/293) of the soil load, but it may be more at the shallower depths.

Table 5.6 Wheel loads (%) transmitted to underground rigid conduits in percent

Depth of backfill over top of tile, H (m)	Trench width at top of drain (m)				
	0.3	0.6	0.9	1.2	1.8
0.3	17.0*	26.0*	28.7*	29.7*	30.2*
0.6	8.3	14.2	18.3	20.7	22.7
0.9	4.3	8.3	11.3	13.5	15.8
1.2	2.5	5.2	7.2	9.0	11.5
1.5	1.7	3.3	5.0	6.3	8.3
1.8	1.0	2.3	3.7	4.7	6.2

*Includes both live load and impact load transmitted to a 0.3 m length of tile. Live loads transmitted are practically negligible below 1.8 m depth. (ASAE, EP 260.4, 1984)

Flexible pipes

As against the clay/concrete tiles, failure of plastic is mostly by deflection. The deflection of a plastic pipe under a soil load is computed using the projecting conditions of a rigid pipe as follows:

$$W_c = C_d\, w\, B_c\, B_d \tag{5. 5}$$

Here all the parameters are as described in the earlier equations. Soil at the sides of the pipe bears the soil load on the pipe. The soil is assumed to have the same stiffness as the pipe.

Flexible pipes must be installed in a trench so that good side support is provided. Bedding conditions should give a load factor of at least 1.5, which can be obtained with a 90-degree V groove. A space of about 8 cm should be provided between the trench wall and the tubing on each side so that granular soil without large voids can be packed to give the needed side support. Since flexible tubing has little load-bearing strength, it can support the soil load only because of the passive pressure exerted by the soil as the pipe deflects. If this pressure was not present, tubing would buckle and fail from the soil load. Deflections up to 20% usually do not cause failure of corrugated HDPE tubing, but failure is likely at 30% deflection. Maximum deflection is not likely to occur until about 4 years after installation.

The load-carrying capacity of flexible pipes is computed from the modified Iowa formula as given in Spangler & Handy (1973):

$$W(c) = 93.16\, Y\, [(EI/r^3) + 0.061\, E'\,]/d(1)\, K \tag{5.6}$$

Here Y is the allowable vertical deflection (1.1 x horizontal deflection), mm, (EI/r^3) is the ring stiffness from the parallel plate test at 20% deflection, KN/m^2; *E'* is the modulus of soil reaction or side support in KN/m^2 (suggested value 345 KN/m^2); d(1) is the deflection lag factor (use as 3.4), and *K* is the bedding angle factor (0.096 for 90° angle or fitted circular groove).

Recommended maximum depths for installing flexible corrugated tubing in loose, fine grained soil is given in Table 5.7 (Fenemor *et al.*, 1979; ASAE, 1984).

Table 5.7 Maximum trench depths for tubing buried in loose, fine-textured soils

Nominal diameter (mm)	Tubing quality (ASTM, 1987)	Trench widths at top of tubing (m)			
		0.3	**0.4**	**0.6**	**0.8 or >**
100	Standard *	3.9	2.1	1.7	1.6
	Heavy- duty**	a	3.0	2.1	1.9
150	Standard *	3.1	2.1	1.7	1.6
	Heavy-duty**	a	2.9	2.0	1.9
200	Standard *	3.1	2.2	1.7	1.6
	Heavy-duty**	a	3.0	2.1	1.9
250	**		2.8	2.0	1.9
300	**		2.7	2.0	1.9
375	**			2.1	1.9

a - Any depth is permissible at this trench width or less; Assumptions: E' = 345 KN/m^2; D(1) = 3.4; K = 0.096; W = 1750 kg/cu m; and Y = 1.1 x horizontal deflection.

*Pipe stiffness = 90 kN/m^2 for 20% deflection and **Pipe stiffness = 124 kN/m^2 for 20% deflection.

NOTE: Differences in commercial tubing, including corrugation design and pipe stiffness and soil conditions may change the assumptions, and therefore may change the maximum depths given above.

Example 5.4

Determine the maximum depth that a 200 mm diameter corrugated plastic tube, standard quality, can be installed in a trench 1.0 m wide at the soil surface and 0.4 m wide at the top of the tubing. Soil is loose, fine-textured and has a unit weight of 1750 kg/m^3. An allowable deflection of 20% is permitted.

Since the conditions in Table 5.6 apply, read for a trench width of 0.4 m, a maximum depth of 2.2 m. Note that 1.0 m width at the soil surface would not apply.

For conditions not covered in Table 5.6, following step-wise procedure is used:

- Calculate the soil load using Eq. (5.5) and compare with the load carrying capacity of the tubing using Eq. (5.6).
- Change the depth until the two loads are equal.
- Solve for C_d, read H/B_d from Fig. 5.7 and solve for *H* (assume ditch conditions to apply).

ENVELOPES

Dieleman & Trafford (1984) used the terms envelope, filter and surround to distinguish between types of envelopes based on their principal functions. An envelope, a generic term used to designate both filters and envelopes, is defined as *'a porous material placed around perforated drain pipes'* (Fig. 5.9). Specific distinction is made on the basis of whether the envelope is used for the filtering function or the hydraulic function.

Filter

A filter is put around the drains to prevent or restrict fine-grained materials from entering the drains along with the groundwater. If fine particles enter the drains, it may clog the system. Ritzema (1994) defined a drain filter *'as a layer or combination of layers of previous materials, designed and installed to provide drainage, yet prevent the movement of soil particles in the flowing water'*. The need to use filters mostly depends upon soils, some soils essentially requiring filters while others may or may not. Some people say that filter is a misnomer because any perfect filter will get clogged over the years.

Envelope

Porous material put around the drains is designated as envelope if *'it is placed to provide a medium of good permeability around the pipe to reduce entrance resistance and thereby improve flow conditions around the drain'*. The need to use an envelope is secondary. If a filter is provided, it also serves the purpose of envelope.

Mechanical/Bedding Function

Besides the filtering and hydraulic functions, filters/envelopes provide bedding support to the pipes in order to prevent excessive deflection and damage to the pipes due to soil load during or after construction (see section on Load on Conduits).

Kinds of Envelope Material

According to the raw material, drain envelopes have been divided into mineral or granular, organic and synthetic or manmade or fabric envelopes.

Fig. 5.9 A synthetic woven envelope (nylon sock) being put on a collector

Granular envelopes

Gravel, brick ballast, coarse sand, pit run sand and artificial aggregates and fine gravel (rounded or angular) are granular or mineral envelopes. Even top soil is included in this group. A combination of these materials such as gravel or sand can also be used.

Organic envelopes

Some of the organic envelopes are straw of flax, rice, wheat and safflower, saw dust, wood chips, peat, building paper, hay, jute cloth (burlap), corn cobs, leather, coconut fiber, volcanic granules, reds, fuel ash etc. It may be noted that performance of organic envelopes is not well tested and therefore, under critical conditions, it is advisable to go for other kinds of envelope materials.

Synthetic envelopes

Amongst these nylon socks, geo-synthetic filters, bonded synthetic fibers, slag, polystyrene foam, cinders, glass wool and fiber glass (made from lime-borosilicate glass) are common. Some new materials such as polypropylene, polyvinylidene and polyethylene have also come into the market. Sometimes a distinction is also made between various materials categorizing them as woven and non-woven materials.

Criteria for Envelope Selection

The selection of envelope depends on following factors.

- Material availability, transportation and cost
- The likely functions (hydraulic, filter, mechanical and bedding) of the envelope material
- Loading of the pipe and envelope
- Handling characteristics during transport and application
- Danger of bio-chemical fouling
- Organic matter and pH of the soil
- $CaCO_3$ content of soil and pH of the water
- The climatic conditions
- Chemical constituents of soil and water

NEED OF AN ENVELOPE

The question whether a drain envelope is needed or not, is often encountered during design of a drainage system. The dilemma is between the increasing cost and the long-term performance of the system. A systematic study can only resolve this dilemma. Moreover, if there are doubts, it is better to use envelop rather than not to use.

Soils and Need of an Envelope

In general heavy clay soils with clay percentage > 40%, coarse soils with 90% of particle sizes larger than maximum drain pipe perforation width do not require filter. As per guidelines given by Long & Brownscornbe (1970), clayey gravels, poorly graded gravel-sand-clay mixtures, inorganic clays of low to medium plasticity, gravelly clays, sandy clays, silty clays, lean clays, Well graded gravels, gravels and mixtures with little or no fines, well graded sands, gravelly sands with little or no fines, inorganic fat clays, organic silts and organic silt-clays of low plasticity, organic clays of medium to high plasticity and peat soils do not require a filter. These may require an envelope if flexible conduits are used. Minimum drain velocity also known as self-cleaning velocity may be ensured if these soils have appreciable fines.

A soil with strong silting tendency must be provided with appropriately designed envelope. Other conditions that necessitate the use of envelopes are: i) the soils that are unstable under prevalent water flow gradients and drain envelope does not adequately protect openings/ perforations of the drain, ii) water velocity in the drain is insufficient to flush the particles out of the drain and iii) the openings of the drain pipe are too large such that bridging of soil particles will not take place. Poorly graded sands, gravelly sands, silty sands, poorly graded sand-silt mixture, inorganic silts and very fine sands, rock flour, silty or clayey fine sands with slight plasticity, inorganic silts, micaceous fine sandy or silty soils and elastic silts require filters (Long & Brownscornbe, 1970).

Depending upon the silting tendency of the soils, one can visualize the need of filtering envelopes (Rycroft & Amer, 1995). Low silting tendency usually means no need of a filtering envelope. Soils with high, medium and no primary and secondary silting are reported in Table 5.8. Soils such as sandy loam, silty loam, loam, sandy clay, loamy clay, medium sand, loamy sand and clay sand, have low primary or secondary or both silting tendencies. Even in soils with silting tendency where water velocity can be maintained at the minimum recommended velocity of 1.4 ft/s (0.42 m/s) to avoid fine sand or silt to accumulate in the system (self-cleaning velocity), the filtering envelope may not be required (http://www.wq.illinois.edu/DG/subsurface.htm). In flat terrain, it may often be difficult to lay drainage systems where velocity of flows would equal or exceed the self-cleaning velocity.

Stuyt *et al.* (2000) indicated the need of drain envelopes in Dutch soils, wherein it emerged that except for the ripened clay soils, the other group's unripen clayey soil, ripened loamy soil, unripened loamy soil, clayey peat, silty sand, very fine sand and moderately fine sand require or the envelope, if provided, can play a better role.

PARAMETERS TO ASSESS THE NEED OF AN ENVELOPE

Clay Percentage

Soil textural analysis of the soil at the drain depth may be used to assess the need of a filter. Experience in Egypt and the Netherlands revealed that soils having > 40% clay, may not require filter but an envelope may be needed to reduce entrance resistance. Aanalysis from four sites in India revealed that clay percentage alone may not be a good criterion to assess the need of an envelope. Other parameters including SAR of the soils need to be taken into account while using this parameter. If SAR >8, even soils with clay content more than 40% may require envelope as dispersion of clay aggregates may result in sedimentation.

Table 5.8 Silting tendency for various classes of soils

Soil type	Primary silting	Secondary silting
No primary and secondary silting		
Coarse sand	-	-
Clay	-	-
Low decayed peat	-	-
Medium decayed peat	-	-
High primary and secondary silting		
Silt sand	+++	+++
Silt	+++	+++
Sandy silt	+++	+++
Medium primary and secondary silting		
Loamy silt	++	++
Fine sand	++	++

- No, ++ Medium, +++ Strong

Plasticity Index (PI)

The plasticity of a soil is defined as the difference in water content expressed as a percentage of the weight of oven-dried soil at liquid limit and at plastic limit.

$$PI = O_{LL} - O_{PL} \tag{5.7}$$

Here, PI is the plasticity index, O_{LL} is the water content of the soil at the liquid limit (%) and O_{PL} is the water content of the soil at the plastic limit (%).

It is an indicator of the stability of the soil and is also used to determine hydraulic failure gradients (Lagacé, 1983). Dieleman & Trafford (1976) recommend the following guidelines for decision making:

$PI < 6$: high tendency to siltation

$6 \leq PI \leq 12$: limited tendency to siltation

$PI > 12$: no tendency to siltation.

Limited data for Indian condition reveal that it may a good indicator to assess the need of a filter.

Coefficient of Uniformity and Particle Size Curve

The sieve analysis as described in Chapter 3 is used to find out the coefficient of uniformity and draw the particle size curves. Uniformity coefficient is determined by using the following equation:

$$Cu = d_{60}/d_{10} \tag{5.8}$$

Such that d_{60} is 40% retained and d_{10} is the 90% retained on the curve. If CU is more than 15, the soil may not require envelope against siltation but check is needed whether an envelope is needed to reduce entrance resistance? The particle size curve is also used in granular envelope design. The data in Fig. 5.10, for a medium textured soil reveal that (i) particle size analysis curve failed to meet the criterion of no – envelope and (ii) the particle size analysis curve crossed both the boundaries when compared with no-problem soil boundaries. Clearly the soil curve is designated as a problem soil likely to result in sedimentation in drains.

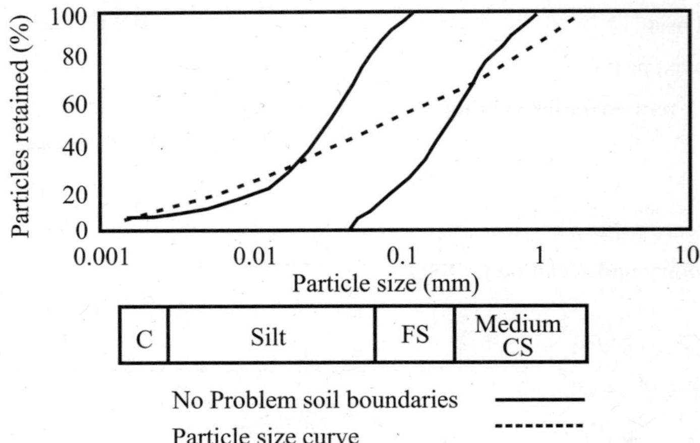

Fig. 5.10 The soil sieve analysis results of soil at drain depth (Dotted line) and no-problem soil boundaries (Solid lines)

Once these parameters are in hand, flow chart given in Fig. 5.11 can be used to assess the potential of soil clogging or need of the filter or the envelope.

Besides what is discussed in the previous paragraphs, one needs to determine iron, calcium carbonate, sulphur and manganese content in the soil and water. That should provide an insight into the chemical clogging potential.

ENTRANCE HEAD LOSS AND RESISTANCE

So far we have dealt with an ideal drain offering no resistance to drain flow. In practice, however the water enters only through the perforations; drain pipes offering considerable resistance to water flow during the entry of water in the drain pipes. Thus, an additional head i.e. an entry head (e) is introduced besides the vertical (v), horizontal (h), radial (r) heads describe earlier in Chapter 4. Thus, the total head, h_t is written as (Fig. 5.12)

$$h_t = h_v + h_h + h_r + h_e \tag{5.9}$$

The envelope provides an easy way to minimize h_e. As such, h_e and associated parameters described in this section are used to assess the performance of drain line or the envelope. The simplest way is to measure h_e/h_t ratio to assess the performance. For this purpose, a network of observation wells is installed at the top of drain pipe, adjacent to the drain (20-40 cm away from drain) and middle of the drain (See Chapter 6). The depth of the water heads below the surface is determined in observation wells (pipes). While h_t is measured at midway between the two adjacent drains, h_e is determined as the difference between the water level in the observation well closest to the drain (at envelope-soil inter-face, or just outside the trench boundary) and in the drainpipe, m (Fig. 5.12). Many times, latter is taken as half the radius of the drain to avoid measurement on the drain, which poses several problems. The envelope performance is rated good, moderate, poor and very poor if h_e/h_t is < 0.20, 0.20-0.40, 0.40-0.60 and > 0.60 respectively (Dieleman & Trafford, 1976, Stuyt *et al.*, 2005).

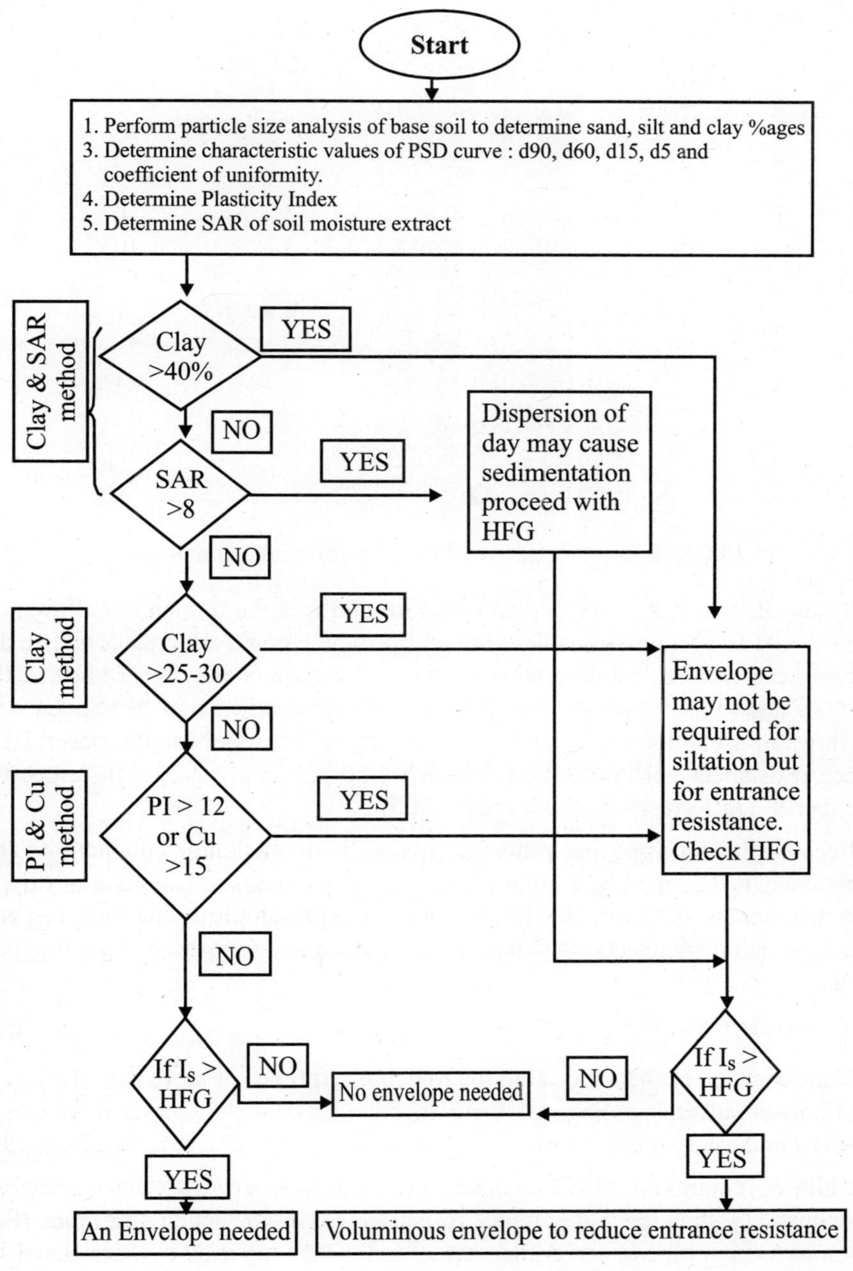

Fig. 5.11 Flowchart to systematically arrive at the need of an envelope

Approach Flow Resistance

Approach flow resistance is also used to test the efficacy of envelope materials used in subsurface drainage. The height of the water head above drain level along with drain discharge is used to calculate the head loss per unit rate of flow caused due to the entrance using the following formula.

$$h_e = q\, R_e \tag{5.10}$$

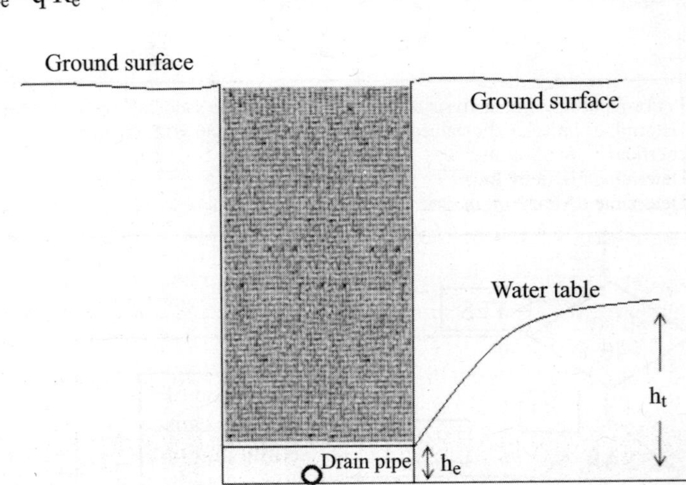

Fig. 5.12 Entrance and total head loss around a drain pipe

Where R_e is the entrance resistance (d/m) and q is the drainage outflow per m length (m^2/d). The value of q is determined from the Hooghoudt equation. The drain performance is rated as good, moderate, poor and very poor if h_e is <0.15, 0.15-0.30, 0.30- 0.45 and > 0.45 respectively. For arid and semi-arid irrigated regions where drain flows of around 4 mm/d for a drain spacing of 50 m are usually observed, the entrance resistance will be <0.75, 0.75-1.50, 1.50-2.25 and >2.25 d/m respectively (Dieleman & Trafford, 1976, Stuyt *et al.*, 2005).

Because the drain spacing and hence implicitly the hydraulic conductivity (*K*) is still included in R_e, therefore comparison between drainage systems and different design features is difficult. To resolve this, a dimensionless entrance resistance factor, r_e, which depends upon the hydraulic conductivity *K* of the surrounding bed is given as:

$$r_e = R_e * K \tag{5.11}$$

If the *K*-value of the material in the trench is ten times or more than the K-value of the undisturbed surrounding soil, the extra head loss due to entrance resistance can be ignored in drain spacing calculations (Smedema & Rycroft, 1983, Ritzema, 1994).

Child & Youngs in 1958 transformed the entrance resistance into an effective drain radius, R_{eff} showing that entrance resistance causes the actual drain pipe (R_o) to function as a drain pipe of lower than actual radius (As reported by Dierickx, 1987). It is calculated by using the following equation:

$$R_{eff} = R_o\, e^{(-2\pi re)} \exp(-2\,\pi\, r_e) \quad (5.12)$$

It can be seen that a 50 mm diameter pipe will perform as a 14.2 mm diameter pipe if r_e is 0.2 but will actually function as a pipe of 44.1 mm diameter, if r_e is 0.02.

CRITERIA FOR GRANULAR ENVELOPE

The USBR Drainage Manual (1978) describes the characteristics of an envelope material as follows:

The envelope should be well graded, free of vegetable matter, clays and other deleterious substances, which could, in time, change the hydraulic conductivity of the envelope. In sieve analysis of the envelope material, 100% should pass the 1.5 inch clear, square screen openings and not more than 5% should pass the No. 50 US standard series sieve. Since few pit run sands and gravels meet these requirements, most envelope materials must be machine sorted.

An envelope material is well graded when all particle sizes from the largest to the smallest are present. To determine whether a material is well graded, coefficients describing the slope and shape of the gradation curve such as uniformity coefficient (Eq. 5.8) and coefficient of curvature are used. Coefficient of curvature (Cc) is defined as follows:

$$\text{Coefficient of curvature} = Cc = (D_{30})^2 / (D_{10} * D_{60}) \quad (5.13)$$

Here D_{10}, D_{30} and D_{60} are the diameters in mm passing the 10, 30 and 60% on the envelope material gradation curve.

To be well graded, the coefficient of uniformity must be greater than 4 for gravels and greater than 6 for sands and in addition, the coefficient of curvature must be between 1 and 3 for both gravels and sand.

In some locations, gradation limits become uneconomical because the majority of pit run material from available sources passes the No. 30 sieve. For such locations, the material passing the No. 200 sieve should be removed. Moreover, a hydraulic conductivity test should be conducted on the remaining sample.

Design

USBR and Soil Conservation Service, USA, have developed envelope design criterion that is being adopted world over (Table 5.9). The criteria for design depend on the base particle size distribution curve. As such, the first step in selecting suitable gravel filter is to determine gradation curves for base material and filter material. On the basis of the design criteria, limits established and summarized as follows should be met by a filter material for a specific base material (USBR, 1978). Commonly, a graded envelope (Natural) is designed as given in Row 2 of Table 5.9:

$$D_{50}/d_{50} = 12\text{-}58$$

$$D_{15}/d_{15} = 12\text{-}48$$

It means that the envelope D_{50} should be in the range of 12 to 58 of the d_{50} of base material and envelope D_{15} in the range of 12 to 48 of the d_{15} of base material. Besides, the general criteria listed in the last row of the Table 5.9 should be met with, which is common to all the envelopes. For other kinds of gravel envelope, follow guidelines given in Table 5.9.

Table 5.9 USBR criteria for envelope design

Uniform envelope (natural)	D_{50}/d_{50} = 5-10	
Graded envelope (natural)	D_{50}/d_{50} = 12-58 D_{15}/d_{15} = 12-48	
Graded envelope (crushed rock)	D_{50}/d_{50} = 9-30 D_{15}/d_{15} = 6-18	
General	$D_{100} \leq 80$ mm	To minimize segregation and bridging during placement
	$D_5 \geq 0.07$ mm	To prevent movement of fines
	$D_{opening} \leq 0.5\ D_{85}$	Opening of drain perforation to be adjusted to filter material used

US SCS (1988) criteria for filter and surround design gives the range of filter/envelope to that of the base material as follows.

SCS criteria for filter gradation	$D_{15} < 7\ d_{85}$ $D_{15} > 4\ d_{15}$ $D_5 > 0.0074$ mm	but need not be smaller than 0.6 mm % passing sieve No. 200 less than 5%
SCS criteria for envelope (surround)	$D_{100} < 38.1$ mm $D_{30} > 0.25$ mm $D_5 > 0.074$ mm Minimum envelope thickness 76 mm	The whole sample should pass the sieve of 38 mm % passing sieve No. 60 less than 30% % passing sieve No. 200 less than 5%

The design of gravel envelope as per SCS criterion (USBR, 1978) is independent of the base soil curve and utilizes only the d_{60} of the base soil to select the filter material. If the d_{60} is 0.04 mm, select the row in which this value in column 1 of Table 5.10 falls. It falls in the first row. Applying the guidelines of Table 5.10, the minimum and maximum values are chosen for various D_m values to design the filters given as follows.

Diameter (% passing)	**Particle size (mm)**	
	Minimum	**Maximum**
D_{100}	9.52	38.1
D_{60}	2.0	10.0
D_{30}	0.81	8.7
D_{10}	0.33	2.5
D_5	0.3	-
D_0	0.074	0.59

Table 5.10 USBR criteria for surround design (USBR, 1978)

Base soil for d_{60} (mm)	Lower limits (mm)						Upper limits (mm)					
	Percentage passing						Percentage passing					
	100	60	30	10	5	0	100	60	30	10	5	0
0.020-0.050	9.5	2.0	0.81	0.33	0.3	0.07	38.1	10.0	8.7	2.5	-	0.59
0.050-0.100	9.5	3.0	1.07	0.38	0.3	0.07	38.1	12.0	10.0	3.0	-	0.59
0.100-0.250	9.5	4.0	1.30	0.40	0.3	0.07	38.1	15.0	13.1	3.8	-	0.59
0.250-1.000	9.5	5.0	1.45	0.42	0.3	0.07	38.1	20.0	17.3	5.0	-	0.59

Thickness of envelope

Bureau of Reclamation, Soil Conservation Service, USA and Edward E. Johnson Co. all recommend a minimum thickness of gravel envelope as 8 cm. No maximum values are suggested except by Edward E. Johnson Co., who suggests a value of 20 cm. BIS standard lays down a minimum thickness of 10 cm gravel around the tile/pipe. For other materials, the thickness requirements are given in Table 5.11.

Table 5.11 Thickness requirement of various envelope materials

Type of material	Thickness (cm)
Organic (wrapping on pipes)	7 cm
Organic (loose such as saw dust, wood chips)	15 cm
Fibre glass type materials*	2.5 cm

*For borosilicate, it equals 2.6 kg/100 m length of pipe

Amount of Gravel Required

The volume of gravel per unit length of drainage pipe is equal to the gross cross-sectional area of the envelope minus the area of the pipe. For a square form of envelope around the pipe, it can be calculated using the following equation:

$$\text{Volume} = (OD/1000 + T/100)^2 - \pi/4\,(OD/1000)^2 \text{ m}^3\text{/m of pipe} \tag{5.14}$$

Where OD is the outside diameter of pipe, mm and T is the thickness of the gravel envelope, cm. This volume is known as neat line estimate and does not include spillage, gravel required for use in difficulties to control the required thickness and stockpile losses. It is assessed that these factors will increase the overall gravel volume by 25%. Experience at Mardan SCARP in Pakistan revealed that gravel requirement was nearly 45% higher than the neat line requirement. It is necessary to assess such increases to calculate the exact requirement of the gravel.

Shortcomings of Gravel Envelopes

When using gravel envelopes, following problems are generally encountered:

- Lack of uniform quality and gradation of gravel
- Segregation during transport and installation
- Difficulties in mechanization
- Unequal distribution around the pipes

DESIGN AND SELECTION OF SYNTHETIC ENVELOPE

Many of the shortcomings of the gravel envelope can be overcome with the use of synthetic envelopes. The design criteria for synthetic material are different than granular material. While a granular envelope is designed, a synthetic envelope is selected on the basis of properties given by the manufacturer. Since, a wide variety of synthetic envelopes is available and new envelopes are being released, it is quite difficult to have a sound envelope selection criterion. Summary of some desirable properties of synthetic envelopes are given in Table 5.12.

Table 5.12 Summary of desirable properties of synthetic materials for drain envelopes

S.N	Parameters	Specifications
1	Material	Polypropylene (PP) polymers are preferred, but nylon (polyamide, PA) and polyester (PET) have been used successfully
2	Permeability	>50 L/m^2 at 2 kN/m^2 > 15 L/m^2 at 200 kN/m^2 >100 m/d at 2 kN/m^2 (Typical values for transmissivity are 0.01- 2.0 x 10^{-3} m^3/min-m) > 10 m/d at 200 kN/m^2
3	Characteristic opening size (COS)	Wet sieving recommended for assessing O_{90}. When other methods are used COS should be converted to standard opening size O_{90}
4	Thickness	Minimum 1 mm preferred, but depends on minimum thickness requirements and standards
5	Mass per unit area	> 200 g/m^2. Commonly available 130-700 g/m^2
6	Tensile strength	> 6 kN/m (wide width tensile strength)
7	Tensile strength of joints	> 6 kN/m (seam strength usually 50- 100% of tensile strength)
8	Sewn-seam strength	> 310- 360 N/m width (for Class B, AASHTO M288-90 which class seems most appropriate for agricultural drains).
9	Trapezoid tearing strength	>110 N (for Class B, AASHTO M288-90)
10	Static puncture strength	>110- 160 N (for Class B, AASHTO M288-90)
11	Grab strength	>360- 400 N (for Class B, AASHTO M288-90)
12	Joint overlap	> 150 mm joining two pipe sections, joined by tape or sewing > 20 mm when wrapped around pipe, joined by sewing or overlap. When pre- wrapped with netting similar to PLM

Measurement of Apparent Opening Size (O_{90})

The dry sieving is a speedy method applicable to most geotextile having a characteristic pore size greater than 100 μm. The pore size distribution is determined by sieving dry spherical glass beads or sand fractions in a commercially available sieving equipment in which geotextile is used as a sieve (Dierickx, 1999). The sieving is carried out at a vibration frequency of 50 Hz with vertical amplitude of 0.75 mm (Fig. 5.13 left). To prepare the sand fractions, following procedure is used:

- Dry the sand at 100ºC – 110ºC for at least 1 h, allow it to cool
- Install the sieves on the pan so that the mesh size increases upwards
- Put the sand on the top sieve, install and screw down the cover clamp to fasten the sieves
- Adjust the amplitude and sieve for 30 min.
- Take off the sieves, collect the sand and store in plastic containers
- Clean both sides of each sieve with the nylon brush so that all sand particles are removed

Following procedure is then used to determine characteristic opening size.

- Put the fastening ring with the cutter diameter on the geotextile sample and cut out a circle of geotextile that matches this diameter.
- Put the lower ring on the pan, then the supporting screen and the geotextile specimen over the screws, and finally the upper ring over the specimen. Fasten the bolts.
- Prepare 50 g of the sand fraction of which it is assumed that 50% will be retained by the geotextile.
- Place the pan and the rings on the vibration plate, fix them with the screws and a wooden strip and adjust the amplitude of vibration on 0.75 mm.
- Pour 50 g of sand in the centre of the ring on the geotextile and let the sieve run for precisely 5.0 min.
- Take off the pan and rings from the sieving plate, collect the sand which is on top of the geotextile and weigh on the analytic balance.
- Record the mass for the particular sand fraction. Put the sand back in its container. Weigh sand which passes into the pan. This sand can also be placed back in same container
- Repeat the sieving procedure with the larger and smaller fractions until 99.0% of a fraction is retained on the filter or until 99.0% of a fraction passes through the geotextile.

The mass of particles which have passed through the test specimen and which are caught in the receiving pan M_1 is determined. The mass of the fraction that is retained on and within the specimen, M_2 is given by:

$$M_2 = 50 - M_1 \tag{5.15}$$

For each size designation of particles used, the mass of particles retained on and within the specimen (M_2) is calculated as a percentage of the total mass of sand sample used (50 g). The percentage of particles retained on the surface and within the test specimen is plotted against the mean size of the particles for each grade on semi-logarithmic paper or on a normal arithmetic paper and a smooth curve is drawn through the points as in the case of sieve analysis of base soil or gravel envelopes. From this curve O_{90} value for which 90% of the particles are retained on and within the geotextile is determined. For 9 samples of an Indian geotextile O_{90} varied from 200 to 460 showing variations within the samples of the same manufacturer. It calls for greater care in the use of synthetic envelopes.

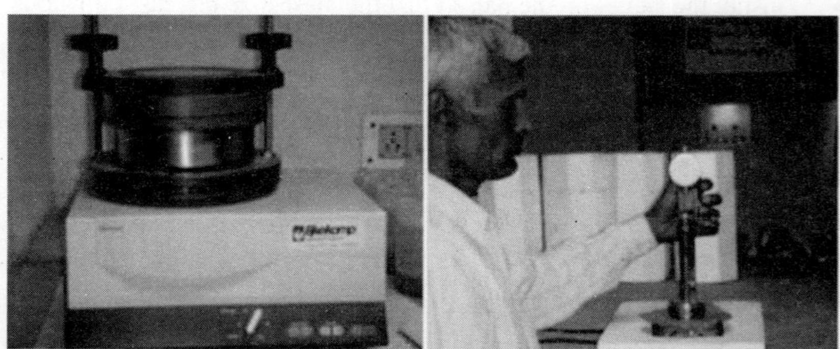

Fig. 5.13 Sieve analysis to assess O_{90} (left) and measurement of envelope thickness (left)

Retention Criteria

The opening pore size of an envelope is expressed as the O_{90} value, which is defined `*as the average diameter of the soil particles, 90% of which is retained by the envelope in a standardized sieve test*`. It is difficult to set a lower limit for detrimental soil invasion into geotextile, but envelopes for which $O_{90}/d_{90} < 1$ are considered inappropriate (Faure, 1991) and are likely to clog (d_{90} – Sieve mesh size through which 90% base soil passes). Experiments in Belgium showed that geotextiles with $O_{90}/d_{90} < 1$ rapidly clog when placed in turbid water whereas no obvious clogging was observed for materials with O_{90}/d_{90} ratio near one or more. Stuyt (1992) investigated the functioning of various geotextiles installed in experimental fields in the Netherlands and found that the O_{90}/d_{90} ratios as proposed by Dierickx (1987) were valid for the problem soils. Most of the applied envelopes had comparatively high O_{90}/d_{90} ratios (4 to 5). For coarser textured soils (d_{90} >200 μm) the ratio may be less than 1 on condition that the O_{90} of the geotextile is at least 200 μm and that the permeability of the synthetic fabric is at least equal to that of the surrounding base soil. The main criteria recommended for synthetic envelope designs are:

- $< O_{90}/d_{90} < 2.5$ for envelope thickness < 1 mm (filter function)
- $< O_{90}/d_{90} < 5$ for envelope thickness < 5 mm (filter function)
- $1 < O_{90}/d_{90} < 3$ for 1-3 mm envelope thickness
- $1 < O_{90}/d_{90} < 4$ for 3-5 mm envelope thickness

Guidelines for design of non-woven synthetic envelopes are given in Table 5.13 (Vlotman,1998). In general, a geotextile sock is recommended for coarse textured soils free of silt and clay. These soils are considered unstable even if undisturbed, so that particles may wash into pipes. The need for an envelope in intermediate soils (clay contents less than 25 to 30%) is best left to a professional contractor or soil and water engineer because soil movement is more difficult to predict.

Table 5.13 Recommended design criteria for synthetic drain envelopes (geotextiles)

Geotextile	Flow type	Criteria	Remarks
Thin ($T_g \leq 1$mm)	-	$O_{90} / d_{90} \leq 2.5$	Retention criterion
Voluminous ($T_g \geq 5$mm)	-	$O_{90} / d_{90} \leq 5$	Retention criterion
In between ($1 \leq T_g \geq 5$mm)	-	Interpolate	Retention criterion
	-	$O_{90} \geq 200$ μm	Hydraulic and anti-clogging criterion
	Dynamic and steady	$K_e > a\ K_s$	Hydraulic criterion if a= 0.1 no safety, a=1 for non-critical conditions, and a=10 for reverse flow
	-	$O_{90} / d_{90} \geq 1$	Anti-clogging criterion
	-	$O_{90} \geq$ 100-200 μm	Anti-clogging criterion

RAJAD Criterion

RAJAD (1995) proposed criteria similar to gravel envelopes to design synthetic envelopes. The criteria depicting in Fig. 5.14, can be used to determine the kind of envelope needed as follows:

- Knowing the clay content, draw a horizontal line
- Add silt content, go vertical and proceed horizontal along silt line
- Similarly, add very fine (VF), fine, medium, coarse (C) and very coarse (VC) sand contents and proceed as in the previous steps.
- Select the envelope as per guidelines given in Fig. 5.14.

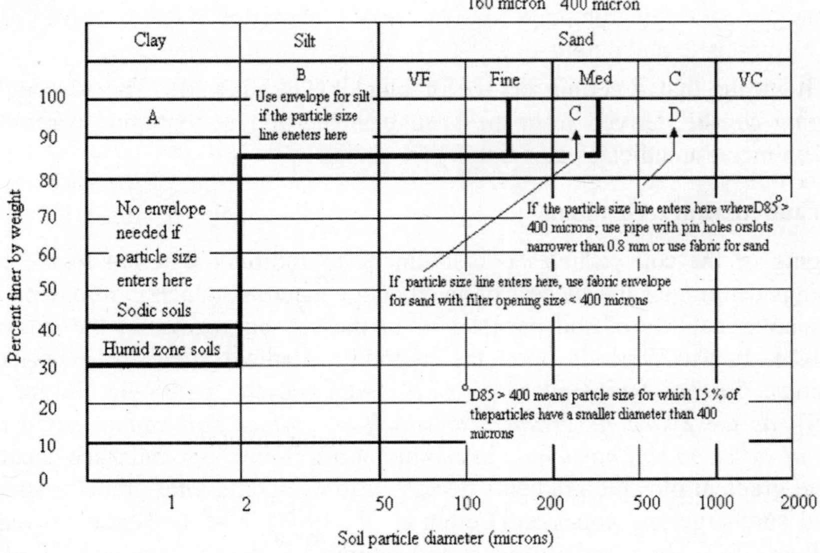

Fig. 5.14 Selection of envelope on the basis of particle size analysis of the soils

On the basis of these guidelines, HOPP (Haryana) uses synthetic envelopes that have O_{90} > 300 microns. The evaluation data reveal that all the three synthetic envelopes performed well (Table 5.14).

Table 5.14 Performance of envelope materials Gohana (Haryana)

Envelope	O_{90} (Micron)	Head loss fraction (h_e/h_t)	Entrance resistance (d/m)	Drain performance
Polypropylene	320	0.11	0.71	good
Polypropylene	366	0.14	0.73	good
Polypropylene	400	0.13	0.49	good

Hydraulic Criteria

The hydraulic characteristics of a particular synthetic envelope material depend on the characteristic opening size, percent open area, or porosity, compressibility, water permeability normal to the plane with and without load, water flow characteristics in the plane with and without load and water penetration resistance. To analyze the clogging potential of the systems, the gradient ratio (GR) is used as per the following equations.

$$GR = i_{es} / i_s \tag{5.16}$$

Here GR is the gradient ratio, i_{es} is the hydraulic gradient of the soil-envelope system, i_s is the soil hydraulic gradient. Severe blocking of the envelope results in immediate increase in hydraulic gradient. Some blocking of the envelop may occur but as long as following condition is met, it may not cause much problem.

$$i_{es} < i_s \quad \text{or} \quad i_{es}/i_s < 1 \tag{5.17}$$

The US Army Corps of Engineering (1977) however, accepted the condition:

$$i_{es}/i_s < 3 \tag{5.18}$$

It means that a certain degree of blocking is allowed. The susceptibility to clogging can be derived from the change in the above-mentioned parameter with time, an increase indicates clogging.

Hydraulic Failure Gradient

Bridging of the soil particles over drain perforations or external to the envelope openings determines the success of an envelope material, which in turn depends upon the relative stability of soil particles or of the soil aggregates at the soil-envelope interface. It also depends upon the hydraulic gradients that are expected at the openings. Samani and Willardson (1981) defined the hydraulic failure gradient (HFG) `*as the hydraulic gradient at which a confined soil cannot resist the drag force of water on soil particles*`. Instability at a soil-envelope interface occurs at this critical gradient plus the gradients necessary to overcome inter-granular stresses due to soil surcharge and cohesion (Luthin *et al.*, 1968). The total gradient required is called as HFG. The plasticity index and saturated hydraulic conductivity of the soil are used to determine the HFG for comparison with the computed exit gradient.

$$HFG = e^{0.332 - 0.132K + 1.07.\ln(PI)} \tag{5.19}$$

Here K is the hydraulic conductivity, m/d, PI is the plasticity index (Eq. 5.7), e is the base of the natural logarithm. An envelope is needed, if the exit gradient exceeds the HFG of the soil.

This method is used only if $0.02 < K < 4.5$ m/d. Once the HFG has been determined, flowchart given in Fig. 5.11 is used to assess the need of an envelope.

TESTS ON SYNTHETIC ENVELOPE MATERIALS

Non-woven Envelope Materials

Based on the guidelines to select a drainage envelope, tests to be conducted on envelope materials are listed in Table 5.15.

Table 5.15 Test procedure and guidelines for synthetic envelopes

Test	What is tested?	Guidelines
Visual appearance and strip width	Width should be measured under zero tension	Should be regular and no open spots; Width should provide a minimum overlap of 25%.
Thickness	Measured to an accuracy of at least 0.02 mm at a pressure of 2 kPa	Average thickness in no case should be less than 3 mm
Mass per unit area	Mass per unit area should be measured to the nearest 0.01% of the total specimen mass (Fig. 5.13, right)	The mass per unit area should be 300 g/m^2 or more
Tensile strength of material and joints	When tension is applied during construction, the fabric must be able to withstand without any tear or without affecting the COS to a greater extent	Tear strength of at least 200 N
Static puncture test	The resistance of the geotextile envelope against puncture by rocks, gravel, and other sharp objects	Pushing a flat plunger through the geotextile and the force applied to puncture the material is measured

ORGANIC ENVELOPES

Design specifications of organic envelopes are based on similar principles as applicable to synthetic envelopes. The life of these envelopes however, is limited and depends upon microbiological activities around the envelope. The microbial activity in turn is governed by temperature, soil chemical properties and presence of oxygen. Rate of decomposition is slow under reducing conditions. In general, the following guidelines apply in the use of organic envelopes.

- The organic envelopes should be avoided in arid and semiarid regions.
- If the performance of an organic material is doubtful, don't go for it and instead use gravel or synthetic envelopes,

QUESTIONS (THEORY)

1. Describe the importance of drainage materials in subsurface drainage. Name and describe some of the commonly used pipe materials. Also name at least five accessories used to assemble the system.
2. Describe the general characteristics of clay tiles.
3. Describe the general characteristics of cement concrete tiles.
4. Differentiate between the various characteristics of polyvinyl chloride and polyethylene pipes.
5. Describe the various kinds of load encountered by drain pipes. Differentiate between ditch conduit and projecting conduit conditions to determine the load on pipes.
6. Differentiate between the static load calculation on tiles and flexible pipes.
7. Name and describe three tests required to test the quality of plastic drain pipes.
8. Discuss the need of providing envelope in subsurface drainage. Differentiate between its role as a filter and an envelope.
9. What do you understand by entrance head loss. How is entrance head used to assess the functioning of drains?
10. Describe the design criterion of gravel envelopes in subsurface drainage.
11. Write short notes on: the shortcomings of gravel envelopes, characteristic opening size of synthetic filters and hydraulic failure gradient.

QUESTIONS (OBJECTIVE)

1. Fill in the blanks

1. HDPE stands for high _____________ polyethylene.
2. Water absorption should not exceed ______% by weight of an individual tile.
3. Smaller diameter corrugated pipes come in the form of_________ .
4. PVC pipes deteriorate when stored in the open because of exposure to __________ radiation from the sun.
5. On per unit weight basis, corrugated pipes have __________ stiffness than plain pipes.
6. The Manning roughness coefficient is ________ for corrugated than plain pipe of the same size and similar perforation pattern.
7. Internal diameter of corrugate pipes is nearly _______times the outer diameter.

8. _________test is conducted to know the quality of extruded PVC.
9. Passage of equipment or vehicles exert _________ load on pipes.
10. Ditch conduit conditions apply when pipes are laid in ___________trenches.
11. Bedding conditions of pipes affect the ______ carrying capacity of the pipes.
12. Burlap (jute cloth) is a kind of___________ envelope.
13. _____________tendency of the soil indicates the need of an envelope.
14. Plasticity index is given by the difference in water content of the soil at the _______ limit (%) and the water content of the soil at the plastic limit (%).
15. The entrance resistance has the unit of_________.
16. Dimensionless entrance resistance is obtained by multiplying the entrance resistance with _________________.
17. Apparent opening size of synthetic envelopes is given by _____.
18. ______ value is the value of pore size at which 90% of the particles are retained on and within the geotextile.
19. HFG stands for hydraulic __________ gradient.
20. Hydraulic failure gradient besides other parameters depends upon _________ index.

2. Write down the correct or most appropriate answer for each question in the box provided

1. Quality of polyvinyl chloride is reduced due to storage

   (A) In the open under shade (B) In the open under sunlight

   (C) In a dark store room (D) None of these

2. Degradation due to ultraviolet rays in polyethylene pipes can be avoided

   (A) By adding carbon black (B) By increasing the wall thickness

   (C) By adding red colour (D) None of these

3. Polyvinyl chloride pipes are affected by

   (A) Exposure to ultraviolet rays

   (B) Becomes brittle at freezing temperatures

   (C) Both A and B

   (D) None of these

4. The hydraulic roughness coefficient of corrugated pipes is

   (A) Higher than smooth pipes (B) Lower than smooth pipes

   (C) Equal to smooth pipes (D) None of these

5. Static load on the pipes is due to

   (A) Overburden soil (B) Passage of equipment or vehicles

   (C) Both A and B (D) None of these

6. In the ditch conduit situation, total load on conduit per unit length is directly proportional to

   (A) Square of the outside diameter of the conduit

   (B) Square of the width of ditch at the bottom of the conduit

   (C) Both A and B

   (D) None of these

7. In the projecting conduit situation, total load on conduit per unit length is directly proportional to

   (A) Square of the outside diameter of the conduit

   (B) Square of the width of ditch at the bottom of the conduit

   (C) Both A and B

   (D) None of these

8. In the case of flexible piping, total load on conduit per unit length is directly proportional to

   (A) Outside diameter of the conduit

   (B) Width of ditch at the bottom of the conduit

   (C) Both A and B

   (D) None of these

9. The functions of an envelope

   (A) Filter Function

   (B) Hydraulic Function

   (C) Mechanical/Bedding Function

   (D) All of these

10. Which is not synthetic filter material

    (A) Glass wool

    (B) Fiber glass

    (C) Nylon socks

    (D) Naturally graded coarse sand

11. Plasticity Index (PI) of soil is defined as

    (A) Water content at liquid limit

    (B) Water content at plastic limit

    (C) Difference in water content at liquid limit and plastic limit

    (D) Sum of water content at liquid limit and plastic limit

12. Lesser the entrance head to total head ratio
    (A) Better the drain performance
    (B) Lower the drain performance
    (C) The ratio has no effect on drain performance
    (D) None of these
13. Entrance resistance has the units of (d and m are day and meter)
    (A) d/m (B) Dimensionless
    (C) m/d (D) m.d
14. Effective radius is always_______ than the original drain radius in the presence of entrance resistance
    (A) Higher (B) Lesser
    (C) Equal (D) None of these
15. Coefficient of curvature is calculated by using
    (A) D_{10}, the grain diameter at 10% passing
    (B) D_{30}, the grain diameter at 30% passing
    (C) D_{60}, the grain diameter at 60% passing
    (D) All of these
16. Coefficient of curvature is directly proportional to the square of
    (A) D_{10}, the grain diameter at 10% passing
    (B) D_{30}, the grain diameter at 30% passing
    (C) D_{60}, the grain diameter at 60% passing
    (D) None of these
17. To design a graded envelope with natural gravel as per USBR criteria (here D stands for envelope and d for base material indicating the diameter of 15 and 50% passing)
    (A) D_{50}/d_{50} = 12-58 (B) D_{15}/d_{15} = 12-48
    (C) Both A and B (D) None of these
18. Apparent opening size is one of the criterion in the selection of synthetic envelopes, it stands for
    (A) Average diameter of the soil particles, 90% of which is retained by the envelope in a standardized sieve test
    (B) Average diameter of the soil particles, 90% of which is passed from the envelope in a standardized sieve test
    (C) Average diameter of the soil particles, 60% of which is retained by the envelope in a standardized sieve test
    (D) Average diameter of the soil particles, 60% of which is passed from the envelope in a standardized sieve test

19. Hydraulic failure gradient depends upon

    (A) Hydraulic conductivity (B) Plasticity index

    (C) Both A and B (D) None of these

20. A layer of 10 mm of sediment in a drain pipe of 80 mm diameter weighs around

    (A) 231 g/m of pipe (B) 462 g/m of pipe

    (C) 693 g/m of pipe (D) None of these

21. Openings in a corrugated PVC pipe are made

    (A) On the top of corrugations

    (B) In the valley between two corrugations.

    (C) Both on the top and the valley

    (D) There is no hard and fast rule

22. Which one of the methods can be used to measure the soil load on the tiles

    (A) Ditch conduit formula

    (B) Projecting conduit formula

    (C) Any of A and B depending upon the conditions

    (D) None of A and B

23. If condition of drains is unknown and loads are calculated both for ditch and projecting conduit formula, then the actual loading situation is

    (A) In which load is less (B) In which load is more

    (C) Any of A and B (D) None of these

■■■■

Chapter 6

CONSTRUCTION AND EVALUATION OF DRAINAGE SYSTEMS

INTRODUCTION

Quality of drainage systems depends upon its design, layout, drainage materials used and its construction. Amongst all the factors, construction of drainage is most paramount in deciding the overall quality of system. Therefore, to construct an effective and efficient system, appropriate construction methods and guidelines need to be adopted. In the past few decades, tremendous progress has been made and construction of field drainage has evolved from a purely manual to a highly automated operation. In India, where subsurface drainage is still in its infancy, development of rapid and low cost techniques is a challenge to the field engineers. Unlike open drains, tile drainage system is an assembly task which can be accomplished through a variety of methods. Besides the pipelines, a number of drainage structures are constructed for the efficient operation and maintenance of the systems. The construction of these structures should also meet the hydraulic and structural design criteria. The aim is to construct a nearly ideal technically sound system at an economic cost that provides enough flexibility in the selection of installation techniques. Most of the times, a good blend of mechanized and manual techniques may be desirable over complete mechanization to use large skilled and unskilled manpower. Moreover, a number of assumptions are made during design and some compromise may be made during selection of drainage materials. Even during construction, some deviations from the ideal solutions and guidelines are imminent. Therefore, monitoring and evaluation of the finally constructed system assumes significance. This chapter describes the general principles of construction and assembly of the systems, important structures required to construct an efficient subsurface drainage system and post-construction monitoring and evaluation procedures.

PRE-INSTALLATION ACTIVITIES

Plans and Records

Ideally, all construction should follow a definite plan prepared in advance by the designer. The base map specifies the layout, location and size of the drainage system. Profile sheets are prepared specifying drain elevations and space. Additional sheets are used showing drain connections, sump, details of appurtenant structures and pump etc. Basic control points are established in the project area, approximately 1 point in 100 ha. Once the layout plan is ready as per the design criteria and basic control points are established, layout of laterals, collectors and related structures is transferred on the ground. All survey work for control should be within desired tolerances, horizontal tolerance being within 1 m in 5000 m and vertical distance (elevation) in mm should be plus or minus 25 multiplied by the square root of the distance in km.

Inspection of Materials

All construction materials should be inspected before and during installation. All materials should be as per the quality for the intended use and should meet the prescribed standards discussed in the previous chapter. Reject any defective or damaged material as otherwise functioning of whole system will be jeopardized.

Storage of Materials

Drainage materials should be protected from damage during storage and handling. The storage area should be dry, well drained, and free of rodents, vegetation, and fire hazards especially for plastic tubing. Since pipes can be harmed by excessive exposure to ultraviolet rays, protect it from sunlight when it is to be stored outside for a long period.

Tree removal

All trees within 18 m of the drain line should be removed. Tree stumps at least up to 0.5 m below ground should be removed. Trees along the collector may not pose major problems if the collectors are used only to transport the water i.e. blind collectors are used.

Safety

Make all arrangements to implement safety standards for persons and machines. Persons working in trenches should be protected from cave-ins (especially when trenches exceed 1.5 m depth). All moving parts of machinery should be protected by proper guards. Persons inspecting the work should not be permitted to come close to the excavating machines.

CONSTRUCTION METHODS

Since the land drainage began, tremendous achievements have been made in the construction technology. Commonly, three methods, namely, manual, mechanical and semi-mechanical methods are used to construct drainage systems in India (Table 6.1).

MANUAL CONSTRUCTION

In principle, there is not much difference in the sequence when laying the system either manually or mechanically. Principally, the construction procedures are quite easier in the later as level control is through the laser beam. Construction sequence comprises the following steps:

- Construction of disposal drain, if needed
- Installation of sump and pump, if pumped outlet

- Laying the collector pipes
- Laying the lateral drains. On small fields, first and the last lateral are located one-half to one-third spacing from the edge of the field. Other drains are then located by dividing the field width by the spacing.
- Blinding/Backfilling and first wetting
- Dressing of disturbed earth/back filled material
- Construction of pump house
- Handing over to the stakeholders

Table 6.1 Methods and approximate rate of drainage installation

Method	Machinery used	Rate of installation	Requirement
Manual	No	20-50 m/d	No problems
Mechanical construction	Trenchers or trench less machines	0.2 km/h - 5 km/h	See Table 6.2 in the forthcoming section
Semi-mechanical construction	backhoes and excavators	30 - 45 m/h	Excavators are available with different bucket width normally ranging from 45 – 90 cm.

Trench Digging

Trenches are dug manually or with the help of backhoes, excavators or trenching machines. General guidelines are:

- Digging should begin from the outlet end. If there is rain or if water accumulates in the drains, it will be easier to empty the drains.
- The digging operation should begin when soils are reasonably dry down to drain depth. Performance of the system is affected if slurry is formed while working under wet conditions.
- The width of the trench should be kept as minimum as possible but should be enough so that no obstructions are encountered while installing the pipes at the desired drain depth. The bottom width should be the outside diameter of the drain pipe plus 130 mm.
- Minimum length of the tile line is twice the spacing. Maximum length should not exceed 300 m if conventional flushing methods are likely to be used to clean the drains. If the length exceeds 300 m, a manhole should be provided at every 300 m.
- Depth of the trench should be slightly more than the depth at which drain is to be placed when gravel or organic envelopes are used.
- As far as possible, maintain the desired grade during the digging operation itself. If a trench is accidentally excavated below grade, the section should be backfilled, tamped and shaped to grade.

- For slope control, line and gauge method is commonly used (Fig. 6.1). Grade stakes are set with Engineers′ level so that tops of the stakes are at the desired depth above the bottom of the ditch. A string stretched over the top of the stakes represents the grade. A L-shaped gauge in upside position is used to decide the bottom depth of the trench. The grades vary from 0.05% for collectors of 455 mm OD to 0.1% for laterals of 80 and 100 mm OD. No reverse grade is permitted.
- Alignments should be in straight sections. Avoid 90° bends. Bends, if necessary, should be provided as smooth curves. Stakes to govern alignment should be provided at 30 m intervals in straight section and 15-18 m on curves.
- The bottom of the trench should be shaped so that about one-fourth of the circumference of the drain remains in contact with the soil.

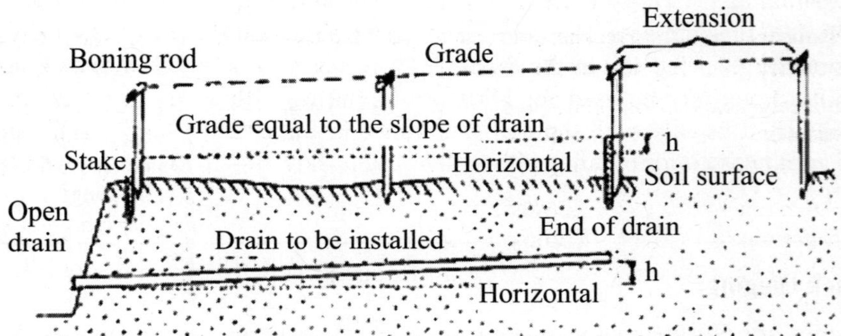

Fig. 6.1 Staking grade stakes during laying of pipe drains

Laying the Tile Line

The following stepwise procedure is used to place the tiles in the prepared trench.

- The envelope to the desired thickness is placed at the drain bottom and suitably compacted. The grade after compaction is rechecked to ensure that it is in line with the desired grade.
- Clay ware or concrete tiles without perforations are laid as close together as possible. No attempt is made to leave gap between the joints. The desired gap of 1.5-3.0 mm between the two tiles will automatically remain for water to enter. In peat and muck soils, the slightly higher gap is kept.
- Cracked tiles should be straightway rejected. In general, broken tiles should not be used. If used, the tiles with broken ends (gap > 5 mm) should be laid with the broken sides facing the bottom. All wide cracks and broken ends should be covered with broken pieces of tiles or other suitable material such as roofing paper or cement mortar. At bends, where space between the two adjacent tiles is more, should be covered with broken tile pieces.
- To avoid fine material from entering the tiles, joints or holes are covered with jute cloth pieces (burlap) or any other suitable materials.

- The ends of all tile lines are tightly closed with a plate, brick or some other suitable structure. It is also done at points where day's work ends.
- As far as possible, manufactured connections should be used.
- Although not mandatory, rechecking the grade after completing each drain line is a useful practice. A slight sag in the grade line is less harmful than a hump. Sag and a hump occurring within a distance of 30 m are not allowed.
- The envelope material to the desired thickness is placed over the tiles. While doing this, one must ensure that no displacement is caused to the drain line.

Blinding, Backfilling and First Wetting

Blinding and backfilling are two similar processes yet are carefully performed in that order. Blinding protects the tiles from displacement during the process of backfilling. In the blinding process top soil or a permeable soil having higher hydraulic conductivity than the soil at the drain depth is placed around the drains. Fine sands and silts should not be used for blinding. Initial backfilling, up to 45 cm depth, is accomplished usually with top soil (blinding). In saline soils, highly saline top soil should not be used for blinding although it might show high hydraulic conductivity.

As a matter of safety, backfilling of the drain is accomplished as soon as the blinding process is over. It protects the trench banks from caving in. For the same reasons, it is a common practice to blind and backfill all drains before suspending the day's work. Wet soils are not used for backfilling. A mound of 150 mm or more from the adjacent ground level is made over the drain line to provide necessary allowance for settlement of soil. Borrow sources are used if soil is insufficient to form the mound. Take care that soils of the borrow pits are comparable to the soil excavated from the trench.

It is not a good practice to flood the drain trench so as to consolidate the backfill. It imposes a sudden high hydraulic gradient in a very loosely packed soil mass, an ideal situation for fine particles to migrate to the drain. If slurry is formed, it may adversely impact the long-term performance of the system. The consolidation of backfill under natural rainfall is most desirable. Such conditions can be created with the help of sprinkle irrigation. The other way to accomplish the same is through the use of irrigation water. In this case, a furrow or a small dyke is made on each side about 1 m away from the drain. The irrigation water is let in on both sides of these dykes or furrows so that consolidation occurs only through seepage of water. Direct access of surface water on the drain line should be avoided.

MECHANICAL INSTALLATION

Since there is an increasing trend to use mechanical installation techniques, the machines and methods used are discussed in the following sections.

Machinery

Two most common types of trenching machines are the bucket-wheel type and the bucket ladder type. The bucket-wheel type has a large wheel mounted on a frame at the rear of the machine. The wheel is moved up and down to keep the machine on grade. Attached to the wheel are excavating buckets, cutting shoe and a shield to keep the loose earth from falling into the trench. In the bucket-ladder type machine, ladder type booms around which the excavating buckets move on an endless chain are used. The excavated material is deposited alongside the trench using a conveyor. The machines are fitted with rubber tires or have tracks (Fig. 6.2). These machines are most suited for digging drains in the depth range of 1 to 3.5 m. Interchange of the digging equipment may be required as per the depth of digging. The trench width can be varied in the range of 12 to 65 cm. The corrugated pipe for field laterals is carried on the machine on a reel and is fed into the trench. Large diameter pipes for collectors are usually laid out and coupled in the field beforehand.

Fig. 6.2 An auto-control corrugated PVC pipe trenching machine (left), in operation in Maharashtra, India (middle), a tractor driven trencher (right)

The continuous trenchers have a wide range of trenching capacities. The majority of these have a series of chain driven cutters that wear very rapidly in stony soils. In such conditions, a bucket wheel trencher is used. Both clay tiles or plastics pipes can be laid using these machines. For gravel envelope, trenching machines are often equipped with two gravel chutes. The first chute places the gravel in the bottom of the trench on which the pipe is placed. The second chute places the gravel on the sides and top of the drain to the desired depth. Light bulk envelope materials are often applied by hand after the passage of the machine. Strip type filters may be fed from a roll and be either wrapped around the pipe as it passes through the machine or placed on the top of the pipe or groove in the trench (Stuyt *et al.*, 2000). Most trenchers have a keel or a V-groove that provides ordinary bedding conditions with a load factor of 1.5 or more (see section on Load on Conduits in Chapter 5). Trenchers normally require a crew of five men, comprising the operator, the assistant-operator (who also does the staking) and three labourers who supply the pipe, put in the envelope material, install the outlet pipe and backfill the trench (using a tractor with blade and trailer). With pre-wrapped plastics pipe, the number of labourers can be reduced.

Self-propelled trench less drain laying machines are categorized as vertical plow and V plow. In the former, which acts like a sub-soiler, large fissures and cracks are formed. The vertical plow is capable to install pipes of 300 mm diameter to drain depths close to 3 m.

The V-plow lifts a triangular wedge of the soil and places it back on the pipe drain after it is installed in one process. The V plow is capable to install pipes of 100 to 125 mm.

The technical capabilities of the principal types of machines are detailed in (Table 6.2). This indicates that the backacter is the slowest machine. The trenching machines have the maximum working capability although it is limited by the efficiency of the support services i.e. the availability of pipes, envelope, spare parts for machine and facilities for on-site maintenance.

The selection of a machine depends upon its technical suitability and upon the total costs involved (HOPP, 1997). The backacter is quite popular because it is technically versatile and relatively less costly. Its disadvantages are that it does not form a uniformly graded bed on which the pipe can be laid and its operation is quite slow. It is most suited for small drainage scheme in hilly areas where the limitations of the grade are overcome by adequate natural slope. The machine works well in stony soils.

Trenchless machines have the potential to work faster and more economically than trenching machines (up to 2-3 or as fast as ≅ 5 km/working day). Here also, continuity and reliability of support services is essential. Trenchless machines can be used in stony soils; a large machine having the capability to push stones up to 1 ton out of its way. Envelopes can also be installed although the inspection may be bit difficult.

Table 6.2 Principal types of drainage machinery and their characteristics

Machine type	Power requirements kW (HP)	Maximum digging depth (m)	Trench widths (m)	Working speed (km/h)
Backacter	30-50 (40-65)	up to 4.0	0.20-0.60	0.2
Continuous crawler trenchers				
Light (8-10 tonnes)	60-90 (80-120)	1.5-2.0	0.20-0.30	0.5-1.0
Medium (13-14 tonnes)	100 (150)	2.5	0.20-0.45	1.0-2.0
Heavy (16-23 tonnes)	150 (200)	3.0-3.5	0.20-0.45	Up to 3.0
Trenchless, winched	50-80 (65-110)	1.2-1.7		1.7-2.0
Trenchless, crawler	100-150 (135-200)	1.6-2.0		Up to 5.0

Source : Schwab *et al.,* (1987)

While using a drainage trencher, construction infrastructure needs to be supported by machines such as an excavator, a bulldozer, tractor and trailers, pick up van, mobile welding generator set, drain cleaning machine and a filter-wrapping machine. The filter wrapping machine, being mandatory for large-scale projects, can be designed locally. Such a machine designed and used in HOPP is capable to handle 2-2.5 km of pipe daily. The wrapping cost per meter length of the pipe is cheaper than the manual wrapping (Mohan *et al.*, 2012a,b). Moreover, a manually operated machine ensures continuous operation even under erratic rural electric supply. The cost of the machine is around Rs. 20,000/-. Similarly, a drain cleaning machine can also be locally assembled/fabricated (Anonymous, 2002).

Construction

From practical standpoint, construction sequence is same as in the case of manual construction. Following additional guidelines may be followed for the smooth construction of drainage systems in waterlogged areas:

- Ideally drains should be installed during the dry season when the ground is able to support the heavy machines such as heavy-wheeled tractors and gravel filled trailers. It may be preferable to install drains even in a cropped land (causes ~5% crop loss) as against the installation under wet conditions. The passage of heavy machinery on wet land causes the greatest damage to soil structure.
- Before proceeding to install any drain line, the row width to install the drain line is smoothed of humps, hollows, stumps or stones using a dozer blade. This process ensures that there is no tipping of the drain laying machine that might affect the grade control.
- Staking the drain begins at 0±0 station near the outlet. Hubs should be set at every 30 m offset to the proper amount and to the correct side of the centerline of the planned trench. Minimum of three stakes are needed to sight a grade but numbers should be increased on bends.
- Depth and grade may be controlled manually by the machine operator keeping a reference mark on the digging part of the machine fixed on a sightline. Modern methods employ laser equipment. Essentially, this consists of an emitter, located at the edge of the field, which establishes a sloping reference plane over the field by means of a rotating laser ray. A receiver on the machine visualizes this plane of reference and transmits the signal via the regulating system to the machine's hydraulics control. The reference plane can be set to a grade with an accuracy of 0.5 cm per 100 m (0.005%).
- Standards for depth control vary somewhat between different countries though there is general agreement that the position of the drain pipe should not be allowed to deviate by more than half the pipe diameter from its designed depth. The maximum permitted deviation in grades is normally half the pipe diameter per 100 m. These standards can be achieved by skilled operators using manual control while machine fitted with laser control can meet even more exacting standards. With trenchless machines working at high speed and/or in non-uniform soils, installation depths may become irregular. Extra care needs to be taken to maintain standards.

Most procedures are automatically controlled in mechanical construction; following rules may ease the operations.

- Technically there is not much difference between laying laterals or collectors. The only difference is in track size of the machine. While wider tracks are used to lay the collectors, smaller tracks are used for laterals. Do change the tracks as it helps to increase the operational speed. Install the collector over its full length leaving construction of the manholes to a later stage, except when bends are encountered.
- In waterlogged lands where pipes are likely to float, one can think of using the perforated collectors to avoid floating (Mohan Lal *et al.*, 2014).
- Start construction of manhole nearest to the sump and move upstream.

- Pump should continue to operate at the sump during the laying of pipeline (If perforated collectors are used) so that the trench is dry when one starts installing the manholes and the laterals. It is useful to operate an additional pump at the last constructed manhole while next manhole is being constructed.
- If the water table is too shallow and the pumping as suggested fails to control the water, a network of shallow point wells (5-6 m deep) around the area of excavation (where sump/manholes or other structures are to be installed) may be used to lower the water table. A gravel filter (instead of nylon socks) around point wells is more effective for dewatering.

Blinding and Backfilling (Mechanized)

Blinding and backfilling process is similar to the one described previously except that in mechanical construction, it is mechanized. The initial backfilling is done with topsoil (blinding) by hand or by means of a blade mounted on the machine or separately pulled by a tractor. For the backfilling operation, graders, scraper blades or even screw augers mounted on the trenching machine are used. In humid temperate climates, fill material is tamped by running the tracks of trenchers on to the trench line.

Construction under Quicksand Conditions

Quicksand is a soil condition in which fine sand in a saturated, fluid condition is buoyed up by hydrostatic pressure from below. Such a situation is occasionally encountered in the deep drainage for salinity control in alluvial soils. This kind of problem is most acute when installing collector pipes which are either too big or too deep. In such cases either the collector should be installed manually or suitable guard plates on the trencher should be used to ensure that the trench remains sufficiently open to allow machine installation of the drains. Under quick sand conditions extensive dewatering is done to keep the trench dry so as to prevent it from collapsing. It can be achieved through conventional well –pointing along the trench or horizontal dewatering, which may prove to be a better alternative. This method involves pumping water from a pipe (usually an ordinary corrugated plastics drain pipe) installed lengthwise below the bottom of the trench. Special machines are available to install these horizontal dewatering pipes at depths exceeding 5 m although some conventional drainage machines can be adapted to install horizontal dewatering pipes up to a 3.0 m depth.

Other recommendations that help in planning and installing pipe drains under quicksand conditions are:

- Install pipe when the water table is at its lowest level usually during the driest season
- Use plastic tubing with a factory wrapped envelope
- Use larger size pipes.
- Use a steeper grade if possible, not less than 0.4%
- Install each line as rapidly as possible without stopping but carefully observing the grade line

- Blind the pipe with coarse material, such as gravel, sawdust, or sod
- Prevent the caving-in of the trench prior to and during backfilling operations; and
- Throw the spoil as far away from the trench as possible. If the grade line cannot be maintained, the trench can be dug without laying the pipe and a solid wooden cradle can be installed on which pipes are placed. Crushed rock can also be placed in the trench and the grade line re-established.

Resolving Connection Problems

Other construction problem is encountered during the installation of the sump, manholes, and during connections of the drain pipes to the manholes or the sump. It is due to formation of slush that makes the soil under the structure unstable. Under such a condition, it is difficult to maintain the required depth or pipes are pushed up due to unstable conditions and/or soil wall gets collapsed. The least cost solution to these problems is to control the system depth. A properly designed and constructed shallow drainage system can provide the same benefits as a deeper system.

Another problem arises at points where the laterals are connected to the collectors. To resolve the problem the connection at the collector may be kept about 0.2 m lower than the lateral pipe near the collector (Fig. 6.3).

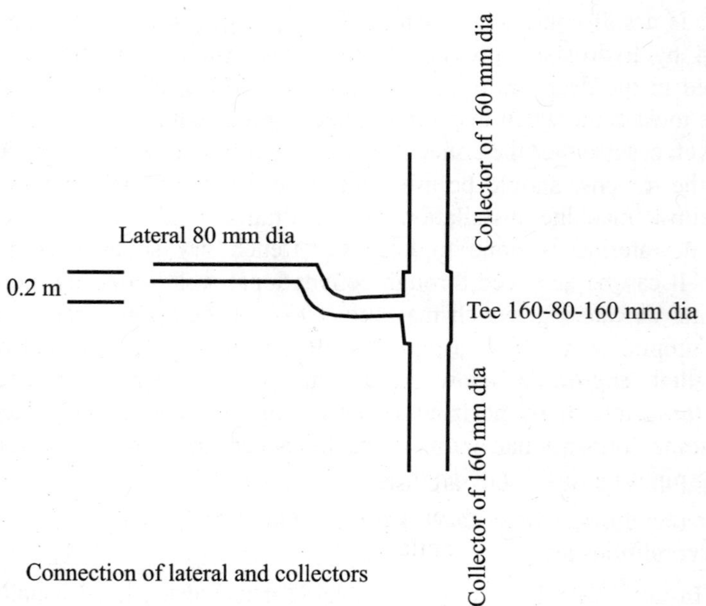

Fig. 6.3 Connection of a lateral with the collector

LAND DRAINAGE FOR PEAT AND MUCK SOILS

Peat and muck soils contain high organic matter, peats containing 50% or more while mucks contain less than 50%. Peat and muck soil are often very wet, but upon drainage become quite productive. Surface subsidence is commonly encountered in these soils because of oxidation and compaction and direct loss of organic matter by erosion and burning. Shrinkage is inevitable following land drainage. The hydraulic conductivity of these soils also decreases with time. In order to secure initial subsidence, the peat and muck soils are drained with open ditches for several years before pipe drains are installed. The pipes are usually laid on the underlying mineral soil, provided the mineral soil is not too deep and is not impermeable. Subsidence can also be reduced by controlling the water table depth at the outlet so as to back up the water in the drains.

DRAINAGE STRUCTURES

A subsurface drainage project cannot simply be constructed by laying out perforated pipes. It must have a number of engineering structures for disposal of the drainage effluent, to maintain the system, maintain the infrastructure of the project area, transport the surface runoff and to minimize or optimize the drainage flow. The description of the structures given in this chapter is based on the structures used in drainage projects in India and structures that have been planned elsewhere to make subsurface drainage systems more effective. Some common structures used in a drainage system are:

- Outlet structure
  - Gravity outlet
  - Pumped outlet
- Manhole/junction box
- Drops
- Surface water inlets
- Relief and breathers
- Crossings: surface drain, canal/distributory/minor/watercourse/road
- Drain bridges
- Closing devices
- Flow management devices

Outlets

Commonly, two kinds of outlets are used. If the natural slope permits, the disposal of drainage water from the collector or main drain into a natural water body is accomplished through a gravity outlet (Fig. 6.4). On the other hand, in flat lands, possibility of having a gravity outlet is minimal. In such cases, a sump is constructed and water is pumped from it to the natural drain. If the water quality permits, pumped water can also be used to irrigate the crops. Down wells, a third kind of outlet, is also a form of gravity outlet. It is useful during drainage of perched water table or in land locked fields.

Fig. 6.4 A gravity outlet for the collector

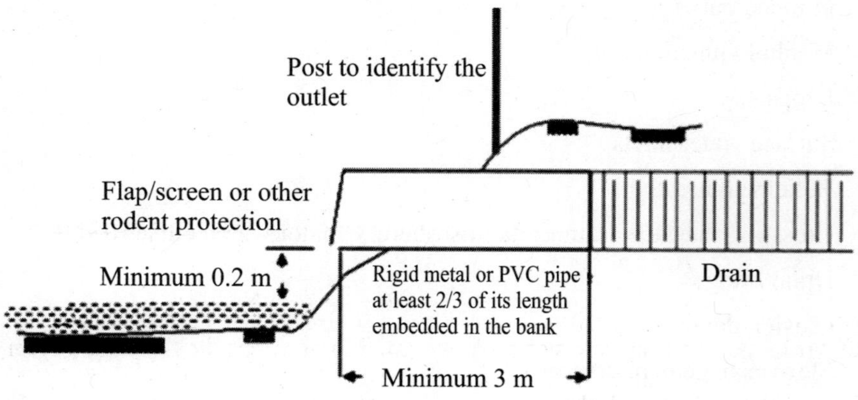

Fig. 6.5 The layout of a gravity outlet in an open drain

Gravity outlet

A gravity outlet is used both in singular or composite systems. For a singular system, pre-cast, reinforced concrete pipes is used as outlet pipe at the end of the drain line (Fig. 6.5). The discharge end is kept at least 20 cm above the normal water level in the drain. To protect this end of the drain from being knocked off or pushed out of alignment, the last 1.5-3.0 m portion is a single length of the pipe. If pipes such as dense concrete, cement or glazed stone-water sewer pipes are not long enough, two or more pipes with cement joints can be used. At least two-third of its length should be embedded into the ridge with the overhanging length discharging at the toe of the ditch slope. The end of the drain should always be protected against erosion and undermining. If the outlet is of corrugated plastic tubing, the exposed portion should be protected against crushing. In general, corrugated thermoplastic conduits should not be used as an outlet pipe as these are likely to be damaged by exposure to sunlight or fire or some earthmoving equipment or livestock.

Details of a typical outlet structure for a composite system are shown in Fig. 6.6. Concrete pipe of socket (bell) and spigot type should be laid with the socket facing upgrade and sealed with mortar. The joints of the concrete pipes and ends of the sleeves for corrugated PVC drainage pipe should be mortared and wrapped with a sheet of plastic film with a minimum thickness of 0.15 mm and minimum 300 mm in length on both the sides of the joints and at the ends of the sleeve. The structure base should be constructed in cement concrete cast-in-place. The head wall should be either masonry with cement mortar or cement concrete cast-in-place. Alignment of the head wall should be parallel to the sloping bank and the concrete pipe should face either perpendicular to the outlet drain or towards the downstream side. Where two or more outlets discharge at the same head wall, the head wall is increased by an amount equal to twice the outside diameter of the largest of the pipes. The outlet structure should always be constructed under dry condition.

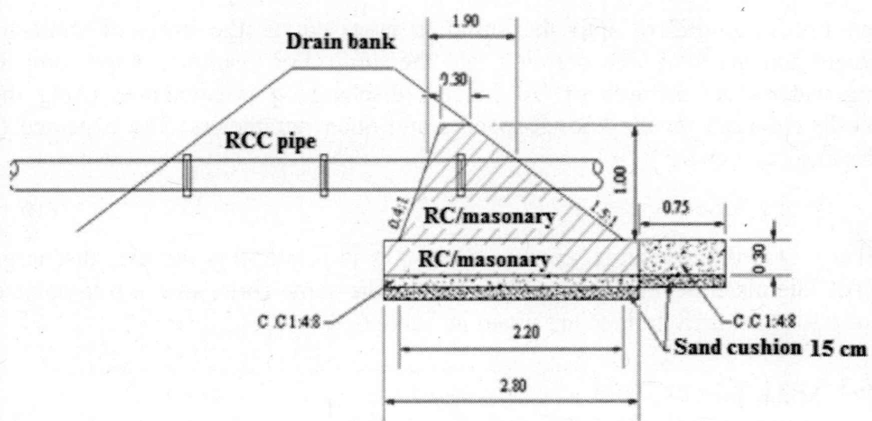

Fig. 6.6 Typical design dimensions of a gravity outlet for a composite drainage system

The rodent guard should always be fixed in the outlet structure (Fig. 6.7). The simplest rodent guard may be a flap hinged at the end of the drain pipe. The flap opens when water flows from the drain pipe to the outlet ditch otherwise it remains closed. The flap also checks back (reverse) flow from the outlet ditch into the drain pipe.

Pumped Outlet: Sump Design

A sump is a prefabricated RCC structure or constructed *in-situ*. To assess the capacity of a drainage sump, due consideration is given to the discharge capacities of the collector/main drain. Due importance is also given to:

- The maximum inflow into the sump
- The type of storage required and its capacity

- The cycles of power operation and discharge levels, and
- The pump and the power unit needed

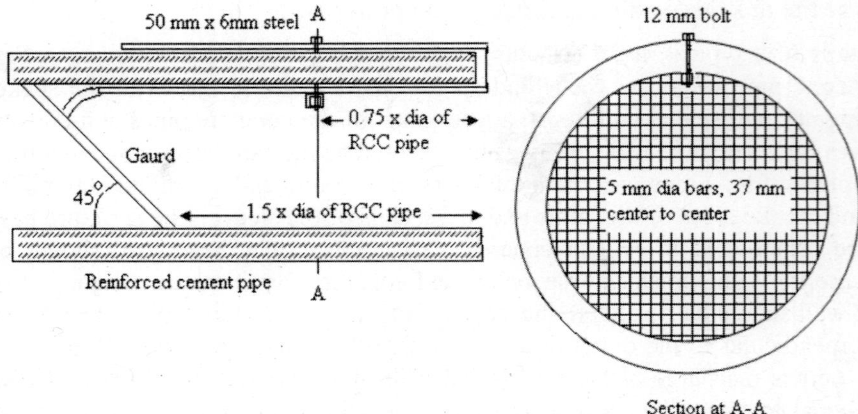

Fig. 6.7 A rodent guard fixed in a gravity outlet

The maximum inflow into the sump is assessed on the basis of drainage requirement and the total area draining into the sump. For example, if the sump is receiving water from an area of ′A′ ha, the discharge *q* is calculated using the Hooghoudts equation for the known spacing and other parameters. The expected *Q* into the sump is given as:

$$Q = c\, q\, A/8640 \qquad (6.1)$$

Where, *Q* is the discharge, cumecs when *q* is in mm/d; *c* is the area discharge factor (All laterals/collectors do not flow full at the same time, so the rationalized values of *c* for total area drained are given in Table 6.3).

Table 6.3 Area discharge factor

Area drained (ha)	Factor, c
16	1.0
16-32	1.0-0.92
32-50	0.92-0.87
50-65	0.87-0.82
65-80	0.82-0.79
80-100	0.79-0.76
100-120	0.76-0.74
120-130	0.74-0.72

The cyclic operation of pump and motor is considered while determining the amount of storage required in the sump. In general, a cycle of 12 minutes at the rate of 5 cycles/h is considered satisfactory. The length of a complete cycle in minutes is equal to the standing time plus the running time. For the motor to have equal on and off times, the storage must be equal to the amount that would run 6 minutes when 12 minutes cycles are used. Therefore, the sump must have a storage capacity, S given by:

$$S = 60*6*Q \ m^3 \tag{6.2}$$

The volume of required storage plus the criterion that the minimum water level should be 0.6 to 1.0 m above the bottom of the sump will determine the actual size of the sump. Let us consider that the pipe drain enters 2.3 m below ground surface and the sump is cylindrical and vertical, the distance between the start and stop elevations (say *d'*) is 0.6 m and the volume of required storage, *S*, is 12.42 m^3, then the diameter, d of the sump is computed from:

$$\pi d^2/4 = S/d'$$

$$d^2 = S/0.785 \ d' = 12.42/0.785*0.6 = 26.37 \tag{6.3}$$

$$d = 5.14 \ m$$

Considering the factor of safety as 25%, d = 6.5 m.

Based on the field experiences, the bottom hole in the sump from where the collector pipe enters the sump should be made at least 10 cm below the designed collector level. It helps to avoid sedimentation in the collector pipe just before the connection (Mohan, 2011). It often happens if the pipe sinks somewhat during backfilling.

Sump/Manholes, Junction Boxes and Drops

The construction of a sump/ manhole/junction box or even drop is almost similar except for their location, size and depth. While a sump is provided at the end of a block from where water is pumped into the disposal drain, manholes are provided at appropriate locations to inspect and to detect damaged sections and to carry out maintenance activities. These are invariably used when the length of the pipe line exceeds 300 m. The manholes and junction boxes are usually provided with silt traps, the bottom being lower than the level of the outlet of the collector (Fig. 6.8). Manholes are usually kept above ground although there are no hard and fast rules for the same. Manholes are provided with appropriate covers. In most cases plastic-coated iron bars are used as steps so that a person can enter the manhole whenever required. Above ground manholes are used to measure drain flow rates as well as take water samples to determine its quality. While the above ground junctions/ manholes are convenient, these are likely to be abused by the farming community to drain excess irrigation or rainwater.

Junction boxes are constructed at major junctions of pipelines, where one drainage collector empties into a bigger one or where sudden change in alignment or grade occurs. These are also provided in collectors adjacent to canals or road crossings and/or where two or more collectors intersect. At some places blind junctions in which lateral and collectors are joined directly using T-pieces or cross pieces are used. The inspection chamber kinds of junction boxes have so far been commonly used. The capacity of the outlet drain should equal the combined capacity of the incoming drain lines and the elevation of the outlet drain flow line should be at the same level or below the flow line of the lowest incoming drain (Fig. 6.9a).

Underground junction boxes should be at least 40 cm below soil surface so as not to hinder the field activities (Fig. 6.9). Moreover, the location of the underground structures should be well documented so that these can be located whenever required for maintenance activities in future. A metal detector can be used to detect the structure if the lid with steel bars is provided at the top of these structures. Whenever considerable drops are encountered on the way, a drop structure is used to mainatian the allowable slope to the drain line. It helps to maintain the drain depth and obviates considerable problems that may be encountered in excavating deeper drains.

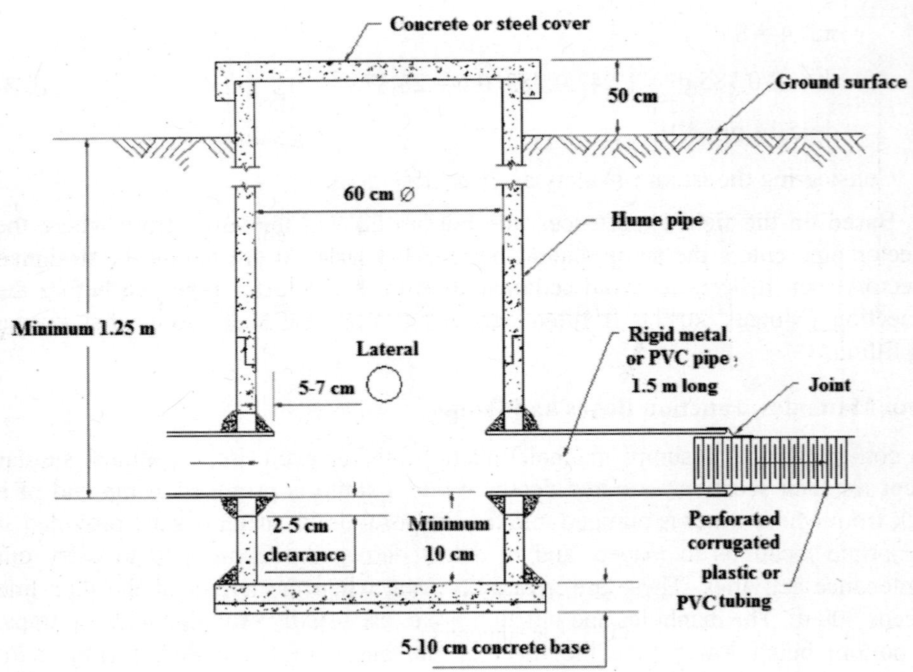

Fig. 6.8 Construction details of a square manhole

Sump/manholes/junction boxes/drops are made of bricks and are square or round in shape. These can also be made of reinforced cement concrete (RCC) rings having an internal diameter of 1.2 m for sump and 0.9 m for manholes. The bottom ring has a monolithic base plate to ease installation under shallow water table (Fig. 6.10). Usually these are 5 m long made of 4 elements of 1.25 m each. Following construction, the excavation around the sump/manhole/junction boxes should be backfilled with local soil. It should be compacted at appropriate moisture content to achieve at least 95% Proctor density. It may be noted that adequate maintenance and supervision is needed to maintain structures like manholes and junction boxes. Therefore, these structures should be provided at places where they are absolutely necessary.

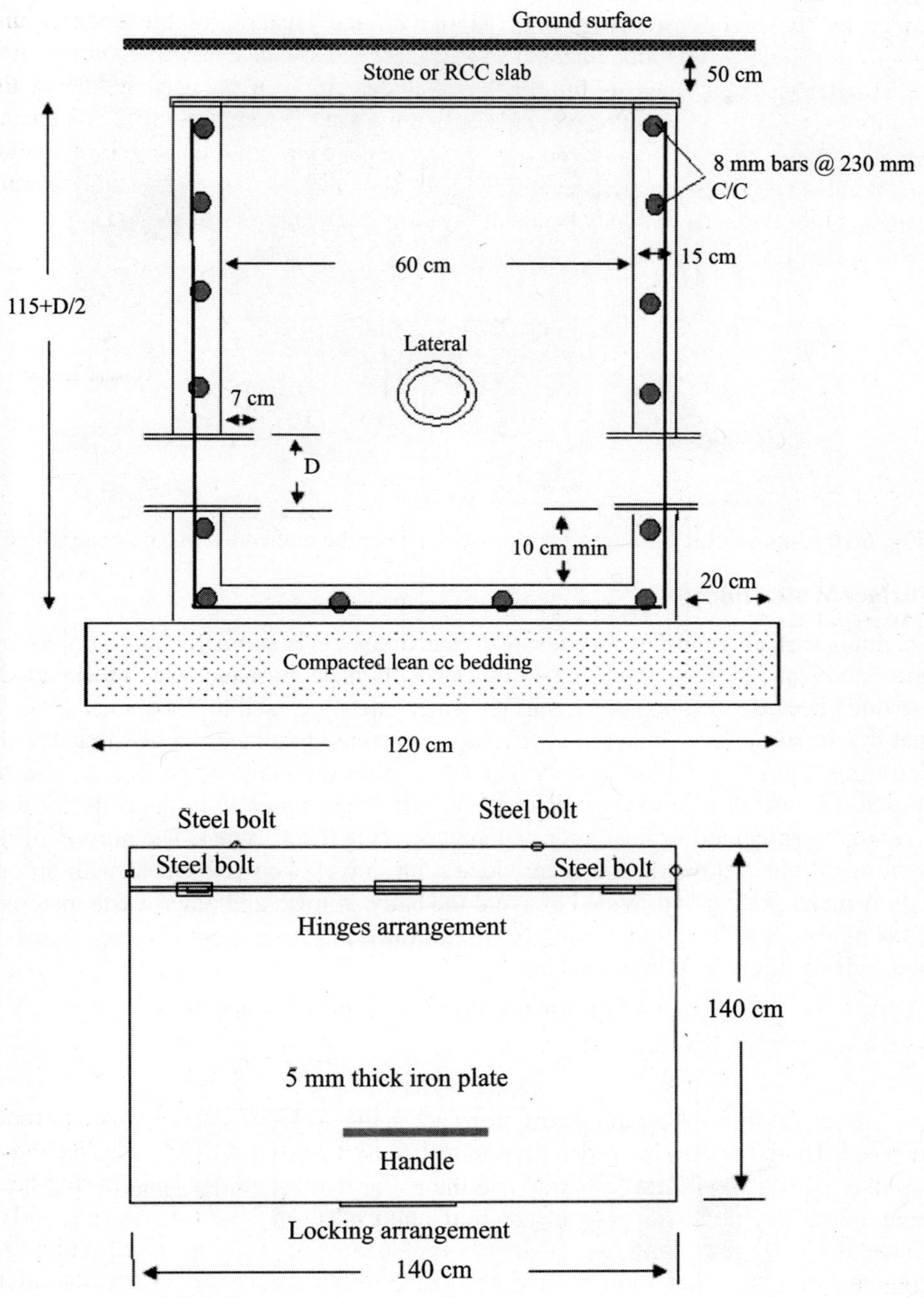

Fig. 6.9 A line diagram of an underground junction box (top) and an iron cover for manhole (bottom)

Fig. 6.10 A monolithic base ring for a manhole. Note the embedded plastic-coated stairs

Surface Water Inlets

To drain surface runoff through subsurface drainage is normally undesirable and uneconomical. At few points where surface drainage is inadequate or cannot be provided because of topography, surface water inlets are used to drain such areas. In that event, subsurface drainage coefficient is increased suitably so as to enable the drain pipes to carry surface runoff. The open inlets are constructed in the collector (Fig. 6. 11, left) or near to the collector line, which is connected to the collector with a siphon to safeguard against poor maintenance (Fig. 6.11, right). The portion of the perforated inlet below ground is provided with gravel. Top is covered with an end cap. A metal grating is provided to avoid the entry of trash and other waste materials if the pipe ends at the ground level. Non-perforated pipe to at least 150 cm is used on both sides of the inlet (USDA, 2008).

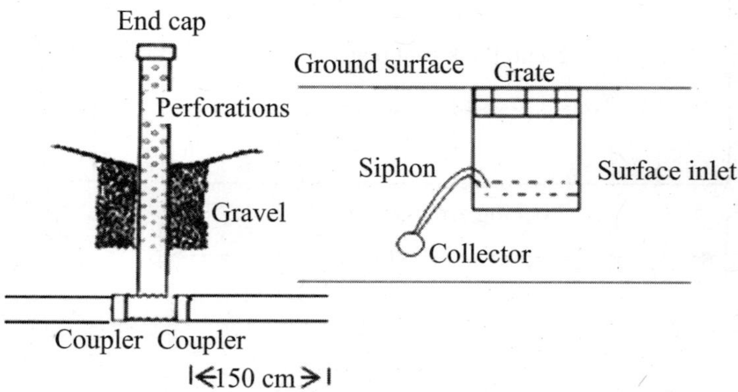

Fig. 6.11 A surface inlet on the collector (left) and near to the collector (right)

The blind inlets are also provided to remove surface runoff although these are also capable of removing subsurface water (Fig. 6.12). Blind inlets remove water at a much slower rate than surface water inlets. The blind inlets are used to drain open fields because in that case farming operations can be accomplished without any hindrance. Moreover, these inlets intercept sediments that may be encountered in drainage of open fields. To construct a blind inlet, a trench is dug around a drain line that ends at about 30 cm above the drain. It is filled with stone, gravel, crushed rock or a combination of these materials. Additional material graded upward (coarse to fine) is filled within approximately 30 cm of the ground surface. The top 30 cm is further covered with topsoil or where silting is a problem, with pea gravel, small stones, or coarse sand. The life of the structure depends upon the quality of construction, the fill material and the amount of sediments that reach the inlet. In general, the use of blind inlets has been found to be unsatisfactory because of its clogging with fine soil sediments. Blind connections are also provided at appropriate locations to permit flushing of the field drains without the need of excavating and dismantling for connections (Fig. 6.13).

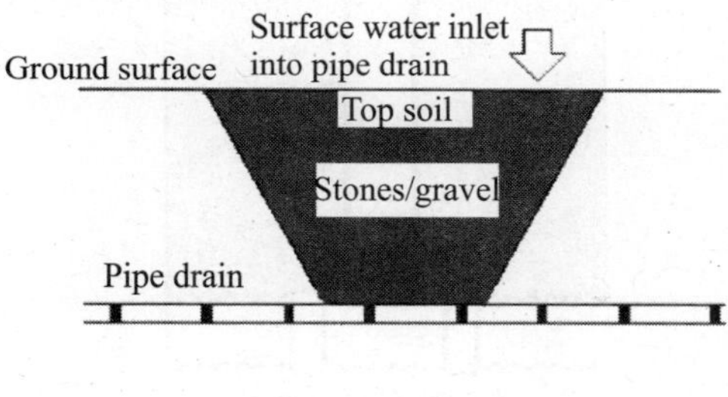

Fig. 6.12 A schematic view of a blind inlet

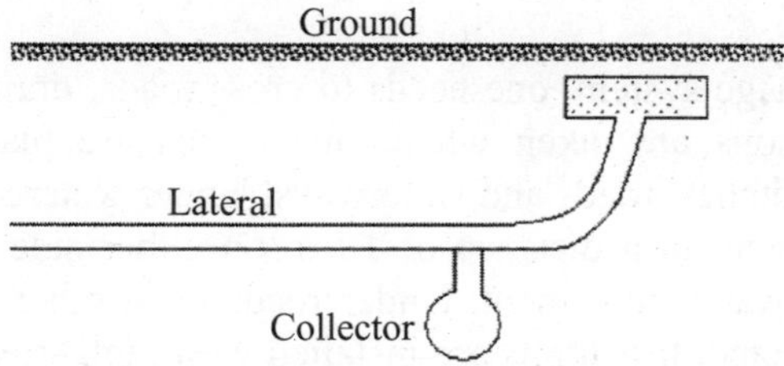

Fig. 6.13 A blind junction for flushing

Relief Wells and Breathers

Relief wells are needed to relieve pressure in the line that might result in blowing out of the tile line. Relief wells are located on long lines at points where tile line might be overloaded for short periods as at the end of steep sections, on lines having surface inlets or where a steep section of a main changes to a flat section. To save cost, relief wells are not provided if the capacity of the flat section exceeds the capacity of the steep section by 25%. The relief well is constructed by providing a T joint at which a pipe extends at least 30 cm above the ground. This can also serve as an entry point for surface run-off provided necessary arrangements are made at the ground level for surface water entry. The exposed end is generally covered with heavy wire mesh. The size of the riser is normally the same as the tile drain but slightly lesser size can also be used.

On the other hand, breathers relieve negative pressures in the drains. These are provided if the drain lengths are more than 400 m or the slope changes from flat to steep especially when pipes flow full in the flat slope. A breather pipe shown in Fig. 6.14 is about 10 cm in diameter for 350 mm or lesser diameter pipes while it is 15 cm diameter for larger than 350 mm pipes. The effect of negative pressure in subsurface drainage has remained unclear so far. As a result, many engineers do not feel the necessity of providing breathers in subsurface drainage.

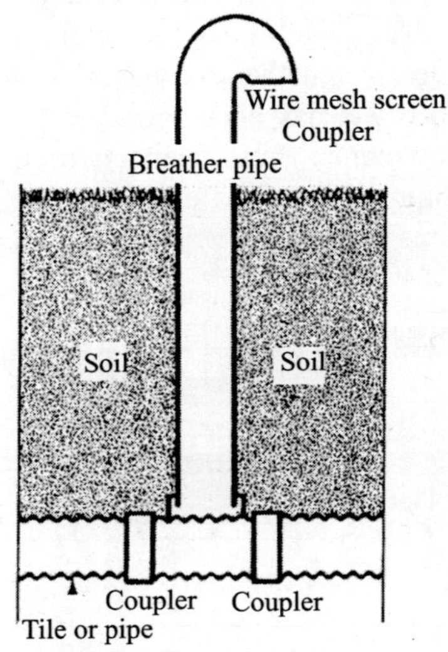

Fig. 6.14 A breather or vent pipe

Crossings

In a large-scale drainage system, one needs to cross roads, drains and water courses etc. Special precautions are taken when pipe drains are placed under irrigation watercourses, open ditches, roads and waterways. Under watercourses or ditches non perforated plastic tubing to a distance of 10 m on either side will be sufficient to prevent any serious water movement. Under roads or at other places where soil or wheel loads are expected, the drains are installed with high strength collar pipe with concrete joints. The load is always assessed before selecting an appropriate kind of pipe. The guidelines for each kind of crossing are available and the designer must follow these guidelines. A road crossing is shown in Fig. 6.15.

Drain Bridges

Undisturbed natural soil has enough strength to support the drain pipes. Drain bridges are used to maintain the level of the drain when a drain crosses a soft spot where the soil has not yet settled *e.g.* a filled-in former ditch. Drain bridges are made of timber blocks on which the drain pipes are laid or the drain is surrounded by a continuous length of solid/rigid pipe (Fig. 6.16).

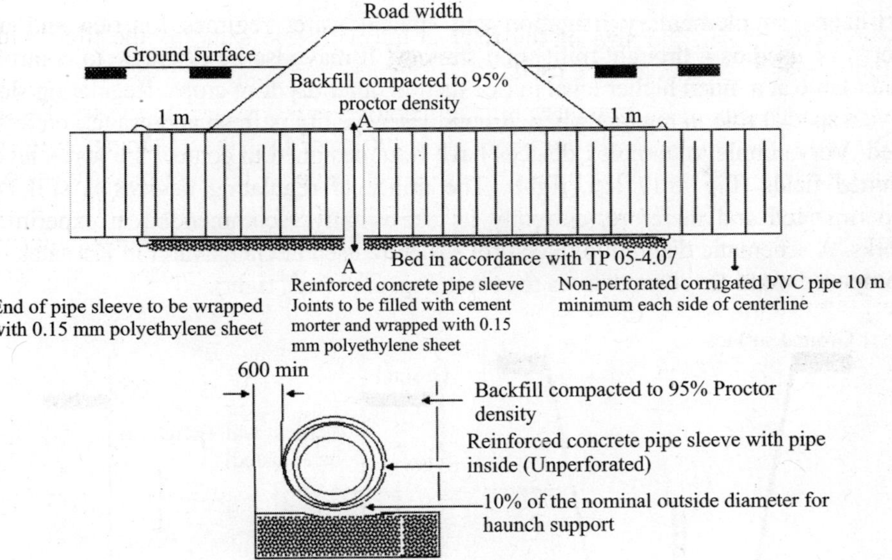

Fig. 6.15 Schematics of a road crossing

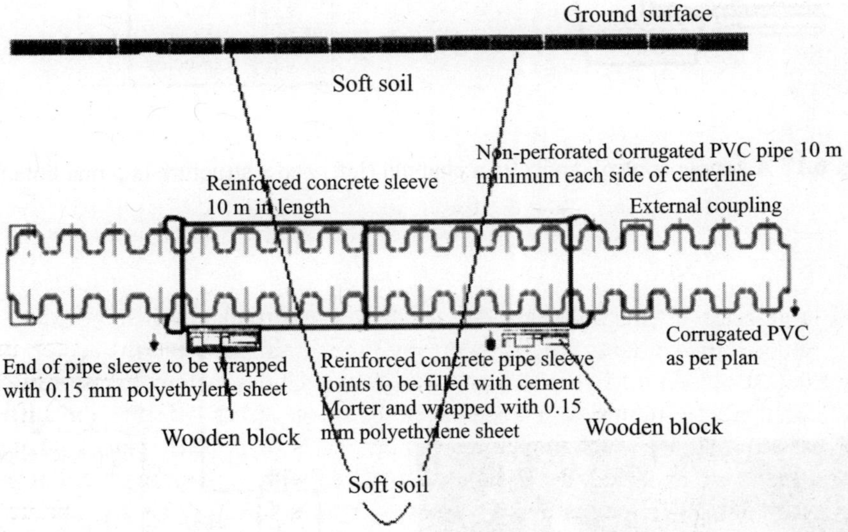

Fig. 6.16 Schematics of a drain bridge

Closing Devices

To prevent soil from entering the open end, each end of the laterals and collectors is capped with a standard cap made for this purpose. If unavailable, a concrete block, a brick wall or blocks of any other material can be used.

Regulating Sluices

There can be several occasions when drainage has to be reduced temporarily or even halted (e.g. environmental considerations, unwarranted and harmful leaching of fertilizers, supplementary irrigation and special water regimes for rice and other crops) or used as a drought mitigation strategy. It may also be desirable to control the water table at a much higher level in rice than in other dry foot crops. Regulating sluices have a special role in regions where groundwater quality is fresh and gravity outlets are used. Very simple to complex devices have been designed to control the water level in drained fields (Fig. 6.17 left, right). The utility of regulating devices is still being experimented and therefore, currently its use is only recommended in experimental works. A schematic diagram of a control structure used at Gangavathi in Karnataka and a proposed multi-level structure is shown in Fig. 6.18 (left, right).

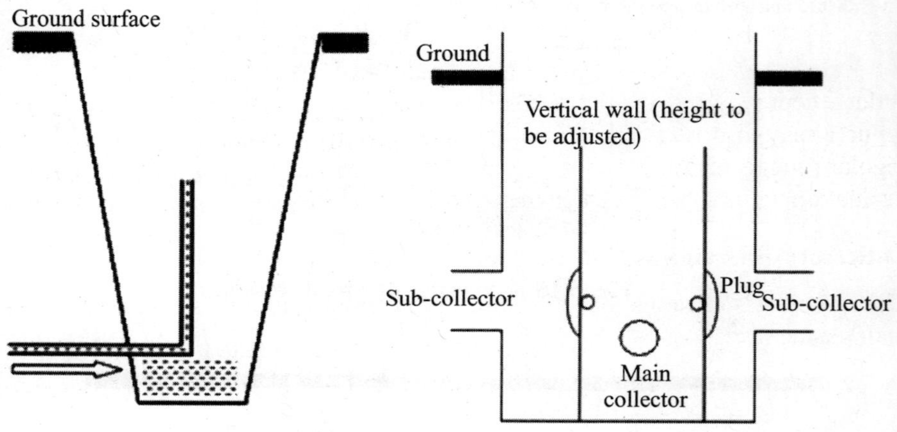

Fig. 6.17 A simple control device in a channel (left) and a structure in a manhole (right)

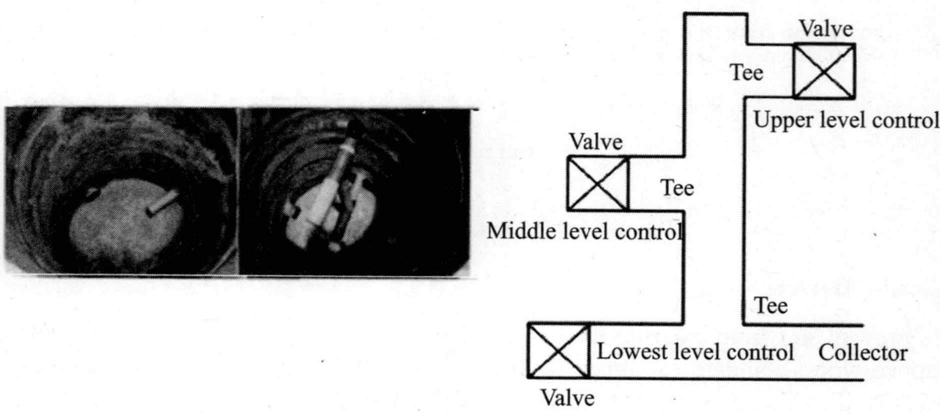

Fig. 6.18 Photos of a manhole in conventional and controlled drainage (left) and a multilevel water table control device (right)

MAINTENANCE OF PIPE DRAINAGE SYSTEMS

Possible Causes of Troubles

Possible causes of failures or partial failures of a SSD system may be related to design of the system such as drain spacing too wide and drain depth too shallow, insufficient capacity of drains and development of perched water table due to layers of low permeability (Gupta & Gupta, 1997). Wrong selection of envelope material or degraded PVC pipes (due to exposure to sun for long period) and selection of improper pipe fittings might also cause the failure of the system. Most causes are related to system construction and may include: pipe installation under wet conditions, pipe joints pulled apart, drain installed without envelope or poor placement, tearing of envelope material and unfolded stitching at seam, improper backfilling and sudden depression in the drain line or installed on reverse grade. Besides, physical failures may also result in several problems. Many problems can be identified with regular/routine inspection. Some problems can be identified during monitoring and evaluation while others might come to the notice through complaints by the farmers.

Indicators of Drainage Malfunctioning

The possible indicators that suggest shortcomings in the system and need for maintenance are:

- Submerged outlet, displacement of outlet flap or rodent guard, soil erosion around the outlet structure-scouring of head wall or undermining.
- Any sudden reduction in drain discharge shortly after rain or irrigation and / or sudden reduction in flow rate in pipes draining in manholes. These conditions might indicate that a pipe or segment of the pipe drain has been completely or partially blocked. A muddy outflow indicates possible direct entry of surface water in the drain pipe or improperly fixed end cap.
- Wet spots that remain wet after the field has dried off reveal clogged drain, overpressure flow in the non-perforated collector resulting in wet area over nearby laterals, inadequate capacity of drain, poor soil physical condition not permitting drainage.
- Appearance of cavity holes either because of sink holes or blow holes over drain lines. Sink holes are formed due to soil washing into the drain pipe at separated joints or broken pipe. A blow hole may be caused by soil washing due to overpressure leakage from drain line at separated joints or broken pipe when its downstream pipe is plugged with soil.
- Poor trafficability in the drained area may often be associated with water table drop at a lower than the designed rate. In such cases, observation wells should be installed at mid spacing and water table should be monitored to investigate the problem.
- Increase in soil salinity is an indicator of the inadequate drainage.
- Growth of shrubs/trees over or near the laterals may cause drain clogging by the intrusion of root hairs and their accumulation in the drain pipe. This may reduce the capacity of the pipes.

- Condition of drainage structures such as damaged manholes, collapsed junction boxes, damaged RCC pipe sleeves in crossings, settlements around structure are other indicators of problems associated with project.

Necessary Requirement for Maintenance

No hard and fast rules are in place to guide as to what constitutes good post-installation maintenance programme. It is because subsurface drainage requires very little attention provided enough quality control is exercised during design and construction. Since out of sight out of mind principle applies to subsurface drainage, to ensure a good maintenance programme, the following points must be kept in view:

Preserve good as built-up maps of the drainage system. The maps should show the location of all ditches and subsurface drains, their size, depth and grade etc. These maps should be updated as and when repairs or extensions are made. In their absence, simple pencil sketches may be kept in record to access the system and its components.

The efficiency of a subsurface drainage system is often governed by surface drainage network. Open ditches should be kept in working condition by removing trees, bushes, and weeds growing in them. Otherwise water movement will slow down resulting in excessive silting. Submergence of the drain outfalls needs to be avoided.

If the spacing is too wide or depth is shallow resulting in improper drainage, it is useful to provide new laterals in the critical areas. A new drain between the two old laterals will reduce drain spacing and improve drainage system performance.

If the drains are not constructed with a good bond with the soil, piping can occur and cause a complete washout of the drain.

A most useful principle of maintenance is to repair the site immediately rather than follow a wait and watch policy.

Identification of Blocked Sections of Drain Pipes

To identify the blocked sections of a drain pipe the drain pipe is excavated at suitable intervals. Any of the following methods is used to know the blocked section:

A rigid torpedo-shaped tip steel rod (process is known as rodding) is pushed into the drain pipe through the pipe opening (an outlet or excavated pit). If the rod fails to pass a particular point in the drain pipe, one can assume a possible blockage at that point. If rodding equipment is not available, a pipe of just smaller diameter than the drain pipe could also be pushed through the opening.

Alternatively, the most commonly adopted technique is to pull a ball attached with a steel wire to find possible blockage. If there is blockage in the pipe line, further investigations are required to know the reasons of the problem.

Cleaning the Drain Line

Deposits of fines and ochre must be removed by flushing the line for its trouble-free long life. Flushing can be done in combination with rodding. Lateral drains are often cleaned with sewer-cleaning rods if the deposits cover a short stretch. Holes are dug down to the drain at intervals of 10 to 25 m, depending upon the size of the drain and the amount of sediments to be removed. It may be convenient to dig the hole below the level of the drain as a temporary sediment basin. Sewer rods, either plain rods or with attached tools such as scoops, hooks and augers can be used. Thick deposits over a long length may be cleaned with a ball of barbed wire or a chain.

Flushing is now commonly employed to clean the drains. Adequate supply of water must be available to flush a drain. In some cases, a large volume of water is used rather than high pressure. With limited water supply, a catch basin, or hole at the upper end of the plugged section is made to serve as a water reservoir. Block off the upper end of the drain and fill the catch basin or hole with water, then remove the block and allow the water to flush suddenly through the drain. This simple procedure of flushing may solve the problem. The cleaning may also be done from the pipe outlet into a ditch (Fig. 6.19, left) or from a man hole or junction box. Currently, pipe cleaning is mainly accomplished using pressurized water jetted from a self-propelling nozzle (Fig. 6.19, right). Most commercially available drain pipe flushing machines have a maximum reach of 250-300 m. The working capacity ranges from 400 m/h under normal conditions to 200 m/h in highly silted pipes. Water pressure at the nozzle is 10-15 atmospheres, which suffices to remove loose deposits but fails to remove deposits that have consolidated due to lack of regular maintenance. For such deposits, flushing is done at a higher pressure. It may be noted that high pressure can also disturb the stabilized soil around the pipe and lead to more entry of fines. Major sediment loads enter the line during initial period after the installation before the soil above and around the line has consolidated. As such, the system needs to be cleaned soon after installation (one to two years). Later a frequency of five to ten years will normally be sufficient. In composite systems with long drain lines, special access points may be provided for cleaning purposes.

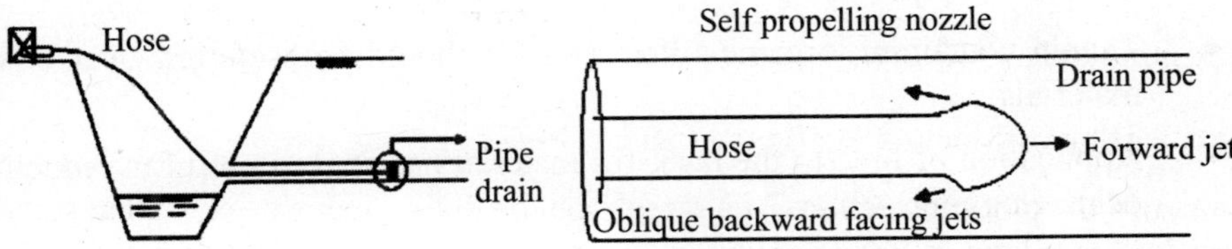

Fig. 6.19 Drain pipe cleaning from outlet end (left) and by means of a jetting nozzle (right)

Ochre Formation and Clogging

Clogging of drains can occur through chemical deposits. These chemicals are mainly the iron compounds and the problem is generally referred as ochre clogging. Ochre is generally a combination of bacterial and chemical sludge, originating from the soil or the groundwater and forms a gelatinous mass that seals the drain openings. Virtually all soils contain appreciable quantities of iron, although the problem is often encountered when soils containing iron sulphate are drained for the first time. Typical soils so affected are lowland peats, recently reclaimed marine soils and soils derived from iron pyrite bearing rocks. If soil is the origin of ochre problem, the problem subsides with time and may disappear in 3 to 5 years. The problem persists if it has its origin in groundwater. As the water enters the drain, the iron is oxidized from ferrous to ferric, an insoluble form. In extreme cases the pipe and envelope can become totally clogged. The potential of ochre clogging in pipe drains can be estimated by analysis of the iron (Fe^{++}) content in the groundwater and its pH (Eggelsmann, 1978). For the same degree of ochre clogging, the iron content is higher in alkali than in acid soil (Table 6.4). It is attributed to the pH in the soil environment.

Table 6.4 The degree of the risk of clogging by iron ochre as related to Fe^{++} content and pH

Fe^{++} concentration (mg/L)		**Iron ochre clogging risk**
pH <7 **Acidic conditions**	**pH >7** **Alkaline conditions**	
<0.5	<1.0	Hardly any
0.5-1.0	1.0-2.5	Slight
1.0-2.5	2.5-5.0	Medium
2.5-5.0	5.0-7.5	Pronounced
>5.0	>7.5	Very pronounced

One of the solutions for soil-originated ochre is to undertake drainage in stages. Initial drainage may be accomplished by ditches, which can later be replaced with pipe drains when the risk of clogging has subsided. In areas where there is only a slight risk of ochre clogging, it can be managed as follows:

- Careful selection of the drainage material, a clay tile is a better alternate than a PVC pipe.
- Tannin containing organic filters (wood) should be preferred over other materials.
- Application of lime to the back-fill material has proved useful in reducing ochre clogging.

To remove the ochre, apply sulfur dioxide gas from the upstream end and after it reaches the downstream end, close the two ends for 24 h. Remove the end plugs so that the dissolved iron will come out along with the drainage water. Safely dispose of the drainage water as it is toxic to animals and fishes.

Self-cleaning of drains with ochre clogging is difficult even at grades up to 5% and therefore, drains must be periodically cleaned.

Other Kinds of Chemical Deposits

Ochre clogging has been extensively investigated being a prominent problem of humid temperate regions. Deposition of other chemicals such as calcium carbonate ($CaCO_3$) and calcium sulphate as gypsum ($CaSO_4.2H_2O$) has not been studied in that detail. The hard and crystalline deposits of these chemicals are likely to buildup comparatively slowly in gypsiferous soils and soils with high calcium carbonate content in arid and semiarid regions. Manganese in groundwater under suitable reducing conditions can also form bacterially enhanced gelatinous black deposits. Sulphur bacteria can also oxidize H_2S to H_2O and elemental sulphur S. Sulphur slime is a yellow to white stringy deposit. Probably more studies are needed to generate evidences as well as to devise ways and means to prevent such deposit in subsurface drainage systems.

Clogging by Roots

Roots of normal field crops rarely enter drain pipes installed at normal depths. Problems may, however, occur where drain lines pass beneath wild bushes, hedges or are near to trees. Roots of trees and other vegetation may enter into the drain and obstruct the flow. Trees growing within 18 m of the drain should first be removed. Sugar beet and alfalfa roots known to enter drains may not be grown although they should not cause much trouble because the roots die, decay and are washed away. The *Prosopis* roots growing wildly on waterlogged saline soils clogged the drainage system at few places in Maharashtra (Fig. 6.20). If it is difficult to remove undesirable vegetation, non-perforated tubing, collar tile, or metal pipe should be installed along the tree line and for at least 30 m distance from the two ends of plantations.

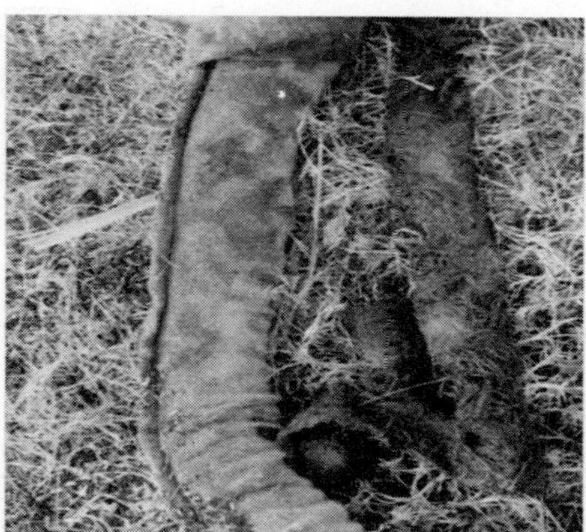

Fig. 6.20. The excavated drainage lateral with *Prosopis* roots

INSPECTION AND QUALITY CONTROL

Inspection is required for many reasons yet it is essential to ensure quality and quantity of works to be accomplished within the allotted funds. Inspection is done by using one or more of the methods such as inspection by observations, inspection by measurement, inspection by testing and inspection by excavation. The method, time and frequency of inspection is decided on the basis of type and nature of work to be inspected. In a drainage project, inspection is required during material selection, layout surveys, pipe installations, construction of drainage structures and site restoration. Inspection is carried out at three stages namely prior to, during and after the construction. Intensive inspection is required for quality control during system construction. At all these stages, inspection by testing essentially requires taking adequate and appropriate number of samples and provision for this must be made in the works. A record of all inspections should be maintained in specified formats for each and every activity to be inspected in accordance with the specifications and standards. These records are important to verify the Bill of Quantities (BOQ), as a proof for settlement of any claim, for the works to be executed on day work basis and last but not the least to prepare/compare the as built drawings.

After the construction, inspection should also be carried out during the defect liability period. Following points should be considered while inspecting at this time:

- Observe and record the defects which are caused due to poor workmanship and inaccurate installation
- Record sink holes, cavities formed due to poor compaction at the time of installation
- Check saturated zones on the laterals as compared to the surroundings
- Take few samples of laterals with and without fabric by excavating them for sediment deposition
- Check effluents quality for sediment coming in the lines at outlet and manholes
- Check sediment deposits in the silt traps of manholes and junction boxes.

The frequency of inspection after the liability period depends upon the age of the system and occurrence of problems. A new system requires frequent observation of its performance particularly during the entire monsoon season and the irrigation season. A 2-3 years old system requires inspection before the onset of monsoon season for necessary repair and after the monsoon season for identification of problem indicators, if any.

TENDER DOCUMENT

A tender document should be prepared in two parts as follows:

- A technical part specifying the works to be carried out
- A legal part specifying rights and duties, liabilities etc., of the parties involved.

Without going into the specificity of the legal part, technical part of the tender document should invariably include the following:

- Layout of the drainage network on 1:5000 or 1:10000 scale showing alignment of all major drains (collectors and larger) with a clear and rigid numbering system. Field drains are not necessarily indicated but first and last line should be shown along with spacing.
- Spacing and depth of laterals, diameter and slopes, elevations of collectors: longitudinal profiles or tables
- For all open drains, indicate
    - Dimensions (cross section)
    - Bed levels and slopes (elevations with respect to a well-defined benchmark e.g. mean sea level): on longitudinal profiles or tables

- Materials and structures
  - For pipe drains: pipes, envelopes, connections between laterals and collectors, outlets
  - For open drains: culverts, passages, bridges etc.
  - Construction specifications
  - Working method
  - Need for dewatering to lay collectors/manholes, laying of pipes
- Working conditions: Time periods when the work cannot be done/ machines would be idle
  - Bill of quantities (Annexure I)
  - Lengths of various pipes to be purchased
  - Lengths of open drains, m^3 of excavations
  - Concrete structures of various types

ORGANIZATIONAL SET-UP

To undertake the activities included in the drainage master plan of the state, a responsive organization is needed. Within the current hierarchy, it should be headed by the Engineer-in Chief/ Director of Agriculture/CAD authority at their respective headquarters. They should spend about 5% of their time on general policy and follow-up activities with the funding/collaborating organizations. A chief Engineer/ Additional Director of Agriculture should respectively be assigned the duties of the Executive Project Directors. They should spend about 20% of their time on the project and prepare cases which have to be decided in the funding organizations. They should also participate in all the meetings of the Steering Committee/Technical Committee.

Overall Supervision

For the overall supervision and guidance of the project, a technical committee with members of a number of organizations involved in drainage should be constituted. The Technical Committee headed by the Engineer-in-Chief/ Director of Agriculture/CAD authority as the case may be, should guide the project on the technical details; meeting at least two times a year.

Engineer-in-Chief/ Director of Agriculture/CAD authority

Chief Engineer/Additional Director of Agriculture

Nominee of the Department of Water Resources

Nominee of the Department of Agriculture

Nominee of two Research Organizations (One Central and one state)

Project Director

Special Invitees (4)

The Subsurface Drainage Project team should be located at some central place around the projects with appropriate support at the project sites. The project teams should be headed respectively by Project Directors (PD), who should not be less than the rank of a Superintendent Engineer/Deputy Director (Agriculture). The tasks of the project director could be:

- Manage the project teams
- Prepare budgets, guard the expenditure and prepare new project proposals
- Liaison with concerned departments and coordinate with the activities of all department involved in the project
- Monitor and evaluate the performance of the various sections of the project, prepare and present the results, to submit Progress Reports.

Appropriate wings with clearly defined responsibilities should be established at the head quarter and/or at project sites for smooth functioning of the project. The organogram of a set-up for a subsurface drainage project is shown in Fig. 6.21.

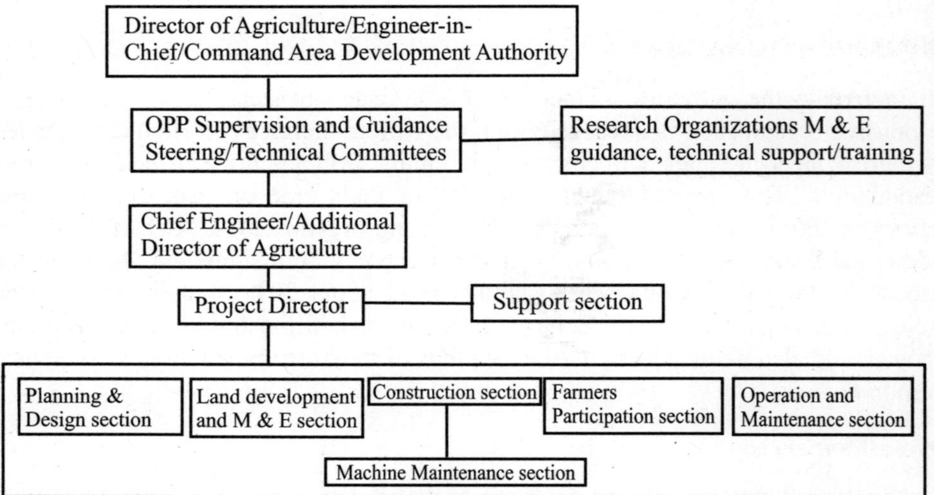

Fig. 6.21 Organogram of subsurface drainage projects

The Planning and Design Section

It should be entrusted with the following activities:

- Delineate the pilot areas of 500-1000 ha each in cooperation with the other sections, including the Department of Agriculture
- Survey and mapping; updating existing topo maps; preparation of new maps
- Determine the design criteria
- Design the pipe drainage system including different layouts, pipe sizes and if necessary filter material, and preparation of the related cost estimates
- Plan improvements of surface drainage systems in consultation with the Construction and Machinery Section

- Formulate the construction schedule
- In consultation with the Farmers Participation Section
  - interact with the farmers about the design, including the location of laterals/collectors/outlets/sumps
  - plan the land acquisition needs for the open drains and sumps, if needed
- Prepare as built maps (showing the actually installed location of the pipes and structures) in cooperation with the Construction Section.

The Construction and Machinery Section

The main task of this section will be to jointly install the pipe drainage system in an effective manner. The Section should be entrusted with the following tasks:

- Survey, check the designs and stake out the planned alignment of the drains
- Preparation of crop compensation and land acquisition plans in cooperation with the Planning and Design and Farmers Participation Sections
- Prepare, manage and monitor the construction schedule
- Install the pipes and the structures (sumps, manholes, connections between collectors and lateral pipes)
- Install pumps on sumps, if necessary, including action needed to energize these pumps including possibilities of installing solar pumps
- Carry out improvements in the surface drainage system, if required
- Supply the data for the as built maps to be prepared by the Planning and Design Section.

The Machinery Section This section should be entrusted with the following tasks:

- Operate and maintain the machinery and pumps during the construction
- To ensure the machinery operate according to the time schedule for construction
- Maintain the stores, keep the stocks of spare parts and POL (Petrol, Oil and Lubricants) and arrange the timely provision of these at the project site.

Land Development, Monitoring and Evaluation Section

This section is entrusted with the following activities:

- Perform the baseline and other socio-economic surveys under the guidance of a Research organization
- Fix monitoring points for Monitoring and Evaluation (M & E) activities
- Monitor the effects of drainage under the guidance of a research organization. This should include

- Monitoring cost of construction
    - o Hydraulic performance of the constructed subsurface systems
    - o Impact monitoring
    - o Environmental monitoring.
- Arrange other on-farm development activities such as land levelling and irrigation and drainage improvement, where applicable
- In cooperation with the Farmers Drainage Societies (FDS) monitor the effects of different designs such as the composite/singular layout.

Monitoring and evaluation should focus on both the social and technical aspects. Monitoring and evaluation should keep records of changes in household consumption patterns, dietary habits, propensity to develop new skills and training at the village household level. It would indicate possible spin-off from the technical activities. Since M&E is a major activity in the project therefore, M&E section should not be involved in the overall Project progress monitoring, which may be part of the Project Administration.

The Farmers Participation Section

The major activities of this section are:

- Appraise the perceptions of farmers toward the project including identification of salient social issues that can potentially impact project outcomes
- Gender issues related to women' role
- Enlist the association of weaker sections of the society
- Provide inputs on social considerations in refining the exact boundaries of the project area
- Involve farmers in monitoring and assessing the social impacts of the project
- Assist in the development of socially acceptable design and layout parameters of the drainage system
- Create awareness of the benefits and motivate farmer involvement in all phases of the project
- Mobilize the farmers participation in construction and O&M of the project
- Ascertain the capacity and willingness of farmers to contribute to project in cash, kind, labour and land.

While performing the above tasks, the project may embark on a social marketing strategy and an operational plan to prepare the beneficiaries to the intricacies of the new technology. This social marketing has to be reinforced by technical training of the users to familiarize them with various equipment that are part of the drainage system. A workable physical model can also be made and demonstrated through a mobile exhibition to show the operations that would take place at the field level. Video films of some old projects can also be shown.

Since, farmers are not well conversant with the benefits of land drainage and are not oriented towards operation and maintenance of the systems, there is a need to train the development agents, members of the association and all the stakeholders to sensitize them towards the benefits of land drainage. This is accomplished by organizing field days, exhibitions and video shows besides few lectures of interest suited to the stakeholders groups. On-site demonstrations of improved practices might help the farmers of the project to replicate practices that impact the productivity.

Skills in conflict resolution and mediation will be a major forte of this section. It should focus on the composition of workable units and ensure the unwilling farmer groups to take interest in the project. While mitigating these conflicts, the discussions may lead to a rethinking on how issues of crop compensation and land acquisition should be presented to the farmer groups. The conflicts during the construction stage will necessarily have to be resolved quickly and efficiently.

The Operation and Maintenance Section

Too often an assumption is made that subsurface drainage does not require any maintenance. This statement is partially correct. Although few drainage systems installed long ago are still functioning efficiently without any major maintenance yet it may not be true for each and every system if appropriate quality control is not exercised. The role of this section should be to see that the completed projects are well maintained and operated. Some of the activities of this section are:

- Train the farmers in operation and maintenance of the system with the assistance of the Machinery Section
- If necessary, operate and maintain the pumps in cooperation with the farmers and Machinery Section
- Maintain and operate pipe cleaning devices where necessary
- Plan and monitor the maintenance of open drains where needed
- To provide inputs to the relevant section for corrections to be made in the system.

MONITORING AND EVALUATION

Drainage systems are designed on the basis of investigations made on small areas. The up-scaled or averaged values may not represent large areas especially when there are large spatial and/or temporal variations in soil parameters. Moreover, some logical assumptions are made and guidelines developed on the basis of experience are used in the designs. Quality of construction is another major factor that might influence the performance of the drainage system. With this view, a representative part of the system is monitored and evaluated to assess its functioning. Both short- and long-term monitoring is desirable at least for first few projects in a region (Table 6.5).

Table 6.5 Objectives of monitoring and evaluation of drainage projects

Short-term monitoring	Long-term monitoring
Hydraulic performance of the system (drain flow, rate of lowering of the water table)	Developing relationships between crop yield and water table, soil salinity etc.
Salt leaching (Trend of soil reclamation)	Salt and water balance
Estimation of hydrological constants (hydraulic conductivity and drainable porosity)	Economic analysis (See Chapter 11)
Sediment deposit in drain pipes	Performance of envelope materials (sedimentation studies)
Crop performance	
Financial analysis (See Chapter 11)	

The data requirement and the frequency of their measurements during monitoring are listed in Table 6.6. For this purpose, strengthening of the observation well network might be required to obtain a representative picture of the depth to water table. Additional observation wells required to monitor the groundwater table and their locations are shown in Fig. 6.22. Besides, observation wells are also installed at the center of the 3 well spread out control plots (non-SSD area). It may be noted that such a rigorous monitoring as suggested in this section is necessary for research orientated projects. In purely applied projects data monitoring can be curtailed to a great extent.

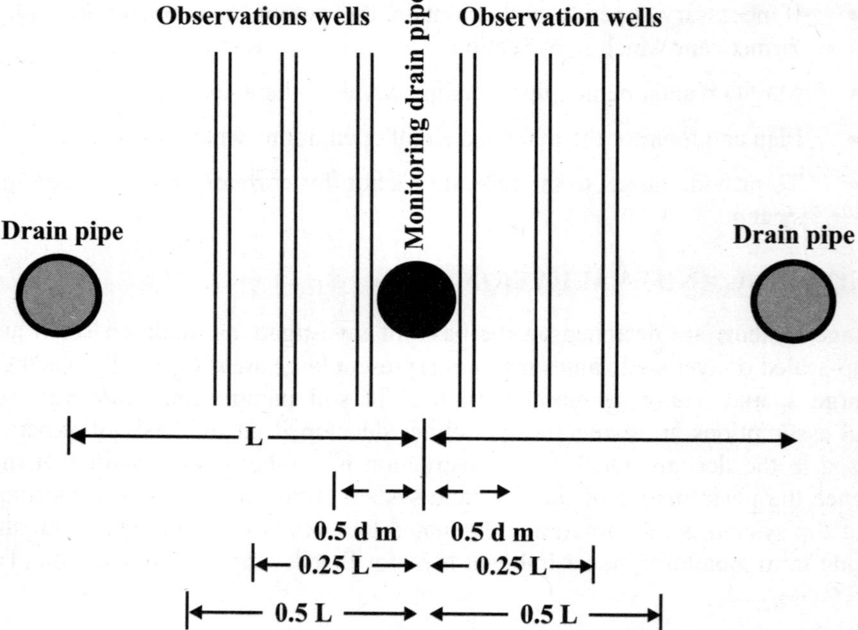

Fig. 6.22 Installation of additional observation wells around monitoring drain (*d* is the diameter of the pipe and *L* is the drain spacing)

Table 6.6 Data required for monitoring and evaluation and their frequency of measurement

Item	Frequency
Meteorological data	
Precipitation-Rainfall	Daily (on project site)
Pan evaporation	Daily (on project site)
Temperature	Daily (nearest observatory)
Humidity	Daily ,,
Sunshine hours	Daily ,,
Wind velocity	Daily ,,
Hydrological data (surface drain flow)	
Irrigation water runoff	Selected irrigation events
Main drain	Daily
Field drain	Selected events
Hydrological data (subsurface drain flow)	
Lateral flow	Daily in the first season, weekly later on
Outlet (collector) flow	Daily
Last outfall	Continuously with stage level recorder
Groundwater table	Daily initially and then weekly
Piezometric head (For artesian pressure)	Daily initially and then as per requirement
Soil water transmission and retention characteristics	
Saturated hydraulic conductivity	Annual
Infiltration characteristics	Annual
Soil moisture retention	Annual
Soil moisture	Selected irrigation event
Sediment deposit in subsurface drains	
Drains with filter	1st or 2nd year seasonal (After Monsoon season and after irrigation season); later on it may be in 3-4 years interval
Drains without filter	
Mechanical analysis of the soil samples of test site	
Entrance head loss at the drain	
Water level in piezometers in the drains and away from the drain	Daily for selected event
Water Quality	
u/s of outlet	Twice a week
d/s of outlet	Twice a week
Soil Chemical properties	
Salinity (EC)	Seasonal (after monsoon season and after irrigation season)
Sodicity (SAR)	

Item	Frequency
Alkalinity (pH)	
Land use and irrigation	
Crop cover distribution	Seasonal (for every crop season)
Crop calendar	,,
Irrigation calendar	,,
Crop yield	‘’
Drainage installation/land development cost	Keep regular records during installation
Cost of crop production and returns	Every season

Source: Gupta (2011)

Monitoring the Water Table

The groundwater table data are used to derive the following information:

- A comparison of the water table behaviour in drained and non-drained areas (Fig. 6.23).
- The relationship between rainfall and water table as well as rate of fall of water table in drained and non-drained areas following a rainfall event (Fig. 6.24).
- The profile of water table between the drains (Fig. 6.25).
- The hydraulic heads at mid-point of the lateral to be used in the hydraulic performance
- Water table contour maps to compare with the pre-drainage situation.

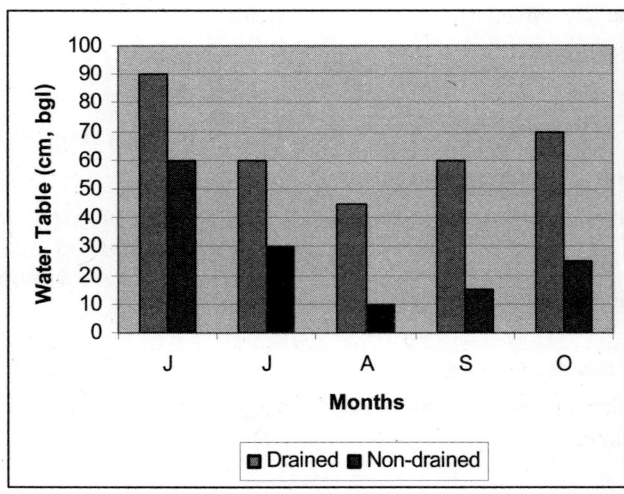

Fig. 6.23 Comparison of water table in drained and non-drained areas during June (J) to October (O)

These data sets are also used to assess whether the rate of fall of the water table is as per the assumption made (30 cm in 2.5 days' time).

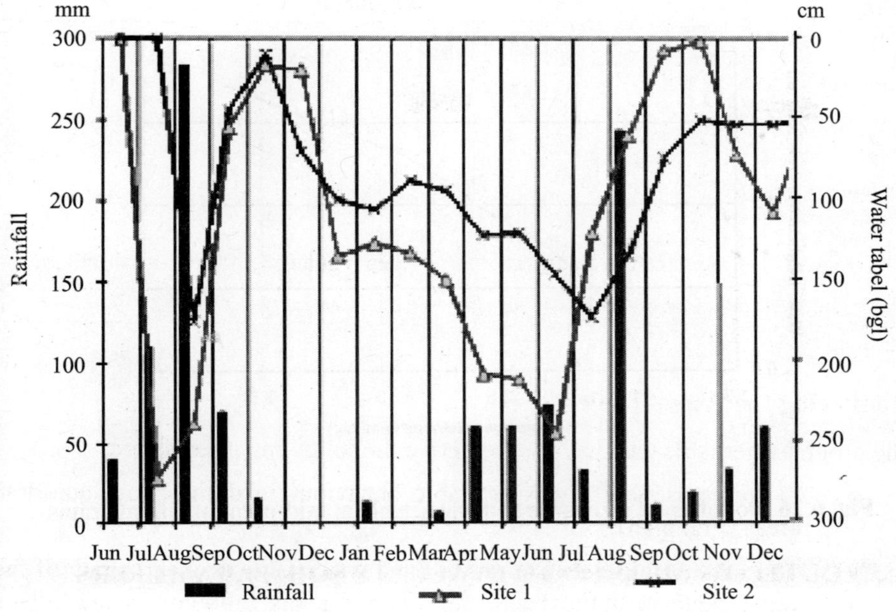

Fig. 6.24 Plot of rainfall and water table in drained areas

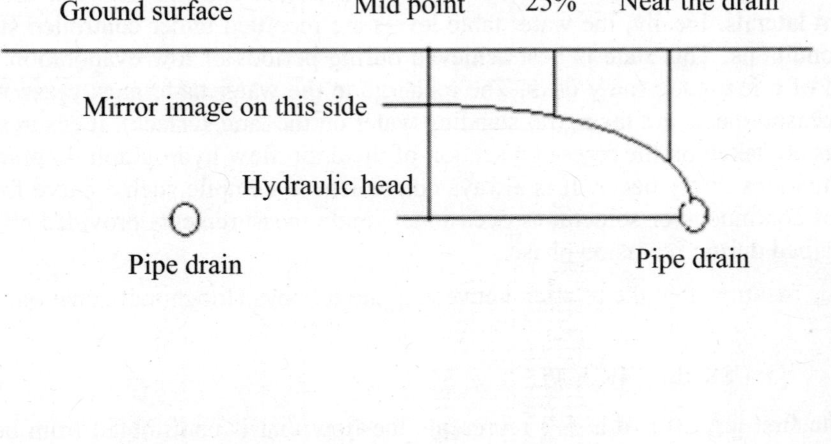

Fig. 6.25 Plot of water table profile between two adjacent lateral

The data on depth to water table at mid-point of the two laterals and drain discharge is also plotted as drain discharge versus hydraulic head curve (Fig. 6.26). These curves are used to determine hydraulic conductivity, drainable pore space and depth to impermeable layer in an inverse mode.

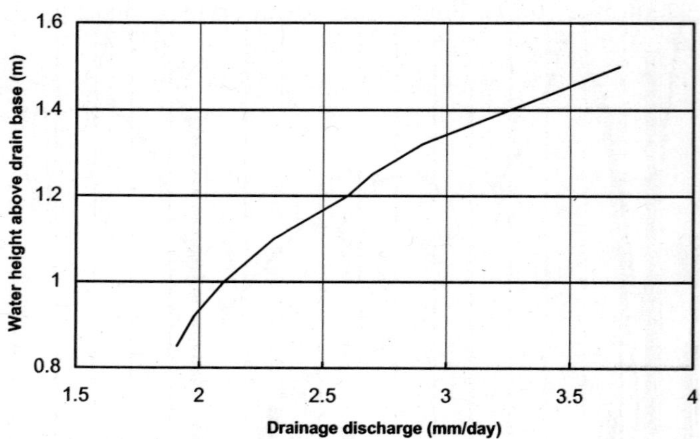

Fig. 6.26 Plot of drain discharge and water table at mid-point of lateral drains

DRAIN OUTFLOW METHOD TO EVALUATE SOIL PARAMETERS

Steady State Equation

The hydraulic conductivity for drainage projects is determined by analyzing the records of drain discharge and water table head at mid-spacing between the two adjacent laterals. Ideally, the water table levels are recorded under controlled steady state conditions. This state is best achieved during periods of low evaporation after the end of a few good rainy days. The recharge to the water table must cease while these measurements are taken (no standing water on the land surface). It ensures that readings are taken on the recession section of the drain flow hydrograph. In practice, such situations rarely occur. It is always convenient to compile such a curve from a series of continuous or sometimes occasional single measurements provided all data are obtained during recession phase.

It is assumed that the relation between q and h obeys Hooghoudt equation, such that:

$$q = (8K_b d_e h + 4K_a h^2)/L^2 \tag{6.4}$$

The first term ($8K_b d_e h/ L^2$) represents the flow that is contributed from below the drain depth while the second term ($4K_a h^2/L^2$) describes the flow contributed from above the drain depth. Equation (6.4) can also be written as follows:

$$q = A\,h + B\,h^2 \tag{6.5}$$

Such that $A = 8K_b d_e/L^2$ and $B= 4K_a/L^2$

Dividing both sides of Eq. (6.5) by h, we also get

$$q/h = A + B\,h \tag{6.6}$$

The shape of plot of q versus h (Eq. 6.5) on a normal arithmetic graph paper will depend upon the relative contribution of the two terms. The plot will become increasingly non-linear with increasing h due to increasing contribution of the second term of the equation (Fig. 6.27). Since data are collected from quite heterogeneous fields, the shape of the curve might vary and one may end up in some peculiar situations. The errors in the measurement of h, depth variation in the drains along the length due to the slope of the drains, error in measurements of the height of the observation wells above the soil surface and hydraulic head needed to compensate the entrance resistance all cause such variations. Such situations need to be tackled with care (Ritzema, 1994). For example, a q-h line might intersect the x-axis at some point. It means that even when there is some positive hydraulic head, the drains are not discharging. Some minimum head is needed for flow to take place. In such cases, it may be advisable to subtract this value from the measured h values. The new h values then are used for further analysis.

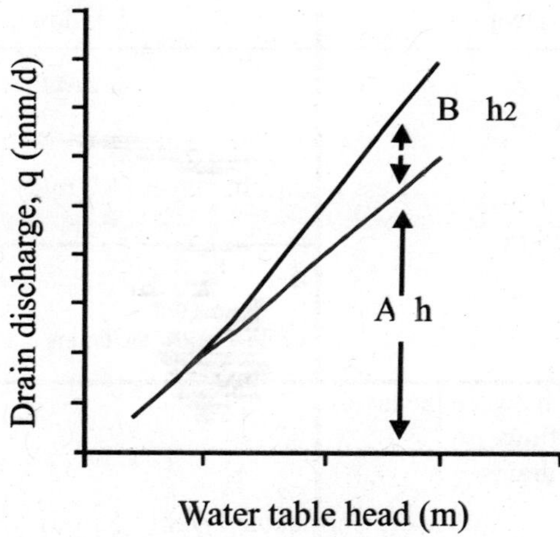

Fig. 6.27 The h-q plot showing non-linearity in the plot with increasing h

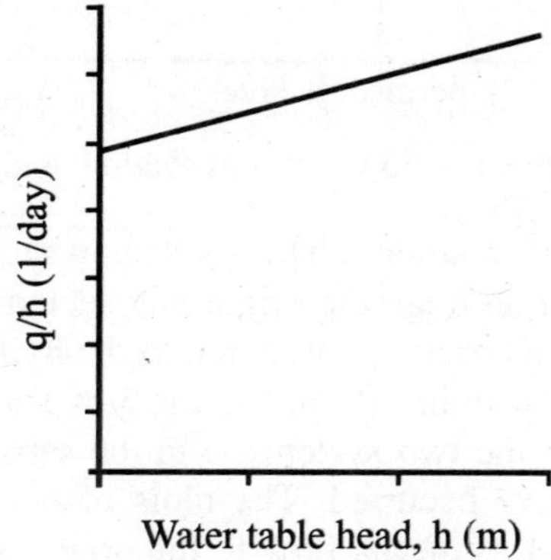

Fig. 6.28 The q/h - h plot with contribution from layer above the drains

If we use Eq. (6.6) instead of Eq. (6.5), q/h versus h graph should be a straight line. Following two situations are anticipated. If there is no flow from above the drains, the straight line should pass through the origin. Slope of this line would be equal to A and hence $K_b d_e$ and if d_e is known, K_b can be calculated straightway. If the flow above the drains is significant, the plot would be straight line, as shown schematically in Fig. 6.28. Slope of this line yields $4K_a/L^2$ while the intercept gives $8K_b d_e h/ L^2$. Thus, it is possible to calculate K_b and K_a. If K_b and K_a are equal, then equivalent depth can also be calculated explicitly and compared with the investigated value.

Example 6.1

At a drainage site, drainage systems were installed at shallow and deep depths. The profile characteristics for the two situations are shown in Fig. 6.29. The hydraulic heads and drain discharges for both the systems were measured and are reported in columns 1 and 2 of Table 6.7. Calculate the hydraulic conductivity and comment upon the differences in the estimated values.

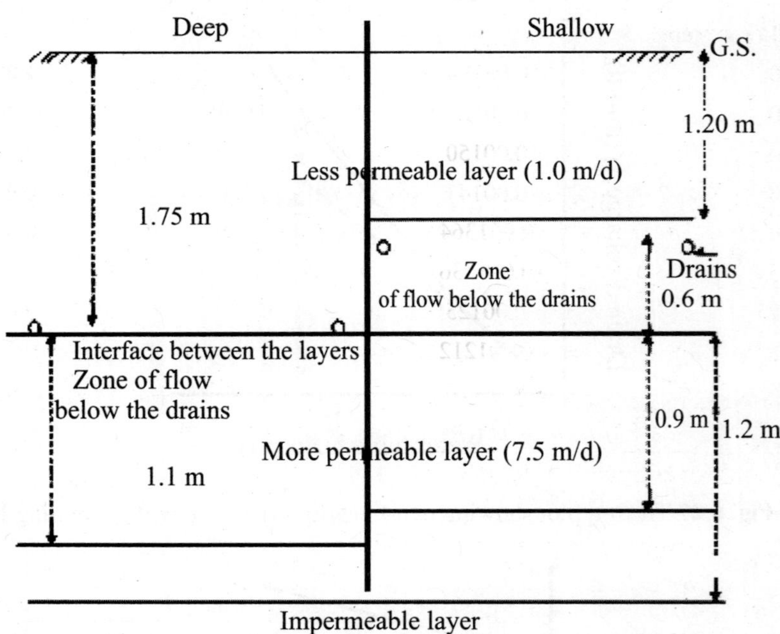

Fig. 6.29 Profile showing effective flow zones in shallow and deep drain systems

Plots of h versus q resulted in straight lines (Not shown). However, the resulting straight lines instead of passing through the origin intersected the X-axis around 15-20 cm right of the origin. On this basis, a correction to the hydraulic heads was made for both the deep and the shallow drains (h′ in column 3 of Table 6.7). The correction estimated at 17 and 20 cm for the two systems is in the same range showing some systematic error that might have occurred. The plots of q v/s h' pass through the origin (Fig. 6.30 and Fig. 6.31). It means flow to the drains in both cases is mainly from below the drains. To double check this observation, the q/h' v/s h' graphs are also plotted, which again resulted in a straight line almost parallel to the X-axis (Fig. 6.30 and Fig. 6.31). It further confirmed that flow in both cases is from below the drains with very little contribution from above the drains (Rawat *et al.*, 2001).

Table 6.7 Hydraulic head and drain discharge data

h(m)	q (m/d)	h' (m)	q/h' (1/d)
Deep system			
0.98	0.0132	0.81	0.0163
0.75	0.0102	0.58	0.0176
0.66	0.0081	0.49	0.0165
0.55	0.0060	0.38	0.0158
0.46	0.0048	0.29	0.0166
0.40	0.0039	0.23	0.0170
0.35	0.0030	0.18	0.0167
Shallow system			
0.620	0.00194	0.420	0.0046
0.570	0.00167	0.370	0.0045
0.530	0.00150	0.330	0.0045
0.520	0.00143	0.320	0.0045
0.500	0.001364	0.300	0.0045
0.495	0.001336	0.295	0.0045
0.470	0.00125	0.270	0.0046
0.450	0.001212	0.250	0.0048
0.445	0.00108	0.245	0.0044
0.420	0.00098	0.220	0.0045

h'= Corrected hydraulic head

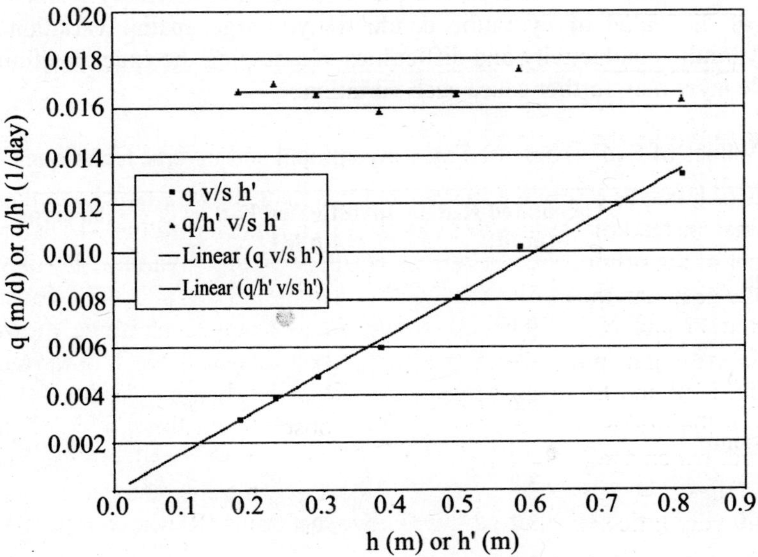

Fig. 6.30 Drain discharge and hydraulic head plots for shallow drains

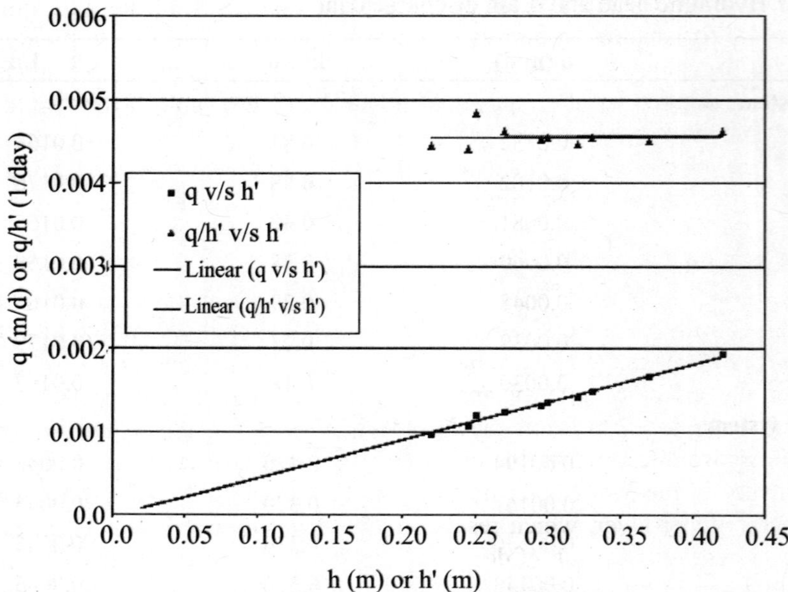

Fig. 6.31 Drain discharge and hydraulic head plots for deep drains

The calculated value of K_bd (6.6 m^2/d) for deep system is much closer to the investigated value of 8.2 m^2/d (Table 6.8). The minor differences in the value of transmissivity can be explained by the uncertainty in the estimated parameter by conventional procedures. For example, the hydraulic conductivity measured by an auger hole represents a very small volume of the soil around the auger. Contrary to this, a large volume of soil is involved in inverse technique, making it a better average value than an auger hole. Moreover, in an uncased auger hole, the dimensions of the auger hole are difficult to maintain and that may result in over-estimation of the value of hydraulic conductivity. Large spatial variation in the values of hydraulic conductivity and difficulties in assessing the exact position of the impermeable layer may further cause such variation.

Table 6.8 Values of K_bd as assessed by conventional and inverse techniques

Relation	Calculated K_bd (m^2/d)	Investigated K_bd (m^2/d)	% error
Deep system			
q-h	6.6		
q-h'	6.6	8.2	20
q/h'-h'	6.6		
Shallow system			
q-h	3.2		
q-h'	3.2	7.4	56
q/h'-h'	3.2		

Contrary to what is stated in the previous paragraph, the transmissivity values of $K_b d$ (3.2 m^2/d) for shallow system is much lower as compared to the value of 7.4 m^2/d obtained with conventional procedures of measurement of hydraulic conductivity multiplied by the equivalent depth to impermeable layer (Table 6.8). This difference may be attributed partly to the difficulty in assessing the exact position of the impermeable layer and partly due to the head loss caused by flow from permeable layer to the drains in the less permeable upper layer. The uncertainties as discussed in deep drains alone cannot be cited as a reason for the wide variation in the $K_b d$ values in the shallow drainage system. If one assumes that about 20% error is expected as a result of uncertainties, about 36% difference is positively due to the placement of drains in less permeable layer, which has altered the hydraulics of flow in the system. In deep drains the water flows through the upper layer under the effect of gravity and horizontal and radial flow remains in the permeable layer up to 1.1 m below the drain level. Contrary to this, in shallow drains, the water flows from less permeable upper layer to permeable lower layer. The water moving radially in the lower layer up to 0.9 m below the interface has to enter the less permeable upper layer, which results in significant loss of head. As the size of pore space decreases as a result of decrease in hydraulic conductivity from 7.5 m/d to 1.0 m/d, some part of pressure head (hydraulic head) is converted to velocity head. It is apparent, therefore, that in such cases, it may always be useful to place the drains at the interface of the layers rather than in the less permeable layer. If it is necessary to place the drains at shallow depth, the reduced transmissivity resulting from such placement must be considered in the design of the system.

Non-steady State Equation

Glover & Dumm as reported by Dumm (1954) equation for the falling water table reproduced as follows can also be used to assess the parameter α, which is equal to (π^2 Kd/f L^2).

$$L^2 = \frac{(\pi^2 \text{ Kdt})}{f \text{ Ln}\left(\dfrac{4h_0}{\pi h_t}\right)} \tag{6.7}$$

The equation can also be written as

$$h_{t+1} = 1.16\, h_t\, e^{(-\alpha \Delta t)} \tag{6.8}$$

Here $\Delta t = (t+1) - t$

Since q and h are linearly related and without making much error by retaining the constant 1.16, we have

$$q_{t+1} = 1.16\, q_t\, e^{(-\alpha \Delta t)} \tag{6.9}$$

Observed q_t or h_t values are plotted on a semi-log paper against time t on the X-axis (Fig. 6.32). If the system obeys the assumptions underlying the non-steady state situations, the observed values will more or less plot as a straight line. The slope of the line will give the value of α, which can then be used for further assessing the unknown parameters. In general, it is recommended that analysis should be done using both the steady and non-steady state equations. It will help to determine all the three parameters. The results are then compared with the values taken in the initial design. Necessary improvements are made in future designs, if necessary.

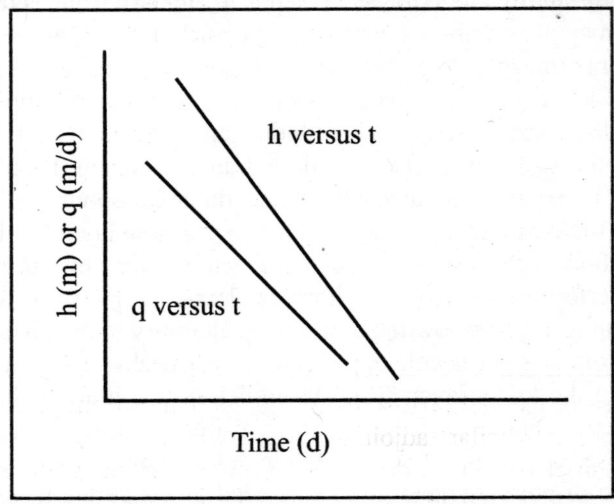

Fig. 6.32 Relations between hydraulic head or discharge with time

CROP CUTTING EXPERIMENT

Land use at the drainage site before and after implementing land drainage should be noted for each season. The data are tabulated (Table 6.9) or mapped to prepare land use maps. Crop cutting experiments are also conducted (HOPP, 2004). These experiments are conducted for at least 2 years before the drainage system is implemented and every season after implementing land drainage for few years. The data are tabulated and analyzed as shown in Table 6.10.

Table 6.9 Seasonal land use under drained lands

***Kharif* crops**	**Percentage**	***Rabi* crops**	**Percentage**	***Zaid* crops**	**Percentage**
Rice		Wheat		Green gram	
-		-		Pulses	
-		-			
-		-			
Other crops		Other crops			
Current fallow		Current fallow			
Permanent fallow		Permanent fallow			
Total	100	Total	100	Total	100

Table 6.10 Performa to monitor crop yield in drained and non-drained areas

Site: Season:

Grid point	Name of the farmer	Crops	Yield (kg)	Yield (t/ha)
1	2	3	4	5
1	A	Wheat		
	A	Oat		
	B	Barley		
2				
N				

The % increase in the yield is obtained by using data for the same crops either before drainage or in similar adjoining plots where drainage has not been implemented yet. Similarly, these data sets are also used to work out change in cropping pattern as well as change in cropping intensity after land drainage has been implemented.

More useful information will emerge if depth to water table and soil salinity (Table 6.11) is also measured at the same grid points. These data sets can then be used to draw relationships between the relative yield and depth to water table and between the relative yield and soil salinity.

Table 6.11 Soil salinity and crop yields in drainage projects

Month and year	Soil salinity (EC_e, dS/m)				Crop yield (t/ha)			
	Season 1	Season 2	Season 3	Season 4	Season 1	Season 2	Season 3	Season 4
2017								
2018								

Note: A similar table for water table is prepared.

Soil and Water Salinity

An important reason to install a drainage system is to facilitate leaching of salts present in the soil profile. The average soil salinity in the drained area before and after drainage will straight way gives a good picture of the performance of the system on this account (Fig. 6. 33). However, a plot of soil salinity against time (the number of seasons/years after the drainage system has been installed) will depict behavior of soil salinity in drained plots. Usually there is a fast decline in initial years. Its comparison with adjoining undrained area can also be made to highlight the benefits of land drainage.

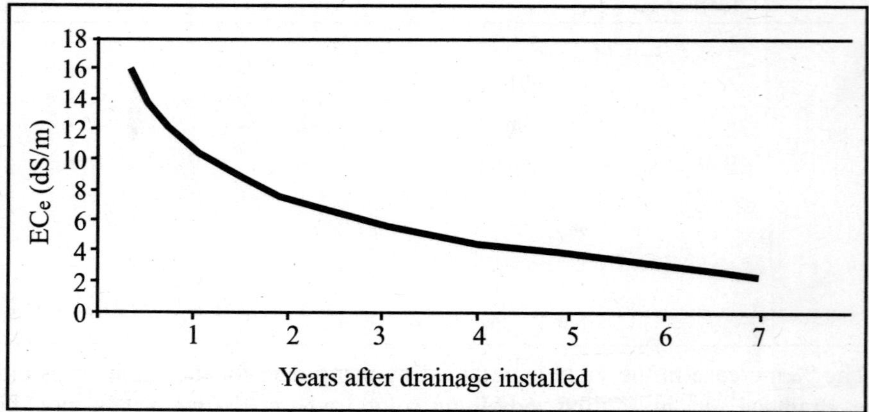

Fig. 6.33 Soil salinity variation in drained plots over the years in Andhra Pradesh

Similar changes in water quality are monitored from observation wells, manholes and sump in the drainage area. A plot similar to soil salinity will give an indication of the water quality before and after the drainage installation. Experiences at various sites in India have shown that there is general improvement in water quality over the years.

MEASUREMENT OF SEDIMENT DEPOSITS

In order to check the condition of the drain pipes, filter and soils, some lateral pipes need to be dug up within one year of installation and every 5 years thereafter. The purpose of this analysis is:

- To check the occurrence of silt deposited in the lateral drains provided with filter or no filter.
- To investigate the reasons of silt deposit in the pipe drains if any, and relate the silt deposit with soil particle size distribution/sodicity of the soil at drain depth.
- To ascertain the need of envelop/filter requirement in future subsurface drainage projects
- To ascertain the need for flushing.

Methodology

Identify the drains on which tests are to be conducted. First sample should be taken at the mid-point of the selected drain line. Soil samples should be taken from the top surface of the filter-envelope as well as 10 cm above it. A piece of pipe about 50 cm should be cut for sediment analysis and the system should be rejoined with a piece of pipe and filled back. Procedure to be followed in the laboratory analysis is:

- Take the sample (pipe + envelope + sediment) and air dry it
- Carefully remove the envelope from the pipe
- Remove any sediments attached on the pipe/filter
- Weight both the pipe and envelop separately
- Wash the pipe and envelope a number of times, air dry and weigh

The soil texture analysis is carried out for the soil drawn from around the drain depth. The results are analyzed for sediment retention in the pipe and envelope. An average depth up to 10 mm of sediment deposit in the pipe is considered acceptable. If depth of silt deposit in the mid length is > 10 mm then another sample at ¼ length on the line measured from d/s end of the lateral (minimum 20 m) should be dug and a similar process repeated. The data are assembled in tabular form (Table 6.12).

Table 6.12 Laboratory results of the excavated pipes/filters

		Weight of sediment in grams[1]			
		Sediments retained in the sample piece		**Soil retained per m of**	
Sample Number and length (cm)	**Site**	**Envelop**	**Pipe**	**envelop**	**Pipe**
A (50 cm)	A				

[1]10 mm of sediment in a drain pipe of 80 mm diameter weighs around 462 g/m of pipe

Field Results

At Dabhou, the quantity of sediments after four years of installation of drainage pipes with nylon netting, gravel, geotextile 90/3, coir mat and without envelope were found to be 200, 240, 122, 954 and 4200 g/m length respectively (Table 6.13). This indicated that minimum sediments were observed in geotextile 90/3. The drain pipes without envelope had the maximum deposits. The result also indicated that the clay percentage in sediment deposits was maximum in drains with geotextile 90/3 followed by gravel, coir mat, nylon netting and drain pipe without envelope. It seems that the geotextile envelope allowed passage of more clay through it while prevented coarser particles, as compared to other enveloping materials. Hence, geotextile 90/3 envelope material performed more effectively in preventing sedimentation (Table 6.13).

Table 6.13 Sediment deposition in exhumed drainpipes with and without envelope

Drain pipe/ Envelope	Drain spacing (m)	Sediments (g/m)	Percent by pipe section
Nylon netting	50	200	1.5
Gravel	50	240	1.7
Geotextile 90/3	50	122	0.9
Coir mat	50	954	6.9
Without envelope	50	4200	30.6

WATER AND SALT BALANCE

Lateral Drain Flow Measurement

Usually water flow in the lateral drain is measured at the man hole using volumetric method. A bucket, a measuring jar and a stop watch are used for this purpose. The drain flow is collected in the bucket while the time is measured with the stop watch. Once the bucket is nearly filled, the time is noted and the volume collected during this period is measured using the measuring jar.

$$\text{Drain flow (L/min)} = \text{Volume of the water in liters/time in min} \quad (6.10)$$

The discharge in L/min is converted in depth equivalent (mm/d) by dividing the area served by the drain, calculated as follows:

$$A=L.S+S.S/2 \quad (6.11)$$

$$q\ (\text{mm/d}) = Q/(1440 * A) \quad (6.12)$$

Where S is drain spacing m; A is the area drained, m^2 and Q is discharge, L/min.

Drain Flow Measurement at the Outlet

The drain flow at the outlet gives an idea of the total outflow generated from the drained area. Both surface and subsurface drainage outlets should be monitored using weirs. Normally a V-notch weir is used to measure the discharges. For each weir the following table is constructed to work out the drain outflows (Table 6.14).

Table 6.14 Calculation of surface and subsurface outflows from the area

Weir number	Average discharge (L/s)	Duration (days)	Drainage area (ha)	Average discharge (mm/d)
1	A	5	20	
2	B	3	30	
3				

Water Balance

All water inflow components such as rainfall, canal/groundwater for irrigation, seepage through irrigation network and other location specific data are monitored regularly. Similarly, all outflow components such as evapotranspiration from cropped and un-cropped areas and the drainage water from surface and subsurface drainage systems are measured. Normally for large drainage projects, natural inflow and outflows are neglected although in small drainage projects such neglection may result in large errors. Neglecting natural inflow and outflow, the water balance equation is written as:

$$I + R + G + S = ET + Sd + Sdd \pm \Delta s \pm \Delta w.f \qquad (6.13)$$

Here I is the irrigation water, cm; R is the rainfall, cm, G is the capillary flow from groundwater, cm; S is seepage contribution from canal network, cm; ET is the evapotranspiration, (cm), Sd is the surface drainage or runoff, cm; Sdd is the subsurface drainage, cm; Δs is the change in soil moisture (cm); f is the drainable pool space and Δw is the change in water table (cm). All these components are shown schematically in Fig. 6.34.

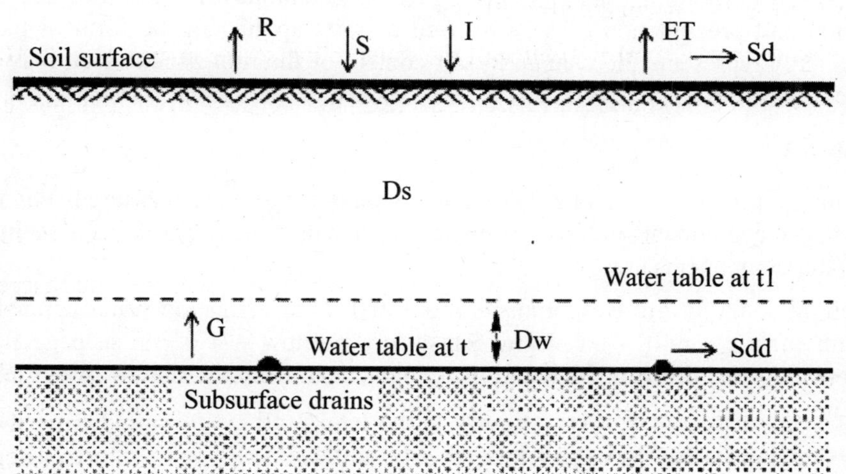

Fig. 6.34 Water balance components in drained lands

Example 6.2

In a drainage project, water balance components were measured and are given as follows:

Rainfall = 50 mm, Evapotranspiration = 55 mm

Irrigation water applied = 40 mm

Surface runoff = 12.5 mm, Subsurface outflow = 31.8 mm

Seepage flow from canal network and upward water movement from the shallow water table were estimated at 1 mm and 13 mm. Assume that there was no change in soil moisture and water table was lowered by 15 cm.

Calculate the water balance of the profile above the drain depth.

Incorporating the values in Eq. (6.13), we have

$$40 + 50 + 13 + 1 = 55 + 12.5 + 31.8 + 0.0 - f(150)$$

Note that water table contribution is negative. Assuming f to be 0.08 (8%), we get

Water balance = 104 - 87.3 = 16.7 mm

Water balance is positive for this period. Therefore, it must have contributed in the change of soil moisture.

Salt balance equation is easily derived from the water balance equation, by multiplying the respective component by its salt content:

$$I{*}Ci + G{*}Cg + S * Cs = Sd{*}Cs + Sdd{*}Csd + \Delta Cr \qquad (6.14)$$

Here C represents the salt concentration and subscript denotes the salt concentration of the relevant component of the water balance equation. Note that R and ET are not included in the salt balance equation because salt content of rain water and ET are quite less and can be neglected. Equation (6.14) also neglects the dissolution and precipitation of salts as well as salts applied in the form of plant nutrients. ΔCr represents the change in salt content of the root zone, which may be negative or positive.

Example 6.3

In a drainage project, water and salt balance components were measured. For the water balance components given in example 6.2, calculate the salt balance, if salinity of the water components are:

Rainfall = 0.0 dS/m, Evapotranspiration= 0.0 dS/m, Irrigation water applied = 0.5 dS/m, surface runoff = 0.6 dS/m, Subsurface outflow = 4 dS/m, seepage flow canal network = 0.5 dS/m, and upward water movement from the shallow water table = 5 dS/m

Incorporating the values in Eq. (6.14), we have

Total salt input = 40*0.5 + 13* 5 + 1*0.5 = 85.5 mm.dS/m = 547.2 kg/ha

Total salt output = 12.5*0.6 + 31.8*4 = 134.7 mm.dS/m = 862 kg/ha

It can be shown that 1 mm.dS/m on 1 ha of land = 6.4 kg of salts

This simple calculation for this period indicates that salt balance is negative or salts are being removed from the profile.

Economic Evaluation

All costs incurred on land drainage and benefits that accrue year-wise need to be documented during field monitoring. This issue is discussed in Chapter 11.

QUESTIONS (THEORY)

1. Write the construction sequence in subsurface drainage project briefly describing each operation.
2. Differentiate between blinding and backfilling of trenches in subsurface drainage.
3. Describe various kinds of machines used in the construction of subsurface drainage systems.
4. Name various drainage structures and describe the construction of gravity and pumped outlets.
5. Differentiate between: Manhole and junction boxes and relief wells and breathers.
6. Describe the need for surface water inlets in subsurface drainage. Differentiate between surface water and blind inlets.
7. What is the role of regulating structures in drainage systems? Explain at least one of such structures.
8. What are the major symptoms of subsurface drainage malfunctioning?
9. Write a note on maintenance of subsurface drainage systems.
10. Discuss the clogging problems of drain pipes.
11. Briefly describe the organizational structure for a drainage organization.
12. Describe the importance of monitoring and evaluation of drainage projects. What additional steps are required to extensively monitor a drainage project?
13. Describe the use of Hoogoudt equation to determine hydraulic conductivity and equivalent depth to impermeable layer using drain outflow data..
14. Describe the use of non-steady state equation to determine hydraulic conductivity and drainable pore space using drain outflow data.
15. Describe the procedure of conducting a crop cutting experiment to assess crop performance in drained lands
16. Describe the procedure to assess sedimentation in drain pipes.
17. Describe the various salt and water balance components in a drained land.

QUESTIONS (OBJECTIVES)

1. Fill in the blanks

1. Backfilling follows the _________ process.
2. A soil condition in which fine sand in a saturated, fluid condition is buoyed up by hydrostatic pressure from below is termed as _______________ condition.
3. Peat soils contain _________ per cent or more organic matter.
4. A blind inlet besides surface water also drains ____________ water.
5. Breathers are provided to relieve ____________ pressures in the drains.

6. Pushing a rigid torpedo-shaped tip steel rod into the drain pipe to check sedimentation is known as _________.

7. As per Hooghoudt′s equation, head versus discharge plot is non-linear because of flow contribution from __________ the drain.

8. As per Hooghoudt′s equation, a discharge/ head versus discharge (q/h versus h) plot should cut the _______ above the origin.

9. In general drainage water quality _________ with years following drainage installation.

10 A ___________ guard is fixed at the end of a gravity outlet

2. Write down the correct or most appropriate answer for each question

1 The construction of the subsurface drainage system should begin from

(A) The laterals (B) The collectors

(C) The outlet (D) Anywhere

2 With increasing size of the pipe in subsurface drainage, the slope of the pipes is generally

(A) Reduced (B) Increased

(C) Remains the same (D) None of these

3 First wetting of subsurface drains is done

(A) By flooding the whole field

(B) By flooding the whole line

(C) Isolating one meter area on both sides of the drain and flooding the area around the dykes

(D) None of these

4 Which machines are used in mechanical installation of subsurface drainage

(A) Backhoes and excavators (B) Trenchers

(C) Trench less plows (D) All of these

5 Rodent guards are provided in

(A) Man holes (B) Outlets

(C) Drops (D) Junction boxes

6 Relief wells are provided in subsurface drainage

(A) On long lines at points where tile line might be overloaded

(B) On lines that have surface inlets

(C) Where a steep section of main changes to a flat section

(D) All of such vulnerable points

7 Regulatory sluices are provided

(A) To maximize salt leaching

(B) To minimize excessive drainage

(C) To minimize nutrient leaching

(D) Both B and C

8 Malfunctioning of subsurface drainage are revealed by

(A) Submerged outlet

(B) Sudden reduction in drain discharge shortly after rain or irrigation

(C) Wet spots that remains wet after the field has dried off

(D) All of these

9 Clogging of drains is caused by

(A) Chemical precipitation (B) Ochre formation

(C) Tree roots (D) All of these

10 Ochre problem is considered more problematic if it is of

(A) Soil origin (B) Groundwater origin

(C) It is of same in both cases (D) None of these

11 As per Hooghoudt equation, a plot of head versus drain discharge curve is a straight line, if

(A) Flow is from below the drains only

(B) Flow is from above the drains only

(C) Flow is from below and above the drains

(D) None of these

12 As per Hooghoudt equation, a plot of head versus drain discharge curve is a straight line, if flow is only from below the drains, the straight line theoretically passes through

(A) Somewhere at x-axis (B) Somewhere at y-axis

(C) Passes through the origin (D) None of these

13 As per Hooghoudt equation, a plot of head versus drain discharge divided by head curve is a straight line, if flow is from above and below the drains, the straight line theoretically passes through

(A) Somewhere at x-axis (B) Somewhere at y-axis

(C) Passes through the origin (D) None of these

■■■■

Chapter - 7

SPECIAL DRAINAGE TECHNOLOGIES

INTRODUCTION

Horizontal pipe drainage, a widely adopted technology to reclaim waterlogged salt affected soils cannot be applied to all sites. To overcome the problems associated with pipe drainage under several settings, a number of alternate techniques have been developed, tested and implemented. These technologies can solve site specific drainage problems at reasonable costs. These can also be integrated with pipe or other drainage techniques. This chapter discusses the basic principles and practices related to some of these technologies (Table 7.1).

Table 7.1 Alternate drainage techniques for waterlogged salt affected lands

Kind of drainage system	Technology
Subsurface drainage systems	Mole drainage
	Drainage for seepage interception (Interceptor drains)
Vertical drainage	Tube well including skimming well drainage
Bio-drainage	Bio-drainage
	Interceptor bio-drainage
Special activities	Sub soiling
	Permeable backfills

MOLE DRAINAGE

Heavy soils, having clay (particles < 0.002 mm) content in the range of 30 to 60%, are characterized by low hydraulic conductivity usually less than 0.03 m/d. On the basis of textural classification, these soils are classified as clay or clay loam and have unfavorable physico-chemical and hydro-physical properties. If natural permeability of these soils is applied to design pipe drainage systems, drains spacing may be around 10 m and in some cases may be even less than 5 m. Such drainage system cannot be economically justified especially because these soils are not suited to high value cropping at least during the first few years of reclamation. These soils are amenable to mole drainage, which is also referred as pipe less drainage. Mole drainage is also used when natural drainage is restricted due to lack of slope or heavy clay sub-soil prevents downward movement of water. In fact, the real purpose of mole drains is to drain water that enters from top and not the groundwater. Mole drainage is an inexpensive and effective method of drainage widely used in the clay soils of temperate regions such as United Kingdom (UK), northern Europe and New Zealand.

The mole drains are unlined channels formed in clay subsoil with a mole plow with a cylindrical foot, often with an expander which helps to compact the channel wall. Typically the foot diameter is 75 mm, the expander diameter 85-100 mm, the leg thickness 25 mm and the leg length 200 mm (Fig. 7.1 left). These dimensions can vary within a narrow range. A mole plough designed at Central Institute of Agricultural Engineering (CIAE), Bhopal had a foot of 100 mm with 110 mm bullet or expander. The foot and expander form the drainage channel while the leg generates the slot with associated soil fissures. The leg fissures are vertical while the others are formed at an angle of about 45° to the direction of travel depending upon the soil conditions (Fig.7.1 right). The fissures extend from the surface of the mole drain and laterally out into the soil. Any surplus water above moling depth can therefore, move rapidly as a preferential flow through these fissures into the mole channel.

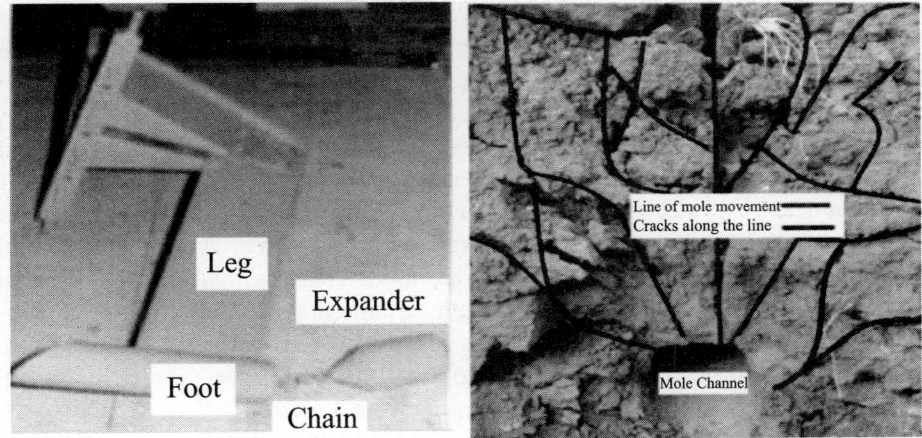

Fig. 7.1 Parts of a mole plough (Left) a view of cracks in a mole drained field (Right)

Soils Suitable for Moling

The success of a mole drainage system depends upon how best the desired water flow path for a particular drainage situation has been achieved? Besides, it also depends upon the stability of mole channels. Soils most suited to mole drainage should have sufficient amount of clay that are free of stones, gravel, or sand lenses (Rycroft and Amer, 1995). The minimum clay content should exceed 30%, preferably in excess of 37%. Clay silt ratio should be more than 0.5, although a ratio as high as 0.62 is desirable. Soils should be fairly consistent and without patches where drains are likely to fail. Many soils with a clay percentage higher than the minimum do not hold stable moles because they shrink and swell excessively between wet and dry cycles or because they are structurally unstable. At the time of moling, the soil at moling depth should be around lower plastic limit (moist enough to be moulded by hand into a ribbon without breaking but not wet enough to become liquid when worked in the hand). The soil above the moling depth should be friable not plastic so that there is adequate traction and the soil cracks well. Conditions to lay mole drains are satisfactory when the soil at mole draining depth has a moisture content of 20-25%.

A month with an overall deficit (Rainfall- evapotranspiration) of 50 mm is suitable for moling. In India, the ideal time for moling is October-November, as the soil starts drying following the monsoon rains. Besides, the mole channel should remain dry for at least a month after installation allowing the soil to ripen before the mole conducts water. It helps in improving the life of mole drains.

Depth and Spacing

Mole drains are usually installed at shallow depths, being in the range of 0.5-1.2 m. The best depth is 60 cm for arable and 55 cm for grasslands. A rule of thumb is that the expander to mole draining depth ratio is 1:7, i.e. a 70 mm diameter expander should have mole depth of approximately 490 mm. The mole drains should be deep enough to remain protected from the dynamic loads exerted by passes of heavy farm machinery and/or from swelling and thawing effect in Vertisols. The drains should also be deeper than the critical depth of the soil, the critical depth being related to soil properties (more plastic the soil, lesser is the critical depth) and the mole plough design (ratio of the tine depth to tine thickness ranging from 2-7).

The spacing of mole drainage is usually based on tradition. The ideal spacing being the one at which all the soil get busted and fractured. This spacing is around 90 cm. Generally, the spacing is in the range of 1.5 – 10.0 m though spacing as large as 15 m are also used. The size of the moles varies widely; larger sizes are generally used at greater depths. A 15 cm mole will usually create a hole of approximately 11.3 cm in diameter. The size should on an average be 75 mm in the range of 60-90 mm, although even 200 mm mole drains have been constructed. A gradient of 1% is preferable though grades as low as 0.2% are used. The maximum permissible grade to safeguard against scouring is 3%. Length of mole drains varies from 20 to 100 m but can go up to 500 m depending on the grade. Greater the risk of channel failure, the shorter should be the length of run to avoid local collapse. Similarly, on steep slopes, length of run may be reduced to avoid channel erosion.

Mole drains are installed using 75 HP wheel or crawler tractors depending upon the size and depth of moles (Fig. 7.2). The speed of operation of a mole plough is less than 1.0 km/h and one can cover about 5-10 ha per working day.

Fig. 7.2 Tractor driven mole plough in operation

The drain out fall or outlet is the most important part of the system. If this fails, the whole system fails. To prevent the collapse of the drain at the outlet, a short PVC pipe is inserted. Mole drains can discharge to open drains or into interceptor drains.

Life

High clay content alone is not a guarantee against failure of mole drains. In a study in Iowa (USA), the mole channels failed early in fields where clay content was higher. In fact, life of a mole drain depends upon several factors, amongst them the consistency being the most important. Well installed mole drains under most suitable condition can work from 10 to 30 years. There are instances when moles have failed in 2-5 years when installed under unfavorable conditions. The mole drainage system installed during 2008-09 at CIAE Bhopal is still functioning without any loss in its efficiency. From practical and economic analysis standpoint, life of mole channels is taken as five years. The cost of mole drainage should be viewed against the life expectancy of mole drains.

MOLE DRAINAGE: INDIAN EXPERIENCES

Mole drains were tried in Digod/Nanta in the Chambal command area in the late seventies under a UNDP project. The mole drains were laid at depths ranging from 0.6-0.8 m and at spacing of 2-5 m covering an area of 17 ha. The cost of the system at that time was assessed at Rs. 200/ha (Lovas, 1972). Relative flow rates from mole vis-à-vis subsurface drains are shown in Table 7.2.

Recently, mole drainage was tried for Vertisols of MP at Bhopal with clay and silt content of 55% and 31% respectively. Mole drains at an average depth of 0.6 m at a longitudinal gradient of 0.20% were installed. Three different spacing (2, 4 and 6 m) each row having a length of 60 m were used. The cost of the system worked out as Rs. 1080, Rs.760 and Rs. 640 per ha for the 3 drain spacing respectively. The drain discharge varied from about 2 mm to 12 mm depending upon the rainfall. The average soybean (JS-335) yield was more in all the spacing as compared to control where no drainage was practiced for all the 4 years of experimentation (Fig. 7.3). Since there was no significant difference in the yields obtained under 2 and 4 m drain spacing (Singh & Ramana Rao, 2014), a mole drain spacing of 4 m is recommended for the Bhopal region (Rao *et al.*, 2009 a,b).

Table 7.2 Drainage discharge from subsurface and mole drains with sub-soiling

Method of drainage	**Drainage discharge (% of the total supply)**		
	July, 1970	**September, 1970**	**February, 1971**
Plastic drains (1.8 m deep with 8-12 m spacing)	31.0	29.1	28.5
Asbestos cement drains (1.2 m deep with 15-20 m spacing)	21.0	20.0	19.8
Mole drains (0.6-0.8 m with spacing of 2-5 m)	12.8	12.0	11.7

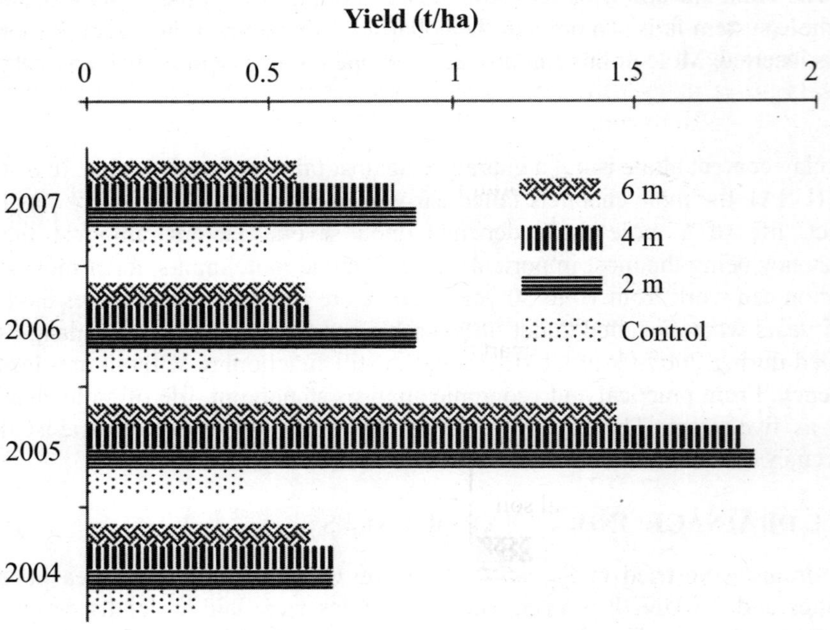

Fig. 7.3 Yield of soybean under different mole spacing and control (no drainage)

Relative Economics of Drainage Technologies

Comparative economics of surface, horizontal subsurface (pipe) and mole drainage was worked out under similar conditions at CIAE, Bhopal. The benefit: Cost (B:C) ratios in surface drainage systems were 1.77, 1.35 and 1.53 and for subsurface drainage system 1.36, 1.44 and 1.55 respectively for soybean, maize and pigeon pea. The B:C ratio for soybean with mole drainage was the highest at 1.91. The results revealed that surface drainage, SSD and mole drainage technologies are technically and economically feasible for soybean, maize and pigeon pea crops in temporarily waterlogged Vertisols.

FAILURES AND REMEDIES

Creep failure or the failure of mole drains at the outfall is common in mole drainage. Drain failure on account of creep in 1-4 years have been reported. To remedy this situation, gravel mole drains are used, which are essentially mole drains filled with gravel. The use of the gravel prevents channel failure in unstable soils. Under some circumstances gravel moles can be installed for lengths up to 200 m to reduce the cost of the subsurface collector system. In this technology clean gravel is introduced down the back of a mole plough, with the expander removed. Gravel mole ploughs incorporate a hopper to allow finely graded gravel to fall into the mole channel. These ploughs have been used successfully in the UK in heavy soils that cannot hold "normal" mole drains. An attempt to test this system in Andhra Pradesh is in the offing. These are relatively expensive costing 70-100 percentage more compared to moles without gravel.

One of the major limitations of mole drainage is in disposal of drainage effluents. When connected with open drains, there is always a fear of water backing up and damaging the moles. On the other hand, the spacing of subsurface drainage in clay soils poses many difficulties and is quite expensive. A way to resolve these problems is to go for a combination of tile and mole drains. With this system, a series of subsurface pipes are first installed to a gradient across the slope at least 150 mm deeper than the proposed mole drainage. Permeable fill is then spread over these pipes so that permeable fill is well above the proposed depth of the mole drains. The moles are then drawn across the pipes. The moles intersect the permeable fill and this provides a rapid connection for the water to flow from the moles into the pipe system (Fig. 7.4). The mole drains are constructed approximately perpendicular to pipe drains which can now be spaced farther apart than the conventional subsurface drainage.

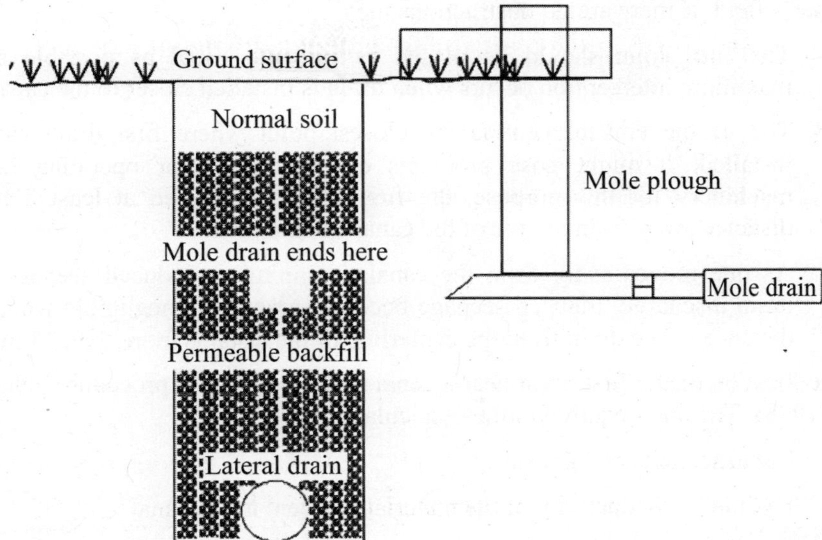

Fig. 7.4 A schematic representation of an integrated mole and subsurface drainage system

DRAINAGE FOR SEEPAGE INTERCEPTION

Interceptor drains are used for the following two purposes:

- To intercept seepage water from canal networks
- To intercept water that seeps down the sloping lands

While interceptor drainage for sloping lands has been discussed in Chapter 4, interception of the seepage water is discussed in this section. Following techniques are used to intercept the canal seepage.

- Deep surface drains (Fig. 1.13, right)
- Pipe drains (Fig. 1.13, right)
- On-farm ponds
- Bio-drains - growing exotic tree species along the canal/seepage surface (Fig. 1.14, right)

While deep surface/pipe drains and farm ponds are discussed in this section, bio- interceptor drains are discussed in the section on bio-drainage in this chapter.

Interceptor Drainage for Canal Network

Open and pipe interceptor drains are used to intercept seepage. Either a single or multiple laterals drain(s) are installed along the canal. These drains, sometimes called seepage drains, prevent development of water logging in the adjoining lands. If the impermeable barrier is at shallow depth, the drains can be placed on or close to the impermeable layer. But if impermeable layer is deep enough, the drains are located above the wet area so as to intercept the major fraction of the flow. Sufficient borings need to be made to locate the impermeable barrier. One cannot play much with the spacing between the first drain and the centerline of the canal especially if it reaches the farmer's field. If there are no obstructions then

- The first drain should be placed as close to canal as possible since maximum interception occurs when drain is installed closer to the canal.
- Toe of the embankment is the closest point where first drain can be installed. It might pose problems during digging or operating heavy machinery for this purpose, the first drain is installed at least 8-10 m distance away from the toe of the canal embankment.
- As one moves away from the canal, it minimizes induced seepage and drain discharge. Induced seepage becomes practically negligible when the distance of the drain from the centerline of the canal is more than 75 m.

The location of the first drain near a canal is estimated by a procedure given by USBR (1978). The data required in these calculations are:

- Canal sections and grades.
- Hydraulic conductivity of the material adjacent to the canal
- Weighted hydraulic conductivity between permissible root zone depth and impermeable barrier.
- Depth to impermeable barrier
- Distance from the centerline of channel to the irrigated land

Assuming that there is no recharge from the precipitation, the seepage from the canal based on Darcy's law for steady state flow is (Fig. 7.5):

$$q_{(c)} = KD.\ d(z)/X \tag{7.1}$$

Here, q_c is the canal seepage in the root zone depth at the edge of the irrigated field, m^3/d- m length; K is the weighted hydraulic conductivity between the root zone depth and the barrier, m/d; D is the average height of the water table between the canal (d+h) and selected root zone depth above the barrier; H, m, d(z) is the depth of the root zone at the edge of the irrigated field and water surface in the canal, m; and X is the distance from the centerline of the canal to the edge of the irrigated field. The q_c can also be estimated from the canal geometry as follows:

$$q_c = K(B + 2d_d)/r$$

such that B is the width of the water in the canal at the normal operating level, m and dd is the depth of flow in the canal, m. The constant r is used to convert K values to seepage loss determined using ponding tests. Commonly used value is 3.5.

If depth of the drain is fixed, then the distance to the first drain is estimated as follows:

$$S = K\frac{(H^2 - d^2)}{2q(c)} + X \tag{7.2}$$

Here, S is the distance from the canal centerline to the first drain, m; H is the vertical distance from the barrier to the selected root zone depth at the edge of the irrigated field, m; and d is the vertical distance between the barrier and the drain, m (Lal, 2005, Fig. 7.5). The spacing for additional parallel drains required to maintain the water level at the desired height can be determined from drain spacing equations discussed in Chapter 4. The intercepted water is either disposed in canal network by gravity at some lower reach or is pumped to irrigate the adjoining lands especially the non-commanded areas, if any.

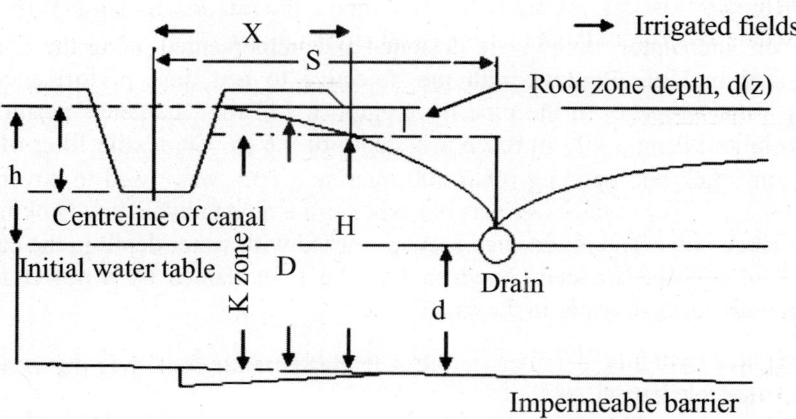

Fig. 7.5 Definition sketch of interceptor drain along a canal

Example 7.1

Determine the distance of the first drain from the centerline of a canal. Assume that the bottom width of the canal is 3.0 m, the canal side slopes are 2:1, average weighted K = 0.5 m/d; d = 6.1 m; H = 7.65 m; root zone depth, d(z) = 1.2 m, X = 18 m, depth of drain, h = 2.75 m, D = 8.25 m and the deep percolation (DC) is 9.4 mm during a period of 14 days between irrigations.

Substituting the value of K, D, d(z) and X in eq. (7.1)

$$q(c) = KD\frac{d(z)}{X} = 0.5 \times 8.25 \times \frac{1.2}{18} = 0.275\, m^3 / d \text{ per } m \text{ length}$$

Substituting above value of q(c) in Eq. (7.2)

$$S = K\frac{(H^2 - d^2)}{2q(c)} + X = 0.5 \times \frac{7.65^2 - 6.1^2}{2 \times 0.275} + 18 \quad = 37.4\ m$$

Calculate deep percolation from irrigation, q(d), for the field width (S-X) above the drain as follows:

$$q_d = DC * (S\text{-}X) / (1000 * \text{irrigation interval})$$

$$= 9.4\ (37.4 - 18) / (1000 * 14) = 0.013$$

Calculate total q = q(c) + q (d) = 0.275 + 0.013 = 0.288 m^3/d per m length

Calculate second estimate of S by substituting q in Eq. (7.2)

$$S = K\frac{(H^2 - d^2)}{2q(d)} + X = 0.5\frac{(7.65^2 - 6.1^2)}{2 \times 0.288} + 18$$

$$= 0.5\frac{21.31}{0.576} + 18 = 36.5\,m$$

Application

Pipe and open interceptor drains were designed and implemented along the Sharda *Sahayak* Canal in Uttar Pradesh with the objective to test their performance in intercepting canal seepage. In the pipe interceptor, a 144 mm diameter perforated pipe was installed along a 400 m reach at a depth of 1.6 m. Geotextile filter of 3.5 mm ± 0.3 mm thick and opening (O_{90}) 300 micron ± 10% was used to cover the perforated pipe. A 10 cm sand layer was placed over the pipe to avoid any choking of the filter in alkali soils. The intercepted seepage varied with water depth in the canal. The amount of seepage arrested shown in Fig. 7.6 is explained by a power form relationship with depth of water in the canal.

$$Q_{drain} = 1837.2\ H^{0.8072} \qquad [R^2 = 0.9448] \qquad (7.3)$$

Here, Q_{drain} is the intercepted seepage flow rate L/h and H is the depth of water in canal, m.

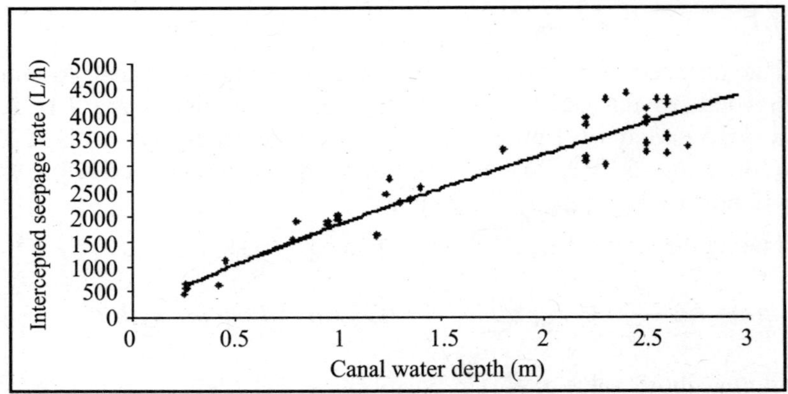

Fig. 7.6 Relationship between seepage and depth of water in canal

Measurement during June 2004 to May 2005 showed that maximum volume of 4825 m^3 was arrested during the month of March, 2005 (Fig. 7.7). Total volume of seepage water arrested during first year and second year of installation were 24084 m^3 and 25781 m^3. Since the outlet remained fully submerged during rainy season, the seepage interception was almost nil during the rainy season (Fig. 7.7). In open interceptor drain, maximum volume of 5867 m^3 was intercepted during the month of December 2004 while minimum 1102 m^3 was intercepted during May 2004. Total volume of seepage water intercepted by open drain was 14122 m^3 in first year and 20844 m^3 in second year. The maintenance of open drains also posed several problems. Based on this experience and the experiences in Rajasthan and Pakistan, it seems that pipe interceptor drain is a better alternative to intercept seepage from the canal network in vulnerable reaches.

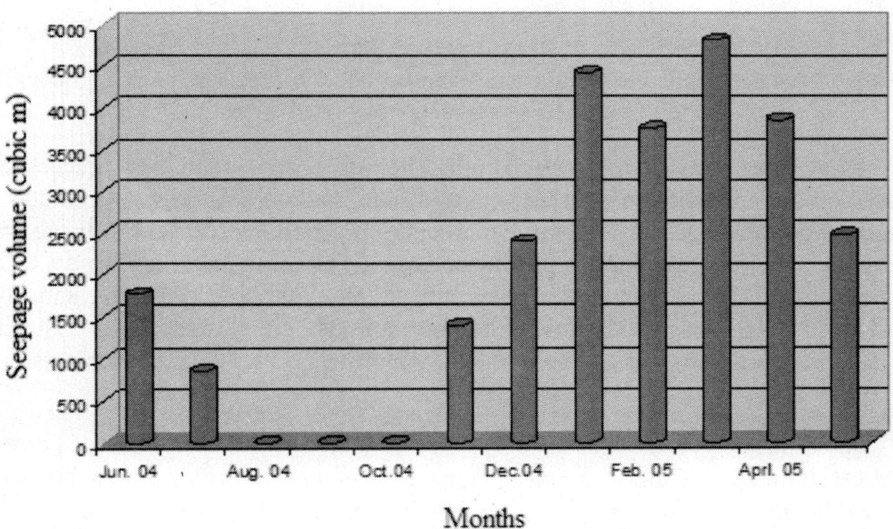

Fig. 7.7 Monthly variations of intercepted seepage volume in pipe interceptor drainage

ON-FARM PONDS FOR SEEPAGE INTERCEPTION

Like a seepage drain, on-farm ponds are also capable of intercepting seepage because of the hydraulic gradients created towards the ponds. A system of this nature was attempted in the Sharda *Sahayak* canal command and found quite effective. In this intervention, a fish pond based multiple water use model was conceptualized and adopted in an area of 1 ha (Sharma *et al.*, 2008). A pond was dug to a 1.75 m depth covering an area of 0.4 ha (25% of the farm area). The seepage water depth measured in the pond varied from 0.10 m to 1.25 m during December 2009 to November 2010 (Fig. 7.8). Since no water from outside sources is added, visually, farmers are convinced that the water in the pond has come from the canal. The water quality is good both for aquaculture and agriculture. While the pond is used for aquaculture, the remaining area under the dykes and remaining portion of the field raised by spreading out the left out bottom soil is put under vegetable, horticulture and arable cropping depending upon the pH of the soil.

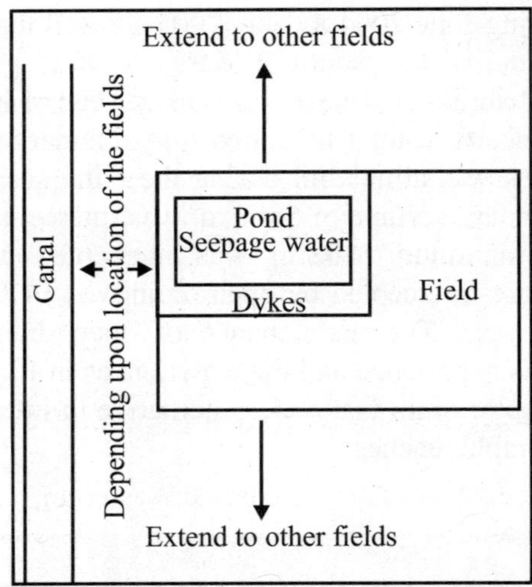

Fig. 7.8 A schematic view of the land shaping model for seepage interception and multiple water use through integrated farming

Rice, wheat, sorghum (forage), mustard, garlic and onion were successfully grown on the embankment and the elevated field beds. If such kinds of models are extended on both sides, it would provide opportunities to intercept the seepage water along the whole reach (Sharma & Gupta, 2013). The most important aspect of this model is that it does not entail any governmental intervention except some subsidy to dig the ponds. Since the guidelines on subsidy are available in watershed development program, the scheme can be implemented without many problems.

VERTICAL DRAINAGE

Vertical drainage is defined as *'the removal of groundwater by means of a borehole (or well) either under pressure (artesian conditions) or through pumping'*. The vertical drainage system practically includes one or more deep/shallow tube well(s). The network of open or closed drains to dispose of the water also constitutes a part of the system. The group of tubewells having sufficient overlap of their individual cones of depressions controls the groundwater table or artesian groundwater conditions. The overlapping of cone of depressions ensures uniform lowering of the water table at all the points. The oldest study on vertical drainage was carried out during the first decade of the 20th century. Few studies in the 20th century were conducted in Punjab, Haryana and Gujarat. All of these have been functional only for a short period due to operational and maintenance problems. Yet, the system performance can be judged from the success achieved in tackling the problems of water logging that was prevalent in sixties in parts of Punjab, Haryana and Uttar Pradesh. Large numbers of tube well structures have lowered the water table. It may be noted that there has been nearly 25 fold increase in tube well structures in Haryana from about 28,000 in 1966-67 to more than 752,000 in 2014-15. Currently on an average 127% of average replenishable recharge is being pumped from these wells making the state as an overexploited state. Similar situation has emerged in the other states as well.

Types of Drainage Systems

Vertical drainage system is a complex system of hydro-technical structures and equipment intended for catching and draining the groundwater. Tube wells with pumping equipment are the initial links of such a drainage system which extract groundwater and influence its regime. Although vertical drainage systems have been categorized in several ways, on the basis of location of wells in the drainage area, well drainage is classified in the following four groups.

Systematic drainage

In this system, wells in the drainage area are located more or less uniformly over the whole area (Fig. 7.9). The wells are located depending upon factors such as (i) access to the well, (ii) convenience in the use of drained water and (iii) disposal of water to the natural drainage system.

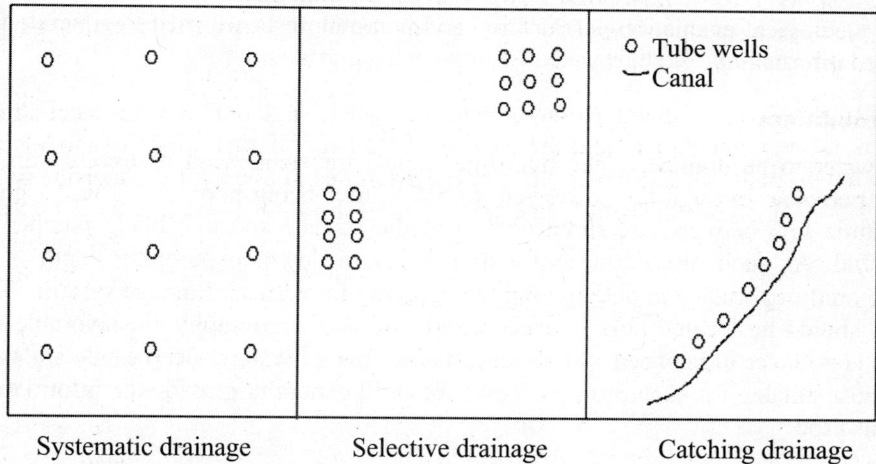

Fig. 7.9 Schematics of kinds of vertical drainage systems

Selective drainage

In this system of drainage, wells are concentrated at unfavorable sites within the area (Fig. 7.9). It may be either a depression or a saline tract within an irrigation command. The excess water in this area might come from irrigation return flows from higher spots within the command.

Head or catching drainage

This system of drainage aims at complete or partial interception of the flow through one or more lines of wells located along the front of flow (Fig. 7.9). This kind of system is also used to intercept seepage from the canal network (Kumar and Singh, 2005).

Shore drainage

It is a linear system of wells along the shores of a river or reservoir to support the groundwater level.

Basic Design Information

For vertical drainage system design, general investigations carried out for subsurface drainage (Chapter 3) should be supplemented at least on three aspects i.e. the aquifer conditions, soil conditions and water and salt balance.

Aquifer conditions

The primary requirement of a successful vertical drainage project is the existence of a thick aquifer capable of supporting an economical well. As such, most area to be drained should be underlain by a good aquifer. The aquifer conditions are largely defined by its area extent, depth and thickness, transmissivity and storage coefficient (specific yield). Neither too shallow nor too deep aquifers in the range of 10-100 m below the ground surface are most suitable. Aquifers in which transmissivity approaches or exceeds 9.29×10^{-3} m^2/s will yield excellent wells. Aquifers with transmissivity 1/10 of this value will permit an uneconomical or at best a marginal well. Geological evaluation, test drilling and pumping tests are used to generate the desired information.

Soil conditions

The water to be drained or the hydro-geological transients such as excess rainfall must percolate through the soil layers to the aquifer being pumped. Thus, a good hydraulic link between the strata nearest to the surface and that being pumped is essential. As such, the permeability of soil layers down to pumping depth is of paramount importance in determining the success of a vertical drainage system. The wells should be located only in areas where soil and stratigraphy are favorable i.e. there is no layer in between, which restricts the flow of water. Deep sandy soils are the most suitable for such projects. Test holes will generally provide the information on this aspect.

Water and salt balance

The information on elements of water and salt balance is essential for the whole area as well as for the regions earmarked for drainage. Separate analysis for the aeration zone, groundwater and surface water should be made. Similarly, monthly and seasonal water and salt balances need to be evaluated. With the information on present salt and water balance, project regime after drainage is chosen and the new salt-water budget for the operative and reclamative periods is determined. This will help to decide the capacity of the vertical drainage system and the intensity of drainage in various regions and seasons.

Site Characteristics for Vertical Drainage

As already discussed, vertical drainage may not be successful at all places. Therefore site conditions must be evaluated before planning a vertical drainage system. Comparative features of sites where vertical drainage systems can work or pose problems are given in Table 7.3.

Table 7.3 Favourable and non-favourable site feature for vertical drainage

Favourable features	Non-favourable features
Good quality aquifers are available	Non-availability of good aquifers; low transmissivity does not allow construction of tube wells of sufficient capacity
Good quality groundwater	Groundwater of high EC, SAR or RSC resulting in problems for its disposal. Tube well components are more prone to wear and tear/corrosion due to salinity resulting in higher maintenance cost
Cost of tube well construction is less and electrical power is available	Lack of fresh water without which possibility of conjunctive use of drainage water my pose problems
The tube wells are installed by the farmers. The government inputs are restricted to priority connections of electricity	High operational and maintenance costs when systems are installed under government set-up
Upward seepage pressure from a confined aquifer is cause of drainage problem and the cost of horizontal subsurface drainage is prohibitive	
The areas to be reclaimed are only waterlogged with no or minimal salinity	
Potential of using drainage water for irrigation or minimum disposal related problems.	
One window system for agriculture and groundwater development is available	

THEORY OF FLOW TO DRAINAGE WELLS

To calculate the total number of tube wells required to create the anticipated drawdown, one must know the discharge of a tube well. Similarly, for proper selection of a pump and a power unit, one must know besides the discharge, the pumping depth. Several equations have been developed to relate well discharge to drawdown or the piezometric surface (or water table), transmissivity and storage coefficient of the aquifers. Both steady and unsteady states have been considered in such developments. The main assumptions in deriving these equations are as follows:

- The aquifer is horizontal and has infinite areal extent. It is homogeneous and isotropic in respect of its hydraulic properties
- The well is pumped (or flows) at a constant rate
- The water table or piezometric surface prior to pumping is horizontal
- The well fully penetrates the aquifer. It is screened, perforated or otherwise open for the entire height of the aquifer.

Steady State Discharge Formulae

The steady state discharge formulae for a water table (unconfined) and a confined aquifer well are respectively given by the equations:

$$Q = \frac{\pi K (h_2 - h_1)}{Ln (r_2/r_1)} \tag{7.4}$$

and

$$Q = \frac{2 \pi K D (h_2 - h_1)}{Ln (r_2/r_1)} \tag{7.5}$$

In these equations Q is the discharge from the well, m^3/d; K is the hydraulic conductivity, m/d; r is the radial distance from the pumped well center with subscripts indicating any two points at two distances, m; D is the thickness of the aquifer, m; and h is the height of the piezometric surface or water table surface from the bottom of the aquifer, m; h_1 is head at distance r_1 and h_2 at distance r_2. Geometry of the well drawdown and the symbols used in Eq. (7.4) and (7.5) are indicated in Fig. (7.10) left and (7.10) right.

In terms of drawdown, S, at a distance r, Eq. (7.4) for a water table well can be written as:

$$S = \frac{Q\, Ln (r_e / r)}{2 \pi K H} \tag{7.6}$$

Here r_e is the radius of influence at which drawdown is assumed to be zero. H is the aquifer thickness. An assumption made in deriving this equation from Eq. (7.5) is that the drawdown is considered to be much smaller than the total thickness of the aquifer.

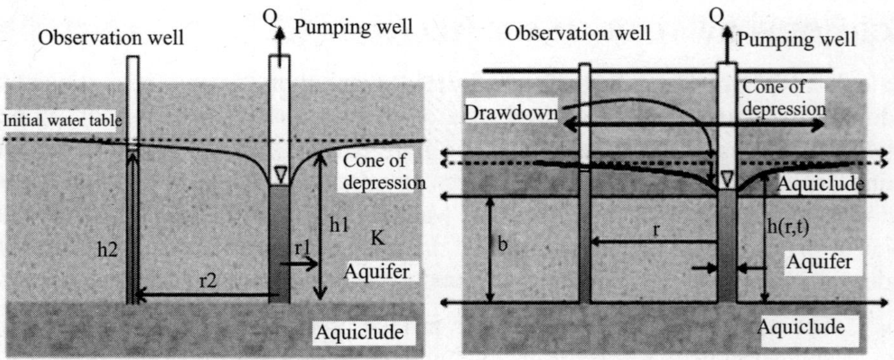

Fig. 7.10 Groundwater pumping under steady state conditions from unconfined aquifer (left) and confined aquifer (right)

A special case of discharge drawdown relationships is when there is a uniform recharge at the soil surface at a rate v (m/d). In that case tube well discharge or radius of influence can be calculated by the following equation.

$$\frac{Q}{K}\text{Ln}\frac{r_e}{r_w}-\frac{V}{2K}(r_e^2-r_w^2)=h_e^2-h_w^2 \tag{7.7}$$

Here r_w is the radius of the well and h_w is the drawdown at the interface of the well. For known v, Q, K and h_w; r_e can be calculated. For a well entirely replenished by vertical flow- through and neglecting the small term r_w^2 in comparison to r_e^2, and writing $r_e^2 = Q/\pi v$, Eq. (7.7) can be simplified to;

$$S=\frac{Q}{2\pi KH}(2.3\log r_e/r_w-1/2) \tag{7.8}$$

The value of 1/2 in Eq. (7.8) is proportionately reduced if a part of the replenishment is by horizontal transfer. It could be neglected if $r_e/r_w > 100$.

The second case is of drainage well, which is pumped for 1/m th part of each day, the drawdown after n days of pumping, S_n is given by (Wesseling, 1973):

$$S_n=\frac{Q}{4\pi KH}2.3\log\frac{n!}{(n-1/m)!} \tag{7.9}$$

In practice, Eq. (7.9) is more practical than the one based on continuous flow as continuous pumping is rarely practiced under field conditions.

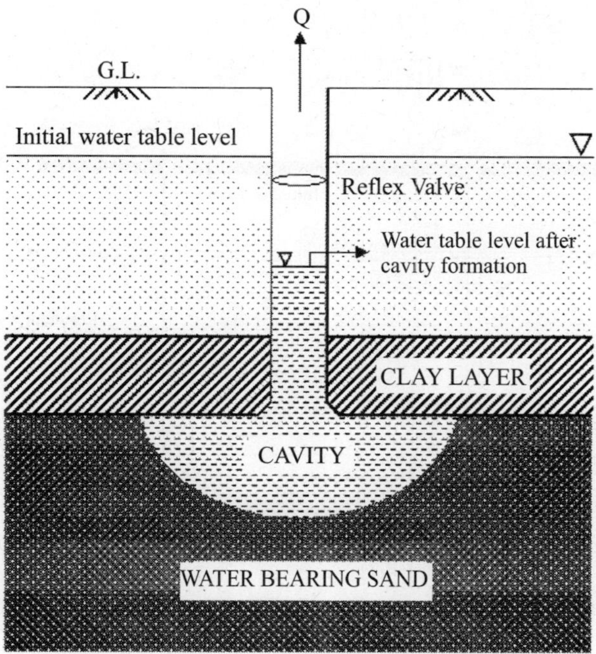

Fig. 7.11 A schematic view of a cavity well

Cavity wells have become quite common in the plains of Punjab, Haryana and Uttar Pradesh (Fig. 7.11). These wells do not require filter (Fig. 7.11). These wells have been instrumental in solving the drainage problem of the waterlogged areas of these regions. It is because clay layer which is essential for the success of cavity well exists at various depths. Thus, there is enough interconnection between the aquifer being pumped and the normal water table in the region. The steady state solution for such a well also known as non-penetrating well is given as:

$$Q = 2\,\pi\,K\,r_0\,(H - h_0) \qquad (7.10)$$

Here H is the piezometric head at r_e and h_0 at a distance r_0 i.e. at the interface of the well. Non-steady state solutions to the non-penetrating wells are given by Chauhan *et al.* (1975) and (Gupta, 1987).

Interference of Wells and Drawdown

Two or more wells located close to each other are said to be interfering, if the cone of depressions intersect each other. Consequently, the sum of the output of all the wells would be less than the total output summed together assuming no interference. The drawdown at any point in the area of influence caused by simultaneous operation of several wells can be worked out by superposing the drawdown caused by each well individually. Thus

$$S_t = c_1S_1 + c_2S_2 + c_3S_3 \ldots\ldots c_n\,S_n \qquad (7.11)$$

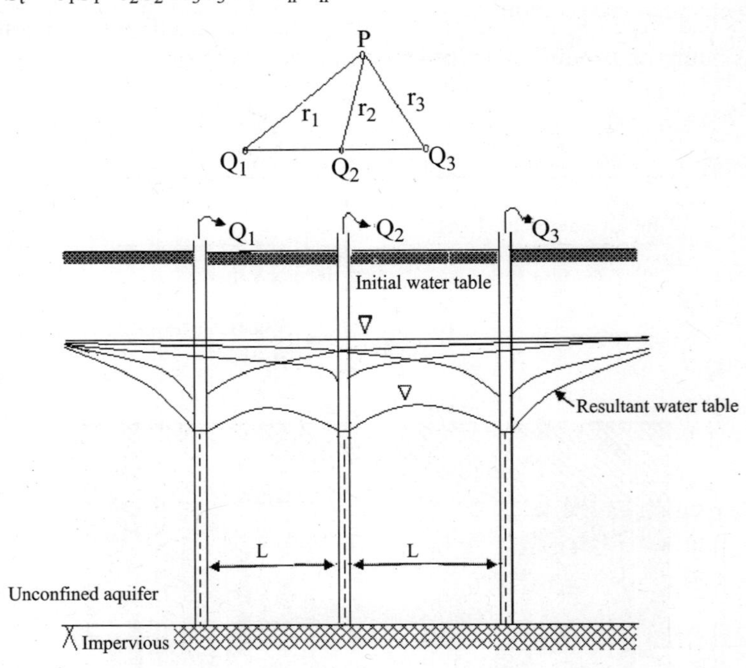

Fig. 7.12 Individual and composite water table situation when three equally spaced unconfined wells pump together

Here S_t is the total drawdown due to n number of wells having drawdown of S_1, S_2 … S_n and c_1, c_2 … c_n are arbitrary constants. Drawdown equations for specific cases can be derived using the above equation. For example, if a number of wells are installed in a straight line (Fig. 7.12) and if the drawdown, of the well is small compared to the saturated thickness of the aquifer, then the hydraulic heads at any point P can be calculated by the equation:

$$h_e^2 - h^2 = \sum_{i=1}^{N} (Q_i/\pi K) \, Ln \, (r_{e.i}/r_i) \tag{7.12}$$

Here Q_i is the constant discharge of i_{th} well, m^3/d; r_i is the distance of the i_{th} well from point p, m; $r_{e,i}$ is the radius of influence of the i_{th} well, m; K is the hydraulic conductivity, m/d; h_e is the static hydraulic head before pumping, m; and h is the hydraulic head during pumping, m.

If all wells have the same constant discharge, Q and have the same radius of influence, r_e, then the equation can be simplified to (Bear, 1972):

$$h_e^2 - h^2 = (Q/\pi K) \, Ln \, (r_e/r_m) \tag{7.13}$$

where r_m is the equivalent distance from point P and is given as follows:

$$r_m = (r_1 \, r_2 \, r_3 \, .. \, r_n)^{1/n} \tag{7.14}$$

Calculation of Number of Tube Wells

While designing a vertical drainage system, it is imperative to calculate the number of tube wells required to control the water table. Number of tube wells required for an area depends upon several factors such as:

- The aim of drainage i.e. whether short-term drainage is required or only seasonal or annual recharge is to be balanced
- Whether groundwater mining is needed or not
- Capacity and number of the existing tube wells
- Proposed measures which might reduce the input to groundwater notably the lining of canals, surface drainage etc.

The well spacing or number of wells for the area may be obtained by the relationship:

$$\text{Number of wells} = \frac{\text{Total drainage volume required}}{\text{Capacity of one well}} \tag{7.15}$$

The recommended capacity of wells for two zones in the state of Haryana is given in Table 7.4 (Bouman, 1984). Similar information from different states can be compiled and used.

Table 7.4 Draft (ha m) norms for minor irrigation structures

Type of well	A zone (Districts of Ambala, Kurukshetra, Karnal, Kaithal, Panipat, parts of Yamunanagar)	B zone (Other districts of Haryana)
Dug well (non-mechanical lift)	0.45	0.45
Dug well with pumping set	1.80	1.35
Shallow tube well	2.50	1.45
Direct irrigation well	15.00	15.00
Augmentation tube well	71.50	71.50

To determine the drainage volume or area irrigated by one well, three scenarios can be considered i.e. control of water table to avoid damage to crops; vertical drainage-cum-irrigation and vertical drainage on the basis of replenishable groundwater recharge (Table 7.5).

Table 7.5 Various scenarios and related information to calculate number of tube wells

Scenario	Explanation	Volume of water (m3)	Number of tube wells
Control of water table to improve aeration and avoid damage to crops*	Any additional rise in water table will be damaging, if water table is already within 2 m of the soil surface. In the monsoon season, it may reach at or near to the surface. For proper aeration, it is recommended that water table should be lowered by 30 cm from the soil surface within a period of 2.5-3.0 days.	$W_p = \frac{A \times 0.3 \times f}{d_p}$ Here w_p is the amount of water to be drained per day, A is the total area to be drained, m^2 and f is the drainable pore space and dp is period within which water table should be lowered.	$N = \frac{W_p \times 24}{C \times n}$ Here C is the capacity of one well, m^3/d, n is the number of hours for which tube wells run per day during the critical period.
Number of tube wells for drainage-cum-irrigation	The area commanded by a tube well is calculated from the information on the capacity of the tube well and related area on crops and their water requirement.	$A_c = \frac{C \times n \times I_n}{I \times 24}$ Here Ac is the area commanded by one tube well, I is the depth of irrigation, m and I_n is the irrigation interval between two irrigations in days.	N = A/Ac Here A is the total area to be irrigated, may be equal to the drained area
Number of tube wells on the basis of seasonal replenishable groundwater recharge	The net replenishable recharge is calculated from groundwater balance (Annexure II). Addition is made if mining is required and subtraction is made for natural drainage, if any	Let Np be the seasonal replenishable recharge in depth units.	$N = \frac{A \times Np \times 24}{C \times n \times n_1}$ n_1 is the average number of days a tube well runs in a year.

***Comment:** The number of tube wells in this case would be too high. As such, vertical drainage is rarely designed on this basis.

Example 7.2

Calculate the number of tube wells required to drain an area of 100 km^2 for the three cases

Case 1: To improve aeration. Here it is proposed to drain the water from the soil surface to 30 cm below the ground surface in 3 days. Drainable pore space is 0.1.

Case 2: Tube wells are used for irrigation wherein depth of irrigation is 5 cm at an interval of 10 days.

Case 3: Net replenishable recharge is 0.8 m. The tube wells are likely to work for 250 days per annum.

The aquifers in the area are suitable for shallow tube wells having discharge of about 12 L/s. Number of hours for which electricity is available is about 8 h/d.

Case 1. This case is illustrated in row 1 of Table 7.5. Using the two equations given in this row, we have

$$W_p = \frac{100 \times 10^6 \times 0.3 \times 0.1}{3} = 10^6 \text{ m}^3/\text{d}$$

Capacity of the tube well in m^3/d is calculated assuming each tube well of 12 L/s capacity as follows:

$$C = \frac{12 \times 1000 \times 60 \times 60 \times 24}{100 \times 100 \times 100} = 1.2*36 \text{ x } 24 = 1036.8 \text{ (Or approximately 1000 m}^3\text{/d)}$$

Number of tube wells for drainage can be calculated

$$N = \frac{10^6 \times 24}{1000 \times 8} = 3000$$

Case 2: This case is illustrated in row 2 of Table 7.5. Using the equations in this row, we have

$C = 1000$ m^3/d (from calculations in case 1)

$N = 10$ days

$I = 5$ cm

Therefore,

$$A_c = \frac{1000 \times 8 \times 10}{0.05 \times 24} = 66666 \text{ m}^2$$

Knowing the total area to be irrigated and the area irrigated by one tubewell, numbers of tubewells are given as:

$$N = \frac{100 \times 10^6}{66666} = 1500$$

It may be noted that number of tube wells required for irrigation are almost half the number of tube wells required for drainage to protect crop (Case 1).

Case 3: This case is illustrated in row 3 of Table 7.5. Using the equations given in this row, we have

For the net replenishable recharge of 0.8 m and putting all other values in the equation

$$N = \frac{0.8 \times 100 \times 10^6 \times 24}{1000 \times 8 \times 250} = 960$$

Irrigation versus Drainage Wells

Technically speaking irrigation and drainage wells are alike having similar components and design techniques. However, a well in the confined aquifer may have little utility as drainage well to control the water table if the confining layer is impermeable. But it can serve as an excellent well for irrigation purposes. The two kinds of wells also differ in some respects mainly related to the site conditions. Whereas an irrigation well is installed at any place with a good aquifer; for a drainage well interconnection between the upper layers and the aquifer being pumped is essential. Within a farm area, irrigation well is generally located at the highest point whereas such a consideration is not necessary for drainage well unless drainage water is used for irrigation. In fact, it may be useful to install drainage well at the lower portion of the field. Another difference in an irrigation and a drainage well is in respect of well spacing. Whereas, interference amongst irrigation wells is avoided, it is useful in drainage wells especially when the system is used to create draw down within a short duration. The interference amongst wells helps to lower the water table uniformly over the whole area.

Well versus Tile Drainage

Well drainage is useful in areas where the pervious substrata are at depths exceeding 5 m. If the groundwater is of good quality, the well water serves as a source of irrigation and will economize the cost of drainage. Since well drainage can lower the water table to greater depths than a tile drainage system, it can play a significant role in saline water regions as the emptied portion of the aquifer gets filled up with fresh water that can be exploited through skimming wells. Deeper depths to groundwater also reduce soil salinization. The cost per ha of a vertical drainage system is less than that of a horizontal drainage system but the operation and maintenance cost is much higher than the tile drainage. Horizontal drainage is useful in areas where artesian pressure in the pumped aquifers is too high or seepage is excessive. Since large amounts of water are contributed by areas outside the project in vertical drainage, horizontal drainage appears to be the best feasible solution to drain small areas. Tile drainage is an easy and most appropriate solution to drain lands where water table frequently reaches the soil surface and drainage is required to lower the water table within a short period to save the crops. If one has to achieve similar results using a vertical drainage system, cost may prove to be prohibitive because number of wells will have to be quite large.

Cost of Vertical Drainage

To make a precise estimate of the cost of vertical drainage system is difficult because large numbers of parameters influence the total cost. Number of tubewells per hectare will widely alter the cost of land drainage on per hectare basis. Even the cost of a single tube well varies widely from one location to another mainly due to differences in labour rates even when standard materials are used with almost similar cost. If the tube well water is used wholly or partially to augment the surface water resources, it is difficult to precisely estimate 'what fraction of the total cost should be charged to drainage'? Depth to groundwater table considerably increases the cost because in such events centrifugal pumps have to be installed in pits or submersible pumping plants need to be installed. When pumps are installed in a pit, cost of a single tube well has been estimated to vary from Rs. 13,000 to Rs. 21,500 when the water table is within 3 m to 35 m below the ground level respectively. A complete estimate of the cost of drainage through wells can be gathered from five schemes prepared by Haryana State Minor Irrigation Development and Tubewell Corporation (HSMITC) under the FAO/UNDP Project. According to these estimates the total cost of vertical drainage varied from 18 lakh to 61 lakh (Table 7.6). In terms of per ha basis, the cost was in a narrow range of Rs. 13,000-14,000 per ha. The annual maintenance cost was estimated at Rs. 2500/ha. The cost of vertical drainage can be reduced substantially, if farmers are encouraged to install tube wells and the water is utilized for crop production either through proper management or by mixing. Following policy issues may help to encourage farmers′ participation in groundwater development:

- Subsidy on tube well construction in saline water belt
- On-demand fresh water supply to leach down the salt or for application of cyclic/blending conjunctive use strategies
- Compensation in water rates in lieu of groundwater use.

Table 7.6 Cost of vertical drainage schemes executed by HSMITC in Haryana

S.No.	Scheme	District	Cost (Rs. Lakh)	Period	Water table decline (m)
1	Narwana area	Jind	19.90	1985-92	1.00-4.00
2	Rori área	Sirsa	21.00	1985-92	1.90-3.00
3	Fatehabad Branch (50 vertical drainage tube wells)	Fatehabad	61.00	1995-96	0.44-3.20
4	Kishangarh Sub Branch/Link channel (50 vertical drainage tube wells)	Hisar	61.00	1995-96	0.95-1.90
5	Jui feeder/canal (15 vertical drainage tube wells)	Bhiwani	18.3	1996-97	0.65-2.18

Case Studies

Few case studies of vertical drainage from Haryana and Punjab and a local adaptation of vertical drainage are described in the following sections.

Haryana Minor Irrigation and Tubewell Corporation, Haryana

A vertical drainage system was installed along Fatehabad branch. Fifty strainer type tube wells with discharge ranging from 10 to 15 L/s were installed during 1995 covering an area of 683 ha. The pumped water quality ranged from 0.8 to 2.0 dS/m and was mixed in the canal network. Initially a sharp decline in water table was noted up to a distance of 300 m from the tube wells. During 1997-2000, a number of tube wells went out of order due to wear and tear and the water table started rising again although it remained below 1.5 m for most of the time. In due course of time scheme failed to operate as most of the tube wells went into disuse. Apparently, the system worked well but required huge operational and maintenance cost, which are rarely available in a governmental set-up. Such issues needs to be considered while planning a vertical drainage scheme.

Department of Irrigation and Drainage, Govt. of Punjab

A pilot vertical drainage system was installed to drain an area of about 700 ha at Gulabewala in Faridkot district of Punjab In the year 1987-88 (Sondhi, 2007). Fourteen tube wells, 30 m deep, were installed in 5 rows. The row-to-row spacing was 1000 m and well-to-well spacing varied from 71 to 94 m. The fall in water table during the operation of the scheme ranged from 0.08 to 0.51 m. The EC of water increased over time with pumping. The increase in EC of the pumped water is attributed to upconing of saline water body situated at deeper depths. Scheme design must take care of such issues by studying the depth wise distribution of groundwater quality during investigations.

Multiple - well point drainage system

A multiple well point system consists of shallow multiple well points to skim the fresh water floating on the saline water, a common feature encountered in southwest Punjab as well as in coastal sandy soils. The well points are pumped centrally or through a centrally located sump. This technology has been tested using a battery of 24 wells, each 6 m deep (3 m blind pipe and 3 m screen) with a line to line distance of 100 m and a well spacing of 50 m. Each well point had a discharge of about 3 L/s (Shakya *et al.*, 1995).

Since the system skims only fresh water, there are no groundwater quality issues. There are no disposal related problems either because the groundwater is used for irrigation (Sondhi, 2007).

BIO-DRAINAGE

Biological drainage or bio-drainage is defined as *'the application of a biological material to absorb, utilize, transport and evapotranspire water stagnating on the land surface or in the root zone to drain the land'*. Quick growing plants are employed as bio-drain materials as these are capable of absorbing and transpiring huge quantities of water. Since water is transpired straight into the atmosphere, there are no disposal problems.

In essence, bio-drainage involves three processes (Fig. 7.13):

- The uptake of water from the soil medium by the roots (absorption)
- Transportation of water from root to plant parts
- Transpiration of water by the plants through the tree leaves.

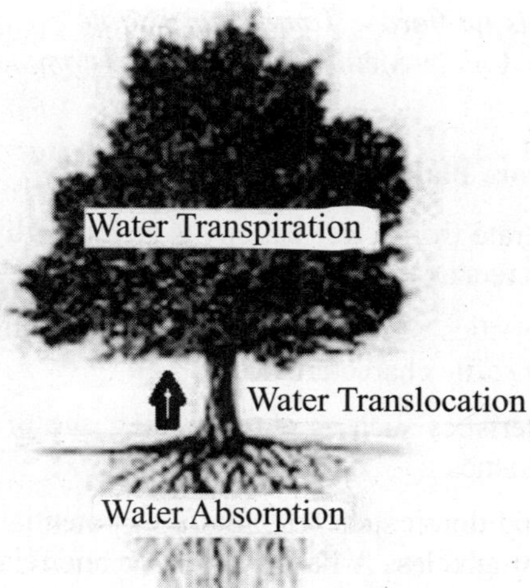

Fig. 7.13 Schematic view of the concept of bio-drainage

Apparently, good bio-drainage friendly plants must be able to absorb, transport and transpire water under stresses of water logging and soil salinity (Gupta, 2005).

Plant Suitability as Bio-drain Material

A critical part of bio-drainage is to select suitable plants that are tolerant to water logging and soil salinity. Dagar (2011) listed several species for block plantations, important amongst them being *Eucalyptus*, *Casuarina glauca*, *Acacia ampleceps*, *Terminalia arjuna* (Arjun), *Pongamia pinnata* (Papri) and *Syzygium cuminii* (Jamun). *Eucalyptus* with the following characteristics fits well into any bio-drainage scheme.

- It has the ability of deep rooting. Thus, it can lower the water table to greater depths. Recent studies however, indicate that maximum lowering of water table is underneath the planted areas only. The zone of lowering (cone of depression) of the water table extends only up to 10 m away from the planted areas.
- Significant amounts of water are used even in summer during which other plants adapt by manipulating stomatal openings.
- It can withstand water logging and alkalinity. Some of its species are also tolerant to salinity. As such, this species can be planted on these soils when economic constraints do not allow the reclamation of soils.
- It can adapt to widely varying climatic and moisture supply conditions. It is also resistant to many diseases.
- It can be easily marketed.

Cloned *Eucalyptus*, known for its luxurious water consumption under excess soil moisture condition, has so far proved to be the most suitable plant for bio-drainage. Toky *et al.* (2011) showed that magnitude of water table depression is in the order of *Eucalyptus hybrid* > *Eucalyptus tereticornis* C-10 > *Eucalyptus tereti-*

cornis C-130 > *Prosopis juliflora* > *Tamarix articulate* > *Callistemon lanceolatus* > *Eucalyptus tereticornis* C-3 > *Meiia azedarach* > *Terminalia arjuna* > *Pongamia pinnata*

Evapotranspiration from Plants

The evapotranspiration rate from a tree canopy is governed by soil, plant, climate and soil water related characters as follows:

- Soil characteristics such as soil texture, hydraulic conductivity, salinity/ sodicity and profile characteristics
- Plant characteristics such as plant species, age of plants, tree density and root characteristics
- Climatic conditions such as rainfall, potential evapotranspiration as measured through class A Pan evaporation and related climatic parameters
- Soil water availability/aquifer related characters such as depth to water table and salinity of the groundwater.

Table 7.7 Rates of tree water use (T) as measured by the heat pulse method

Serial number	Species	T (mm/ annum)	E_{pan} (mm/ annum)	T/E_{pan}	Soil texture	Mean water table depth (m)	Mean water table salinity (dS/m)
1	*Eucalyptus camaldulensis*	1334	2032	0.66	Sand	1.16	9.0
	Eucalyptus camaldulensis	1680	–	0.83	–	–	–
2	*Eucalypus camaldulensis*	460	2430	0.189	Clay	3.0	6.2
3	*Eucalyptus camaldulensis*	307	2025	0.152	Sandy clay	4.5	6.2
4	*Acacia nilotica*	1248	2650	0.471	Silty loam	1.3	20
5	*Acacia nilotica*	2225	2650	0.84	Silty loam	2	1.5
6	*Eucalyptus camaldulensis*	1181	1646	0.72	Sandy loam	1.5	5
		1090	–	0.66	–	3	5
7	*Eucalyptus microtheca*	1048	1618	0.65	Loam	1.5	3
8	*Eucalyptus camaldulensis*	340	1350	0.252	Clay	2	6
	Casuarina cunninghamiana	350	–	0.259	–	–	–

[1.] Marshall *et al.* (1997); Akilan *et al.* (1997); [2.] Fraser *et al.* (1996); [3.] Salama *et al.* (1994); [4 and 5]. Khanzada *et al.* (1998); [6, 7 and 8.] ACIAR Project 9316 (1996); – not reported

The average annual evapotranspiration from plants is assessed to be 3-6 times more than pasture/field crops. The water use by tree plantations is not less than 1.5 times that of agriculture crops or about 1.25 times of Class A pan. In view of this and to be on safer side, it has been suggested that tree water use may be taken as 1.0 times the Class A pan, if groundwater is of good quality. Age of the plant however, has a significant effect on tree water use. Under north Indian conditions, evapotranspiration rates for irrigated *Eucalyptus* increased from 53 to 106 cm per annum between 2-6 years old trees having an increment of about 14 cm per year. Thereafter, rates stabilized in the range of 138-141 cm per annum (NAAS, 2015). It shows that the benefits of bio-drainage over annual crops flow only after 6 years of plantation. The rates of tree water use along with some soil and water characteristics are listed in Table 7.7. Such data sets are used to design bio-drainage systems.

Bio-drainage System Design

The first and foremost task is to assess the groundwater balance that will determine the required capacity of the bio-drainage system. For this purpose all inputs to the water balance equation such as rainfall, seepage due to canal network, seepage from other sources and deep percolation losses from cropped land are assessed. Similarly, all output components such as losses from the bare soil including capillary contribution from shallow water table and draft of the groundwater from pumps and natural vegetation are also assessed. Some components such as groundwater inflow into and groundwater outflow from the area including contribution from below layer due to piezometric pressure can be neglected for a long-term water balance. All values are taken in depth units ($m^3/m^2 = m$). If the net input so calculated is positive (Net input = Input – output), then considering this value and plant's capacity to bio-drain the water, area to be put under plantations can be calculated. Following step wise procedure is used.

1. Select T/E_{pan} ratio for the plants proposed to be planted (Table 7.7 or other location specific information)
2. Calculate annual water use by the trees multiplying by E_{pan} of the area.
3. Per cent area to be planted is found out by dividing the seepage rate calculated using water balance equation by annual water use calculated in step 2.

If the percent area to be planted to trees is more than the desired, either change the bio-drainage plant or integrate conventional techniques with bio-drainage.

In canal-irrigated areas, groundwater imbalance is due mainly to the additional seepage and deep percolation losses. In such cases, it may not be necessary to work out the complete water balance equation. Bio-drainage should take care of additional seepage and deep percolation losses in the command to ensure that there is no rise in water table due to introduction of canal irrigation. Kapoor (1997) and Heuperman *et al.* (2002) suggested the use of the following equation for this purpose:

$$P = \frac{(R * Af * Ir) C}{A_{pan}} \qquad (7.16)$$

Here P is the total area under plantation (ha); C is the total culturable area (ha) of the command; R is the fraction of the irrigation water that goes to recharge the groundwater (dimensionless), Af is the intensity of irrigation (0.6–2.0); Ir is the gross irrigation requirement (mm); and A_{pan} is the amount of water transpired by one ha of plantation (mm).

Example 7.3

The replenishable groundwater recharge in an area is assessed at 10 cm. Calculate the area to be planted with *Eucalyptus* to balance this recharge. Soils are sandy clay in nature. Take annual E_{pan} of the area as 1950 mm.

If we plant *Eucalyptus camaldulensis* in these kind of soils, the likely T/E_{pan} is 0.152 (Table 7.7).

Therefore, ET from the plants = 1950 * 0.152 = 296.4 mm/annum

Area to be planted = 10*10/296.4 = 0.34 ha or 34% of the area to be drained.

Bio-Drainage and Seepage Interception

Bio-interception has been used as a means to intercept seepage from canal network since the introduction of canal irrigation in India. Complete length of the canal reach causing acute seepage or affected by water logging needs to be covered under bio-drainage as a bio-drainage belt in pockets or blocks will not be very effective (Fig. 1.14 right). Width of the area over which plantation work has to be done is quite critical and should be assessed carefully. To decide this, canal seepage rate needs to be assessed first. For this purpose, actual measurements of canal seepage, mathematical models (analytical or numerical) or water balance approach are commonly used. For large canals, steady state seepage rate is used to assess the width of bio-drainage belt. Since evapotranspiration rate (ET) of plantations types used in establishing is known, the input and output is matched as per the following equation:

$$W \times L \times ET \times T = Q \times T \tag{7.17}$$

Such that

$$W = Q/(L.ET) = q/ET \tag{7.18}$$

Here L is the length of the canal reach, m; W is the width of bio-drainage belt, m, ET is the crop evapotranspiration of the tree species, m; T is the time period; and q is seepage rate per unit running length of the canal, m^3/m.d while Q is the seepage m^3/d from the whole canal reach

To use Eq. (7.18), most critical period is selected. It can either be the winter season or the rainy season. If the ET for such a period is ET_c and seepage is q_c then the expression to calculate width of bio-drainage belt can be written as follows:

$$W = q_c/ET_c \tag{7.19}$$

Example 7.4

Calculate the area of bio-drainage belt to intercept canal seepage along a canal length of about 1000 m built in fill. The seepage rate is 9.0 L/m per hour and the evapotranspiration rate during most critical period is 6 L/d from an area of 2.25 m^2.

Seepage rate to be intercepted from one side of the canal = (0.009 m^3/m per hour)/2 = 1.25 x 10^{-6} m^3/m-s.

Average evapotranspiration rate per unit area = (0.006 m^3)/(24x60x60) *2.25 m^2 = 5.0 x 10^{-3} x = 3.086 x 10^{-8} m^3/m^2-s.

Width of bio-drainage belt using Eq. (7.19) is given as

$$W = \frac{q_c}{ET_c} = \frac{1.25 \times 10^{-6}}{3.086 \times 10-8} = 40.5\, m \cong 40\, m$$

The area under bio-drainage belt, A = W x L = (40 x 1000) m^2 = 4 ha

Bio-drainage for Shallow Water Table

To control shallow water table, strips of trees can be planted and considered to represent the pipe. It has been suggested that a steady state pipe drainage design equation can be used to estimate the appropriate spacing between the tree strips (Stirzaker *et al.*, 1997). Another suggestion in line with this proposal is to design conventional pipe drainage system and replace alternate drains with bio-drainage strips (Fig. 7.14). One may agree with Chauhan's (2000) observation that these are conjectures and nowhere such kinds of systems have been field tested.

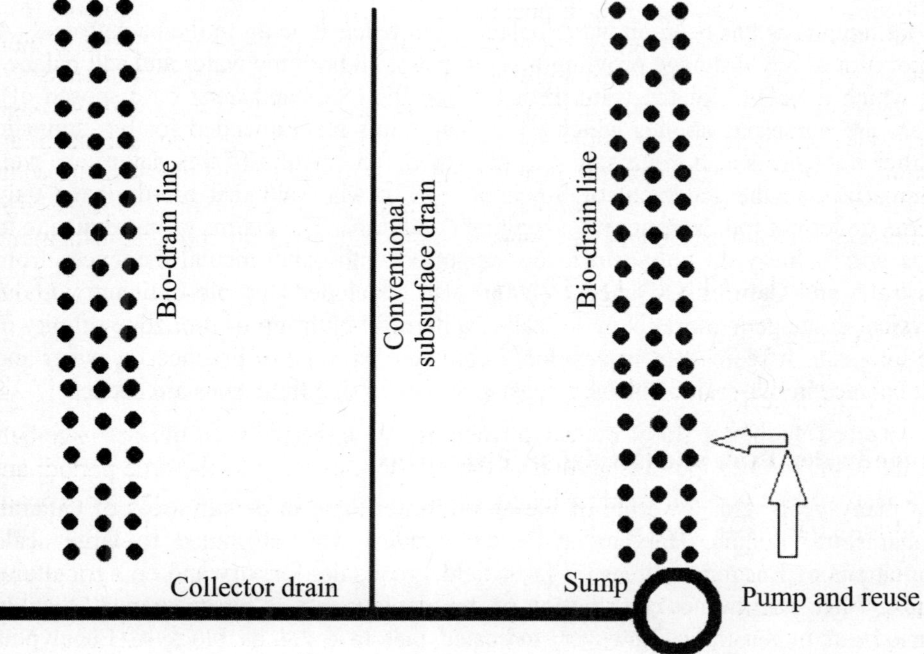

Fig. 7.14 Schematic diagram showing integration of conventional pipe and bio-drainage

Bio-drainage: Benefits and Limitations

The benefits of bio-drainage emerge from the following:

- Simple technology, can be easily implemented by the community

- Eco-friendly, no disposal related problems
- Alternate land use with good potential of returns

Limitations of bio-drainage emerge from the following:

- Limitations in selection of plant type
- Does not play any role in ensuring salt balance
- Loss of land for arable cropping
- Loss of water brought into the command for irrigating crops
- Loss of crop yield due to shading effect, competition for soil moisture between the plants and crops and damage by birds
- Emergence of new diseases/insects/pests
- Cause of lowering of water table beyond desirable depths
- Benefits will emerge slowly, minimum being in the initial years of plantations
- Several issues including design criteria are in developing stage

Salt Balance

So far, emphasis has been on water balance, salt balance being in the background (A major aim of any drainage programme is to maintain both the water and salt balance for which a fraction of the water brought into the command must be disposed of). There are numerous studies which show that plants recommended for bio-drainage neither transpire salt nor are salts bio-harvested. The results of this nature are well summarized in the statement of Smedma (1997), who said that bio-drainage only merits under not too extreme water logging conditions. The claims of bio-drainage to cope with salinity do not seem to be supported with experimental evidences from Australia and California. Denecke (2000) also concluded that bio-drainage without physical drainage is more likely to lead to a harmful build-up of root zone salinity in the long run. It seems that to develop a complete package of practice for water and salt balance in bio-drained lands, intensive and extensive field trials are needed.

Groundwater Table and Eucalyptus Plantations

Not many years ago, lowering of the groundwater table in certain areas of Gujarat, Maharashtra, Punjab, Haryana and Uttar Pradesh was attributed to large scale plantations of Eucalyptus trees on farm fields, roadside, forestry and on agricultural lands. What was the net contribution of the plantations was never known? A study carried out in Australia, however, indicated that in a 2.4 ha block of Eucalyptus plantation with 4 species and 31 provenances, the groundwater was lower by 3 m compared to general groundwater table around the plantations. Closely spaced trees lowered the water table to a greater extent than the ones spaced at larger distances. This effect can be easily explained on the basis of number of plants per unit area. The interesting observation, however, was that the effect of lowering of the water table was completely lost within a short distance from the plantation boundaries (Heuperman, 1999). This shows that the water table decline is more or less of local nature. The author made a simple calculation which revealed that a 1 ha plantation of *Eucalyptus* trees cannot transpire more water than 1/3 of an average discharge of a shallow tube well in Karnal region estimated at 10-12 L/s (Table 7.8).

Table 7.8 Discharge from plantations as affected by number of trees and transpiration capacity

Number of trees	Discharge (L/s) with a transpiration capacity*(L/d. tree)			
	30	40	50	60
2500	0.87	1.16	1.45	1.74
3000	1.04	1.39	1.74	2.08
3500	1.21	1.62	2.03	2.43
4000	1.39	1.85	2.32	2.78
4500	1.56	2.08	2.61	3.12
5000	1.74	2.32	2.90	3.48**

*Expected capacity of a good transpirer 2-3 years after transplanting

**The discharge is around 1/3 of an average irrigation well in the Indo-Gangetic plain of Haryana.

Case Studies

Bio-drainage rate up to 5.5 m/annum for *Eucalyptus tereticornis* and up to 4.2 m/annum for *Bambusa arundinacea* have been reported through a lysimetric study conducted at CSSRI, Karnal (Chhabra & Thakur, 1998a; Chhabra & Thakur, 1998b). No salt uptake by the trees was observed yet the salt remained lower in the profile as compared to unplanted lysimeters. The evapotranspiration rates under field conditions were much less than the lysimeteric studies. For a six years old plantation, annual total consumptive use of water was 2200 mm under high density and 1300 mm in low density *Eucalyptus* plantation, which seems to be quite reasonable.

The parallel strip plantations of *E. tereticornis* were raised on *killa* lines at Puthi and Baas villages in Hisar district. The plantations lowered the groundwater table underneath the plantation strips more than the adjacent fields. When plants were two years and three months old, the average draw down under plantation was 18 cm. It increased to 85 cm in the next three years. Impact on water table was noticed even beyond a distance of 66 m from the edge of outer strips (Dagar, 2011). The technology is now being extended to cover more areas in Jhajjar district of Haryana (Fig. 7.15, left)

A bio-drainage belt in an area of 1.2 ha (400*30 m^2) was established during March 2004 along the Sharda *Sahayak* canal command to intercept seepage. The measured average daily water uptake of the Eucalyptus plant varied from a low of 2.63 L/d-plant in September, 2004 to 21.76 L/d-plant in April, 2005. It may be noted that the transpiration capacity of the plants is much less in the initial year of plantation. The total depth of water annually extracted by the bio-drainage belt was 880.8 mm from three years old plantation belt (Lal, 2006).

In another trial, bio-drainage proved to be an attractive option to intercept seepage by planting trees alongside a water course. *Acacia nilotica*, *Dalbergia sissoo*, *Sesbania grandiflora* and *Casurina equisetifolia* intercepted 86, 84, 72 and 72% of the canal seepage (Patil *et al.*, 2005).

Most spectacular results have been reported from bio-drainage in Indira Gandhi *Nahar Pariyojana*, (IGNP) Phase II in Rajasthan. The water table, which has come near the surface, is now more than 10 m deep. The area is now being cleared of bio-drainage trees and the land is being allotted to stakeholders for cultivation. The condition of allotment stipulates that crops will be cultivated with groundwater using sprinkler irrigation so that water table remains under control (Fig. 7.15, right).

Fig. 7.15 A view of bio-drainage at Paharipur in Jhajjar district (left) and bio-drained land allotted to stakeholders for cultivation in IGNP (right)

SUITABILITY OF DRAINAGE METHODS

It may be noted that no single drainage method can have universal application at all the sites. One needs to investigate and arrive at a sound economic solution to the problem. Some of the situation under which one or the other method of drainage would be applicable is given in Table 7.9. In general, an integrated solution needs to be implemented as discussed in the following section.

INTEGRATED DRAINAGE SYSTEMS

For large areas only one solution of drainage will be applicable. As such, discussion on merits and demerits of different kinds of drainage systems is immaterial. We are not in the process of selecting a common prescription to the whole area or all the sites. When, site specific, cost effective and eco-friendly solutions are sought, it will in all probabilities be an integrated system with a good blend of surface, subsurface, vertical, mole and bio-drainage systems (Gupta, 2014). An integrated system will only be a cost-effective and eco-friendly solution. A system that integrates vertical, bio-interceptor, bio and subsurface drainage is shown in Fig. 7.16. More systems can be integrated in a similar fashion as per need of the sub-areas within the area to be drained.

Table 7.9 Suitability of drainage methods for specific situations

Situation/aquifer condition	Preferred method of drainage		
	Tube well	**Pipe**	**Bio-drainage**
Pervious strata at deeper depth (> 5 m)	Y	*	Y
Groundwater quality good for irrigation	Y	*	*
Water table to be lowered to depths exceeding 2.5 m with no shallow impermeable layer	Y	*	*
Water table to be lowered in short duration to remove aeration problem	*	Y	*
Drainage of small farms within a watershed	*	Y	*
Hard pan is encountered at 2-5 m depth	*	Y	*
Limited funds for initial investment	Y	*	Y
Limited funds for operation and management	*	Y	Y
Prevention of water logging and soil salinization (water table 2-3 m depth)	Y	*	Y
Reclamation of waterlogged saline lands	*	Y	*

Y = Yes and * = No. Source: Modified from IMTPC

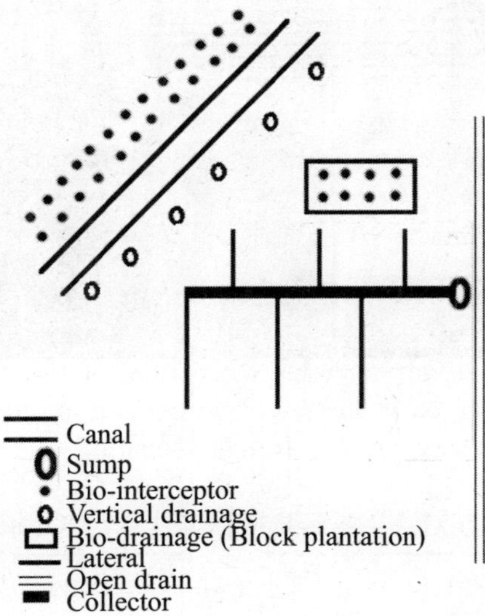

Fig. 7.16 A schematic representation of bio-interceptor, bio, vertical and subsurface drainage

SPECIAL ACTIVITIES TO IMPROVE DRAINAGE EFFECTIVENESS

Sub-soiling

Sub-soiling, also called ripping, chiselling or aerating, is the process of breaking up of the dense horizons below the plough layer and to loosen up the soil. An implement known as sub-soiler (Fig. 7.17 top), a single or multi-blade equipment, is used to loosen and break up the soil at depths below the level of a traditional disk harrow or rotavator. Typically, a sub-soiler is operated at depths of 30-50 cm to improve permeability of the soil if permeability of the profile is low. When a sub-soiler is pulled through the soil, it loosens the compacted layers by lifting and cracking them, creating a network of interconnected pores (Fig. 7.17 bottom). Many of these pores extend from depth of loosening up to the soil surface. They can, therefore, act as pathways for root penetration and for transmission of water and air. The best conditions for sub soiling occur during the dry season, when the conditions are optimum for maximum bursting and it is easier to work with farm implements. The results of sub-soiling have not been consistent in terms of yield increase. It is believed that sub-soiling alone may not be as useful as a combination of sub-soiling and pipe drainage. The main disadvantage of this technique is that sub-soiling is not a permanent operation. It has to be repeated say after every 5 years depending upon the farm mechanization and soil conditions. Another limitation of sub-soiling is that the operation requires a high draft tractor, which may not be available with the farmers. This limitation can be circumvented with popularization of custom hiring concept, yet the need and benefits of sub-soiling must be evaluated before the operation is taken up.

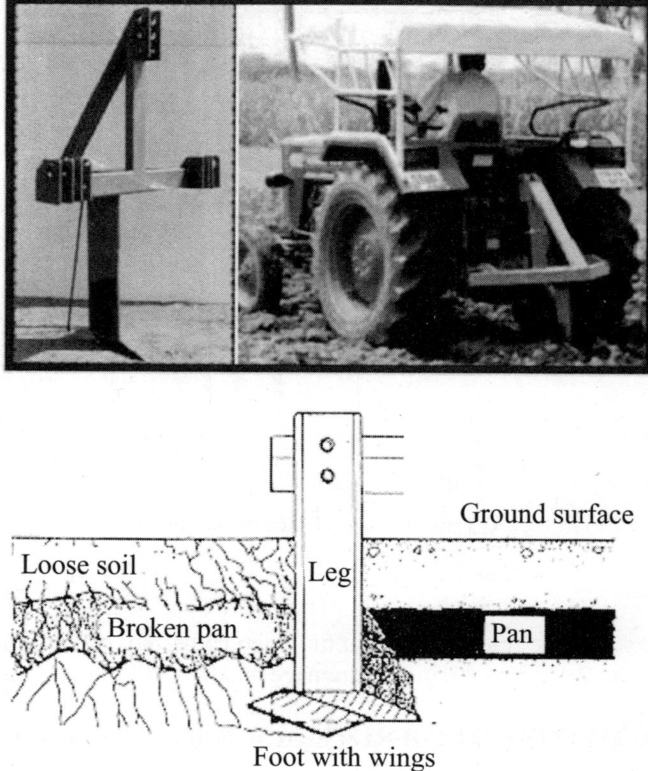

Fig.7.17. A view of a sub-soiler and sub-soiling (Top) and loose soil and broken pan behind a sub-soiler (Bottom)

Permeable Backfills

One of the major difficulties with land drainage is to ensure the movement of water to drains especially in heavy textured soils. Although, it is difficult to improve the whole field, yet the performance of drains can be considerably improved by backfilling the trenches with a more permeable fill than the excavated soil. In the first place, it provides an easy entry route for water from the surface layers to the drains and secondly improves the general performance of the drain by providing a permeable surround around the drain. The recommended permeability of this backfill varies from a low value of 5 times to a high value of 1000 times the permeability of the original soil. The results of experiments where this technique has been adopted are not consistent and in many cases the improvements are not commensurate with the extra expenditure. Many times the improvement is short-lived and the expenditure made on the material is wasted.

Chemical methods of structuration and stabilization have also been adopted in achieving results similar to the one expected from a permeable backfill. In these cases, soil conditioners are used to improve hydraulic conductivity. Because of their prohibitive cost, the use of these conditioners is advocated only to the backfilling material. The commonly used soil conditioners are bituminous emulsions, polymers, rubber solutions, cement, lime and gypsum. Gypsum is most suited to sites where soils are poorly permeable because of high sodium content of the exchange complex. There is no evidence as yet proving the usefulness of these additives at economical cost.

QUESTIONS (THEORY)

1. Describe various soils features that support mole drains.
2. Write a short note on mole drainage. Describe few of the remedial measures that can be taken to avoid failure of mole drains.
3. Briefly describe the methods of seepage interception in canal networks.
4. Name and describe the systems that characterize the various vertical drainage layouts.
5. List the favourable and non-favourable features of sites for vertical drainage.
6. Why is it necessary to have interfering wells in vertical drainage?
7. Differentiate between an irrigation and drainage well on the basis of its location and other characteristics.
8. Describe the various criteria used to decide the number of tube wells required in vertical drainage
9. Calculate the number of tube wells required to drain an area of 5 km^2. The aquifers in the area can sustain shallow tube wells of 8 L/s discharge. Number of hours for which electricity is available is about 10 h/d. It is proposed to grow a crop of wheat in the area. Depth of irrigation is 6 cm applied at an interval of 21 days.
10. Write a short note on the characteristics of plants suitable as bio-drain material.

11. How will you proceed to determine the number of plants required to drain a given area?
12. Describe some of the limitations of bio-drainage.
13. Write a short note on the concept of integrated drainage.
14. Write short notes on skimming wells, sub-soiling and permeable backfills.

QUESTIONS (OBJECTIVE)

1. Fill in the blanks

1. Mole drainage is also known as ________ drainage.
2. Mole drains are ____________ channels formed with a mole plough.
3. Expander in the mole plough helps to __________ the channel.
4. Vertical drainage is accomplished by __________.
5. For vertical drainage, tube wells are installed in __________ aquifers.
6. Unlike irrigation wells ________ between cone of depression of adjoining wells is necessary in drainage well to have uniform drawdown.
7. Most promising tree for bio-drainage is ________ species.
8.. Bio-drainage neither transpires salts nor are salts ___________.
9. A hard pan is encountered at 2-5 m depth from the ground surface. The most appropriate drainage method is ______________ drainage.
10. Sub-soiling is also known as _________.

2. Tick mark correct/most appropriate answer (A, B, C or D) from the following

1 Mole drainage is also known as

(A) Pipe less drainage (B) Vertical drainage

(C) Interceptor drainage (D) None of these

2. The components of a mole plough are

(A) Leg (B) Expander

(C) Foot (D) All of these

3. Soils suitable for moling should have clay content exceeding

(A) 20% (B) 30%

(C) 60% (D) None of these

4. At the time of moling, the moisture content at the moling depth should be nearly equal to

(A) Liquid limit (B) Saturation

(C) Plastic limit (D) None of these

5. Vertical drainage is also known as
   (A) Tube well drainage (B) Horizontal subsurface drainage
   (C) Both A and B (D) None of these
6. A well in the __________ aquifer is most suited for vertical drainage
   (A) Confined (B) Unconfined
   (C) Both A and B (D) None of these
7. Vertical drainage is most suited if
   (A) Groundwater is of good quality
   (B) Availability of good aquifers
   (C) Farmers willingness to install tube wells
   (D) All of these
8. Drawdown in a water table well is ____________ proportional to discharge
   (A) Directly
   (B) Inversely
   (C) Does not depend upon discharge
   (D) None of these
9. Steady state discharge in a cavity well is
   (A) Inversely proportional to the well radius
   (B) Directly proportional to the well radius
   (C) Does not depend upon the well radius
   (D) None of these
10. Which is not a part of mole plough?
   (A) Leg (B) Expander
   (C) Disc (D) Foot
11. A plant transpires 15 L/d water per day. How much area should be planted to take care of recharge of 40 m^3/d. Assume plant density as 4000 plants/ha.
   (A) 0.67 ha (B) 2.67 ha
   (C) 0.38 ha (D) None of these
12. A chiseler is used for
   (A) Construction of mole drains
   (B) To fill the excavated drains
   (C) Sub-soiling
   (D) None of these

13. Which is not used to intercept canal seepage
    (A) Pipe drainage (B) Open deep drains
    (C) Bio-interceptor (D) Mole drains
14. General lowering of water table in many parts of India is attributed to
    (A) Horizontal pipe drainage (B) Vertical drainage
    (C) Expansion of irrigation (D) Surface drainage
15. The total volume of water to be drained to lower the water table by 30 cm in 3 days over an area, A can be written as A*0.3*f/3, here f is
    (A) Drainable pore space
    (B) Moisture percentage at wilting
    (C) Field capacity moisture percentage
    (D) None of these
16. A good material for bio-drainage is a plant that
    (A) Deep rooted with good water absorption
    (B) Efficient water translocation from roots to the leaves
    (C) Efficient transpiration
    (D) All of these
17. A good material for bio-drainage is a plant
    (A) That can tolerate poor aeration
    (B) That can tolerate salt stress
    (C) That can transpire luxuriously
    (D) All of these
18. In general bio-drainage is designed to maintain
    (A) Water balance (B) Salt balance
    (C) Both water and salt balance (D) None of these
19. Sub-soiling is also known as
    (A) Ripping (B) Chiseling
    (C) Both A and B (D) None of these
20. These are used in heavy textured soils to improve flow to drains
    (A) Permeable backfills
    (B) Chemical methods of structuration and stabilization
    (C) Addition of amendment to back-fills
    (D) All of these

■■■■

Chapter 8

SALT AFFECTED SOILS: RECLAMATION AND MANAGEMENT

TERMINOLOGY

Electrical Conductivity (EC)

The total salt content of soil solution or irrigation water is represented in milligrams per liter (mg/L), millimole per liter (mmol/L) or electrolyte conductivity expressed in dS/m at 25ºC. Earlier, it also used to be represented by TDS (total dissolved salts), which is the sum of the anions and cations present in units of ppm (Table 8.1). Electrolyte conductivity is by far the simplest representation and is a measure of the ease with which current will flow through the soil water or the irrigation water. EC is reciprocal of the electrical resistivity, the resistance in ohms of a conductor, metallic or electrolytic, which is 1 cm long and has a cross sectional area of 1 cm^2. Hence,

Table 8.1 Conversion factors for soil/water salinity

Component	Multiply by	To get	Remark
TDS, mg/l	1.0	TDS, ppm	
Salinity, dS/m	1.0	Salinity, mmho/cm	
Salinity, mmho/cm	1000	Salinity, μmho/cm	
Salinity, dS/m (3-30)	0.36	Osmotic pressure, atm.	
Salinity, dS/m (< 5 dS/m)	640	TDS, mg/l	
Salinity, dS/m (>5 dS/m)	800	TDS, mg/l	The value of the factor 800 increases with increasing EC and approaches near unity at very high salinity levels.
mg/L	1/equivalent weight	meq/L	As an example 80 mg/l of calcium is equal to approximately 4 meq/L of calcium, the equivalent weight of calcium being 20.04 mg/meq.

mg/l = milligrams per litre; meq/l = milliequivalent per litre; ppm = parts per million; dS/m = deciSiemens per meter; μS/cm = microSiemens per centimeter; mmho/cm = millimhos per centimeter; TDS = total dissolved solids.

electrical conductivity was formerly expressed in reciprocal ohms per cm or mhos per cm and now in SI units it is expressed in dS/m. It is equivalent to the earlier unit expressed in millimhos/cm (mmhos/cm) (1dS/m = 1 mmhos/cm). Milliequivalent per liter is the milliequivalent weight of the anion or compound (1/1000 part of the equivalent weight) in a liter of solution. Various conversions for soil/water salinity are given in Table 8.1.

The equivalent weight of the element or compound indicates the combining capacity of an element or radical with hydrogen. It is expressed as the weight in grams of the ion or compound that combines or replace 1 gram of hydrogen. It is evaluated by dividing the atomic weight with valency of the ion or compound. The equivalent weights of some of the ions and compounds are given in Table 8.2. The milliequivalent weight is 1/1000 part of the equivalent weight. In the range of 0.1 to 5.0 dS/m, following relation holds true:

$$1 \text{ meq/L} = 10^* \text{ EC (dS/m)} \tag{8.1}$$

Table 8.2 Equivalent weight of various ions, salts and amendments used in land reclamation

Ions/Salts	Equivalent weight	Salts	Equivalent weight	Salts/amendments	Equivalent weight
Calcium	20.04	Magnesium chloride	47.62	Sodium carbonate	53.00
Magnesium	12.16	Magnesium sulphate	60.19	Sodium bicarbonate	84.01
Sodium	23.00	$MgCl_2 \, 6H_2O$	101.7	Potassium chloride	74.56
Potassium	39.10	$MgSO_4 \, 7H_2O$	123.2	Potassium sulphate	87.13
Sulphate	48.03	Magnesium carbonate	42.16	Potassium carbonate	69.10
Chloride	35.46	Sodium chloride	58.45	Potassium bicarbonate	100.11
Carbonate	30.00	Sodium sulphate	71.03	Sulphur	16.03
Bicarbonate	61.01	Sodium carbonate	53.00	Sulphuric acid	49.04
Calcium chloride	55.50	Sodium bicarbonate	84.01	Aluminum sulphate	111.07
$CaCl_2 \, 2H_2O$	73.5	Sodium chloride	58.45	Iron sulphate (ferrous)	139.01
Calcium sulphate	68.07	Sodium sulphate	71.03	Calcium carbonate	50.04
$CaSO_4 \, 2H_2O$	86.09				

This relation can be used to assess the EC of the water if meq/L of various cations is known such that:

(Sum of the cations in meq/L)/10 = EC (dS/m) (EC is < 5.0 dS/m) (8.2)

Within practical limits of measurement, sum of anions and cations in the water is equal and therefore, even the sum of anions can be used in the above relation.

The EC is easily determined using EC bridge (Fig. 8.1, left). EC of the soil saturation extract is usually written as EC_e. For quick appraisal of soils, EC is also determined in 1:2 soil water extract (EC_2) or 1:2.5 soil water extract ($EC_{2.5}$), which is prepared by dissolving 10 g of dry soil with 20 g or 25 g of water. For heavy textured soil even 1:5 (EC_5) soil water extracts have been used. A calibration curve relating EC_2 or EC_5 and EC_e is used to estimate EC_e (Gupta & Gupta, 2014). EC_e can also be determined using the following relation:

$$EC_e = \frac{\Theta_k}{\theta_{sp}} * EC_k \qquad (8.3)$$

Here Θ_k and Θ_{sp} are the soil moistures contents in soil-water paste (1:k) and at saturation for a given soil. EC of the soil solution at field capacity (EC_{fc}) is usually taken as twice that of EC_e.

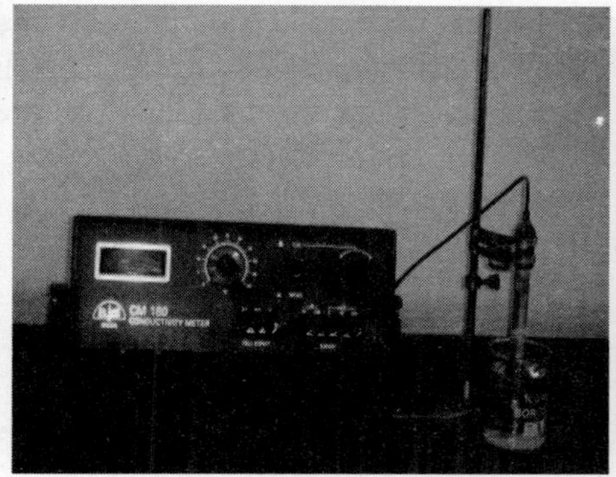

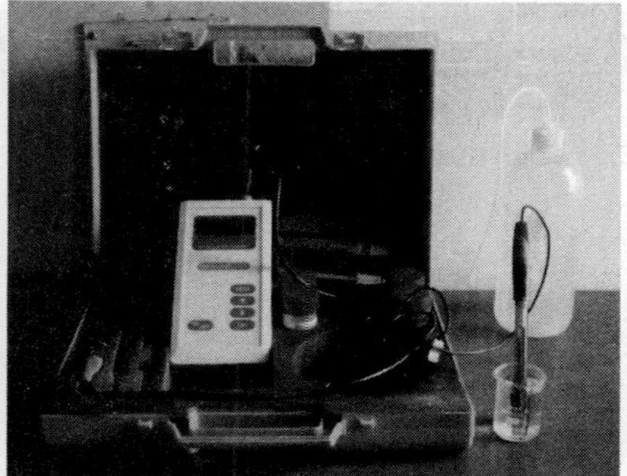

Fig. 8.1 The EC bridge (left) and a pH meter (right)

Example 8.1

4 g of sodium chloride and 2 g of calcium chloride is dissolved in 2 L of water. Express the salt concentration in ppm and also determine its EC. Calculate the concentrations of Na and Ca in the solution in meq/L.

Amount of sodium chloride dissolved = 4*1000/2 = 2000 milligrams per litre

Amount of calcium chloride dissolved = 2*1000/2 = 1000 milligrams per litre

Salt concentration in mg/L = 2000+1000 = 3000 mg/L or ppm

EC = ppm/640 = 3000/640 = 4.69 dS/m or mmhos/cm

Concentration of sodium in the solution, meq/L = 2000/58.46 = 34.22

Concentration of calcium in the solution, meq/L = 1000/55.50 = 18.02

Note: 58.46 and 55.5 are the equivalent weights of sodium chloride and calcium chloride respectively (Table 8.2).

Example 8.2

The TDS of a saturation extract of the soil is 750 ppm. Calculate the approximate EC of the soil saturation extract. What is the amount of salts present in mg/L? What will be its EC_2, if saturation soil percentage on weight basis is 30%?

Since ppm = mg/L, amount of salts present is 750 mg/L.

EC of the soil saturation extract is given by the Eq. (8.2) = 750/640 = 1.17 dS/m

To assess EC_2 (k =2), applying Eq. (8.3), we have

$EC_2 = EC_e * \theta_{sp} / \theta_2 = 1.17 * 200/30 = 7.8$ dS/m (Note: 100 g of soil will have 200 g of water in the determination of EC_2, while only 30 g at the saturation level).

Soil Reaction, pH

The pH is a measure of the effective concentration of the hydrogen ions in the irrigation water or the soil water. The log of the inverse of the hydrogen ion concentration in the saturated paste of the soil is expressed as soil pH_s. It is used as an index of the acidity or the alkalinity of the soil. A pH of 7.0 is neutral. Soils are acidic below pH 7.0 and alkaline above pH 7.0. From practical stand point, however soils having some acidic and some alkaline reactions around the neutral pH are treated as normal soils. Only at a pH_s of 8.5 or above soil is characterized as alkali (Fig. 8.2). When rapid determinations are needed, soil water suspensions (1:k) are used to determine pH and designated as pH_k.

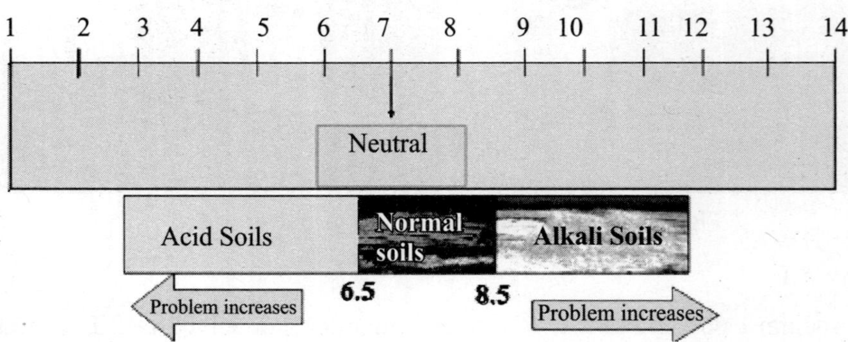

Fig. 8.2 The pH scale indicating acid, normal and alkali soils

A pH meter with glass electrode is used to measure the pH (Fig. 8.1, right). Soil pH_2 and pH_s are significantly correlated, the former being more than the later. The pH in some cases is used to evaluate gypsum requirement as is explained in a forthcoming section.

Sodium Adsorption Ratio, SAR

Ratio of the monovalent and the divalent cations in the soil water is used as an index of the sodium adsorption ratio, SAR. As the potassium on the exchangeable complex is comparatively less than the sodium, it is neglected in the calculations. On this basis, the sodium adsorption ratio is given as

$$SAR = \frac{Na}{\sqrt{((Ca + Mg)/2)}} \tag{8.4}$$

Here concentrations of the ions in the soil solution are in meq/L. If concentrations are in mmol/L, then SAR is given as:

$$SAR = \frac{Na}{\sqrt{(Ca + Mg)}} \tag{8.5}$$

Based on the last definition given by Eq. (8.5), the units of SAR are $(mmol/L)^{1/2}$.

Although SAR of the irrigation water (SAR_{iw}) and the SAR of the soil solution (SAR_{sw}) are related yet no firm indication can be given of the SAR_{sw} because of the following factors:

- Evapotranspiration increases the salt concentration as well as the SAR of the soil solution such that saturation extracts can be 1 to 3 times more concentrated than the irrigation water applied.
- Rainwater has a dilution effect on soil water.
- There may be precipitation (loss of Ca) or dissolution (gain in Ca and Mg) due to interaction between the applied water quality and the soil constituents. Since ESP build-up is due to SAR_{sw}, SAR_{iw} alone might not be an indicator of true hazard of high SAR water.
- The SAR in the root zone varies with depth and time and cannot be assigned a single value as in the case of irrigation water.

On this basis, recent developments have shown that if the RSC is more than 2.5 meq/L, Adj. SAR (Ayers and Westcot, 1976) and Adj. R_{Na} comes into play to assess precipitation of calcium from the water or dissolution of native calcium of the soils (See chapter 10). Current opinion is also not in favour of Adj. SAR as it is observed that it overestimates the impact of high SAR (Ayers and Westcot, 1985).

Exchangeable Sodium Percentage (ESP)

Each soil has certain capacity to adsorb cations (the positively charged constituents of the dissolved salts) such as calcium (Ca^{++}), magnesium (Mg^{++}), sodium(Na^{+}) and potassium(K^{+}). Depending upon the concentration of these ions in the soil solution, they get adsorbed. Once the relative concentration of these ions changes in the soil solution, the cations are replaced by each other although the capacity of the soil remains the same. The total capacity of a soil to absorb cations is known as cation exchange capacity (CEC) and is expressed as c mol (p^{+})/kg of soil or meq/100 g of soil. Exchangeable sodium percentage is the amount of adsorbed sodium, expressed in percent of the CEC of the soil as:

$$ESP = \frac{\text{Sodium on the exchange complex (c mol (p}^{+}\text{)/kg soil)}}{\text{CEC (c mol (p}^{+}\text{) / kg soil)}} \times 100 \tag{8.6}$$

It is also expressed as follows:

$$\text{ESP} = \frac{\text{Exchangeable sodium (meq / 100 g of soil)}}{\text{Cation exchange capacity (meq/100 g soil)}} \times 100 \qquad (8.7)$$

The exchange capacity of the soils is dominated by Na^+, Ca^{2+} and Mg^{2+} with small amounts of K^+. If the exchangeable cations concentration of these ions is known then CEC can be determined by adding the exchangeable cations. ESP, then can be determined using either eq. (8.6) or eq. (8.7). More accurate laboratory procedures are used to assess soil CEC (Gupta *et al.,* 2012).

According to USDA (Richards, 1954), if the irrigation water of relatively constant composition is used and drainage conditions are good, the ESP of the soil varies only slightly from season to season or year to year after the steady state conditions have reached. It may happen only after prolonged irrigations or after few seasons. Under the Indian monsoon conditions, where waters are significantly invariant such a situation might occur after a long period during which some variations in ESP may be noticed.

According to the USDA, the steady state ESP of the soil can be assessed using the SAR of the irrigation water as per the following equation:

$$\text{ESP} = \frac{100\,(-0.0126 + 0.01475 * \text{SAR})}{1 + (-0.0126 + 0.01475 * \text{SAR})} \qquad (8.8)$$

In deriving this relation, it is assumed that there is no precipitation, no dissolution of salts and no absorption of soluble salts by plant roots. The actual ESP of the soils is generally higher than given by Eq. (8.8).

Example 8.3

The exchangeable cations concentration in a soil is as follows:

Ca^{2+} = 3.5 meq/100 g, Mg^{2+} = 1.5 meq/100 g, Na^+ = 19.0 meq/100 g, and K^+ = 0.1 meq/100 g.

Calculate the exchangeable sodium percentage. Make any assumption that is required in this calculation.

Assumption: Other exchangeable cations are present in negligible quantities.

Cation exchange capacity (CEC) = 3.5+1.5+19.0+0.1 = 24.1 meq/100 g

The exchangeable sodium percentage (ESP) is then given by Eq. (8.7) as follows:

ESP = (19.0/24.1) * 100 = 78.8

Soil Saturation Extract

Saturation extract is the solution extracted from a soil at its saturation moisture percentage. To obtain soil saturation extract, first a soil saturation paste is prepared by slowly adding water to the soil. At saturation, the soil paste glistens as it reflects light, flows slightly when the container is tipped and the paste slides freely and cleanly from a spatula for all soils except those with a high clay content. Soil is kept as such overnight. Next day, paste is examined as discussed previously and more water is added, if necessary. Vacuum pump is used to obtain the soil saturation extract from the saturated soil paste.

SALT AFFECTED SOILS

All soils contain some amount of soluble salts, which are essential for the healthy growth of plants. If the quantity of soluble salts in a soil exceeds a certain value, the growth and/or the yield of most crops is adversely affected. Similarly, if the sodium on the exchange complex of a soil exceeds the desirable value, the crop productivity is affected. *'Soils which contain excess soluble salts or high exchangeable sodium percentage to adversely affect the plant growth, are called salt affected soils'*. These soils are known by different local terms, such as *Kallar* or *Thur* in Punjab and Haryana, *Usar* or *Reh* in Uttar Pradesh, *Luni* in Rajasthan, *Khar* or *Kshar* in Gujarat and Maharashtra, *Chouddu* or *Uppu* in Andhra Pradesh, *Choppan* in Karnataka etc.

Salts present in saline soils are mostly chlorides and sulphates while bicarbonates and carbonates predominate in alkali soils. The salts either originate from local geologic deposits but most often they accumulate in the soils as a result of irrigation, capillary rise from the groundwater, accumulation of runoff from adjoining areas and/or tidal ingress in coastal areas.

Classification of Salt Affected Soils

As per the USDA classification (Richard, 1954), salt affected soils are grouped into three classes' namely saline soils, alkali soils and saline-alkali soils. Later, US salinity Laboratory changed the terminology and designated alkali soils as sodic soils. It was to highlight that adverse effect in this soil are due to high exchangeable sodium. Now, the two terms are used synonymously to describe alkali problem. Henceforth in this chapter, term alkali is used to designate these soils.

Alkali soils

These soils have high ESP exceeding 15. It is reflected in the high pH_s of these soils (> 8.5). The soluble salt content given by the EC is variable but is less than 4 dS/m (Table 8.3). As such, these soils have dispersed soil structure and low infiltration rate than similar normal soils. Following an irrigation or rain event, water stagnates for longer period particularly in the lower spots. Upon drying, these soils develop 1-2 cm wide cracks. These soils are also termed as black alkali soils because the organic matter present in the soil solution of highly alkali soils gets dissolved and is deposited on the soil surface through capillary action causing dark black surface. The cropped fields have spotty growth. Crops show scorching and leaf burn injury typical of sodium toxicity in highly alkali soils. Yellowing of leaves is noticed mainly resulting from poor soil physical properties. In extreme cases, the soils remain barren with practically very little or no vegetation. Depending upon the ESP, degree of alkali hazard varies from none to slight to extremely high (Table 8.4). Based on EC_e and ESP of the soils, categorization of salt affected soils can be seen in Fig. 8.3.

Table 8.3 Classification of salt affected soils on the basis of chemical characteristics

Class	EC_e (dS/m)	ESP[§]	pH_2[$]	Indian Local Names
Saline Soils	> 4*	< 15	< 8.5	*Thur, Uippu, Lona, Shora, Soula, Pokhali, Khar and Kari*
Saline-alkali soils	> 4	>15	> 8.5	*Usar, Kallar, Karl, Chopan, Bari, Reh, Choudu and Kshar*
Alkali	< 4	>15	> 8.5	*Usar, Rakkar, Bara, Chopan, Kari*

*Present literature sometimes gives this value as 2.0 dS/m. [§]Sometimes SAR is used instead of ESP. For Indian conditions, only ESP should be used for characterization. Current research suggests that if pH_s is measured, the value is taken as 8.2.

Table 8.4 ESP and degree of alkali hazard

Approximate ESP	Alkali hazard
Up to 15	None to slight
15 – 30	Slight to moderate
30 – 50	Moderate to high
50 – 70	High to very high
> 70	Extremely high

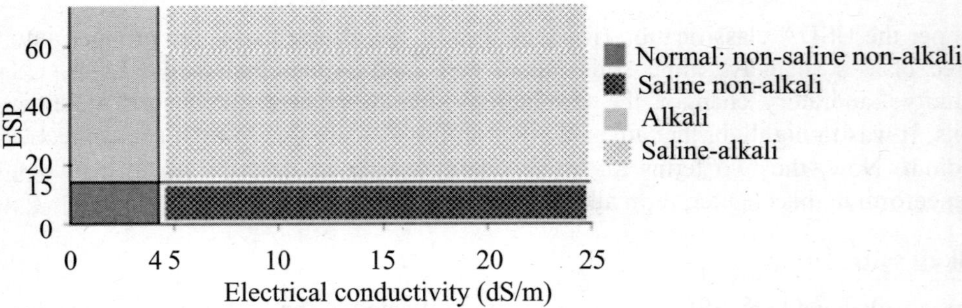

Fig. 8.3 Categorization of salt affected soils based on EC_e and ESP of the soils.

Factors affecting plant growth in alkali soils

Plant growth, yield and/or quality of crops in alkali soils are affected by one or a combination of the following factors, the degree of damage depending upon the amount of exchangeable sodium, pH, and nature and kind of crops grown, macro and micro-environmental conditions and management interventions (Gupta *et al.*, 2002).

- High ESP and pH due to high concentrations of HCO_3 and CO_3
- Poor water and air permeability as a result of high dispersion of soil aggregates and clay particles
- Toxicity of Na, HCO_3 and CO_3 ions

- Excess of Na causing imbalance in the uptake of other ions such as K and Ca nutrition due to antagonistic effect.
- Decreased solubility and availability of micronutrients like Zn, Fe and Mn due to high pH and presence of $CaCO_3$ and soluble CO_3 and HCO_3 ions
- Presence of hard $CaCO_3$ layer at or around 1.0-2.0 m depth, which act as a physical barrier for the movement of water, salts and roots of crops and trees.
- Hard crust on the surface layer which greatly hinders the seedling emergence and reduces the germination percentage and thus plant population
- Poor drainage causing aeration problems
- Low organic matter content.

In most situations, a combination of the above factors affects crop production and productivity in alkali soils.

Saline soils

These soils have EC of the saturation paste extract (EC_e) exceeding 4 dS/m and have a variable ESP in the low range (<15) with pH_s < 8.5 (Table 8.3). These soils show patchy growth of crops and sometimes, visible signs of salt injury such as tip burn or chlorosis (pale yellow colour) of leaves can be seen. Soluble salts, which are invariably present in large quantities, form a white efflorescence on the soil surface. For this reason, these soils are also called white *kallar*. Since the neutral salts present in this type of soil can absorb moisture from the soil surface, these soils look wet although the water is not available to the plants. Salt content of these soils vary from season to season. Salts are leached during the monsoon season, when salt tolerant vegetation appears on these soils. Visually these soils at this time can be identified by the vegetation. Depending upon the degree of EC_e, degree of the salinity hazard is expressed in Table 8.5.

Table 8.5 EC_e and degree of salinity hazard

Soil salinity class	EC_e (dS/m)	Effect on crop plants
Non-saline	0 - 2	Salinity effect negligible
Slightly saline	2 – 4	Yields of sensitive crops restricted
Moderately saline	4 – 8	Yields of many crops restricted
Strongly saline	8 – 16	Tolerant crops give satisfactory yields
Very strongly saline	> 16	Only a few very tolerant crops yield satisfactorily

Following three distinct groups of saline soils are identified.

- Inland saline soils having shallow water table, groundwater quality being saline in nature. These soils suffer from water logging and salinity.

- Saline soils developed from irrigation, being irrigated over the years with saline groundwater. These soils normally have deep water table. However, in certain situations the water table may be shallow. Theses soils generally occur in arid and semiarid regions.
- Coastal saline soils where the problem is due to ingress of sea water resulting in high salinity and temporary flooding. The climate is generally humid to semiarid/arid.

Factors affecting crop production in saline soils

In saline soils plant growth is adversely affected due to the following factors:

- Excess soluble salts present in the soil solution depress the water potential of the medium that restricts the water uptake by plant roots. A simple relationship between the osmotic pressure (OP) in atmosphere and electrolyte conductivity of the saturation extract in dS/m at 25^0C can be approximated by the simple relationship:

$$OP = 0.36 * EC_e \tag{8.9}$$

The data in Table 8.6 clearly brings out this effect that soils are unable to take up the water with increasing soil solution salinity (Kelly and Rengasamy, 2006 as reported in Rengasamy, 2010).

Table 8.6 Percentage of available soil water not taken up by plants in different soil types due to osmotic pressure of soil water salinity in relation to laboratory measured soil salinity and gravimetric field soil moisture (With permission from Rengasamy)

Laboratory measured soil salinity $EC_{1:5}$ ($dS\ m^{-1}$)	**Gravimetric field soil moisture(%) below which osmotic pressure due to salinity is**		**Percentage of available soil water not taken up by plants due to osmotic pressure (>1000 kPa) of soil water salinity**			
	(>1000 kPa)	**(>1500 kPa)**	**Sand**	**Sandy loam**	**Clay loam**	**Clay**
0.25	4.5	3.0	25	0	0	0
0.39	7.0	4.7	50	0	0	0
0.50	9.0	6.0	70	25	0	0
0.72	13.0	8.7	100	50	15	0
1.00	18.0	12.0	100	81	40	4
1.11	20.0	13.3	100	94	50	12
1.25	22.5	15.0	100	100	63	22
1.50	27.0	18.0	100	100	85	40
1.64	29.5	19.7	100	100	98	50
1.64	29.5	19.7	100	100	100	58
2.00	36.0	24.0	100	100	100	76
2.33	42.0	28.0	100	100	100	100

- High concentrations of some specific ions (e.g. Na^+, Cl^-, SO_4^{--}, Mg^{2+}, B) can be toxic and can induce physiological disorders
- High salinity may affect several metabolic processes such as CO_2 assimilation, protein synthesis and respiration
- In waterlogged salt affected soils, some elements such as aluminium, iron and manganese may be produced and accumulate to toxic levels
- Imbalance in the uptake of ions.

Usually, not one but a combination of the above factors are in operation to affect crop production and productivity in waterlogged saline lands.

Saline -alkali soils

These soils are formed due to a combined process of salinization and sodification. Saline alkali soils have pH_s greater than 8.5, EC_e greater than 4 dS/m at 25°C and ESP greater than 15 (Table 8.3).

Waterlogged Salt Affected Soils

The lands, which are both water logged (See Chapter 1) and salt affected are termed as waterlogged salt affected lands. As such, there are additional combinations of the problems such as waterlogged alkali and waterlogged saline soils. Interventions that take care of both the excess water and excess salts are used to reclaim these soils.

FORMATION OF SALT AFFECTED SOILS

Salts in soils originate from the weathering of parent rocks during soil formation. A close relationship exists between the kind of salts in the soil and the chemical composition of rocks from which these soils are formed. Soils formed from salt bearing rocks usually contain more salts. These salts may stay in the soil or are carried downward with the percolating water as a result of rainfall or applied irrigation water. This kind of soil salinity is usually referred as primary salinity. Depending upon the percolation, salts may accumulate at a certain depth depending upon rainfall and other ago-climatic conditions, may precipitate or continue to be transported in solution, ultimately ending up in the groundwater and to the sea. Soil salinization may also occur through the processes of accumulation of salts applied through some outside source. Even the naturally occurring salts resting at some depth may return to the soil surface with water acting as a carrier. The soil salinity developed through these processes is termed as secondary salinity. Depending on the salt composition of the irrigation water, the ESP of the soil solution may also increase resulting in alkalization or sodification of the soils. The two terms as already discussed are used synonymously. Apparently the sodification process does not occur in isolation and is a consequence of soil salinization (Smedema and Rycroft, 1983). Secondary soil salinization and alkalization processes are anthropogenic in nature and can be traced to mismanagement of water. Some of the processes of soil salinization/alkalization listed in chapter 1 are reproduced as under:

- Weathering of rocks and the salts brought down from the upstream to the plains by rivers and subsequent deposition along with alluvial materials

- Secondary soil salinization through capillary rise from subsoil salt beds or from shallow brackish groundwater
- Lack of natural leaching due to topographic situation, especially in arid and semiarid regions.
- Indiscriminate use of irrigation water of variable qualities including poor quality groundwater
- Ingress of sea water along the coast
- Salt laden sand blown by sea winds

While the first four processes are common to all settings, the last two are common in coastal regions.

Table 8.7 Extent of salt affected soils in India (ha)

State	Saline	Sodic	Coastal saline*	Total
Andhra Pradesh	0	196609	77598	274207
Andaman & Nicobar Island	0	0	77000	77000
Bihar	47301	105852	0	153153
Gujarat	1218255	541430	462315	2222000
Haryana	49157	183399	0	232556
Karnataka	1307	148136	586	150029
Kerala	0	0	20000	20000
Madhya Pradesh	0	139720	0	139720
Maharashtra	177093	422670	6996	606759
Odisha	0	0	147138	147138
Punjab	0	151717	0	151717
Rajasthan	195571	179371	0	374942
Tamil Nadu	0	354784	13231	368015
Uttar Pradesh	21989	1346971	0	1368960
West Bengal	0	0	441272	441272
Total (ha)	1710673	3770659	1246136	6727468
Total (M ha)	1.71	3.77	1.25	6.73

Source: Adapted from Mondal *et al.* (2011). There is mismatch between the figures quoted since the extent as high as 3.1 M ha for coastal saline soils are being reported.

EXTENT AND DISTRIBUTION OF SALT AFFECTED SOILS

Salt affected soils occur extensively in different agro-ecological and soil zones of the country, particularly in the arid, semiarid and the sub-humid coastal regions (Agarwal *et al.*, 1979). Exact extent of the area afflicted with salts is unknown as different organizations and authors have arrived at different estimates. The differences are attributed to the methodology of obtaining data, characterization guidelines used and several other factors. According to a current estimate, the salt affected soils occupy about 6.73 M ha of land in India (Table 8.7, Mondal *et al.*, 2011). The States most affected are Uttar Pradesh, Gujarat, Rajasthan, West Bengal and Andhra Pradesh (Fig. 8.4).

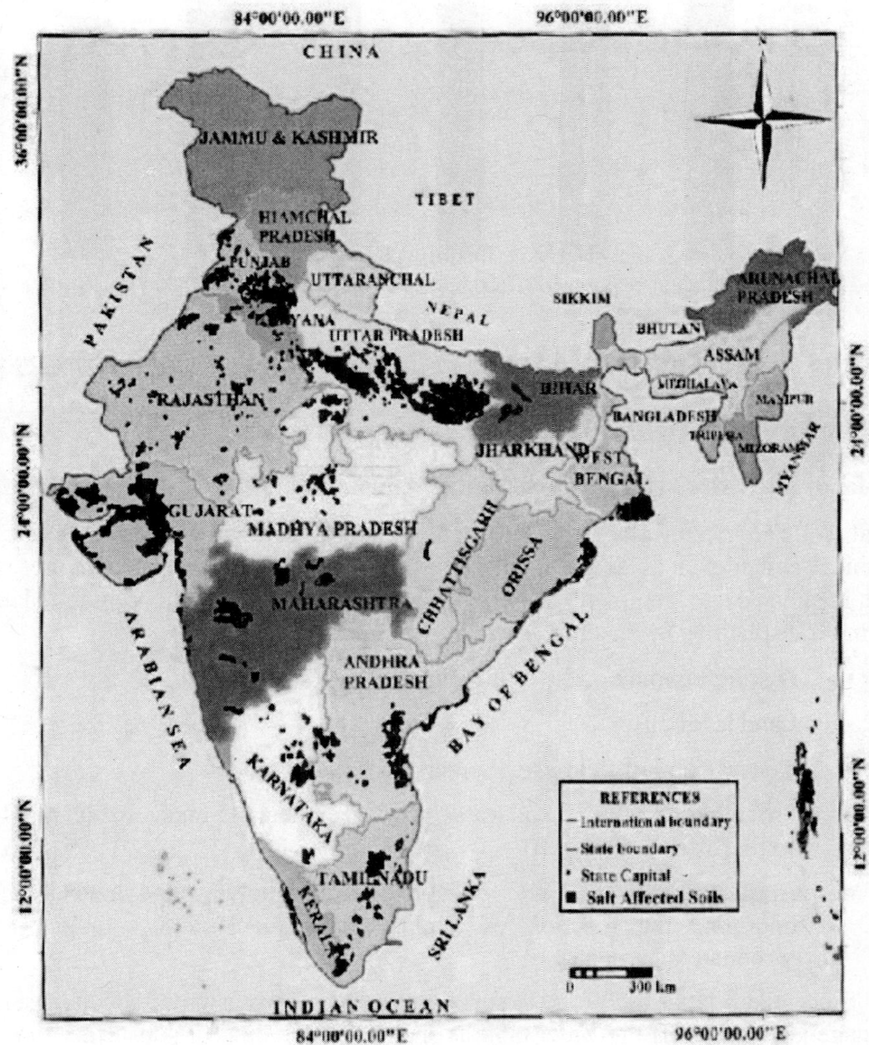

Fig. 8.4 Distributions of salt affected soils in India (Source: Mondal *et al.*, 2011)

Soil Salinity and Yield Losses

The yield loss in salt affected soils can be as high as 100% when such soils present a desolate look and remain barren. In other cases, the yield loss may vary from a low value of about 10% to around 75%. Data from one of the project sites investigated for subsurface drainage in Haryana revealed yield loss varying from 35 to 52% (Fig. 8.5).

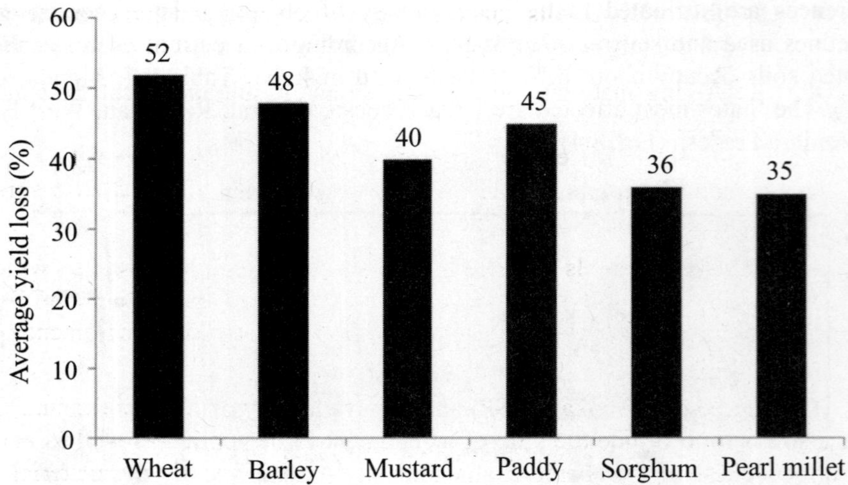

Fig. 8.5 Yield loss in various *rabi* and *kharif* crops at Jhajjar (Average of 2000-2004)

RECLAMATION OF ALKALI SOILS

On-farm Activities before Application of Amendments

Alkali soils are chemically degraded lands with poor physical condition. Before any chemical amendment is added to these soils, on-farm development activities must start in advance to complete the tasks well in time before the optimum date of sowing/transplanting of the crop. The activities include:

- Dyking/bunding of the fields
- Land leveling
- Construction of field irrigation and drainage channels
- Provision of a source of water to ensure adequate and reliable supply of water (often a tube well)
- Application of a heavy irrigation before applying the amendment to minimize the reaction of applied amendment with soluble sodium carbonate/bicarbonate

Since alkali lands are prone to surface stagnation, provision of adequate surface drainage is essential. If the water table is shallow, some kind of subsurface drainage may also be required to facilitate leaching.

Application of Amendments

For cropping in alkali lands, pH/ESP of the soil must be brought down to acceptable level, for which chemical amendments are primarily used. These amendments neutralize the pH of the soil, by replacing the exchangeable sodium by calcium.

Classification of chemical amendments

The chemical amendments are broadly classified into three categories:

- Soluble calcium salts, viz. calcium chloride, gypsum and phospho-gypsum which directly supply soluble Ca for the replacement of exchangeable Na.
- Acidifying materials, viz. pyrites, sulphur, sulphuric acid, and industrial wastes (distillery spent wash, dairy waste) which due to their acidulating action and/or increasing the partial pressure of CO_2 supply Ca indirectly by dissolving calcite natively found in alkali soils. The Ca so mobilized is used to replace Na from the exchange complex
- Organic materials viz. rice husk, FYM and green manure etc., which help to improve soil structure as well as dissolve the native $CaCO_3$ albeit weakly. The reclamation efficiency with these kinds of amendments is usually low.

The degree of soil alkali (ESP) and the availability of *Kankar* (impure $CaCO_3$) in the soil help to decide the kind of amendment to be applied as well as determine the effectiveness of the chemical amendments. Acids or acidifying material will be effective only if sufficient amounts of $CaCO_3$ are present in the alkali soil. A major factor that determines the use of a particular amendment is its cost. Mined gypsum being cheaper has been the most preferred amendment for the reclamation of alkali lands. The chemical reactions that occur in soils following application of a soluble calcium salt, sulphur and sulphuric acid are presented in the following sections.

Calcium sulphate

The commercial di-hydrate form of calcium sulphate generally known as gypsum ($CaSO_4.2H_2O$) reacts with exchangeable sodium and replaces it with calcium as per the following reaction:

$$\underset{\text{Alkali soil}}{2\,\text{Na - soil}} + CaSO_4 2H_2O \longrightarrow \underset{\text{Normal soil}}{\text{Ca - soil}} + Na_2SO_4 \downarrow + H_2O \quad (8.10)$$

Sulphur

Mineral sulphur is converted to H_2SO_4 through the process of biological oxidation by *Thiobacillus thiooxidens* bacteria. It being a very slow process, the relative efficiency of mineral S in land reclamation process is generally low.

$$2S + 5O_2 + 2\,H_2O \longrightarrow 2H_2SO_4 \quad (8.11)$$

$$CaCO_3 + 2H_2SO_4 \longrightarrow CaSO_4 + CO_2 + H_2\,O \quad (8.12)$$

$$2\,\text{Na–soil} + CaSO_4 \longrightarrow \text{Ca–soil} + Na_2SO_4 \downarrow \quad (8.13)$$

Sulphuric acid

Sulphuric acid constitutes one of the major chemical amendments in land reclamation programs in Russia. It is applied directly to the soils. It reacts rapidly with soil lime [as per Eqs. (8.12) and (8.13)]. as it does not have to go through the oxidation process. However, it is highly corrosive and quite difficult to handle at the field level.

DETERMINATION OF GYPSUM REQUIREMENT (GR)

Gypsum requirement (GR) is the equivalence of gypsum which should be added to reclaim an alkali soil. It depends upon the initial ESP and the desired final ESP of the alkali soil. The difference is an index of the exchangeable sodium that must be replaced. The quantity of gypsum needed also depends upon the depth of the soil to be reclaimed. Both the final level of ESP and the depth of reclamation depend on the crops to be grown in the reclaimed land. Quantity of gypsum also depends upon the nature of the soil, specifically its cation exchange capacity (CEC).

GR is determined in the laboratory or by several other simple means. Theoretically, for every one meq of Na to be replaced, one needs one meq of soluble Ca. The meq of sodium to be replaced is calculated by the following equation:

$$\text{Sodium (meq) to be replaced/100 g of soil} = [\text{CEC (initial ESP} - \text{final ESP)}]/100 \qquad (8.14)$$

Here CEC is in meq/100 g. Final ESP is usually taken as 15.

$$\text{Weight of the soil to be reclaimed per ha} = 10000 * \rho * d/100 = 100 * \rho * d \text{ kg/ha} \qquad (8.15)$$

Here ρ is in kg/m^3 and d is the depth of the soil to be reclaimed, cm. It may suffice to reclaim the top 15 cm of the profile with appropriate crop selections. It is presumed that once this top layer is reclaimed, it will promote good seed germination, plant growth and help to initiate a self-reclamation process.

The equivalent weight of calcium sulphate ($CaSO_4.2H_2O$) is its molecular weight divided by the valency of the basic element present in it. Thus, one milli equivalent weight of CaSO4.2H2O contains 86 mg of calcium sulphate [(40 +32 +16*4 +2*2 +2*16)2 = 86 mg]. The milliequivalent weight of Ca in calcium sulphate is its atomic weight divided by its valency (40/2 = 20 mg). The amount of Ca required in milligrams is the amount in milli-equivalent times the equivalent weight. Thus, to replace 1 meq of sodium 0.02 g Ca /100 g soil or 0.086 g of pure $CaSO_4.2H_2O$/100 g or 0.86 g of pure $CaSO_4.2H_2O$/kg of soil will be required. Combining with Eq. (8.14) and Eq. (8.15), we get

$$\text{GR (kg/ ha)} = [\text{CEC(initial ESP} - \text{final ESP)}] * 0.86 * \rho * d / 1000$$

$$\text{GR (t/ ha)} = [\text{CEC (initial ESP} - \text{final ESP)}] * \rho * d * 0.86 * 10^{-6} \qquad (8.16)$$

The final equation takes care of variation in soils and crop characteristics through the parameters of CEC and ρ of soils and final ESP and depth d for the crop salt tolerance characteristics and rooting zone. If the purity of gypsum is not 100%, values calculated with Eq. (8.16) should be divided by the purity of gypsum being used.

Example 8.4

A soil having CEC of 20 meq/L has an initial ESP of 45. The proposed final ESP is 25. Calculate the gypsum requirement of the soil to reclaim 20 cm depth of the soil, if purity of gypsum as 80%. Assume an appropriate value of bulk density.

Assuming that the bulk density is 1350 kg/m^3 and using Eq. (8.16)

$$GR\ (t/ha) = [CEC\ (initial\ ESP - final\ ESP)] * \rho * d * 0.86 * 10^{-6}$$

$$= 20\ (45\text{-}25) * 1350 * 20 * 0.86 * 10^{-6} = 9.3\ t/ha$$

Since the gypsum is 80% pure, the quantity of gypsum needed is

$$= 9.3/0.8 = 11.6\ t/ha$$

Laboratory Method

Schoonover method is used to work out the gypsum requirement (Schoonover, 1952). Saturated gypsum solution of known calcium concentration is prepared by adding 5 g of chemically pure gypsum ($CaSO_4$-$2H_2O$) to one liter of distilled water by shaking it vigorously for about 10 min on a mechanical shaker or 1 h by hand. It is filtered through Whatman filter paper no. 1. Five gram of air dry soil is taken in a 250 ml conical flask to which 100 ml of the saturated gypsum solution is added. The mixture is shaken for 5 min on a mechanical shaker or 50 min by hand. After filtering the contents through Whatman filter paper no.1, 5 ml of aliquot is transferred into 100 ml porcelain dish. Calcium and magnesium concentration is determined. A blank is run using 5 ml of the saturated gypsum solution prepared earlier.

$$GR\ (meq/100\ g\ soil) = 2\ [(Ca_b) - (Ca_f)]$$

Here Ca_b is the calcium concentration in blank, meq/L and Ca_f is calcium concentration in the filtrate, meq/L.

$$GR\ for\ 30\ cm\ soil\ layer\ (t/ha) = GR\ (meq/100\ g\ soil) * 3.54 \quad (8.17)$$

Abrol *et al.* (1988) recommended that in rice-wheat rotation, it may suffice to reclaim 15 cm of the soil profile. As such, gypsum requirement can be halved to what was proposed by Schoonover. A further modification was also proposed. It was postulated that GR estimated from Schoonover method also includes the gypsum required to neutralize the soluble sodium carbonate, commonly present in alkali soils. Therefore, it was proposed that the sample of the soil (5 g) should be leached with 20 ml of 60% ethanol. It will leach down the soluble sodium carbonate and therefore, will not add to the gypsum requirement of the soil. Rest of the procedure is same as proposed by Schoonover.

pH versus GR Graph

The absolute quantity of exchangeable sodium percentage of an alkali soil will continue to be the basis to accurately determine the quantity of gypsum required for reclamation. In practice, pH of the saturated soil paste (pH_s) or pH of 1:2 soil water suspension (pH_2) may also be used since both are fairly correlated with each other and with ESP of the alluvial soils of the Indo-Gangetic plains. This method is particularly useful where skilled staff and laboratory facilities to determine the exchangeable sodium and gypsum requirement are not available. A graph between pH_2 and GR of the soils for different soil textures is shown in Fig. 8.6. This graph is a quick and fairly good method to estimate GR of alkali soils of the Indo-Gangetic plains (Abrol *et al.,* 1988). However, this graph is not universally true for all kinds of soils. For example, pH_s and ESP of black cotton soils (Vertisols) are not well related and therefore, use of such a graph will not be advisable for such soils.

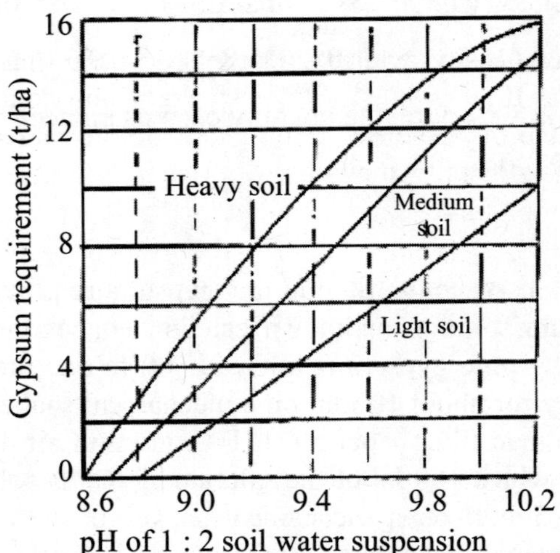

Fig. 8.6 pH_2 versus GR graph for alluvial soils of Indo-Gangetic plains

Use of Mined Gypsum as an Amendment

Mined gypsum is in the form of boulders. Before, it can be used as an amendment, it is essential to ground it so that it could be uniformly spread over the soil surface. Besides, powdered gypsum solubilizes faster in water compared to boulders since for the same amount of gypsum, surface area of the powdered gypsum is more. As per the Bureau of Indian Standards, agricultural grade gypsum should pass through a 2 mm sieve and at least 50% of it should pass through 0.25 mm sieve. Mined agricultural grade gypsum being 70% pure, the GR calculated as described in the previous section needs to be divided by 0.7 to assess the actual gypsum requirement.

Dose of amendments

In many reports, it has been mentioned that actual GR is more than the calculated GR mainly because there is no one-to-one replacement of sodium by calcium. Moreover, whole of applied gypsum may not come in solution because of low solubility especially when the amount of water added is inadequate. Current studies however, indicate that only 50% GR as calculated by Eq. (8.16 or 8.17) is sufficient if appropriate crop selection and management are adopted. For this reason, transplanted rice and wheat cropping pattern has emerged as the best to reduce investments on gypsum. High tolerance of transplanted rice to soil ESP and its shallow root system makes it the best crop for the first season after application of amendment. The dissolution of soil $CaCO_3$ in the presence of rice roots also helps to reduce the soil exchangeable sodium. The GR calculated with the help of Fig. 8.6 should not be reduced to half since it is a plot of actual GR. If sensitive crops like pulses are cultivated in the initial years, one needs to add relatively higher amounts of amendments.

After the gypsum is applied, mix the gypsum in a shallow depth using disc harrow, limiting the depth of mixing to about 10 cm. Deep mixing will reduce the efficiency of gypsum to reclaim the shallow depth where rice roots will stay initially.

Other Amendments

Being the cheapest and an easily available material, gypsum has been extensively used as a soil amendment. If amendments other than gypsum are used, their quantity is determined with the help of Table 8.8. The factors are the ratios of the equivalent weights of the given amendment to the equivalent weight of gypsum. For example; the equivalent weights of sulphur and gypsum are 16 and 86 respectively. As such 1 unit of gypsum is equal to 1*16/86 =0.186 or 0.19 units of sulphur as given in the Table 8.8. It may be noted that limestone, a commonly available mineral that also contains calcium, is not used to reclaim alkali soils because it is not soluble at high pH. However, it can be used to reclaim acid soils, being the most preferable amendment (See Chapter 9)

Table 8.8 Equivalent quantities of some common amendments for alkali soils reclamation

Amendment	Relative quantity
Gypsum ($CaSO_4$ $2H_2O$)	1.00
Calcium chloride ($CaCl_2$ $2H_2O$)	0.85
Sulphuric acid (H_2SO_4)	0.57
$^+$Sulphur (S)	0.19
$^+$Pyrite (FeS_2: 30%S)	0.63
Iron sulphate ($FeSO_4 \cdot 7H_2O$)	1.61
Aluminum sulphate ($Al_2(SO4)_3 \cdot 18\ H_2O$)	1.29

* These quantities are based on 100% pure materials. If the material is not 100% pure, necessary correction must be made. Thus, if the sulphur is only 80% pure the factor applied will be 100 x 0.19/80 = 0.238.

Example 8.5

Calculate the sulphur requirement for the reclamation of an alkali soil which has a pH_2 of 10.0. The soil is sandy loam in texture. Assume sulphur as 80% pure.

For such a case, first of all gypsum requirement is calculated. Since the data given allows the use of graph (Fig. 8.6), the GR (t/ha) for this case is calculated as 13 t/ha for a medium textured soil.

Sulphur requirement = 13*0.19 = 2.47 (from Table 8.7)

If sulphur is only 80% pure, then

Sulphur (t/ha) = 2.47/0.8 = 3.08 or 3.1 t/ha

On-farm Activities during and after Application of Amendments

Once the gypsum has been uniformly broadcasted over the area following activities should follow:

- After 7-10 days of leaching, rice crop is transplanted. It will take care of the additional leaching requirement as the rice crop requires standing water

- Follow improved cropping practices such as higher plant population, nutrient management and shallow and frequent irrigation in rice-wheat cropping.
- Apply Zn, a micro-nutrient, @ 20 kg/ha to rice crop. Application of organic manure along with amendments helps in efficient reclamation.

Leaching

Leaching alkali soils after amendment application helps to dissolve the amendment so that it can reach the exchange sites. It also helps to leach down the reaction products from the root zone. From reclamation point of view process of leaching is grouped in two categories namely the flushing and leaching. Flushing in alkali land reclamation is used only before the amendment application and not after the amendment has been applied.

Flushing

Process of flushing the salts is quite different than leaching in many ways although the aim in both cases is to reduce salts in the root zone. In flushing, salts are removed by surface washing of the soil as compared to pushing of soluble salts below the root zone in leaching. In the former, salts move with runoff or irrigation water as the water moves down the sloping land surface. The water with washed away salts is removed through a surface drain provided at the tail end. Flushing method is useful under the following situations:

- Soil permeability is low
- Soils have impermeable layer near to the soil surface
- If a crust of highly hygroscopic salts exists and it prevents use of leveling or other heavy equipment because of low bearing capacity. Flushing in that case will permit better drying of the surface soil.

A combination of flushing and leaching has been successfully used for the reclamation in several projects of *Khar land* Development Board, Gujarat and is useful in highly saline alkali soils of relatively finer texture encountered in central India. For effective flushing, water is kept ponded on the land surface for 2-3 days. Land is then puddled for effective dissolution of soluble salts. The salty water is then drained through the surface drainage system. Two flushing with an interval of about 7 days in between the flushing is recommended. Flushing is impractical in land locked fields. Besides, this technique as a means of reclamation has limited application due to its high water requirement, inefficient salt removal, need of an efficient drainage system and other technical constraints.

Leaching

'Leaching refers to the process of removal of soluble salts or exchanged products by the passage of water through the soil'. It is accomplished by ponding water on the soil surface. It is recommended to pond the water for about 7-10 days. Cultivation of rice will take care of additional leaching, since water is kept in lowland transplanted rice recommended during the first year of reclamation. For more details on this issue, refer the leaching section in saline soils of this chapter.

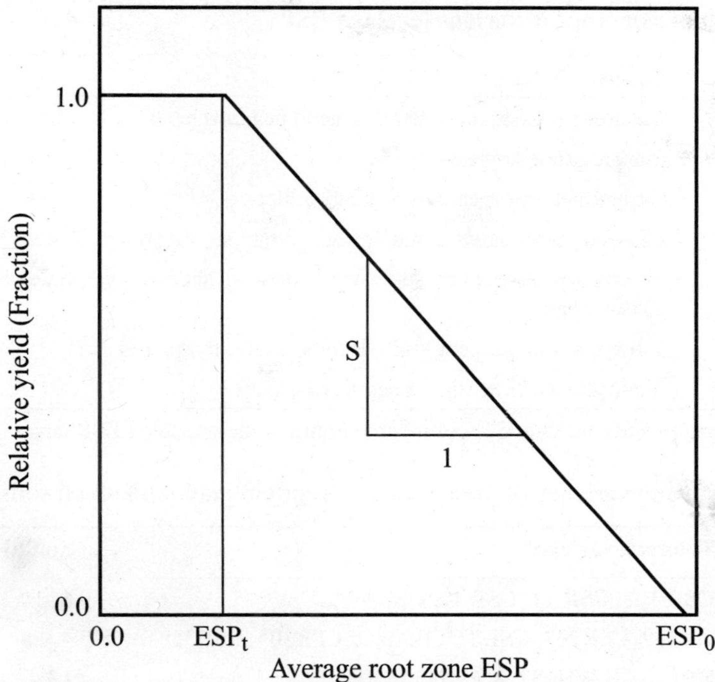

Fig. 8.7 Schematic description of response function of soil ESP and crop yield

ALKALI TOLERANCE AND CROP SELECTION

Crops differ widely in tolerance to alkali environment of the root zone. Crop production function in an alkali environment can also be described by a piecewise linear model of Maas and Hoffman (1977). According to this model described in Fig. 8.7, the productivity is not affected until the ESP exceeds a particular value designated as ESP threshold (ESP_t). Beyond this ESP, yield declines linearly ultimately becoming zero at ESP_0 (Gupta & Sharma, 1990). Each crop has a distinct production function. Such production functions help to decide which crops can be grown depending upon the likely ESP of the root zone at a given time. Since the reclamation is a slow process, alkali tolerance crops should be grown initially to counter the high ESP during the initial years of reclamation. Crops that can withstand excess moisture conditions are generally more tolerant to alkali conditions as well. For example transplanted rice, sugar beet and Karnal grass are most tolerant and can withstand high ESP. It is for this reason reclamation is initiated with rice. It may also be noted that even crop varieties respond differently to alkali conditions. Therefore, selection of crops, cropping sequences and their varieties is of practical and great economic importance to manage alkali lands. Although the crop tolerance to soil alkali might vary with the stage of growth, variety, environmental factors and level of management, their relative tolerance is summarized in Table 8.9. Tolerant varieties within a crop give more production and better economic returns than other varieties under stress conditions (Misra, 1996). Such varieties have been identified and evolved through selection and hybridization in crops like rice, wheat, mustard and gram (Table 8.10).

Table 8.9 Tolerance of crops to various levels of ESP

Range of ESP	Crops
10 – 15	Safflower, mash, peas, lentil, pigeon pea, *urd* bean
15 –20	Bengal gram, soybean, maize
20 – 25	Groundnut, cow pea, onion, pearl millet
25 – 30	Linseed, garlic, guar, lemon grass, pamarosa, sugarcane, cotton
30 – 50	Wheat, *raya*, sorghum, sunflower, *berseem*, shaftal, *senji*, blue panic, gutton panic
50 – 60	Barley, *sesbania*, para grass, Rhodes grass, matricaria
60 - 70	Transplanted rice, sugar beet, Karnal grass

Note: Relative crop yields are only 50% of the maximum in the indicated ESP range

Table 8.10 Promising varieties of important crops for cultivation in alkali soils

Crop	Tolerant varieties	Soil pH_2
Rice	CSR 10*, CSR 11, CSR 12, CSR 13*	9.8 -10.2
	CSR19, CSR23*, CSR27*, Basmati CSR30*, CSR36*	9.4-9.8
Wheat	KRL 1-4*, WH157, KRL 210, KRL 213	<9.3
	Raj3077, KRL19*	<9.3
Barley	DL200, Ratna, BH97, DL348	8.8-9.3
Indian mustard (Raya)	Pusa Bold, Varuna	8.8-9.2
	Kranti, CS52*, CSTR330-1	8.8-9.3
	CST609-B 10, CS54*, CS56*	8.8-9.3
Gram	Karnal Chana 1	<9.0
Sugar beet	Ramonskaaya 06, Maribo Resistapoly	9.5-10
Sugarcane	Co453, Co1341	<9.0

*CSSRI, Karnal varieties released by Central Varietal Release Committee

Improved Agronomic Practices

Seed rate

Germination of crops suffers in alkali soils due to formation of a hard crust or other adverse physico-chemical properties. High mortality rates further depresses the plant population. This together with poor tillering adversely affects the yield of the crops. Increasing the plant population either by reducing the plant-to-plant and/or row-to-row distance in transplanted crops or increasing the seed rate in seeded crops helps to improve crop stand. In general, seed rate should be increased by about 25% over and above the seed rates recommended for normal soils. In rice, increasing the number of seedlings per hill also helps to achieve higher yields. In general 4-6 seedlings should be planted per hill as against the practice of 2 seedlings per hill in normal soils

Puddling

Puddling in lowland rice cultivation is performed mainly to reduce the infiltration rate to conserve water and to control weeds. In alkali soils, infiltration rate is naturally low and weed problem may not be encountered at least during the first 2-3 years. Therefore, puddling can be avoided in these soils unless the soil is too hard to transplant the rice seedlings. It helps to reduce the cost, save energy and reduce environmental pollution.

Age of seedlings

Somewhat older rice seedlings, 35 to 40 days of age, are preferred in alkali soils instead of the usually recommended 30 days old seedlings. Still better results are obtained when rice seedlings are prepared in normal soils as it provides healthy seedlings for transplanting. Similarly, to establish tree species, planting older seedlings grown in normal soil is beneficial.

Continuous cropping

Since most alkali soils have sufficient amount of salts in the profile, leaving the land fallow for a long time results in the upward movement of salts. Once the reclamation process begins, the land should be cropped continuously without long gaps between the crops. Otherwise it may result in resodification of the reclaimed lands.

Time of first irrigation to wheat

Compared to 21 days in a normal soil, the first irrigation to wheat in alkali soils should be given at 30 days after sowing. The maximum yield of wheat was recorded when 5 irrigations were given at different crop growth stages or when irrigation was given at 1.0 IW/CPE ratio (Ratio of irrigation water to cumulative pan evaporation).

Nutrient Management

High pH, high exchangeable sodium, high $CaCO_3$, negligible to low organic matter content and inhospitable soil physical environment of alkali soils affect the transformation and availability of native and applied fertilizer nutrients. Alkali soils, in general, are deficient in Ca, N and Zn, but are rich in available P and K. Following practices are recommended for efficient nutrients management:

Because of low available N status and slow transformation of organic and inorganic-N in salt affected soils, added N exhibits poor use efficiency and as a result crops grown in these soils generally respond to higher levels of N than those raised in normal but otherwise similar soil and climatic conditions. Therefore, general recommendation has emerged that apply 20 to 25% more nitrogen than recommended for normal soils in 3 splits to rice-wheat.

Application of Zn plays an ameliorative role in alkali soils. Field experiments show that application of 10 kg $ZnSO_4$ per ha to each crop of rice and wheat or 20 kg $ZnSO_4$ per ha to rice increased the grain yield significantly over no application of Zn.

Integrated nutrient management

Disproportionate and inappropriate continuous application of chemical fertilizers has lowered the fertility status and caused nutrient hunger in soils. Integrated nutrient management comprising in a good blend of organic and chemical fertilizers helps to sustain crop production in reclaimed lands. Besides, farmyard and green manure increase organic carbon and also increase the availability of plant nutrients. FYM also helps to meet a part of the zinc requirement of the crops. Evidences exist to show that application of farmyard manure and/or green manure helps to build-up nitrogen status of the soils and over the years increase crops yield.

Application of green manure enhances organic matter (OM) content of the soil, increases partial pressure of CO_2, lowers pH, enhances solubility of native $CaCO_3$ and adds considerable amount of plant nutrients to the soil. *Sesbania*, a green manuring (GM) crop, being tolerant to high ESP and water logging is an ideal crop. It is grown during summer months (May-June), a lean period in the rice- wheat cropping sequence. Normally, 45-50 days old *Sesbania* crop, when it has attained a height of 1.5 to 1.8 m, is incorporated into the soil. Longer growth period renders the plants hardy that make the decomposition difficult. A fifty days old *Sesbania aculeata* is able to supply about 4.2 t/ha dry matter and 90 kg N, accumulated mostly through biological N fixation, 11 kg P and 90 kg K.

THE PACKAGE (ALKALI LANDS)

The complete package of alkali land reclamation and management is summarized as follows:

- On-farm land development including dyking, leveling, construction of field irrigation and drainage channels and arranging assured supply of water.
- A heavy irrigation (pre-amendment application) followed by application of adequate quantities of chemical amendment
- Leaching for 7-10 days if rice is taken as the first crop
- Cultivation of crops as per tolerance, the preferred cropping sequence is rice-wheat (high and medium tolerant respectively). Select salt tolerant varieties within the selected tolerant and medium tolerant crops. Green manuring crop, *dhaincha* can be grown after wheat, if possible.
- Use improved agronomic practices such as higher plant population in transplanted crops (Lesser plant to plant and row to row distance and more number of seedlings per hill in rice), 25% higher seed rate in case of wheat or other crops, avoid puddling in initial 2-3 years unless necessary to transplant rice, use rice seedlings of higher age preferably grown in normal lands and avoid prolonged fallowing in between the crops
- Apply 25% more nitrogen to crops, apply micro-nutrients as per requirement, application of zinc is absolutely necessary to rice crop@ 20 kg Zn/ha and practice integrated nutrient management.

PHYTO-REMEDIATION OF ALKALI LANDS

Phyto' means plant and phyto-remediation is a generic term for all technologies that use plants to remediate soils, sediments and water contaminated with organic and inorganic contaminants. Phyto-remediation therefore, is defined *'as the use of plants to remove, detoxify or immobilize environmental contaminants in soil or water through the natural biological, chemical or physical activities and processes of the plants'*. Synergy of the chemical amendments (application of 25% of GR) and the salt tolerant crops or their varieties can be utilized to phyto-remediate an alkali soil. The basic principle is that during cultivation process roots, fertilizers added and organic matter applied produce acids which react with native $CaCO_3$ present in the soils. It releases Ca to lower the ESP of the soils albeit slowly. Similarly CO_2 produced by the respiring roots increases the partial pressure of CO_2 which also helps to dissolve the $CaCO_3$.

The actual strategy to be adopted also depends upon the initial ESP of the soils. For highly alkali soils (pH > 10.0), cultivation of salt tolerant grasses is an attractive proposition. Karnal grass (*Diplachne fusca*) is an excellent choice for this purpose as it is able to tolerate high ESP and water logging stresses. Para grass (*Brachiaria mutica*) is an equally good alternative. Field experiments proved that after growing Karnal and para grasses for 1 year, satisfactory yield of rice can be obtained. After 2 years, even good yield of wheat is possible. Besides, during these years, good yield of Karnal and para grass is obtained. Both grasses can be used as fodder. Although slow and less efficient, it can be adopted by resource poor farmers who cannot afford high investments on reclamation program. For medium degree of alkali soils (pH around 9.6), reclamation can begin with salt tolerant rice varieties such as CSR 10 of CSSRI, Karnal. In the first year yield of wheat may be affected but from second year onward, salt tolerant wheat varieties such as KRL 210 and KRL 213 give good yield.

RECLAMATION OF SALINE SOILS

Saline soils are often encountered under two entirely different set-ups:

- Saline soils in association with shallow water table in irrigation commands and/or coastal regions
- Saline soils resulting from irrigation with saline water, have deep water table and are encountered in arid and semiarid regions.

The basic principle of reclamation of saline soils is to remove the salts through the process of leaching. But in the first kind of soils where water table is shallow or natural subsurface drainage is restricted, a subsurface drainage system needs to be installed (discussed thoroughly in previous chapters) to control the water table and to facilitate the percolated soil solution to move out of the soil profile. Soils with deep water table can be reclaimed through leaching and management alone. Since drainage design, construction, operation and maintenance have been discussed in the previous chapters, processes of leaching and management are included in this chapter.

Local Solutions

Scraping of top soil

In order to get quick returns, farmers resort to remove the salts that accumulate on the soil surface using manual or mechanical means. They scrap about 5 cm of the soil and keep it as a heap in one corner of the field. Generally speaking, scarping may or may not provide even the short-term relief. Even if some relief is received, it will be quite temporary say for 1-2 years. The practice has limited utility in areas having shallow water table as illustrated in Fig. (8.8 left). The reduced depth to water table accentuates the process of capillary action, increases the evaporation rate and quickly makes the soil salinized. The cost of the operation and ultimate disposal of salts will continue to be major problems. Therefore, scraping as a means of land reclamation does not find much favor.

Addition of normal soil to raise the level

Realizing the problems encountered in scraping, farmers now adopt a quite different approach. They raise the surface of their fields by adding a layer of normal soil. Although the cost of such an operation is enormous, yet it may obviate some of the problems encountered in scraping. The water table is now deeper by the amount with which field level has been raised. The problem of soil disposal is also resolved (Fig. 8.8 right).

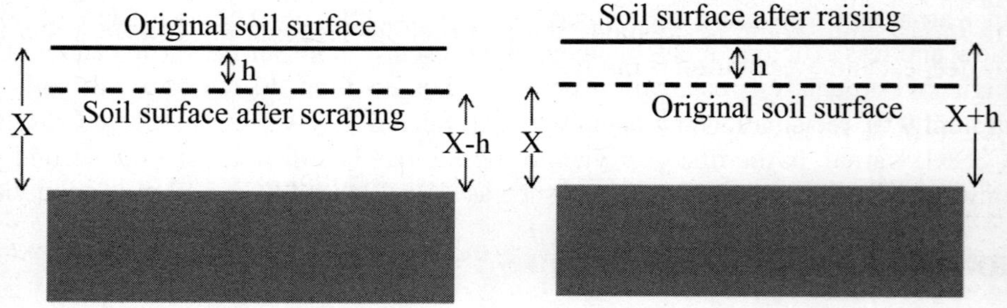

Fig. 8.8 Groundwater depth in relation to scraping (left) and after raising the field surface (right)

Leaching

The only reasonable way to accomplish reclamation seems to be leaching. *'Leaching of saline soils is a process in which water of low concentration is applied to displace the soil solution of relatively high concentration. The application of the excess water with the aim of pushing the salts below the root zone is defined as leaching'*. For efficient leaching following steps are practiced.

On-farm activities before leaching

After the subsurface drainage is constructed, fields should be dyked/bunded and leveled. Inadequate leveling results in spotted desalinization and may cause delay or failure of the reclamation program. Since leaching is one time operation, fields should be laid as per the cropping and irrigation requirement. If need be smaller plots with temporary bunds can be constructed within the overall layout plan of the farm (Fig. 8.9). Tillage of land before and in between leaching (in case of intermittent leaching) is beneficial because it enhances infiltration, reduces the time of reclamation and improves the leaching efficiency.

Fig. 8.9 Fields being filled with water for leaching

The depth of reclamation depends largely upon the potential land use. It is 0.6 m for shallow rooted crops or pastures to more than 2 m for trees. For diversified cropping, the depth of reclamation usually varies from 1 to 1.5 m. It may be noted that these are the minimum recommended depths and appropriate measures against re-salinization should be adopted. The best time for leaching is when the water table is deepest and evaporation is minimal.

Amount of water required for leaching

In an ideal soil system without bypass, without dissolution or precipitation of salts, without salt diffusion or dispersion, the salt concentration of the soil water after one pore volume (1 PV) of water has percolated below a given depth should drop to the concentration of the applied water. One pore volume equals one pore space in the given volume of soil per unit area (θ. d m^3) such that θ is the soil porosity and d is the depth of the soil. The actual amount of water required under practical leaching, however, depends upon many factors, important among them are:

- Amount of salt initially present in the profile
- Desired level of salt in the root zone
- Soil texture
- Type of salts present
- Soil depth to be reclaimed
- Efficiency of the drainage system, if provided
- Method of leaching

As a thumb rule, a unit depth of water (D_w/D_s =1) passing through the root zone of a medium textured soil will remove about 80% of the salts from the given depth of the soil. Field data, however, show a wide range depending upon the soil texture and other factors, the values varying from 0.3 cm to a high value of 2.58 cm (with continuous leaching).. The final soil water salt concentration is normally determined by the tolerance of the crop to be grown soon after reclamation. Crop production function of any crop with respect to soil salinity can also be described in the same manner as shown for ESP (Fig. 8.7). Most comprehensive data on salinity tolerance of crops have been published and are often used to frame guidelines for crop selection in reclaimed lands (Maas & Hoffman, 1977). The crop tolerance data given in Table 8.11 provides limited information on some crops for some selected stations in India. The relative tolerance of maize (Indore) and barley (Sampla) from this table depicted in Fig. 8.10 highlights and reveal that barley is much more tolerant to salinity than maize. If the level achieved after leaching is insufficient to cultivate the proposed crop(s), either a more tolerant crop should be selected or additional water must be applied to leach the salts. Most favorable conditions must be created during the germination and initial establishment stages, because crop productivity at the end of the season depends upon the initial crop stand.

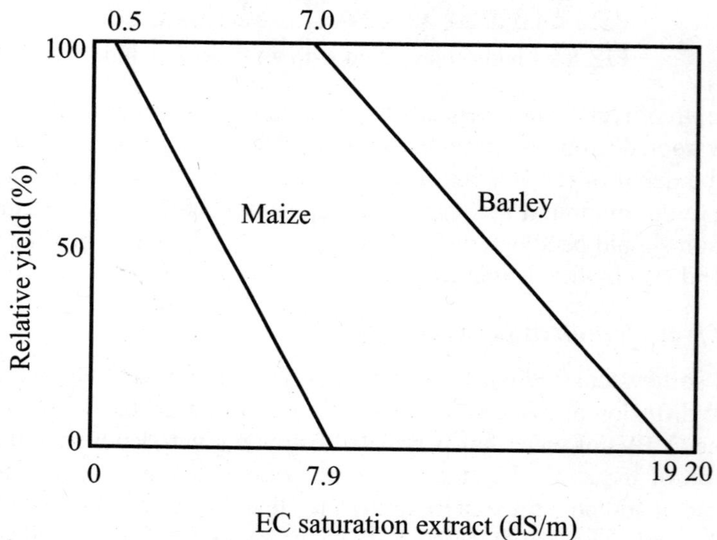

Fig. 8.10 Relative tolerance of maize and barley to soil salinity

The data given in Table 8.11 or those given by Maas & Hoffman (1977) are used to determine the amount of salts to be leached expressed as a fraction (Lf) of the amount of salts present in the profile at the beginning of leaching. Mathematically,

$$Lf = 1 - \frac{\text{Threshold salinity of the crop}}{\text{Initial salinity of the soil}} \tag{8.17}$$

If some yield reduction is acceptable, then threshold salinity in eq. (8.17) is replaced by the salinity at which acceptable yield reduction (EC_d) is likely to occur. It can be calculated by the data reported by Maas & Hoffman (1977) or given in Table 8.10 as per the following relation:

$$EC_d = \frac{\text{Desired yield reduction (\%)}}{\text{Slope (\%)}} + \text{Threshold EC} \tag{8.18}$$

Since most crops notably wheat, beans, alfalfa and even sugar beet are sensitive to salinity at germination stage; these may not grow fairly well if average salinity guidelines are used to decide the soil salinity at the germination stage (Canada

Table 8.11 Crop tolerance to soil salinity

Sites	Soil type	Crop	Threshold salinity (dS/m)	Slope (%)
Sampla	Sandy loam	Wheat	4.0	29.0
		Mustard	6.0	75.0
		Barley	7.0	19.0
Karnal	Sandy loam	*Moong* bean	1.8	20.7
		Mustard	3.8	6.9
		Sorghum	2.2	10.6
Agra	Sandy loam	Wheat	8.2	19.8
		Mustard	6.1	20.7
		Berseem	3.5	12.5
		Potato	4.4	16.1
		Tomato	1.3	6.5
Dharwad	Black clay	Wheat	2.3	20.5
		Safflower	2.8	20.7
		Sorghum (W)	2.1	3.9
		Italian millet	6.1	50.0
		Sataria	1.12	2.19
Indore	Black clay	*Berseem*	2.0	11.22
		Safflower	2.80	5.00
		Maize	0.50	7.90
Chakuly	Clay	Rice	0.60	2.70

Department of Agriculture, 1977). In such cases, it may be prudent to use the threshold salinity levels for the germination stage.

The Lf values so calculated with Eq. (8.17) or Eq. (8.18) are used on the y-axis of the leaching curves to determine the ratio of depth of water to be applied to depth of the soil (see next section). Since the depth of soil to be reclaimed is known, the depth of water required for leaching can be calculated.

Leaching curves

Leaching curve is a useful tool in planning reclamation programmes. Leaching curve is developed by conducting a simple, straightforward replicated test similar to the one used to evaluate infiltration rate. Two concentric cylinders are used in this test (Fig. 8.11). The water level in the inner cylinder is kept at a constant level and water infiltrated is measured till the pre-determined amount of water has entered the soil. In this test, 3 or 4 depths of water are passed and are taken as the treatments. Soil samples in the inner cylinder are taken before and after the completion of the test to evaluate moisture content and EC of different layers (Gupta & Gupta, 1997).

Fig. 8.11 A view of the set-up to prepare a leaching curve

The amount of water drained below a given depth, D_w is found by subtracting from the total depth of water infiltrated, the amount of water evaporated and the amount of water absorbed in the soil layer. Thus, D_w/D_s can be calculated. The EC_{eq} is commonly used to make the leaching curve independent of the salinity of irrigation water and the existing evaporation and drainage conditions. It is obtained by measuring the soil salinity in the 0-5 cm layer in the treatment where largest amount of water has entered the profile. It is expected that the salinity of this layer is practically a constant minimum that can be achieved upon leaching the soils. Current EC_e of different layers is then used to calculate $(EC_e - EC_{eq}) / (EC_o - EC_{eq})$.

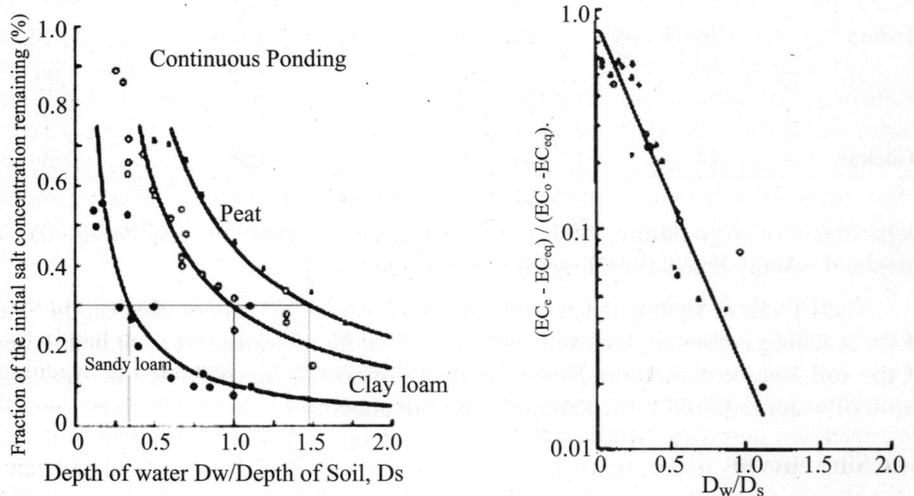

Fig. 8.12 Leaching curves for different soil texture groups (left) and for a medium texture soil (right)

The leaching curve is a dimensionless plot of $(EC_e - EC_{eq}) / (EC_o - EC_{eq})$ versus D_w / D_s, such that EC_o is the initial soil EC_e and D_s is the depth of the soil. The data are plotted either on an arithmetic paper (Fig. 8.12, left) or on a semi-log paper (Fig. 8.12 right). These curves are used to determine one time leaching required for reclaiming the land for the test soil and conditions under which leaching tests have been conducted. The data in (Fig. 8.12, left) suggest that for the same fraction of initial soil concentration remining, water required to leach a light textured soil is relatively less than a heavy textured soil.

Some thumb rules have also been proposed to assess the water required for leaching. The simplest equation is given as (Hanson *et al.*, 2006):

$$D_w/D_s = k\ EC_{in}/EC_{fi} \tag{8. 19}$$

Here k is a constant that varies from 0.1 for coarse textured soils to 0.3 for heavy textured soils and to 0.45 for organic soils and EC_{in} and EC_{fi} are the initial and final EC of the soil depth being considered, D_s. D_w and D_s have the same meaning as indicated before.

Continuous and intermittent leaching

Leaching efficiency can be improved by creating unsaturated conditions during leaching and/or by reducing the average rate of water movement in the soil profile. This can be achieved through intermittent ponding of water on the soil surface. As against the continuous ponding throughout the period of leaching, water is ponded and allowed to percolate down in cycles in intermittent ponding until the leaching is complete. The intermittent leaching minimizes the loss of water through large pores and provides greater opportunities for diffusion of salts from immobile or less mobile pockets to the mobile zone. It has emerged that as much as 30 to 70% water required for leaching in continuous ponding mode can be saved through intermittent leaching. Intermittent leaching is useful in soils with large bypass (with cracks or structured soils), poor soil permeability, deep groundwater table and during periods of low evaporative demand. On the other hand, continuous ponding may give reasonably good results in soils with good permeability, unstructured soils, areas that have shallow saline groundwater and during periods of high evaporative demand. Under certain conditions, benefits ascribed to intermittent leaching may not be obtained. For example, it may happen in soils where surface layer is the restricting layer for water infiltration. It will cause unsaturated flow condition in the soil. Such conditions naturally prevail in alkali soils. As a result, continuous ponding gives equally good results as expected with intermittent leaching in alkali soils.

Leaching efficiency (LE) in field soils can also be enhanced by controlling water application rate. Sprinkler irrigation system for this purpose is an handy tool to apply water at pre-determined rates. In the Indian context, it appears that leaching by sprinkling may not play an important role unless sprinklers are purchased and used for irrigation purposes. Moreover, data to date suggest that similar conditions can be created through intermittent leaching. Therefore, usefulness of investment in sprinklers for leaching is questionable.

Practices to Enhance Leaching Efficiency

Leaching is the only practical means to reduce soil salinity and bring saline soils under the plough. To reclaim the soils with limited amount of water or to achieve early reclamation, a number of practices can be used as summarized in Table 8.12.

Table 8.12 Practices to improve leaching efficiency

Practice	Effect
Sub-soiling	• Improves infiltration rate • Removes/reduces macro channels
Tillage with inversion	• Improves infiltration rate of the soil • Physically moves the surface layer with high concentration of salts to lower depths
Chemical amendment	• Improves infiltration rate • Controls surface crusting • Maintains soil structure
Breaking hard pan, if present	• Improves water movement through the hard pan • Reduces chances of perched water table build-up
Irrigation practices	• Improve LE by maintaining high moisture content • Situations resembling intermittent leaching can be created

CROPPING IN SALINE SOILS UNDER RECLAMATION

Halophytes are most tolerant to soil salinity. Glycophytes are usually less tolerant but have a wide range of salt tolerance. For example, sugar beet and barley are more tolerant to salts than pulses or other cereal crops. Qualitatively crops are categorized in 4 groups depending upon their tolerance to soil salinity (Table 8.13). The ratings given in the literature may be treated as guidelines for the relative salt tolerance as minor variations might occur between tables given by various workers. The relative tolerance of crops is used as a selection criterion to cultivate crops during reclamation. The crop must fit to the degree of soil salinity after the basic reclamation measures such as land drainage, field leveling, *bunding* and leaching have been implemented. Suitable crops and cropping sequences also need to be identified for cultivation during intervening periods before the lands are fully reclaimed. The crops during this period are selected on the basis of salt tolerance, tolerance to water logging, water requirement-water supply scenario and economic value. Moreover as already explained, it would be prudent to select crops and irrigation strategies to minimize salt stress at the sensitive growth stages.

LAND MODIFICATIONS AND PLANTING MANAGEMENT

When it is difficult to fully reclaim a saline soil, then one may adopt practices that help to live with the salts. It can be achieved by irrigation management and/or by adopting certain crop bedding techniques. These may help to move the salts away from the primary root zone as the goal is to ensure salt free zone near to the germinating seeds and plant roots. For example; salts in a furrow irrigated field accumulate on the ridges while the area at the bottom of the furrow and around it has the minimum accumulation of salts (Fig. 8.13, left, Fig. 8.13, right). Thus, seeds can be planted in the furrows to take advantage of a wet and less saline zone. Moreover, salts in the furrows are leached more efficiently than those on the ridges. Studies reporting higher yields in trench method of planting sugarcane than flat sowing attest to

Table 8.13 Crop groups based on response to salt stress

Sensitive groups		**Resistant groups**	
Highly sensitive	**Medium sensitive**	**Medium tolerant**	**Highly tolerant**
Lentil	Radish	Spinach	Barley
Mash	Cowpea	Sugarcane	Rice (transplanted)
Chickpea	Broad bean	Indian mustard	Cotton
Beans	Vetch	Rice (direct sowing)	Sugar beet
Peas	Cabbage	Wheat	Turnip
Carrot	Cauliflower	Pearl millet	Tobacco
Onion	Cucumber	Oats	Safflower
Lemon	Gourds	Alfalfa	*Taramira*
Orange	Tomato	Blue panic grass	Karnal grass
Grape	Sweet potato	Para grass	Date palm
Peach	Sorghum	Rhodes grass	*Ber*
Plum	Minor millets	Sudan grass	Mesquite
Pear	Maize	Guava	*Casurina*
Apple	Clover, *berseem*	Pomegranate	*Tamarix*
		Acacia	*Salvadora*

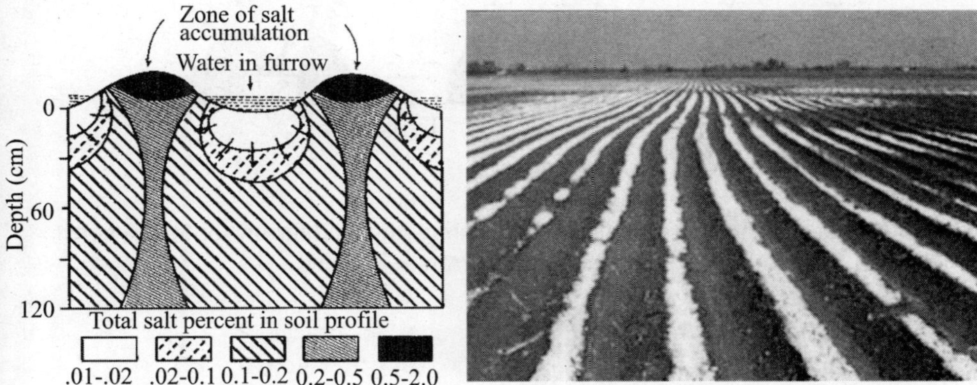

Fig. 8.13 Salt distribution in a furrow irrigation method in saline soils

the advantage of this strategy. It is attributed to less concentration of salts in the trench moving to the top of adjoining ridges.

In a double-row bed system, salt accumulate toward the middle of the bed (Fig 8.15, top). Thus, plants planted away from the middle towards the shoulders of the furrow encounter relatively low salt stress. Another modification to this method is to have a notch at the center of the ridge and to plant the crop halfway between the wet furrows (Fig. 8.14, bottom). Both these strategies require uniform application of water in furrows. Otherwise, salts might accumulate closer to one side of the bed near to furrow receiving less water. Thus, the plant growth along that furrow may be adversely affected.

In alternate furrow irrigation in a single-row bed systems, salts are pushed across the bed from the irrigated side of the furrow to the dry side. Thus, planting towards the shoulder of the wet furrow is helpful (Fig. 8.15). Care must be taken to ensure that enough water is applied to wet all the way across the bed to prevent build-up in the planted area. This method of salinity management can harm the plant if large amounts of rainfall fill the normally dry furrow and push the salts back across the bed toward the plants. There are many other configurations that may help in managing the salts (Fig. 8.16). In drip irrigation salts move to the fringes of the wetting front leaving most of the area under the drip salt free (Fig. 8.17). It is because of this reason; higher yields under relatively more saline soils and irrigation with relatively more saline water are obtained in drip as compared to surface irrigation methods.

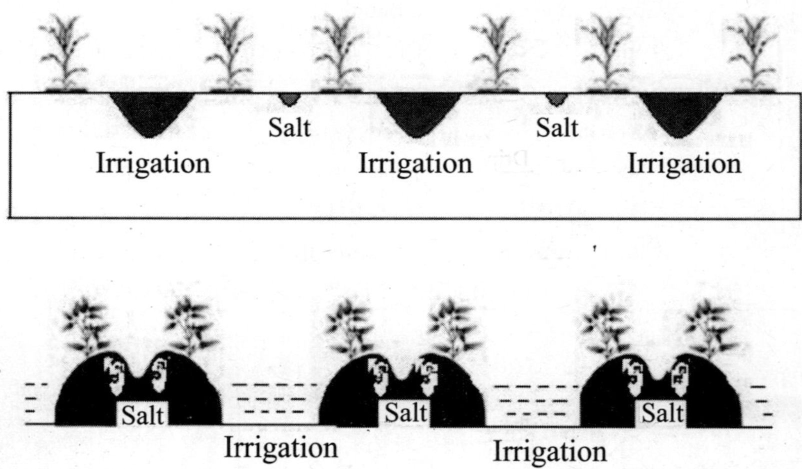

Fig. 8.14 Salt accumulation in double row planting (top) and furrow top with a notch (bottom) where each furrow is irrigated

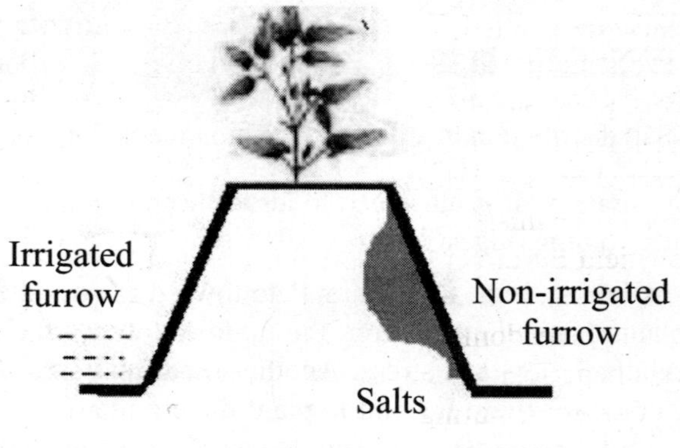

Fig. 8.15 Salt accumulate on the dry furrow side in alternate furrow irrigation

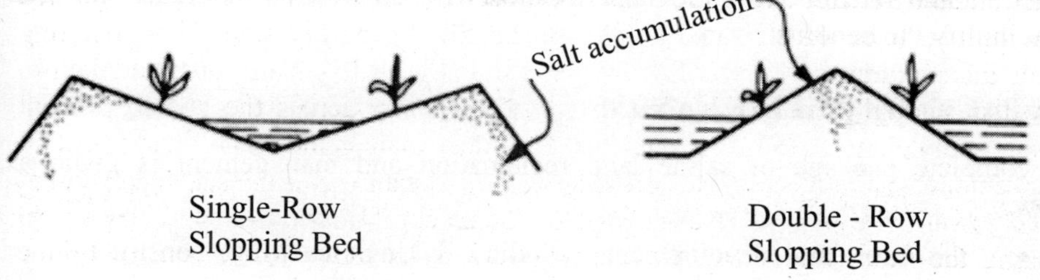

Fig. 8.16 Planting on the sides of slanted ridges to minimize salt stress on growing plants

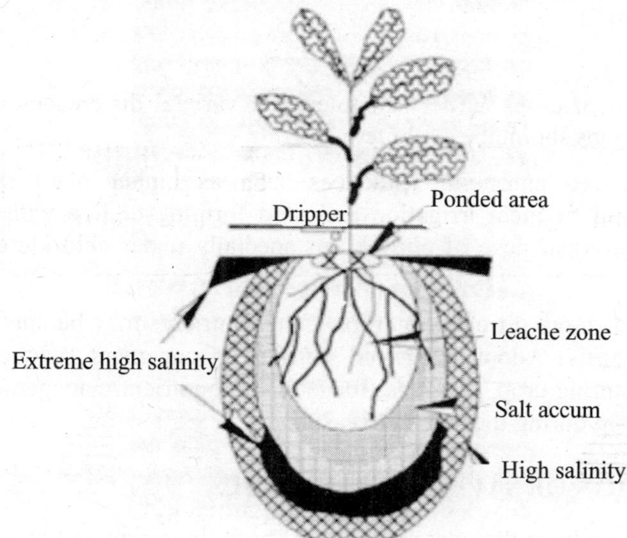

Fig. 8.17 Salt accumulation zones at wetting front in a drip irrigation system

Cultural Practices for Higher Productivity

Few improved cultural practices, if adopted, help to achieve higher productivity in saline soils without incurring much cost. Two main practices are to increase the seed rate by 25% to improve plant stand and application of 25% more nitrogen compared to recommended dose for normal soils. Uptake of P by plant roots in saline soils may also be adversely affected by competitive uptake of Cl^- ions as both these anions are absorbed by the same mechanism. Thus, in saline soils, additional application of phosphorus enhances yield by directly providing the phosphorus and by decreasing the uptake of toxic elements like Cl^- and F^-. Additional application may not be needed in soils with sulphate dominated salinity Application of phosphorus along with FYM is still more beneficial as this combination markedly alleviates salt injury symptoms in crops. Green manuring during fallow periods is recommended to provide crop cover and supplement nutrients and organic matter which invariably helps in improved crop yields. Crops grown in saline soils must be irrigated more frequently compared to crops grown under non-saline conditions so that the plants are not subjected to excessively high soil moisture stress. The total amount of water applied should remain about the same as otherwise application of higher salt load might nullify the benefits.

THE PACKAGE (SALINE LANDS)

The complete package of saline land reclamation and management is given as follows:

- On-farm land development including dyking, levelling, construction of field irrigation and drainage channels for surface drainage and arranging assured supply of water.
- Implementing subsurface drainage for waterlogged saline lands to control water table
- Leaching
- Cultivation of crops as per salt tolerance, varietal differences within the selected crops should be exploited
- Use improved agronomic practices such as higher plant population, shallow and frequent irrigation and land forming to live with the salts. Apply additional dose of phosphorus specially under chloride dominated salinity
- Apply 25% more nitrogen to crops. Other nutrients may be applied as per soil test basis. Adopt integrated nutrient management practices. Grow green manuring crop, *dhaincha* for integrated nutrient management and as a cover crop during the summer season.

RECLAMATION OF SALINE-ALKALI SOILS

Saline-alkali soils have the characteristics of both the saline and alkali soil and therefore needs to be treated as such. Initially, these soils have good soil structure and adequate water movement through the soil profile. It is because of high electrolyte concentration that helps to keep the soil flocculated. As the leaching process in these soils proceeds, soil structure slowly deteriorates that prevents water to infiltrate into the profile. Before this situation develops, an amendment must be applied to correct the sodium problem, followed by further leaching to remove salts. Flushing has been found quite effective in removing the salts from medium to heavy textured saline-alkali soils of central India. Flushing not only removes the salts but also lowers the amendment requirement of these soils (Table 8.14, Arora *et al.*, 2015). As explained earlier, it is because of leaching of soluble sodium carbonates and bicarbonates present in the soil solution. These salts, if not leached, react with the amendment to reduce its efficiency in lowering the exchangeable sodium. Thereafter, all management steps of alkali land reclamation namely cultivation of rice-wheat crops for initial few years, cultivation of salt tolerant crop varieties within the selected crops, application of 25% more nitrogen, application of $ZnSO_4$,and integrated nutrient management are adopted to reclaim these soils.

Table 8.14 Effect of flushing on gypsum requirement of a saline-alkali soil in Uttar Pradesh

Location	pH_2	EC_2 (dS/m)	Gypsum requirement (meq/100 gm)	
			Schoonover method	After washing
Kausambhi	9.2	0.74	3.70	2.90
Rasoolpur	8.9	2.43	1.60	1.10
Baratikhera	10.1	9.81	5.20	3.10
Lalganj	10.0	1.28	5.80	3.40
Sandila	10.3	2.50	5.30	3.70
Dhaura	10.1	4.04	4.50	2.20
Vishnupur	9.8	2.25	5.80	4.80

QUESTIONS (THEORY)

1. What do you understand by EC and pH of soils? Give the relationship of EC with ppm and mg/L.
2. What is SAR of the soil solution? Calculate the SAR of the soil solution, if Na, Ca and Mg are 25, 10 and 2 meq/L. Comment on the soil with respect to its SAR.
3. Differentiate between saline and alkali soils. When a soil is termed as saline-alkali?
4. If a soil of CEC 22 meq/L has an initial ESP of 40. The proposed final ESP to be achieved is 20. Calculate the gypsum requirement of the soil to reclaim 30 cm depth of the soil. Assume an appropriate value of bulk density and purity of gypsum as 80%.
5. List the causes of low productivity in alkali soils.
6. List the causes of low productivity in saline soils.
7. If initial and desired final ESP is known, derive the formula to determine the gypsum requirement.
8. Describe the laboratory method to determine gypsum requirement.
9. What do you understand by crop tolerance to salts? How this character is used to manage salt affected soils.
10. Discuss the role of scraping of salts in land reclamation. What are its limitations?
11. What do you understand by leaching curves? Explain, how these curves are prepared and used in land reclamation programme.
12. Differentiate between leaching and flushing. Discuss some of the problems associated with flushing.
13. Write short notes on continuous and intermittent leaching clearly bringing out the differences and advantages and limitations of each method.

14. Write a short note on land modification for salinity management.
15. Briefly describe the reclamation of a saline-alkali soil.

QUESTIONS (OBJECTIVE)

1. Fill in the blanks

1. 1 dS/m is equal to ____ mmhos/cm
2. 1 mg/l divided by __________weight is equal to 1 meq/l.
3. Sum of the cations/10 is approximately equal to ______ of water (EC <5.0 dS/m)
4. Equivalent weight of sodium is ______ than the equivalent weight of calcium.
5. If concentration of ions is in mmol/l then ________of the water is Na/ $(Ca+Mg)^{1/2}$.
6. ESP of the soil is 20 but EC of the soil is unknown. It can be an ________ or saline_____ soil depending upon the EC of the soil.
7. Osmotic Pressure (OP) of the soil water in atmosphere is given as: OP = _____ *EC_e . Here EC_e is the EC of the soil saturation extract in dS/m.
8. For calcareous soils ______ forming amendments can also be used for reclamation.
9. Chemical formula for gypsum is _________.
10. Pyrites is an _________ forming amendment.
11. For the same degree of reclamation, amount of water required for leaching is ______in a heavy textured than a light textured soil.
12. For the same degree of reclamation and soil type, amount of water required for leaching is more in _________than intermittent leaching.
13. In drip irrigation salts accumulate at the outer fringe of the ______ front.
14. Initial infiltration rate in a saline alkali soil is _______ because high electrolyte concentration keeps the soil flocculated.
15. Transplanted rice is _________tolerant to alkali conditions than direct seeded rice.

2. Tick mark correct or most appropriate answer (A,B,C or D) from the following

1. For quick appraisal, EC of the soil is measured using

   (A) 1:2 soil water extract (B) 1:2.5 soil water extract

   (C) 1:5 soil water extract (D) Any of these

2. EC of the soil solution at field capacity is about______ times the EC of the soil solution at saturation

   (A) One time (same in both cases)

(B) Twice (Two times)

(C) Thrice (Three times)

(D) None of these

3. The saturation extract of the soil has an EC of 4 dS/m, what is the salt concentration in mg/l?

(A) 2.56 (B) 256

(C) 2560 (D) None of these

4. pH is the log of the

(A) Inverse of the hydrogen ion concentration

(B) Hydrogen ion concentration

(C) Inverse of the *hydroxide ions (OH)* concentration

(D) None of these

5. Sodium adsorption ratio, SAR is defined as SAR= Na/√(Ca+Mg). In this relation cation concentration is expressed in

(A) meq/L (B) mmol/L

(C) mg/L (D) None of these

6. Sodium adsorption ratio, SAR is defined as SAR= Na/√((Ca+Mg)/2). In this relation cation concentration is expressed in

(A) meq/L (B) mmol/L

(C) mg/L (D) None of these

7. The exchangeable cations concentration in a soil Ca^{2+} = 3.6 meq/100 g, Mg^{2+} = 1.4 meq/100 g, Na^{+} = 19.2 meq/100 g, and K^{+} = 0.2 meq/100 g. What is the approximate CEC of the soil?

(A) 24.4 meq/100 g (B) 24.2 meq/100 g

(C) 19.2 meq/100 g (D) None of these

8. The exchangeable cations concentration in a soil Ca^{2+} = 3.6 meq/100 g, Mg^{2+} = 1.4 meq/100 g, Na^{+} = 19.2 meq/100 g, and K^{+} = 0.2 meq/100 g. What is the approximate ESP of the soil.

(A) 78.7 (B) 79.3

(C) 1.0 (D) None of these

9. If a soil having cation exchange capacity of 28.3 meq/100g has Na= 12.8 meq/100 g and Ca = 5.4 meq/100g on its exchange complex then what is the soil ESP?

(A) 18.2 (B) 45.2

(C) 64.3 (D) None of these

10. ESP of the soil is given as

(A) (Exchangeable sodium /cation exchange capacity)*100

(B) (Exchangeable calcium /cation exchange capcity)*100

(C) (Exchangeable (sodium+calcium) /cation exchange capcity)*100

(D) (Exchangeable (sodium - calcium) /cation exchange capcity)*100

11. Alkali soils are also known as

(A) Saline soils (B) Saline-sodic soils

(C) Sodic soils (D) None of these

12. H_2SO_4 is an effective amendment only in soils in the presence of

(A) $CaCO_3$ (B) NaCl

(C) KCl (D) None of these

13. A soil has a pH of 8.5, EC_e is less than 4 dS/m and ESP is more than 15, the soil is categorized as

(A) Saline (B) Saline-alkali

(C) Alkali (D) None of these

14. A soil has a pH of 8.1, EC_e is more than 4 dS/m and ESP is less than 15, the soil is categorized as

(A) Saline (B) Saline-alkali

(C) Alkali (D) None of these

15. The EC of the saturation extract of a soil is 6 dS/m. What is the osmotic pressure in the soil?

(A) 6 atmosphere (B) 2.16 atmosphere

(C) 24 atmosphere (D) None of these

16. Which kind of the amendments can be universally applied in all kinds of soils for quick reclamation?

(A) Soluble calcium salts (B) Acidifying materials

(C) Organic materials (D) None of these

17. If EC_e of the soil is 1 dS/m. EC_2 of the soil will be

(A) More than 1 (B) Less than 1

(C) Equal to 1 (D) None of these

18. If EC_e of the soil is 1 dS/m. EC_2 of the soil will be _____ times if the saturated moisture content by weight is 40%

(A) 2 times (B) 5 times

(C) 1 time (will remain same) (D) None of these

19. If the CEC of the soil is expressed in cmol(p+)/kg of the soil, it will be equal to

(A) meq/kg of soil (B) meq/g of soil

(C) meq/mg of soil (D) meq/ 100 g of soil

20. Which is the most tolerant to alkali conditions

(A) Rice (transplanted) (B) Rice (direct seeded)

(C) Wheat (D) Maize

21. Which is most tolerant to soil salinity

(A) Wheat (B) Maize

(C) Barley (D) Tomato

22. Amount of water required to leach the salts depends upon

(A) Initial soil salinity (B) Soil texture

(C) Soil depth to be reclaimed (D) All of these

23. A waterlogged saline soil has

(A) EC_e > 4dS/m (B) Depth to water table <1.5 m

(C) Both A and B (D) None of these

24. In the relation osmotic pressure, OP = 0.36 *EC_e, if ECe is in dS/m, what are the units of OP

(A) Atmosphere (B) meter of water

(C) millibar (D) None of these

25. Intermittent leaching may not give additional advantage over continuous leaching in

(A) Normal soils (B) Saline soils

(C) Alkali soils (D) All of these

■■■■

Chapter 9

RECLAMATION AND MANAGEMENT OF ACID SOILS

INTRODUCTION

Whereas salinity and sodicity problems are associated with arid, semiarid and sub-humid regions, soil acidity is associated with sub-humid and humid regions posing severe limitations to agriculture. While acid soils occupy about 3950 M ha area globally, problem of soil acidity in India is spread over 89.94 M ha (Table 9.1). Within this figure, 58.9 M ha has pH in the range of 5.5 – 6.5 (moderately to strongly acidic), 24.81 M ha in the range of 4.5 – 5.5 and remaining with pH < 4.5. It is also assessed that 48 M ha of acid soils are under cultivation of which 25 M ha have pH less than 5.5 while about 23.0 M ha are slightly acidic with pH ranging between 5.6 and 6.5 (Bhaumik & Roy, 1964). An area of 10.2 M ha has been identified for amelioration on priority (Table 9.1). The low agricultural productivity of acid soils is largely attributed to adverse chemical, physical and microbiological properties of these soils. Besides, loss in soil fertility, excessive presence of toxic elements, intensive cropping and poor management practices lead to low productivity in acid soils. This chapter addresses the issues related to reclamation and management of acid soils in India. The write-up is also extended to briefly cover the potential and actual acid sulphate soils.

DEFINITION, CHARACTERIZATION AND FORMATION OF ACID SOILS

Soil acidity is identified by the soil reaction (pH). The reaction is alkaline (Not alkali as in Chapter 8) at pH more than 7.0, neutral at 7.0 and acid below 7.0. *'Soils with pH values below 7 or where the concentration of H^+ ions exceeds that of OH^- ions are known as acid soils'*. On the basis of pH, acid soils are classified in the following five categories:

(1) Slight acidic (pH range 6.6 to 6.1)

(2) Medium acidic (pH 6.0 to 5.6)

(3) Strong acidic (pH 5.5 to 5.1)

(4) Very strong acidic (pH 5.0 to 4.6)

(5) Extremely strong acidic (pH 4.5 or lower)

Sometimes, soils are also categorized as ultra-acidic if the pH goes below 3.5. Soil pH in the range of 6.5-7.0 is not likely to impact the productivity much.

Table 9.1 State wise distribution of acid soils and prioritized area for reclamation (M ha)

State	Area affected (pH < 6.5) (M ha)	Prioritized area for reclamation (M ha)
Assam	4.66	1.5
Bihar	2.36	-
Chhattisgarh	10.84	-
Goa	0.30	-
Himachal Pradesh	1.78	0.1
Jammu & Kashmir	1.57	-
Jharkhand	6.77	0.4
Karnataka	3.31	-
Kerala	3.76	2.1
Madhya Pradesh	11.72	-
Maharashtra	0.54	0.3
Manipur	2.19	*
Meghalaya	2.24	*
Mizoram	2.05	*
Nagaland	1.65	*
Odisha	8.67	1.8
Sikkim	0.60	*
Tamil Nadu	4.85	-
Tripura	1.05	*
Uttaranchal	3.48	-
West Bengal	4.76	2.0
Total	**89.94**	**10.2 (all regions)**

Source: Panda *et al.* (2009), nfsm.gov.in/Presentations/Soil_Rice_22may2010.ppt (Opened on 07.04.2014, Ministry of Agriculture, 2011), - Not identified, * all north-eastern hill region put together 2.0 M ha

Formation of Acid Soils

Several factors contribute to the formation of acid soils. In the high rainfall regions, soluble basic salts such as those of Ca^{2+}, Mg^{2+}, Na^{+} and K^{+} from the surface soil are leached away and insoluble acidic residues composed chiefly of oxides and silicates of iron, silicon, aluminium accumulate in pretty high amounts. The nature of these compounds is acidic and their oxides and hydroxides react with water to release hydrogen ions in the soil solution causing colloidal exchange sites to be dominated by H^{+} and Al^{3+} ions.

$$CO_2 + H_2O \longleftrightarrow H_2CO_3 \quad (9.1)$$

$$H_2CO_3 \longrightarrow H^+ + HCO_3^- \tag{9.2}$$

The hydrogen ions replace the Ca^{2+} ions held by soil colloids. The released Ca^{2+} ions combine with HCO_3 to form calcium bicarbonate that gets leached resulting in acidification of the soils.

Besides, other causes that produce acidity in the soil are:

- Weathering and dissolution of parent materials especially when soils are formed on the acidic rocks.
- Humus and organic acids, alumina-silicate minerals, carbon dioxide produced within the soil and presence of hydrous oxides
- Vegetation cover with plants like conifers, the foliage leaves of which lack alkali elements and their mineralization process is very slow
- Application of acid forming fertilizers in the soil, such as ammonium sulphate, which undergoes biological transformation in the soil and forms acid forming nitrate ions. While the hydrogen ions replace Ca^{2+} and Mg^{2+} on the colloidal complex, the nitrate ions forms calcium and magnesium nitrate and are leached

$$NH_4^+ + 2O_2 \longleftrightarrow NO_3^- + H_2O + 2H^+ \tag{9.3}$$

- Sulphur also produces acid forming sulphate ions through oxidation. Divalent cations of soluble salts usually have a greater effect on lowering the soil pH than monovalent metal cations.
- Human interference in developing permanently submerged areas into cultivable land or for improving drainage in submerged or saline lands
- Continuous removal of lime and other base elements by crops and accumulation of acids contained in the manures
- Microbial action
- Acid rains, the problem is likely to increase with increasing environmental pollution.

In India, it is believed that acid soils are encountered in regions that receive more than 750 mm annual rainfall except in temperate regions where acid soils can develop even if the rainfall is scanty. The regions with annual rainfall of 1350 mm may have acid soils with pH value 5.0 or even less than that.

KINDS OF SOIL ACIDITY

Theoretically speaking, pH measurement is not a correct criterion to evaluate total soil acidity, since total soil acidity is the sum of the three kinds of acidity namely the active, the potential or exchangeable and the residual acidity. Whereas active acidity is the amount of hydrogen and aluminium ions in the soil solution or free H^+ ions, potential acidity represents the amount of hydrogen and aluminium ion adsorbed on cation exchange capacity (CEC). Aluminium and hydroxyl ions present in non–exchangeable form with organic matter and clay constitute the residual acidity. Most widely used method to measure active acidity is pH of the saturated paste or pre-decided diluted soil water mixture of 1:2, 1:1.5, or 1:2.5 or 1:1 ratio using a pH meter. Both active and residual acidities are limited, yet active acidity adversely impacts the plant roots and has adverse effects on the microbes around the rhizosphere. Moreover, active and potential acidities are in equilibrium, it becomes necessary to neutralize both kinds of acidities. As such, determination of exchangeable hydrogen in soils constitutes the basis of more accurate and quantitative measurement of soil acidity.

EXTENT AND DISTRIBUTION

In India, acid soils occur under varying topography, geology, climate and vegetation conditions such as:

- Greater and Higher Himalayas - mountain + valleys
- Middle and lower Himalayas and sub-Himalayas
- Peninsular region - Subtropical and tropical monsoon (Laterite, lateritic, red and yellow soils)
- Eastern plains - Assam valley, Eastern Bihar and Lower Gangetic plain of West Bengal.
- Coastal Plains - Kerala, West Coast, Deltaic area of Sunderbans.

The acid soils are mostly distributed in Assam, Manipur, Tripura, Meghalaya, Mizoram, Nagaland, Sikkim, Arunachal Pradesh, West Bengal, Jharkhand, Odisha, Madhya Pradesh, Himachal Pradesh, Jammu & Kashmir, Andhra Pradesh, Karnataka, Kerala, Maharashtra and Tamil Nadu (Table 9.1, Fig. 9.1). In northeastern region 95% soils are acidic, and nearly 65% soils suffer from strong acidity (pH <5.5). Considering the magnitude of the problem, soil acidity management in eastern and northeastern regions of India is important (Chaitanya *et al.*, 2017).

CAUSES OF LOW PRODUCTIVITY IN ACID SOILS

Low productivity of acid soils mainly result from low pH and is attributed to the following physico-chemical and biological changes:

- General toxic effects of low pH in the root zone, root membrane being damaged by H^+ ions at low pH
- Adverse effect on the permeability of the plasma membrane for cations
- Imbalance between uptake of basic and acid constituents through roots
- Enzymes being sensitive to pH changes, low pH adversely impacts the enzymatic processes
- Adverse effects on soil physical properties
- Reduced availability of P because of the increased P fixing capacity of soils at low pH

- Reduced availability of many micro-nutrients such as copper, boron, molybdenum and zinc etc.
- Phyto-toxicity of certain elements such as aluminium, manganese and iron as a result of their increased solubility
- Deficiency of bases such as calcium, magnesium and potassium. It also affects the plant nutrition as Ca and Mg are two important secondary nutrients
- Adverse effects on biological activities such as nitrogen fixation in leguminous plants.

All the above constraints mostly in combination with one another cause sub-optimal productivity of crops raised in acid soils. The loss in yield may range from about 10% to more than 50% (Table 9.2).

Table 9.2 Degree of acidity and its impact on crop productivity

pH	Loss in production (%)
>6.5	Nil
5.5-6.5	Up to 10
4.5-5.5	10-25
3.5-4.5	25-50
<3.5	>50

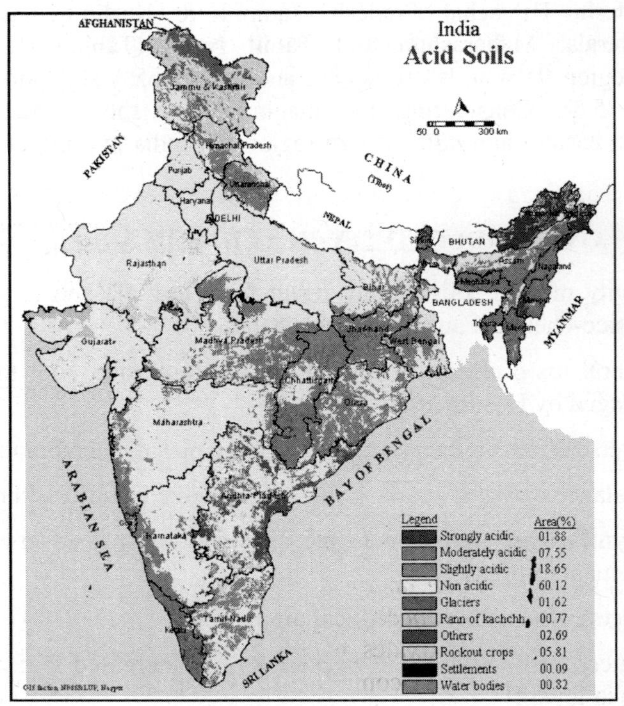

Fig. 9.1. Distribution of acid soils in India (Sharma and Sarkar, 2005)

MANAGEMENT OF ACID SOILS

Slightly acidic soils

These soils are relatively easy to manage for crop production. Simple agronomic/ cultural interventions such as application of organic matter, tank slit, conservation tillage, contour and strip cropping, intercropping of cereals with legumes or oilseeds and *in-situ* rainwater harvesting help to improve production in such upland soils. The phosphorus fixation in these soils can be circumvented or managed by the application of relatively large quantities of organic matter and/or mixture of rock phosphate and super phosphate.

Highly Acid Soils

The real challenges lie in management of acidic soil with pH < 5.5. The management requires raising the soil pH through application of suitable amendments, to reduce the toxicity of aluminium, manganese and iron, improve the nutrient supplying capacity of soil and to introduce improved agronomic and engineering practices. Cultivation of acid tolerant crops/varieties is also an option if liming is uneconomical or unavailable or to take advantages of their synergy with other options. The reclamation and management steps for these soils are listed in Table 9.3.

Table 9.3 Reclamation and management steps for reclamation of acid soils

Steps	Activities
Reclamation	
Land management	Farm layout, irrigation and drainage
Amendment (lime) application	Assessing lime requirement and its application including industrial wastes
Management	
Engineering and agronomic interventions	Drainage, water management, seed rate, application of organic matter, tank slit, conservation tillage, contour and strip cropping, intercropping of cereals with legumes or oil seeds and *in-situ* rainwater harvesting
Nutrient management	Rate of application, integrated nutrient management and synergy of amendments and organic inputs
Crop management	Cultivating acid tolerant crops, adopting a long-term cropping sequence, crop diversification
Varietal selection	Selection of acid tolerant crop varieties

LIMING

'Liming, the application of an alkaline compound of calcium (Ca) or magnesium (Mg) or both, is a widely accepted practice to reclaim/manage acid soils'. Carbonates, oxides and hydroxides of calcium and magnesium are referred as agricultural lime because of their common use in agriculture. A good liming material should have the following characteristics:

- It should have high neutralizing value

- It should be locally available
- It should be easy to ground
- It should be a low cost material

The common liming materials used for reclamation of acid soils are:

1. Calcic limestone ($CaCO_3$) (ground limestone)
2. Dolomite ($CaCO_3$. $MgCO_3$)
3. Quick lime (CaO) known as burnt limestone
4. Hydrated (slaked) lime [Ca $(OH)_2$]
5. Coral shell lime
6. Marl or chalk ($CaCO_3$).
7. Slags obtained as by-products from iron and steel plants, commonly used slags are blast furnace slags, basic slag and electric furnace slag.
8. Press mud obtained from sugar mills using carbonation process. Sulphitation press mud should not be used in these soils.
9. Miscellaneous sources of lime, such as, wood ash, ground oyster shells, by-product of lime resulting from paper mills, tanneries, water softening plants, and by product $CaCO_3$ form fertilizer factories using gypsum process (such as Sindri Fertilizer Factory, Bihar, India).

More than 90% of the lime used in agriculture for reclaiming acid soils is generally in the form of calcium carbonate, some in calcium and magnesium carbonates and much smaller quantity in the form of calcium oxide or calcium hydroxide. Stromatolytic limestone, a poor grade lime, which contains 28-32% CaO, 12% MgO and 0.5% P can also be used to lime the acid soils. The algal deposits in it contribute to its P content (Panda & Mishra, 1970).

It may be noted that all Ca containing salts may not be useful as amendments in acid soils. Although gypsum ($CaSO_4$) or calcium chloride ($CaCl_2$) can be applied to supply calcium ions to the soils but it is worth considering their effect on soil acidity. These amendments may produce H_2SO_4 and HCl in the treated soils.

Chemical Reaction of Liming Materials

The primary use of applying agricultural lime is to raise the pH of the acid soils and reduce the concentration of aluminium (Al) in soil solution. The following reactions occur with application of amendments.

Lime

When lime is applied, it reacts with the soil water as follows:

$$CaCO_3 + H_2O \text{ (in soil)} \longrightarrow Ca^{2+} + 2OH^- + CO_2 \text{ (gas)} \quad (9.4)$$

The OH^- ions produced, besides increasing the pH, react with soluble Al^{3+} to form insoluble $Al(OH)_3$ as follows:

$$Al^{3+} \text{ (soluble)} + 3OH^- \longrightarrow Al(OH)_3 \text{ (insoluble)} \quad (9.5)$$

OH^- ions react with H^+ ions to form water eliminating H^+ ions. The rate of this reaction depends on the amount of hydrogen ions (acidity) present in the soil.

The Ca^{2+} brought into the solution replaces H^+ and Al^{3+} on the soil colloidal complex. The adsorption of the calcium and magnesium ions reduces the percentage acid saturation of the colloidal complex and corresponding increase in the pH of the soil solution.

Dolomite

Similar reaction occur when dolomite ($Ca.Mg(CO_3)_2$) is applied to the soil. Here Mg entering the solution also takes part in replacing H^+ and Al^{3+} on the soil exchange complex.

Calcium oxide

Calcium oxide (quick lime) used for agricultural applications is rarely in pure form. It is found in various combinations depending on the source of the parent material. The minerals that may be present include magnesium (MgO), silica (SiO_2), iron (Fe_2O_3), and aluminium (AlO_3). In the presence of water (soil), calcium oxide in its pure form ionizes readily to form

$$CaO + H_2O \longrightarrow Ca^{2+} + OH^- \quad (9.6)$$

It forms a medium to strong base or alkali. This effectively neutralizes acid soil solutions and the calcium replaces aluminium on the cation exchange complex. The other reactions chain is similar to lime. It may be noted that calcitic or dolomitic limestone reacts slowly while burnt lime and hydrated lime react faster and bring about changes in soil pH within a few days.

Factors Affecting LR

The lime requirement principally depends on pH, amount and type of clay, organic matter, cation-exchange capacity and buffering capacity. Several workers have reported that greater the acidity, and buffering and cation-exchange capacity of the soil, greater is the lime requirement. It means that the soils with low clay content will need less lime. A fine textured acid soil requires much larger quantity of lime than a sandy soil or a loamy soil with the same pH value. Soils containing 2:1 clay lattice (montmorillonite) require more lime than soil containing 1:1 clay mineral lattice (kaolinite) even when organic matter and pH values are similar in both the soils. The acid soil containing higher amounts of organic matter requires more lime than the soils with low organic matter. Besides all these, lime requirement also depends upon the depth of the soil to be reclaimed, crops to be grown, fineness of liming material, efficiency of the liming material and economic returns likely to be achieved through the lime application.

Lime Requirement (LR)

As the gypsum requirement is calculated to lower the pH of alkali soil, the lime requirement is assessed to raise the soil pH of an acid soil from less than 6 to a value in the range of 6.0-7.0. For this purpose representative soil samples are drawn and analytical procedures are used to estimate LR. Procedures are mainly based on soil-lime incubation, soil based titration and buffer solutions (of specific known pH). Commonly used methods rely on buffer solution principle since lime requirement (LR) of soils not only take into consideration the current soil pH but also its buffering ability to resist change in pH (Table 9.4). When buffer solutions, a solution having pH ≥7.3, are added to the soil, the buffer pH is depressed in proportion to the original soil pH and its buffering capacity. The resultant pH is then used to calculate the lime requirement. A large drop in buffer pH indicates a low pH soil with a great potential acidity and a high lime requirement.

Exchangeable Al is also used as an index to assess the lime requirement especially in highly weathered soils with pH < 5.0, low CEC where exchangeable Al is equal to or exceeds the total exchangeable bases. Field trials reported by Sharma and Sarkar (2005) revealed that the yields can be optimized by applying the limestone 1 to 2.5 equivalents of exchangeable Al^{3+}. Potential acidity requires titration of the soil with a base. A number of methods are available but are not included in the chapter being beyond the scope of this book.

Table 9.4 Commonly used methods based on buffer solution principle to determine LR

Method	**Conditions under which used**	**Remarks**
Shoemaker *et al.* (1961) single buffer method	Soils with pH < 5.8, organic matter < 10%, and appreciable quantities of soluble Al^{3+}.	Overestimates the lime requirement for soils with low lime requirement
Mehlich (1976) buffer	Highly weathered soils	-
Modified Woodruffs buffer method (As reported by Brown and Cisco, 1984)	Successfully used to determine LR of acid soils in Odisha	Field trials reveal that one half of the lime requirement (LR) dose determined by Woodruffs buffer (pH 7.0) is good enough for maize and cotton grown in lateritic soils

Lime requirement by Shoemaker *et al.* (1961) method

In this method, the soil sample is equilibrated with a buffer solution of pH 7.5, whereby the reserve H^+ is brought into solution. It results in the depression of pH of the buffer solution. The pH of the soil-buffer solution is then used to determine lime required to raise the pH to a desired value. The reagent required is extractant buffer, which is prepared by dissolving 1.8 g paranitrophenol, 3 g potassium chromate, 2g calcium acetate, 53.1 g calcium chloride dehydrate ($CaCl_2$. $2H_2O$) and 2.5 mL triethanolamine in one liter of distilled water. The pH is adjusted to 7.5 with NaOH.

A test whether this procedure is applicable requires 10 g soil sample to which 25 mL of distilled water is added. It is shaken intermittently for half an hour and soil pH is recorded. This method is applicable only if pH < 5.8. If yes, proceed further as described in the following paragraph.

Weigh 5 g soil in a 50 mL beaker. Add to it 5 mL of distilled water and 10 mL buffer solution. Stir continuously for 10 minutes or intermittently for 20 minutes. Determine the soil pH. Lime requirement is determined on the basis of soil - buffer pH ready reckoner given in various publications or using Fig. 9.2 or derived equation given therein. Here the three curves are for three desired pH levels of 6.0, 6.4 and 6.8. Depending upon the desired pH, a curve or equation is selected to determine equivalents in the form of agricultural lime. The values calculated are divided by 2.43 to express the lime requirement in tons/acre.

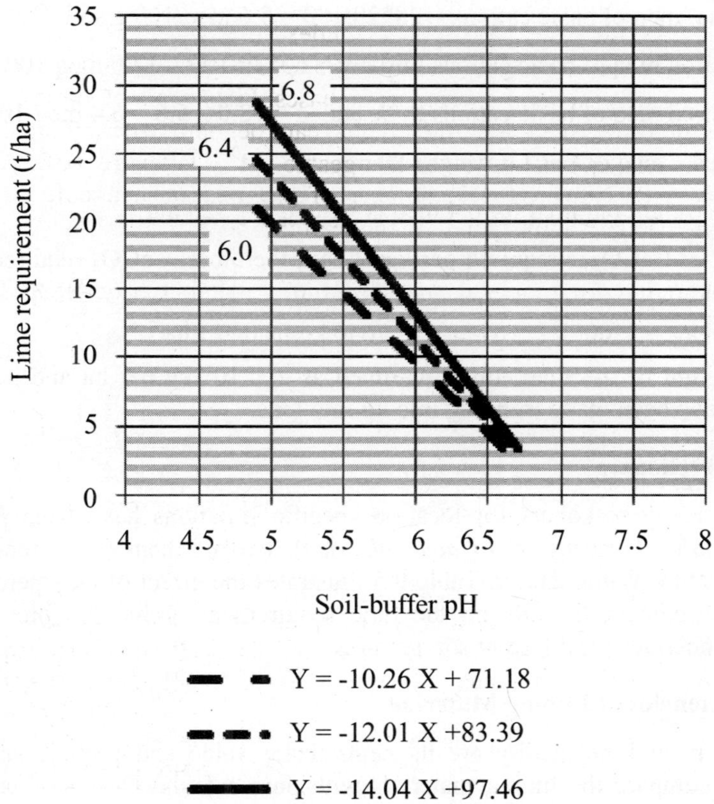

Fig. 9.2 Determination of lime requirement using Shoemaker *et al.* (1961) method

Base saturation to determine LR

Percent base saturation (BS) is defined as the 'percentage of the CEC occupied by the basic cations Ca^{2+}, Mg^{2+} and K^+ rather than the acid cations H^+ and Al^{3+}'. It is mathematically written as:

$$BS (\%) = [(Ca^{2+} + Mg^{2+} + K^+)/CEC] \times 100 \quad (9.7)$$

Depending on soil pH, the soil's base saturation is only a fraction of soils' CEC. Therefore, sandy soils low in CEC need to be limed more frequently but at lower rates of application than clayey soils. Higher lime rates are needed to reach an optimum pH on high CEC soils due to their greater abundance of acidic cations at a given pH. Theoretical lime requirement can be assessed if we know the current base saturation and desired base saturation and the CEC. The methodology is illustrated through Example 9.1.

Example 9.1

A soil having CEC of 20 cmol(p+)/kg and 28% base saturation is proposed to be limed for raising the BS to 60%. If the depth of the soil to be reclaimed is 15 cm, calculate the theoretical lime requirement with $CaCO_3$.

Initial meq of basic cations = 28*20/100 = 5.6 meq/100 g soil

Desired meq of basic cations initially = 60*20/100 = 12.0 meq/100 g soil

Desired meq of basic cations to be added = 12.0 – 5.6 = 6.4 meq/100 g soil

Since 1 meq of $CaCO_3$ weighs 50 mg

The weight required for 6.4 meq is = 50*6.4 = 320 mg

3200 mg of $CaCO_3$ is required per kg of soil, therefore, $CaCO_3$ required for 15 cm soil layer is

$CaCO_3$ in kg = $3200*2.2*10^6*10^{-6}$ = 7040 kg/ha or 7 t/ha

Herein the weight of 0-15 cm soil is assumed as $2.2*10^6$ kg per ha and necessary conversions have been made from mg into kg into tons.

Use of ready reckoners

A number of ready reckoners for location specific situations have been prepared (*https://www.nabard.org/english/ld_acidsoils.aspx*), two of them being reported in Tables 9.5 and 9.6. While data in Table 9.5 illustrates the effect of clay percent and organic matter percent in soils on the lime requirement, Table 9.6 presents the generalized guidelines for different soil textures.

Relative Efficiencies of Liming Materials

The key factors in lime quality are its neutralizing value and particle size. It is customary to compare the liming materials with pure calcium carbonate taking its neutralizing value as 100%. Calcium carbonate equivalent (CCE) of an agricultural liming material is expressed as:

$$\text{CCE of a liming material} = \frac{\text{Molecular weight of calcium carbonate}}{\text{Molecular weight of liming material}} \times 100 \quad (9.8)$$

It is also referred as neutralizing value (NV) or acid neutralizing capacity. CCE values of different liming materials vary from a low value of about 30% in wood ash to a high value of 179% for burnt lime as compared to 100% of the pure calcite (Table 9.7). Besides, one needs to consider the purity of the amendment to arrive at the actual amount. For example; when the purity of lime or other amendments is less than 100%, actual lime requirement is determined by dividing the theoretical requirement by the purity of lime/amendment.

Table 9.5 Ready-reckoner to assess lime requirement to achieve a pH of 5.5 (t/ha)

Soil pH	Organic matter, 2%				Organic matter, 4%				Organic matter, 6%				Organic matter, 8%			
	Clay (%)				Clay (%)				Clay (%)				Clay (%)			
	10	15	20	25	10	15	20	25	10	15	20	25	10	15	20	25
4.5	3.56	3.83	4.11	4.39	3.72	4.00	4.28	4.56	3.89	4.17	4.44	4.72	4.05	4.33	4.61	4.89
4.6	3.15	3.42	3.70	3.98	3.32	3.59	3.87	4.14	3.49	3.77	4.04	4.32	3.66	3.93	4.21	4.49
4.7	2.74	3.01	3.29	3.60	2.91	3.18	3.34	3.62	3.08	3.36	3.63	3.91	3.24	3.52	3.80	4.08
4.8	2.25	2.53	2.81	3.09	2.42	2.70	2.98	3.26	2.59	2.87	3.14	3.42	2.76	3.03	3.31	3.59
4.9	1.93	2.22	2.49	2.75	2.10	2.38	2.66	2.93	2.27	2.54	2.82	3.10	2.43	2.71	3.00	3.27
5.0	1.52	1.80	2.08	2.35	1.69	1.97	2.24	2.52	1.86	2.13	2.41	2.69	2.02	2.30	2.58	2.86
5.1	1.11	1.40	1.68	1.94	1.29	1.56	1.84	2.12	1.46	1.73	2.01	3.29	1.62	1.90	2.18	2.46
5.2	0.71	1.00	1.27	1.56	0.90	1.16	1.43	1.71	0.06	1.33	1.61	1.90	1.22	1.50	1.78	2.06
5.3	0.31	0.59	0.86	1.13	0.48	0.76	1.03	1.31	0.64	0.92	1.20	1.48	0.81	1.09	1.37	1.64
5.4		0.18	0.44	0.72	0.07	0.34	0.62	0.90	0.23	0.51	0.82	1.10	0.40	0.68	0.96	1.23

Table 9.6 Generalized guidelines of lime requirement in different soil groups

Soil texture	Targeted soil pH	
	From 4.5 to 5.5	From 5.5 to 6.5
Sand and loamy sand	0.6	0.9
Sandy loam	1.1	1.5
Loam	1.7	2.2
Silt loam	2.6	3.2
Clay loam	3.4	4.3

Table 9.7 Calcium carbonate equivalent (CCE) of liming materials

Material %	CCE*
Pure calcite	100
Calcitic lime	75-100
Dolomitic lime	75-109
Hydrated lime	120-136
Burned lime	179
Pel-lime (finely ground ag-lime)	90-95
Fly ash**	43-44
Wood ash	30-40

** CCE values for fly ash may vary due to variations caused by source of coal, collection procedures and other factors.

Neutralizing effect of an amendment is always greater with small particles because of increased total surface area exposed to soil acidity. As such, lime requirement also depends upon an index of the fineness of lime. Although fine lime less than 60 mesh is considered to be 100% effective, BIS standard stipulates a mesh size of 80. To determine the fineness index, two sieves of 8 and 80 mesh are used to separate a sample of amendment into three particle sizes: Fine - less than 80 mesh, Medium- less than 8 mesh but greater than 80 mesh and Coarse- greater than 8 mesh. The percentages of these three components are multiplied by factors of 1.0, 0.4, and 0.1 respectively and added together to give the fineness factor. For example, if a liming agent has a particle size distribution of A% fine, B% medium, and C% coarse, the particle fineness of the material is calculated as:

$$\text{Fineness} = (A \times 1) + (B \times 0.4) + (C \times 0.1) \tag{9.9}$$

Effective calcium carbonate equivalent (ECCE) is the measure of the effectiveness of liming materials and is calculated as the product of the CCE and the fineness value divided by 100. For example, if the CCE value is 80 and the fineness value is 75%, then:

$$\text{ECCE} = (80 \times 75)/100 = 60\% \tag{9.10}$$

From practical point of view, liming material, 90% of which passes through a 10 mesh sieve and 50% through a 60 mesh sieve have given the best results. Sarkar (2005) further suggested that at least 20% of the liming material should pass a 100 mesh sieve (Sarkar, 2005).

Method and Time of Lime Application

During application of liming agents, following points are taken into consideration:

(i) The liming agents are used in powdered state as discussed in the section on relative efficiencies of liming materials.

(ii) Liming materials should be in direct contact with clay/organic exchange complex so that H^+ ions of exchange complex may be easily displaced by Ca^{++} ions.

(iii) Liming agents are applied to soils at least one month before sowing the crops or are applied and thoroughly mixed in soils after harvesting the crops.

For arable crops, lime should be uniformly broadcasted and mixed thoroughly into the soil (0-15 cm) using a disc plough. If mixing depth is more than 15 cm, proportionately more lime is required to reach the targeted soil pH (Somani *et al.*, 1996). Although, rates of lime application can be reduced proportionately with decreasing depths, but in that event frequency of liming needs to be increased. For horticultural crops, the depth can be increased depending upon the rooting behaviour. To avoid drifting of the fine powder, the moisture content of the amendment at the time of application may be in the range of 7 to 10%.

For crops sown in line, furrow or bed application is preferred. It will not only reduce the amount of lime to be applied but will also minimize the drifting of the dust (Fig. 9.3). Yields of crops were higher when 0.25-0.50 t/ha limestone was applied in furrows every year as compared to broadcasting at higher rates.

Fig. 9.3 Application of lime in acid soils: furrow application (left) and bed application (right)

Frequency of application

Lime should be applied more frequently in split applications in regions having high rainfall and soils of coarse textured nature. For example, lime applied once at optimum dose was effective only for two years in regions having rainfall of about 3,000 mm per annum. It may last for 3 to 4 years in areas having rainfall around 2,000 mm per annum. Sen (2003) and Patiram, (2007) observed that limestone application every two years is able to sustain productivity even if the rate of lime application is half of the recommended dose. Based on such experiences, a technology package to ameliorate 25 M ha of critically degraded acid soils has been developed. According to the package, band placement/incorporation of lime is proposed @ 1/10 of lime requirement along with recommended level of fertilizers every year. The grounded material is broadcast /applied in furrows at the time of planting of an upland crop preceding rice. It may also be applied to rice crop during puddling (0.2-0.4 t/ha), if the soil is highly acidic (soil pH < 5.5). Results have proved that it is a more economical, practicable and effective way of reclaiming and managing acid soils.

Impact of Liming on Crop Yield

Liming has a positive profound effect on crops yield because application of lime reverses most of the soil processes that adversely impact the yield in acid soils. For example, liming agents reduce soil acidity and stabilize pH of the soils as per the following reaction:

$$\text{Acid-clay} + Ca(OH)_2 \rightarrow \text{Ca-clay} + H_2O \quad (9.11)$$

It also eliminates active acidity and minimizes residual acidity. It ameliorates toxicity of Al, Fe and Mn. Lime reduces the solubility of iron and aluminium such that less phosphorus is held in insoluble and unavailable forms, making it easily available to the crops. Besides, it increases plants uptake of Ca and Mg (depending upon the type of amendment used), enhances the availability of nitrogen and other nutrients, stimulates microbial activity and encourages N_2 fixation. It also improves the physical conditions of the soil by causing aggregation of soil particles, decreasing its bulk density, and increasing its infiltration rate.

Indian Council of Agricultural Research (ICAR) formulated a Network Project on Acid Soils for optimizing liming dose for different crops in acid soil areas throughout the country. Sharma and Sarkar (2005) reported that application of lime @ 0.2-0.4 t/ha increased the yield of rape seed mustard, wheat, green gram, maize, pigeon pea, field pea, black gram, groundnut by 14 to 52% over farmer's practice. The recommended application of fertilizers (100% NPK) also increased yield by 15 to 99% although application of lime with 100% NPK resulted in 49-189% higher yield over farmer's practice. The yields recorded with 50% NPK + lime was equal or slightly higher to the yield with 100% NPK (Sharma and Sarkar, 2005). Application of lime through paper mill sludge @ 0.2 LR increased the yield of groundnut by 17-36%, green gram 5-21%, cabbage 15-16% and cauliflower 22% over farmer's practice (Jena, 2008). Beg (2016) observed that crops yield in acid soils of Chhattisgarh were 16 to 42% higher when management practices were adopted than without them (Table 9.8). Kumar *et al.* (2014) reported that furrow application of lime (control, 0.2, 0.4 and 0.6 t/ha) in rice-bean (*Vigna umbellata*) cropping sequence increased the growth and yield resulting in benefit : cost (B:C) ratio of 2.29. The performance of maize crop with and without liming shown in Fig. 9.4 shows a perceptible improvement following lime application. In another experiment, liming of soil and their combined application with FYM resulted in increasing root length, cob length, diameter and seed weight per cob respectively (Behera and Pattanayak, 2016). Lime amelioration of an acid soil resulted in 32% higher yield compared to the yield of 2.2 t/ha due to inorganic nutrition only. Combining organic amelioration (FYM) with inorganic amelioration and inorganic nutrition yielded 26% higher grain yield in maize crop.

INDUSTRIAL WASTES AS AMENDMENTS

Steel Slag

Industrial wastes like basic slag and blast furnace slag are the by-products of steel industries. Steel slag is alkaline in reaction, its pH being in the range of 8-10. The pH can have values of 12 or higher, if the free CaO content is high. Steel slags contain Ca and Mg in the form of silicates, ferrites, aluminates, oxides and some free CaO and MgO (Table 9.9). Free Ca in slag reacts rapidly with water to form $Ca(OH)_2$, which in turn reacts rapidly with soil acidity. The less soluble silicate compounds react more slowly and provide long-term buffering of soil pH. Considering the efficiency of limestone as 100%, efficiencies of basic slag is assessed at 90-110%. Field experiments in Jharkhand, where soil pH varied from 4.8-5.5 and basic slag was applied @ 0.4 t/ha, revealed that percent yield increase over farmer's practice was from 11.2 to 23.8 in wheat, from 8.15 to 42.5 in gram, from 5.9 to 8.1 in mustard and 6.42 to 8.56 in black mustard (*rai*). The main disadvantage of basic slag is its hardness that poses hurdles in grounding to optimum particle size of 60 mesh.

Table 9.8 Performance of crops with and without acidity management in acid soils of Chhattisgarh

Crops	Without management	With management*	Increase in yield (%)
Rice	1.85	2.33	26
Wheat	1.19	1.51	27
Maize	2.08	2.48	19
Chick pea	0.52	0.74	42
Pigeon pea	0.39	0.53	36
Soybean	0.67	0.88	16
Tomato	12.01	14.55	21

*With management means application of lime with either fertilizer or manure or both

Fig. 9.4 Effect of liming on maize crop, without lime (left) and with lime (right)
Source: http://www.kiran.nic.in/acid_soil.html.

Linz-Donawitz (LD) slag, containing 29% Ca, 21% Fe and 5% Mg, is also a by-product of the iron and steel industry. LD slag from SAIL, Rourkela had pH 10.49, Ca content 16.82%, Mg content 2.13% and neutralizing value 58.8%. The low P content of the LD slag discouraged its use in agriculture as a phosphate fertilizer, but the total high Ca and Mg contents makes the LD slag a potential liming agent.

Table 9.9 Chemical characteristics of steel slag

Constituent	Amount (% by weight)
CaO	24-50
SiO_2	10-19
Al_2O_3	1-3
MgO	2-10
Fe (FeO or Fe_2O_3)	10-40
MnO	5-8
TiO_2	0.5
P_2O_5	1-7
Free CaO	2.1
Metallic Fe	0.5-1.0

Paper Mill Sludge (PMS)

Paper mill sludge, as the name suggests, is a byproduct of the paper industry. The chemical and physical properties of the paper mill sludge vary according to nature of manufacturing processes. The analysis of Emami paper mill, Balasore revealed that the paper sludge had pH (1:2.5) of 8.23; EC 0.20 dS/m, average Ca and Mg 7.2 meq/100g, CaO equivalent 30.0%, water holding capacity 119.0%, organic carbon (OC) 25.0%, nitrogen (Total) 0.87%, phosphorous 0.20%, potassium (Total) 0.14% and sodium (water extract) 0.10%. PMS has 65-79% neutralizing value. Field studies revealed that higher net returns are obtained from rain fed acidic upland when PMS @ 50% LR was applied along with 100% NPK.

Press Mud

Press mud obtained from carbonation plants of sugar mills contains calcium and can be used to manage acidity in the soils. The pH of the press mud varied in the range of 6.5 to 6.92 and OC 16-18%. The neutralizing value of 22-24% indicates that it cannot be compared with $CaCO_3$ or PMS as a liming material. However, it can be used as an acid soil amendment because of high organic carbon content that improves soil physical properties. It may be useful to apply this product in conjunction with other amendments. For example; the pod yield in control was 1.76 t/ha, increased by 52-95% in lime treatments, but highest yield of 3.64 t/ha was obtained when PMS and PM were applied in ratio of 75:25 (Jena, 2013).

Fly Ash

Fly ash, a byproduct of the thermal power plants, is a low- density amorphous ferro-alumino-silicate. It helps to improve soil pH and availability of nutrients. Studies in USA have shown that use of fly ash as soil ameliorant in place of lime helps to reduce CO_2 emissions, thus contributing to minimize global warming.

Adverse Impact of Excess Lime

Each crop has an optimum soil pH range in which nutrient availability is maximized. Over liming harms the crops especially in sandy soils low in humus because of the following reasons:

1. Imbalance of nutrients in the soil system especially altering the ratio of Ca and Mg
2. Boron deficiency
3. Iron, manganese and zinc deficiency
4. Reduced availability of phosphorus to a critically low level
5. Reduced potassium uptake
6. The increased incidences of disease infestation.

Such injurious effects may be reduced by application of large amounts of compost manure, green manure crops, phosphorus fertilizers, boron or a mixture of minor elements.

ENGINEERING AND AGRONOMIC MEASURES

Humid regions, having the most acid soils, receive annual rainfall in excess of 1000 mm, much of which occurs as heavy rains in short spells. Under monsoon climatic conditions, the intensity of rainfall is usually much higher than the infiltration capacity of soils, resulting in huge runoff. Even the paddy fields may be unable to store the water as excessive storage in paddy fields may also result in production losses. The excess water therefore, must be disposed of with minimal erosion. Proper soil and water management helps to check leaching of bases and enhances decomposition of organic matter. Moreover, even in humid regions there are large breaks between two rain events or there may be early withdrawal of monsoon rains, a most appropriate option is to store the drainage water and reuse it to meet crop water requirement. On-farm ponds and raised and sunken bed systems have been found to work well as drainage measures. If subsurface drainage needs to be provided, control structures in tertiary and secondary drains may be provided to maintain the water table depth at or above 90 cm to prevent rapid oxidation, minimize soil acidity and land subsidence. Ridge and furrow tillage, conservation tillage, contour and strip cropping, bench terracing, raising of field bunds for *in-situ* rain water harvesting, runoff water harvesting and recycling, compaction of surface layers with heavy rollers/heavy tractors also help to manage acid soils (Kar *et al.*, 2012).

Iron toxicity is a problem in some parts of Assam, Meghalaya and Odisha in low land rice. Improving drainage facility by excavating deep drains around fields, checking lateral seepage of water carrying Fe by constructing check embankments across the slope with provision of diversion weirs and liming have been successfully used to check iron toxicity. Application of 60 kg K_2O per ha has given encouraging results.

Improved agronomic practices such as increasing the seed rate, mulching the seed lines with straw mulch to avoid crust formation, addition of tank silt and clay to fields to improve water retention and incorporation of legumes in the cropping pattern have also been suggested to get higher yields from acid soil (Kar *et al.*, 2012).

Nutrient Management

All the major plant nutrients (nitrogen, phosphorus, potassium, sulphur, calcium, manganese and also the trace element molybdenum) in acid soils may be unavailable or only available in insufficient quantities. Plants express deficiency symptoms despite the fact that adequate fertilizers are added to the soils. The relationship between soil pH and availability of plant-nutrient, decreasing with decreasing pH, is well known. There is a marked stimulating effect of liming on production of nitrate nitrogen in soil. Besides, available calcium and phosphate increase with the added amount of calcium carbonate. Available nitrogen and potassium also increase but up to the level of 2,242 kg calcium carbonate per ha. The content of extractable iron, soluble plus exchangeable manganese and available copper decreased with an increase in the added amount of calcium carbonate. Conjunctive use of lime and fertilizers pays dividends as the synergy of the two improves yields of crops by 49 to 189% over the farmer's practice. A few nutrient management guidelines are as follows.

Nitrogen and potassium

Nitrogen (N) should be applied at higher rates than recommended to compensate leaching losses. Split application of N is highly beneficial. Introduction of acid tolerant N-fixing legumes in the cropping pattern helps to minimize the need for nitrogen fertilizers as well as to improve the efficiency of fertilizer nitrogen. Split application of potassium fertilizers also helps in higher recoveries of K.

Phosphorus

Phosphate fertilization is essential to sustain crop production in acid soils because of phosphorus fixing ability of these soils. Application of rock phosphate is recommended in such situations. A low pH (below 5.0) facilitates release of phosphate from rock phosphate. The effectiveness decreases with increasing pH above 5.0. Besides, it also contains 30 to 50% CaO and therefore, provides calcium to the soil as well as helps in raising the soil pH. Rock-phosphate should be applied well in advance to allow its dissolution. Avoid application of water soluble fertilizers like TSP, SSP and DAP.

Integrated nutrient management

Integrated nutrient management calls for application of plant nutrients in organic and inorganic forms. Organic forms include FYM, green manure and plant residue. Regular application of well decomposed organic matter in acid soils is essential to prevent sudden fluctuations in soil pH as it improves the buffer capacity of soils. Moreover, it increases the availability of P and reduces the toxicity of Fe and Al in acid soils. Organic inputs improve aggregate stability and physical properties of the soils. Application of lime along with organic manures is more efficient having synergic relation (See Table 9.8). Fertilizers respond well when farmyard manure (FYM) is applied to each crop in Al-toxic acid soils. Judicious use of lime and organic manures or residue management can help to overcome micro-nutrient deficiencies.

CROP MANAGEMENT: ACID TOLERANT CROPS

Crops differ in their tolerance to low pH having large inter- and intra-genic differences. As such, different plant species (even varieties within a species) grow best at different pH ranges. Rice is able to endure a fairly acidic reaction (pH = 5.0). Flooding rice fields helps to create favourable conditions (increase in pH and availability of P, Si and K) for growth of rice. Minor millet (*Panicum miliare*) and finger millet (*Eleusine coracana*) are also quite tolerant to acidity. On the other hand, Bengal gram, lentil, groundnut, sorghum, maize and field peas are medium tolerant. Pigeon pea, soybean and cotton are sensitive. Optimum pH values for various crops are given in Fig. 9.5. The sensitivity of crops increases as one move towards the right. Some crops are able to grow in a wide range of pH whereas others have a narrow range. For example, rice and tea can be grown in a wide range of 4.0 to 6.0 pH.

Thus, selection of crops in favour of acid-tolerant crops is the best management option to get sustainable yields. Based on the response of crops due to liming, the crops have been grouped into high response group (pigeon pea, soybean, cotton), medium response group (gram, lentil, groundnut, maize, sorghum, wheat, pea) and low response group (barley, minor millet and paddy). The best responses are received in legume crops, cotton, maize, sorghum, wheat and linseed etc. Therefore, in limed lands, one should adopt a long-term cropping pattern depending upon the frequency of lime application. In the initial seasons, high response crops should be cultivated. As the time elapses, these crops are replaced with medium response crops. Towards the later years when the effect of liming has considerably reduced, one can opt for low responsive crops.

Commercial cultivation of horticultural, plantation, vegetables and cash crops is also practiced in the Western Ghats and in the northeast. The relative tolerance of horticultural and vegetable crops grown in the Northeast region of India along with their acidity tolerance is given in Table 9.10 and Table 9.11. Optimum soil pH for some spice crops range from 4.5 to 7.5 being 6.0-6.8 for black pepper, 5.5 to 6.5 for ginger, 6.0 to 7.5 for turmeric, 5.7 to 6.0 for cardamom and 4.5 to 6.0 for tree spices (Hamza *et al.*, 2013).

Table 9.10. Relative tolerance of fruit crops to soil acidity

Slightly tolerant to acid soil (pH 6.8-6)	**Moderately tolerant to acid soil (pH 6.8-5.5)**	**Highly tolerant to acid soil (pH 6.8-5.0)**
Mango	Pineapple	Strawberry
Citrus	Orange	Gooseberry
Banana	Litchi	Bel
Papaya	Passion fruit	Elephant apple
Guava	Jackfruit	Plum
Cashew, Carambola, Apple, Peach, Kiwi fruit		

Source: http://www.kiran.nic.in/acid_soil.html

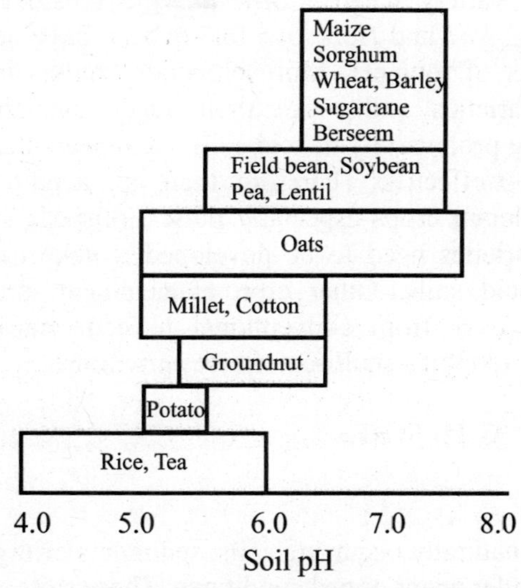

Fig. 9.5 Crop tolerance to soil pH

Table 9.11 Relative tolerance of vegetable crops to soil acidity

Slightly tolerant to acid soil (pH 6.8-6)	Moderately tolerant to acid soil (pH 6.8-5.5)	Highly tolerant to acid soil (pH 6.8-5.0)
Beet	French bean	Potato
Broccoli	Lima bean	Sweet potato
Cabbage	Carrot	Chow-chow
Cauliflower	Cucumber	Yam
Celery	Brinjal	*Dioscoria* spp
Spinach	Knolkhol	
Chinese cabbage	Pea	
Leek	Chilli	
	Pumpkin, Radish, Summer squash, Winter squash, Tomato, Colocasia	

Source: http://www.kiran.nic.in/acid_soil.html

Crops varieties/cultivars/genotypes of various crops also differ in their tolerance to soil acidity (Sarkar, 2005; Kar *et al.*, 2012). Although some work on the relative tolerance of varieties has been conducted, yet the results are not spectacular because there are a number of factors that affect the performance of crops in acid soils. For example, within acid soils, crops may suffer because of general impact of low pH, Al toxicity, iron toxicity and low availability of phosphorus etc. Amongst the three rice varieties namely *Ranjit*, *Siyal Sali* and *Mahsuri*, *Mahsuri* was observed to be tolerant to iron toxicity while the other two were categorized as sensitive for northeastern India. Amongst the 24 varieties tested for Al toxicity, *Disang* variety was found to be comparatively tolerant variety whereas *Joymati*, the sensitive one. In Himachal Pradesh varieties *Bragg*, *Ph1* and *Harasoya* for soybean have been found promising against general impacts of low pH. More elaborate studies in various states are needed to identify varieties and make their seeds available to the farming communities to enhance productivity in acid soils. It appears that this option is most easy to adopt and is cost effective. Therefore, there is a need to broaden researches on breeding of acid tolerant crops especially pulses, oilseeds and cereals. Besides, multi-stress tolerant varieties need to be developed to take care of various stress factors operative in acid soils. Other crop management interventions are crop diversification, switch over from conventional to economically important non-conventional crops and adopting multi-enterprise agriculture.

SPECIAL CASES OF ACID SOILS

Acid Sulphate Soils

Acid sulphate soils are naturally occurring soils, sediments or organic substrates (e.g. peat) that are formed under waterlogged conditions. These soils contain iron sulphide minerals (predominantly as the mineral pyrite) or their oxidation products. In an undisturbed state below the water table, acid sulphate soils are benign. However, if the soils are drained, excavated or exposed to air by lowering the water table, the sulphides react with oxygen to form sulphuric acid, the soil pH going below 4. Release of the sulphuric acid from the soil can in turn release iron, aluminium, and other heavy metals (particularly arsenic) within the soil. Once mobilized in this way, the acid and metals can create a variety of adverse impacts: killing vegetation, seeping into and acidifying groundwater and water bodies, killing fish and other aquatic organisms and degrading concrete and steel structures. In India, acid sulphate soils are mostly found in Kerala, Odisha, Andhra Pradesh, Tamil Nadu, and West Bengal. The most severe problems are encountered in the low-lying (below sea level) *bunded* rice fields in the Kuttanand region of Kerala, where pH is as low as 3.2 (Mathew *et al.,* 2003).

Acid sulphate soils are categorized as potential acid sulphate soils if sulphidic minerals contained in these soils have not yet been oxidized. Potential acid sulphate soils are poorly drained soils with high pyrite content. The pH of these soils is neutral or slightly acid in the field. When these soils are drained or altered enough for cultivation, sulphides begin to oxidize and the soil becomes strongly acidic. Unfavorable properties of potential acid sulphate soils are:

- Most potential acid sulphate soils are situated in coastal areas and therefore, water logging and soil salinity are perpetual problems because of tidal influences
- Strong acidification upon drainage
- Low accessibility and trafficability because soils have not ripened and consist of soft mud
- High permeability
- Engineering problems arise when dikes are constructed on soft mud or acidity from oxidizing dike material may attack steel and concrete structures.

The formation of actual acid sulphate soils consists of two main processes: formation and oxidation of pyrite. The combination of sulphate from sea water, organic matter from plants and the presence of Fe^{2+} together with anaerobic conditions caused by exclusion of atmospheric oxygen by the excess water result in the formation and accumulation of pyrite in tropical coastal wetlands. Although pyrite is formed in a number of steps, yet the overall reaction equation is:

$$Fe_2O_3 + 4\ SO_4^{2-} + 8\ CH_2O + 1/2\ O_2 = 2\ FeS_2 + 8\ HCO_3^- + 4\ H_2O \qquad (9.12)$$

Usually, these materials remain in reduced conditions because of submergence. Under humid or moist aerobic conditions, sedimentary sulphide minerals can oxidize chemically, but it is a slow process. Various micro-organisms are adapted to oxidize sulphides either directly through sulphur transformations or by facilitating (catalyzing) such rate-limiting reactions. The following reaction equations describe pyrite oxidation and formation of sulphuric acid that causes soil pH to reduce below 4.0.

$$4\ FeS_2 + 15\ O_2 + 2H_2O = 2\ Fe_2(SO_4)_3 + 2\ H_2SO_4 \qquad (9.13)$$

The formation of these soils in some cases can also be traced to tectonic uplift or oceanic regression. The strong acidity in acid sulphate soils makes them generally unproductive. Their low productivity is due to one or more of the unfavorable factors such as low soil-pH, aluminium toxicity, soil salinity, phosphorus deficiency, ferrous iron (Fe^{2+}) toxicity, acidification of surface water, N-deficiency, engineering problems and H_2S toxicity.

Managements of Acid Sulphate Soils

A distinction must be made between potential acid sulphate soils, which are not yet oxidized but contain pyrite in the soil material and actual acid sulphate soils, which are oxidized and acidified.

There are two strategies almost opposite to each other to reclaim and use potential acid sulphate soils:

The first strategy comprises completely draining the land to facilitate oxidation. It is followed by flushing the acidity out of the soil. For flushing, even saline or brackish water can be used. Although the problem is solved once for all but the process is expensive and has environmental consequences resulting from the disposal of acid drain water.

In the second strategy attempt is made to live with the problem. In this case pyrite oxidation is reduced by maintaining a shallow groundwater table. The difficulties in doing so result from the high water requirement and re-acidification of the soil.

For actual acid sulphate soils, the easiest and the most effective way to manage is to leave them as such without disturbing. Such a strategy includes the decision to drain a potential acid sulphate soil without exposing the iron sulphide layer beneath. For this purpose one must know the distribution and depth of sulphide layer from the soil surface, the acidity stored in the sulphide layer and drainage system of the area. A unique case is the low-lying paddy lands in acid-sulphate soils of coastal Kerala where the land elevation is 1-2 m below the mean sea level. The protective dikes built around the cultivated fields frequently breach, causing inflow of huge quantities of outside water to the paddy fields that needs to be drained. Also, lands are purposely kept under water fallow during the non-cultivated season to reduce acidity hazard. This water is drained before the start of cultivation. Wide, shallow drains are constructed to quickly drain the surface water from the low-lying lands. Deep, narrow drains are likely to expose the iron sulphide layer resulting in leaking of sulphuric acid into waterways. Flap gates are used for water control in areas with shallow inundation and at least some tidal influence in creeks or canals. In the rainy season, water is discharged at low tide, or irrigation water is applied if needed. Practically speaking, this solution almost limits the use of the area to cultivate rice. To improve productivity of lowland rice additional amelioration measures include: leaching and shallow drainage, submergence, liming, addition of manganese dioxide, nitrogen, phosphorus and potassium application including use of rock phosphate and cultivation of resistant varieties. Application of MnO_2 along with lime reduces the sulphide formation that also prevents the H_2SO_4 formation. Many times the application of these practices may not be economically and financially attractive. The best course therefore, is to neutralize acid sulphate soils by the alkalinity of seawater, an ultimate natural process of management. Another attractive proposition is selection of tolerant varieties, a simple and economical solution for the small and marginal farmers. Tolerant rice varieties can give grain yields of nearly 3 t/ha when iron toxicity is severe and over 6 t/ha when it is mild (Attanandana and Vacharotayan, 1986). Ponds for fish and shrimp farming can also be constructed in the acid sulphate soils but require careful management.

Acid Sodic Soils

Acid sodic soils are soils having pH less than 7 but exchangeable sodium percentage greater than 15. These soils are degraded soils, an advanced stage of soil development resulting from the washing out of salts. These have been reported in deltaic regions of West Bengal. No special management practices are required as lime can be used as an amendment in these soils. It helps to reduce the ESP. Other practices for reclaiming acid soils should be adopted.

QUESTIONS (THEORY)

1. What is an acid soil? List various factors/processes that adversely affect the productivity in acid soils.
2. Briefly describe the various factors that are responsible for the formation of acid soils.
3. What do you understand by liming? Name the various liming materials explaining the chemical reaction of at least one amendment.
4. Explain the concept of relative efficiency of soil amendments for acid soils.
5. Name various buffer methods to determine lime requirement. Explain the use of one of them for calculating LR.
6. A soil has cation exchange capacity of 18 meq/100 g and base saturation of 25%. Calculate the lime requirement per ha for the top 15 cm layer if the base saturation is to be raised to 55%. Assume appropriate values, if needed.
7. Write short notes on (i) Crop tolerance to soil acidity, (ii) Acid sulphate soils and (iii) Industrial wastes as amendments to reclaim acid soils

QUESTIONS (OBJECTIVE)

1. Fill in the blanks

1. The pH of acid soils is _______ than 7.0.
2. Soil acidity is a major problem of _______ climatic regions
3. ________ acidity of the soils is due to hydrogen and aluminium ions in the soil solution.
4. ________ acidity of the soils is due to hydrogen and aluminium ions adsorbed on the soil exchange complex.
5. The common liming materials used for reclamation of acid soils is ________.

6. Chemical formula of quick lime is ________.
7. Neutralizing effect of an amendment ________ with increasing fineness of the amendment in practical range.
8. Rice is ______ tolerant to acid pH than maize.
9. Acid sulphate soils mainly occur in the state of ________ (India).
10 Acid sodic soils have pH less than 7.0 and ESP ________ than 15.

2. Tick mark correct or most appropriate answer (A, B, C or D) from the following

1. Soils with a pH less than 7.0 are categorized as
   (A) Acid (B) Saline
   (C) Sodic (D) None of these
2. The soil acidity is of the forms
   (A) Active (B) Exchangeable
   (C) Residual (D) All of these
3. Which is not the cause of low productivity in acid soils?
   (A) Toxic effects of aluminium
   (B) High fixation capacity for P resulting in imbalance in plant nutrition
   (C) High pH
   (D) None of these
4. Which acidity is due to Al and H ions on the soil exchange complex?
   (A) Active (B) Potential
   (C) Residual (D) All of these
5. Amount of liming material required to reclaim acid soils depends upon
   (A) Clay content (B) pH
   (C) Organic matter (D) All of these
6. Which is the most acidity tolerant crop
   (A) Tea (B) Wheat
   (C) Maize (D) Cotton
7. The pH of acid sulphate soils is
   (A) <4 (B) 6
   (C) 7 (D) >7
8. The calcium carbonate equivalent value is highest in case of
   (A) Calcium carbonate (B) Quick lime
   (C) Hydrated lime (D) Wood ash

9. Effective calcium carbonate equivalent value depends upon
   (A) Calcium carbonate equivalent value
   (B) Fineness of lime material
   (C) Both A and B
   (D) None of these
10. Best results are obtained when
   (A) Lime is integrated with inorganic fertilizers
   (B) Lime is integrated with organic manures
   (C) Lime is integrated with inorganic and organic fertilizers/manures
   (D)Lime alone
11. Base saturation per cent is defined as the % cation exchange capacity occupied by
   (A) Ca, Mg, K (B) H, Al
   (C) All of A and B (D) None of these

■■■■

Chapter 10

CHARACTERIZATION AND MANAGEMENT OF SALINE WATER

INTRODUCTION

Conceptually, *'water quality refers to the characteristics of water that influence its suitability for a specific use'*. For example, good quality river water can be used successfully for irrigation but may pose problems for municipal use because of its sediment load. Treatment to remove the sediment is essentially required before the water is released for domestic use. Similarly, water from the snowmelt is of excellent quality for municipal use but may prove to be too corrosive for industrial use without treatment to reduce its corrosion potential. Quality of water is governed by its physical, chemical and biological characteristics. To evaluate water for irrigation, emphasis is on the chemical and physical characters of the water, other factors being not so important. In recent times even microbiological properties have received importance because of the use of sewage water for irrigation.

From chemical characteristics point of view, almost all irrigation water contains some salt. It is only when the amount of salt in irrigation water exceeds certain limit then the water is considered unfit for irrigation or special practices needs to be adopted for its use. *Per se*, even the saline water is not harmful to the plants as such but when used for irrigation over a period of time, salts accumulate in the root zone to adversely affect the plant growth and soil physical properties. Based on the experiences of using water of different qualities, problems that developed have been documented and scientifically analyzed. This cause and effect relationship between a water constituent and the problem encountered have been framed into criteria that helps to evaluate water quality or its degree of acceptability for the intended use. This chapter describes the physico-chemical properties of water to assess its quality and dwells upon the problems likely to develop with the continuous use of such waters. Ways and means to overcome these problems are included for long-term sustainability of irrigation.

ORIGIN OF SALTS IN IRRIGATION WATER

Rain water in general qualifies as fresh water resource containing very little amount of salts (EC = 0.005 to 0.05 dS/m). The salinity of the river/canal water is generally more than the rainfall (EC = 0.1 to 0.45 dS/m). The increased salinity in the rivers/ canals waters is part of the ongoing geochemical processes. The runoff while flowing into the streams through overland flow from the watersheds picks up salts from rocks and soils that contain salts. The pickup of salts is more when the stream flows through soils or rock formations that contain soluble salts. The salt pick up continues as the stream moves down and into the ocean. Salts may also be added on its way through seepage and irrigation return flow from irrigated lands, groundwater contribution as base flow to the stream and human interference in the form of diverting wastes from industrial/urban development. All these factors together not only determine the quantity of salts present in the water but also govern the nature of the salts present in the water.

The other course of water is to travel to form groundwater and into the oceans. Although the source for both the surface and groundwater is rainfall, groundwater is usually more saline than the surface waters. Although all factors that influence salinity of the stream water are operative in the case of groundwater as well; there are several additional factors that control the salinity of the groundwater.

- Climate of the area including duration and frequency of rain, humidity, intensity and duration of sunshine, wind velocity etc., in as much as these affect the quantity and frequency of recharge and the losses due to evapotranspiration, all influence the net balance between annual evaporation, precipitation and runoff.
- The rate of local rock weathering with resultant formation of water soluble salts from the decomposed products.
- The rate at which salts are transported in the region by streams, winds, rains from outside sources and deposited in the soil.
- The rate of groundwater circulation.
- The interconnection between surface water bodies and the groundwater.
- Hydrogeological conditions.

EXTENT AND DISTRIBUTION OF SALINE WATER

Groundwater quality surveys are being conducted in different states by various organizations. Groundwater potential and distribution of good, marginal and poor quality water for several states covering arid and semiarid regions reveal that 25 to 84% waters are marginal or poor in quality in these states (Table 10.1). These waters have little application for human consumption or for industrial use. It is the agricultural sector, where such waters can be profitably used to tide over the impending water crisis (Agarwal *et al.*, 1979).

Central Soil Salinity Research Institute, Karnal compiled the information, whatsoever was available, to prepare a map of groundwater quality for irrigation on a 1:6 million scale (Gupta *et al.*, 1994). Four legends were used in mapping the quality i.e. good water ($EC_{iw} < 2$ and $SAR < 10$), saline water ($EC_{iw} > 2$ and $SAR < 10$), high-SAR saline water ($EC_{iw} > 4.0$ and $SAR > 10$) and alkali water (EC_{iw} variable, SAR variable and $RSC > 2.5$). Though large spatial variations are encountered, high salinity groundwaters are mostly encountered in arid parts of northwest states like Rajasthan, Haryana and Punjab (Fig. 10.1). Associated with salinity, the groundwater in some pockets may contain toxic levels of B, F, NO_3, Se and Si, etc. The alkali waters are found mainly in the semi-arid parts of India where the annual rainfall varies from 500 to 700 mm. CGWB (1997) approximated that the total area underlain with the saline groundwater ($EC > 4$ dS/m) is 193438 km^2 with annual replenishable

Table 10.1 Groundwater potential and per cent distribution of marginal and poor quality groundwater in the arid and semi-arid regions of India

State	Utilizable potential (M ham/yr)	Net draft (M ham/yr)	% distribution of groundwater		
			Good	Marginal	Poor
Andhra Pradesh	3.66	0.74	68	25	7
Gujarat	2.03	0.69	70	20	10
Haryana	0.88	0.61	37	8	55
Karnataka	1.30	0.18	62	12	26
Madhya Pradesh	5.95	0.49	75	10	15
Punjab	1.31	0.93	59	22	19
Rajasthan	1.83	0.43	16	16	68
Tamil Nadu	2.69	0.99	*	*	*
Uttar Pradesh	9.27	2.68	32	22	46
Total	28.92	7.74			

* not available

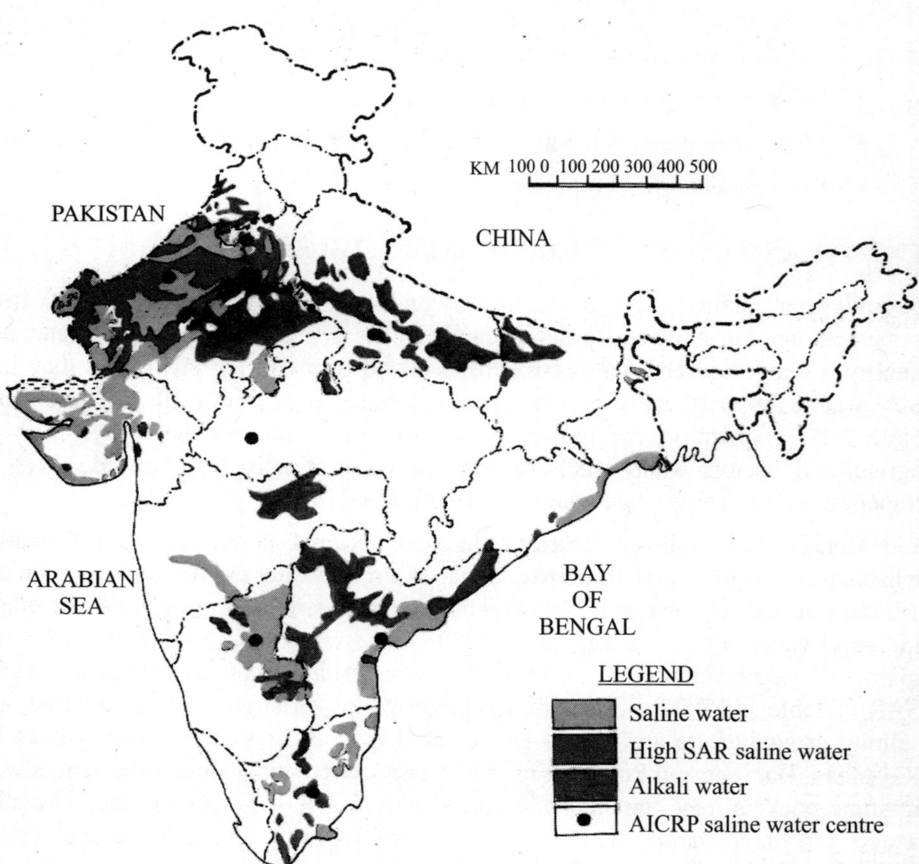

Fig. 10.1 Groundwater quality for irrigation (First Approximation)

recharge of 11765 million m^3/annum. Recently, CGWB (2010) has prepared a water quality map based on EC of the water demarcating the area with groundwater having EC > 3.0 dS/m. Since the drinking water supply guidelines state that water up to 3 dS/m can only be supplied if no other source of water is available, this provides a good insight to areas where groundwater is unfit for drinking and poses few problems for irrigation as well. This CGWB publication also contains maps related to presence of various ions such as fluoride, iron, chloride, nitrate, Arsenic etc., which are briefly discussed in relevant sections of this chapter. The statewise districts affected by water quality are given in Table 10.2.

Table 10.2 State-wise districts affected by salinity (EC > 3 dS/m) problem

State/ Union Territory	District affected in parts by groundwater salinity (EC > 3 dS/m)
Andhra Pradesh	Anantapur, Kurnool, Kadapa, Nellore, Prakasam, Guntur, Mahabubnagar, Nalgonda, Krishna, Khammam, Warangal, Medak, East Godavari, Srikakulam, Visakhapatnam, Vizianagaram
Delhi	North West, West, South West
Gujarat	Ahmedabad, Amreli, Anand, Bharuch, Bhavnagar, Banaskantha, Dohad, Porbandar, Jamnagar, Junagadh, Kachchh, Mehsana, Navsari, Patan, Panchmahals, Rajkot, Sabarkantha, Surendranagar, Surat, Vadodara
Haryana	Bhiwani, Fatehabad, Gurgaon, Hisar, Jhajjar, Kaithal, Mahendergarh, Rewari, Rohtak, Sirsa, Sonepat
Karnataka	Bagalkot, Belgaun, Bellary, Davangiri, Gadag, Gulburga, Raichur
Kerala	Palakkad, Ernakulam
Madhya Pradesh	Bhind, Indore, Jhabua, Sheopur, Ujjain
Maharashtra	Akola, Buldhana, Jalna, Jalgaon, Nasik, Satara
Punjab	Firozepur, Faridkot, Bhathinda, Mansa, Muktsar, Sangrur
Rajasthan	Ajmer, Alwar, Barmer, Bharatpur, Bhilwara, Bundi, Bikaner, Churu, Chittaurgarh, Dhaulpur, Dausa, Ganganagar, Hanumangarh, Jaipur, Jaisalmer, Jalore, Jhunjhunun, Karoli, Nagaur, Neemuch, Raja Samand, Sirohi, Sikar, Swai Madhopur, Tonk, Udaipur
Tamil Nadu	Dharmapuri, Pudukkottai, Thoothukkudi, Coimbatore, Dindigul, Ramanathanpuram, Salem, Karur, Namakkal, Perambalor, Thiruvannamalai, Vellore, Villupuram, Cuddalore
Uttar Pradesh	Agra, Hathras, Mathura
West Bengal	Haora, Medinipur, S- 24 Parganas

CHEMICAL CHARACTERISTICS TO CHARACTERIZE WATER QUALITY

Some of the potential hazards and related important constituents/characteristics of irrigation water are listed as follows:

- **Salinity hazard**: Total concentration of soluble salts; Electrical Conductivity (EC)
- **Sodicity hazard**: Relative proportion of sodium to other cations; Sodium Adsorption Ratio (SAR) and adjusted SAR
- **Alkalinity hazard**: pH, carbonate and bicarbonate concentration as related to the concentration of calcium plus magnesium; Residual Sodium Carbonate (RSC) and adj. R_{Na}.
- **Specific ion toxicity hazard**: Ionic composition; major constituents (Sodium, Magnesium, Chloride, Bicarbonate, Carbonate, Silica, Nitrate) and minor constituents (Boron, Lithium, Fluorine) and other micro-toxicants including heavy metals.

Table 10.3 Laboratory determinations to evaluate irrigation water quality problems

Water parameter	Symbol	Unit[1]
Salinity/Salt content		
Electrical Conductivity	EC_{iw}	dS/m
Total Dissolved Solids	TDS	mg/L
Cations and Anion		
Calcium	Ca^{++}	meq/L
Magnesium	Mg^{++}	meq/L
Sodium	Na^{+}	meq/L
Carbonate	CO^{--}_3	meq/L
Bicarbonate	HCO_3^-	meq/L
Chloride	Cl^-	meq/L
Sulphate	SO_4^{--}	meq/L
Nutrients[2]		
Nitrate-Nitrogen	NO_3-N	mg/L
Ammonium-Nitrogen	NH_4-N	mg/L
Phosphate-Phosphorus	PO_4-P	mg/L
Potassium	K^+	mg/L
Miscellaneous		
Boron/Fluoride or other toxic metals		mg/L
Acid/Basicity	pH	–
Sodium Adsorption Ratio[3]	SAR	$(mmole/L)^{1/2}$
Residual Sodium Carbonate [3]	RSC	meq/L

[2] NO_3 -N means the laboratory will analyze for NO_3 but will report the NO_3 in terms of chemically equivalent nitrogen. Similarly, for NH_4-N, the laboratory will analyze for NH_4 but report in terms of chemically equivalent elemental nitrogen. The total nitrogen available to the plant will be the sum of the equivalent elemental nitrogen. The same reporting method is used for phosphorus.

[3] Calculation procedures are discussed in subsequent sections.

Sometimes, sodicity and alkalinity hazards overlap and a clear distinction needs to be made between the two kinds of hazard. To assess the kind and degree of hazards, laboratory determinations are made using gravimetric, volumetric and colorimetric methods (Table 10.3, Gupta *et al.*, 2012) to evaluate the water quality problems.

SALINITY HAZARD: ELECTRICAL CONDUCTIVITY (EC)

The general principle of assessing EC of soils and irrigation water is discussed in Chapter 8. Based on EC, waters have been categorized as listed in Table 10.4. Important conversion between salinity units are given in Table 8.1.

Table 10.4 Classification of waters based on EC

Water class	EC (dS/m)	Type of water
Non-saline	<0.7	Drinking/irrigation water
Slightly saline	0.7-2.0	Irrigation water
Moderately saline	2.0-10.0	Drainage/groundwater
Highly saline	10.0-25.0	Drainage/groundwater
Very highly saline	25.0-45.0	Very saline groundwater
Brine	>45.0	Sea water

Example 10.1

Prepare a saline water of EC 4 dS/m using various salts.

Using meq/L = 10 * EC, we have to prepare a solution of 40 meq/L

If Na_2SO_4 (Equivalent weight 71.0) is used then dissolve 2.84g Na_2SO_4 per liter (40* 71/1000)

If $CaCl_2$ (Equivalent weight 55.5) is used then dissolve 2.22 g $CaCl_2$ per liter (40* 55.5/1000)

If $CaCl_{2.}$ $2H_2O$ (Equivalent weight 73.5) is used then dissolve 2.94 g $CaCl_2$ $2H_2O$ per liter (73.4*40/1000)

Example 10.2

Prepare a saline water of EC 10 dS/m using sodium chloride.

Since meq/L = 10 * EC is not applicable, EC being more than 5, we follow the other conversions

Salt present in saline water = 10*800 = 8000 mg/L

If we divide this value with equivalent weight of the salt to be added, say sodium chloride, we have

8000/58.5 =136.7 meq/L

Dissolution of 58.5 mg of sodium chloride in 1 liter gives 1000 meq/L. Therefore,

NaCl to be dissolved in 1 liter = (58.5/1000)* 136.7 = 7.99 or 8 g.

Compare with the approximate relation (meq/L = 10 * EC), we need to dissolve only 5.85 g of NaCl (58.5*100/1000).

Example 10.3

To prepare a salt water of, say, 5 dS/m having Na:Ca ratio as 50:50 with a single anion chloride.

Since the relation, meq/L = 10 * EC can be used. Therefore, meq/L = 50. For Na and Ca to be in 50:50 ratio, we have to add 25 meq/L of Na and 25 meq/L of calcium salt. We prepare the following table.

Salt	Concentration (meq/L)	Quantity to be added per liter
NaCl	25	58.5*25/1000= 1.46 g
$CaCl_2$	25	55.5*50= 1.388 g
or $CaCl_2.H_2O$	25	73.5*50= 1.838 g
Total	50	

Therefore, either 1.46 g of NaCl and 1.3888 g $CaCl_2$ or 1.46 g of NaCl and 1.838 g of $CaCl_2.H_2O$ should be dissolved in 1 liter of water.

Harmful Effects of Saline Water

Some of the visual symptoms on crops irrigated with saline water are: plants look stunted; leaves are smaller but thicker and have often-dark green colour as compared to plants growing in a salt free soil irrigated with good quality water. There is decline in crop growth that gets reflected in crop yield. It happens because of the following reasons.

- First and the most important is that the water availability to the plants decreases with increasing amount of salts in the soil water. Although the soil may contain sufficient moisture, it will not be utilizable by the plants. The osmotic pressure of the soil solution increases with the increase in salt concentration. Thus, plants are unable to extract water as readily as they can from a relatively non-saline soil solution. It may be noted that matric and osmotic stresses are additive in nature. As such osmotic stress affects the availability of water in the same fashion as the matric stress. It is why excessive salts in root zone can induce early wilting and the effects are almost similar to those of drought.
- Secondly, at high concentration, absorption of an individual ion (e.g. Na^+, Cl^-, SO_4^{--}, Mg^{2+}, B) may prove toxic to the plants and can induce physiological disorders.
- Thirdly, preferential absorption of one ion may also retard the absorption of other essential plant nutrients necessary for the normal growth of plants.

For example, high concentrations of chloride may depress the uptake of phosphorus.

- Finally, some elements like lithium, boron etc are toxic to plants even though they may be present only in very less quantities.

It is believed that the adverse effects are usually due to combined effects of these factors although one might dominate the others (Gupta and Gupta, 1997).

SODIUM/ALKALI HAZARD

pH

The pH, a measure of the acidity or alkalinity of water, is of interest as an indicator but is seldom of any real importance by itself. The main use of pH is quick evaluation of the possibility that the water may be abnormal. If an abnormal value is found, this should be a warning and the water needs further evaluation and possible corrective measure. The normal range for irrigation water is 6.5 to 8.4. Within this range the crops do well. Irrigation waters having pH outside this range may still be satisfactory but other problems of nutrition or toxicity become suspect. High pH above 8.5 is often associated with high bicarbonate (HCO_3^-) and carbonate (CO_3^{2}) concentrations, which might turn the water alkali in nature. The measurement procedure is same as discussed in Chapter 8.

Sodium Adsorption Ratio, SAR

Ratio of the monovalent and the divalent cations in the water is used to assess the sodium hazard of water. As potassium is comparatively less in waters than sodium, it is neglected in the calculations. On this basis, sodium adsorption ratio is given by Eq. (8.4) or Eq. (8.5).

Concentrations of these ions in water samples are typically reported in mg/L. To convert Na, Ca, and Mg from mg/L to meq/L, one needs to divide the concentration by 22.9, 20 and 12.15 respectively. To convert meq/L to mmol/L, divide by the charge or the valency of the ion.

If the SAR of the water is more than 15 $(\text{mmol/L})^{1/2}$, the water may pose problem with its continuous use for irrigation by affecting soil permeability. Since in heavy textured soil SAR of even 10 might create problems, some classifications of water on the basis of SAR treat such water of medium hazard (Table 10.5)

Table 10.5. Classification of water based on SAR values

SAR	Sodium hazard of water	Remarks
1-9	Low	Cautious use on sodium sensitive crops
10-17	Medium	Amendments (such as gypsum) and leaching needed
18-25	High	Generally unsuitable for continuous use
>26	Very high	Generally unsuitable for use

Adj. SAR

Ayers & Westcot (1976) proposed the concept of adj. SAR, which can be calculated by the following relation:

$$\text{Adj. SAR} = \text{SAR}\,(1 + \text{SI}) \qquad (10.1)$$

Here SI is Langelier Saturation Index (SI). The positive values of SI indicate that $CaCO_3$ will precipitate while negative values indicate that $CaCO_3$ will dissolve as per the following equation:

$$\text{SI} = pH_a\,(\text{actual pH of water}) - pH_c\,(\text{calculated pH}) \qquad (10.2)$$

The pH_c is a theoretical parameter which is the calculated pH of the irrigation water in contact with lime and in equilibrium with soil CO_2. It is given by the following equation:

$$pH_c = (pk_2 - pk_c) + p(Ca) + p(Alk) \qquad (10.3)$$

Since the values of pH_c above 8.4 indicate a tendency to dissolve lime from the soil and values below 8.4 indicate a tendency to precipitate lime from the applied water, Bower (1961) replaced pH_a by 8.4 in eq. (10.2) and p(Ca) by p(Ca+Mg) in Eq. (10.3) assuming that Ca and Mg behave in a similar fashion. In Eq. (10.3) pk_2 and pk_c are the negative logarithms of the second dissociation constant of H_2CO_3 and solubility product of $CaCO_3$, respectively both corrected for ionic strength. Also p(Ca+Mg) and p(Alk) are the negative logarithm of the molal concentration of (Ca+Mg) and equivalent concentration of titratable bases (CO_3+HCO_3). The final equations for pH_c and Adj. SAR are thus written as follows:

$$pH_c = (pk_2 - pk_c) + p(Ca+Mg) + p(Alk) \qquad (10.4)$$

$$\text{Adj. SAR} = \text{SAR}\,(1 + 8.4 - pH_c) \qquad (10.5)$$

pH_c is calculated using Eq. (10.4) and Table (10.6), which relates it to the concentration values from the water analysis. The Table values of ($pk_2 - pk_c$) are determined using the sum of (Ca + Mg +Na) in meq/L in column 1, p(Ca+Mg) using the sum of (Ca + Mg) in meq/L in column 1 and p (alk) using the sum of (CO_3 + HCO_3) in meq/L in column 1. These are then substituted in the pH_c Eq. (10.4) (Ayers and Wescot, 1976). The calculation procedure is illustrated in the example 10.4. While several authors agreed with the contention to replace Ca by (Ca+Mg), Suarez (1981) argued that magnesium carbonate does not precipitate. Magnesium precipitation in irrigated soils is only under agriculturally unusual conditions (Suarez, 1981). While carbonate completely precipitates the calcium, bicarbonate may precipitate calcium only partially under certain conditions increasing the proportion of the sodium in the soil solution but in no case precipitation approach completeness. The tests on eq. (10.5) proved that the equation over predict the sodium hazard and should be multiplied by a factor of 0.5 to evaluate more correctly the effect of HCO_3 on calcium precipitation (Ayers & Westcot, 1985). *In view of this, the concept of adj. SAR has little relevance although it is being reported as a matter of routine.*

Table 10.6 Values of pk_2-pk_c, p(Ca+Mg) and p(Alk) for calculation of pH_c

Concentration	pK_2 - pK_c	p(Ca+Mg)	p (Alk)
0.05	2.0	4.6	4.3
0.10	2.0	4.3	4.6
0.15	2.0	4.1	3.8
0.20	2.0	4.0	3.7
0.30	2.0	3.8	3.5
0.50	2.1	3.6	3.3
0.75	2.1	3.4	3.1
1.00	2.1	3.3	3.0
1.25	2.1	3.2	2.9
1.50	2.1	3.1	2.8
2.00	2.2	3.0	2.7
2.50	2.2	2.9	2.6
3.00	2.2	2.8	2.5
4.00	2.2	2.7	2.4
5.00	2.2	2.6	2.3
6.00	2.2	2.5	2.2
8.00	2.3	2.4	2.1
10.00	2.3	2.3	2.0
12.50	2.3	2.2	1.9
15.00	2.3	2.1	1.8
20.00	2.4	2.0	1.7
30.00	2.4	1.8	1.5
50.00	2.5	1.6	1.3
80.00	2.5	1.4	1.1

Alkalinity Hazard: Residual Sodium Carbonate (RSC)

If irrigation water has an appreciable carbonate and bicarbonate content, calcium and magnesium carbonate will precipitate with the increase in concentration of the soil solution through evapotranspiration. On the assumption that their precipitation is complete, a concept of residual sodium carbonate (RSC) has been proposed to evaluate the quality of bicarbonate containing water and defined it, in meq/L as:

$$RSC = (CO_3^{2-} + HCO_3^{1-}) - (Ca^{2+} + Mg^{2+}) \qquad (10.6)$$

Richard (1954) recommended that both RSC and SAR be used to evaluate the sodicity hazard. Subsequent investigators showed that general reliability of the RSC concept for evaluating the sodium hazard of bicarbonate water is questionable. Recent developments have shown that if the RSC is more than 2.5 meq/L, Adj. R_{Na} comes into play to assess precipitation of calcium from the water and/or dissolution of native calcium of the soils. In several textbooks, Adj. SAR comes into play is also mentioned but as already pointed out Adj. SAR is obsolete and has been discussed only as a matter of routine.

Adjusted Residual Sodium

To assess the dissolution and precipitation of calcium, Suarez (1981) proposed an equation and designated it as adjusted RN_a. It is given as follows:

$$\text{Adj. } R_{Na} = \frac{Na}{[(Ca_x + Mg)/2]^{1/2}} \tag{10.7}$$

where Ca_x is the value of Ca in the water that takes care of EC_{iw} and HCO_3/Ca ratio and the estimated partial pressure of CO_2 in the surface few mm of the soil (Table 10.7). The Ca_x values represents meq/L of Ca expected to remain in the soil solution at equilibrium. It seems that with certain modifications this concept can be well utilized to predict soil ESP under Indian monsoon conditions.

Example 10.4

Water analysis of a given water sample is

Ca = 2.54 meq/L Mg = 1.34 meq/L

Na = 6.87 meq/L EC = 1.1 dS/m

CO_3 = 0.38 meq/L HCO_3 = 4.21 meq/L

Find RSC, SAR, Adj. SAR and adj. R_{Na}

$$RSC = (CO_3 + HCO_3) - (Ca + Mg) \quad meq/L$$

$$= (0.38+4.21) - (2.54+1.34) = 4.59 - 3.88 = 0.71 \text{ meq/L}$$

$$SAR = \frac{Na}{\sqrt{(Ca + Mg)/2}}$$

where Na, Ca and Mg are in meq/L

$$= \frac{6.87}{\sqrt{(2.54 + 1.34)/2}} = 4.93$$

$$\text{Adj. } SAR = \frac{Na}{\sqrt{(Ca + Mg)/2}}[1 + (8.4 - pH_c)$$

Sum of Ca + Mg +Na = 2.54+1.34+6.87 = 10.75 meq/L

($pk_2 - pk_c$) obtained from the Table 10.6 using sum of Ca + Mg +Na = 10.75 meq/L is 2.3

Sum of Ca+Mg = 2.54+1.34 = 3.88 meq/L

p(Ca+Mg) is obtained by using sum of Ca+Mg = 3.88 meq/L in column 1 of Table 10.6 as 2.7

Sum of CO_3 + HCO_3 = 0.38 + 4.21 = 4.59 meq/L

Table 10.7 Calcium concentration (Ca_x) expected to remain in near-surface soil-water following irrigation

Ratio of HCO_3/Ca (meq/L)	Salinity of applied water (EC_{iw}) (dS/m)											
	0.1	**0.2**	**0.3**	**0.5**	**0.7**	**1.0**	**1.5**	**2.0**	**3.0**	**4.0**	**6.0**	**8.0**
0.10	8.31	8.57	8.77	9.07	9.31	9.62	10.02	10.35	10.89	11.32	12.01	12.56
0.15	6.34	6.54	6.69	6.92	7.11	7.34	7.65	7.90	8.31	8.64	9.17	9.58
0.20	5.24	5.40	5.52	5.71	5.87	6.06	6.31	6.52	6.86	7.13	7.57	7.91
0.25	4.51	4.65	4.76	4.92	5.06	5.22	5.44	5.62	5.91	6.15	6.52	6.82
0.30	4.00	4.12	4.21	4.36	4.48	4.62	4.82	4.98	5.24	5.44	5.77	6.04
0.35	3.61	3.72	3.80	3.94	4.04	4.17	4.35	4.49	4.72	4.91	5.21	5.45
0.40	3.30	3.40	3.48	3.60	3.70	3.82	3.98	4.11	4.32	4.49	4.77	4.98
0.45	3.05	3.14	3.22	3.33	3.42	3.53	3.68	3.80	4.00	4.15	4.41	4.61
0.50	2.84	2.93	3.00	3.10	3.19	3.29	3.43	3.54	3.72	3.87	4.11	4.30
0.75	2.17	2.24	2.29	2.37	2.43	2.51	2.62	2.70	2.84	2.95	3.14	3.28
1.00	1.79	1.85	1.89	1.96	2.01	2.09	2.16	2.23	2.35	2.44	2.59	2.71
1.25	1.54	1.59	1.63	1.68	1.73	1.78	1.86	1.92	2.02	2.10	2.23	2.33
1.50	1.37	1.41	1.44	1.49	1.53	1.58	1.65	1.70	1.79	1.86	1.97	2.07
1.75	1.23	1.27	1.30	1.35	1.38	1.43	1.49	1.54	1.62	1.68	1.78	1.86
2.00	1.13	1.16	1.19	1.23	1.26	1.31	1.36	1.40	1.48	1.54	1.63	1.70
2.25	1.04	1.08	1.10	1.14	1.17	1.21	1.26	1.30	1.37	1.42	1.51	1.58
2.50	0.97	1.00	1.02	1.06	1.09	1.12	1.17	1.21	1.27	1.32	1.40	1.47
3.00	0.85	0.89	0.91	0.94	0.96	1.00	1.04	1.07	1.13	1.17	1.24	1.30
3.50	0.78	0.80	0.82	0.85	0.87	0.90	0.94	0.97	1.02	1.06	1.12	1.17
4.00	0.71	0.73	0.75	0.78	0.80	0.82	0.86	0.88	0.93	0.97	1.03	1.07
4.50	0.66	0.68	0.69	0.72	0.74	0.76	0.79	0.82	0.86	0.90	0.95	0.99
5.00	0.61	0.63	0.65	0.67	0.69	0.71	0.74	0.76	0.80	0.83	0.88	0.93
7.00	0.49	0.50	0.52	0.53	0.55	0.57	0.59	0.61	0.64	0.67	0.71	0.74
10.00	0.39	0.40	0.41	0.42	0.43	0.45	0.47	0.48	0.51	0.53	0.56	0.58
20.00	0.24	0.25	0.26	0.26	0.27	0.28	0.29	0.30	0.32	0.33	0.35	0.37
30.00	0.18	0.19	0.20	0.20	0.21	0.21	0.22	0.23	0.24	0.25	0.27	0.28

Adapted from Suarez (1981); Assumes a soil source of calcium from lime ($CaCO_3$) or silicates; no precipitation of magnesium, and partial pressure of CO_2 near the soil surface (pCO_3) is 0.0007 atmosphere; Ca_x, is in meq/L.

p(alk) is obtained by using the sum of $CO_3 + HCO_3 = 4.59$ meq/L in column 1 of Table 10.6 as 2.3

$$pH_c = 2.3+2.7+2.3 = 7.3$$

$$\text{Adj. SAR} = 4.93\,[1+(8.4-7.3) = 4.93[2.1] = 10.35$$

$$\text{Adj.}\,R_{Na} = \frac{Na}{\sqrt{(Ca_x + Mg)/2}}$$

Since EC is 1.1dS/m and HCO_3/Ca ratio is 4.21/2.54 = 1.66.

Ca_x value from Table 10.7 corresponding to the nearest values of HCO_3/Ca and EC_{iw} is 1.43.

$$= \frac{6.87}{\sqrt{(1.43 + 1.34)/2}} = 6.87 / 1.18 = 5.82$$

Note the difference in the values of adj. SAR and Adj. R_{Na}.

Harmful Effects of Alkali/sodic Waters

Alkali or sodic water when applied continuously to the soil, adversely affects crop growth as a result of the following factors that come into play.

- The increased exchangeable sodium percentages (ESP) resulting from long-term use leads to breakdown of soil structure due to swelling and dispersion of clay particles. Fine textured soils remain dispersed and puddled when wet and hard when dry. Thus, less air and water enters the soil resulting in poor environment for root growth. Internal drainage is restricted resulting in temporary water logging. These soils do not easily attain proper soil moisture conditions for cultivation.
- A thin crust is often formed at the soil surface, which acts as a barrier to water penetration and seedling emergence.
- Compacted topsoil and sub-soil creates problems for root penetration and cultivation.
- Adverse crop growth may also be due to direct toxic effects of sodium in many crop plants.
- Increase in soil pH reduces the availability of a number of plant nutrients like nitrogen, zinc, iron, etc. It also results in decreased availability of calcium, magnesium and potassium.

In most cases, the yield of crops is adversely affected by a combination of the factors described above.

TOXICITY HAZARD

Toxic waters have variable EC, SAR and RSC but have excess of an individual ion that causes toxicity to plants that result in reduced crop growth/yield. A toxicity problem is different than the problems due to salinity or sodicity as the toxicity occurs within the crop as a result of the uptake and accumulation of certain constituents from the irrigation water. The constituents of concern are: chloride, sulphate, sodium, nitrate, boron, fluoride, or heavy metals such as selenium, cadmium, lead and arsenic etc. High concentration of these ions may also cause imbalance in the uptake of various ions or even reduce the uptake of plant nutrients. Moreover, these ions may affect the micro-irrigation systems or cause foliar injury to plants. The adverse effects may occur even at low salinity or sodicity. Toxicity problem accompanying salinity or sodicity complicates the issue with simultaneous development of salinity and infiltration problems.

Chloride is essential to plants albeit in very low amounts. It can cause toxicity to sensitive crops at high concentrations (Table 10.8, Maas, 1980). High sodium and chloride concentrations cause more problems when applied with sprinkler irrigation (Table 10.9). Leaf burn under sprinkler from both sodium and chloride are known to reduce crop yields.

Table 10.8 Chloride classification of irrigation water

Chloride (ppm)	Effect on crops
Below 70	Generally safe for all plants
70-140	Sensitive plants show injury
141-350	Moderately tolerant plants show injury
Above 350	Can cause severe problems

Table 10.9 Susceptibility ranges for crops to foliar injury from saline water containing sodium and chlorides

Ion	Na or Cl concentration (mg/L) causing foliar injury			
	Low	Medium	High	Excessive
Na concentration	<46	46-230	231-460	>460
Cl concentration	<175	175-350	351-700	>700

Excessive bicarbonate concentrations tend to precipitate calcium in the soil as per the following reaction:

$$2HCO_3 + Ca = CaCO_3 + H_2O + CO_2 \quad (10.8)$$

Accordingly, the proportion of sodium relative to Ca increases in the soil. This increases the SAR of the soil solution resulting in higher adsorption of sodium on the soil exchange complex. Besides it can also prove to be problematic for drip or micro-spray irrigation systems when calcite or scale builds-up causing reduced flow rates through orifices or emitters.

Boron is another element that is essential in low amounts, but toxic at higher concentrations (Table 10.10). In fact, there is a very thin margin between the essentiality level and toxicity level of boron. Toxicity on sensitive crops can occur at concentrations less than 1.0 ppm. It affects the metabolic activities of the plants.

Sulphate in irrigation water like boron has beneficial effects from fertility point of view. Nitrogen in irrigation water (N) is also largely a fertility issue, as nitrate-nitrogen (NO_3-N) can be a significant N source. Waters high in N result in excessive vegetative growth as well as can impair product quality in crops such as barley and sugar beets. However, these problems can be overcome by nutrient and irrigation management.

Table 10.10 Boron sensitivity of selected crops (mg/L*)

Sensitive		Moderately sensitive	Moderately tolerant	Tolerant
0.5-0.75	0.76-1.0	1.1-2.0	2.1-4.0	4.1-6.0
Peach	Wheat	Carrot	Lettuce	Alfalfa
Onion	Barley	Potato	Cabbage	Sugar beet
	Sunflower	Cucumber	Corn	Tomato
	Dry Bean		Oats	

*Maximum concentrations tolerated in soil water or saturation extract without yield or vegetative growth reductions.

Without going into the details of each and every toxic constituent, it may suffice to add that the problem of toxic elements in water has assumed greater importance only now as the problem is spreading quite fast. Besides naturally occurring toxic waters, surface and groundwater is being constantly polluted by industries directly dumping their wastewater in surface sources of water, leakage from industrial ponds and landfills, septic tanks, mining impoundments and municipal landfills. Besides, chemicals such as herbicides and pesticides used in agriculture are getting into the water resources through runoff after a rain or irrigation or through seepage into the groundwater. The area underlain with toxic elements has been increasing rapidly as more and more analysis results are pouring in. Since the toxic chemicals are a primary cause of cancer, heart disease, stroke, premature aging and birth defects, toxic waters have been in news recently and would continue to dominate the water quality scene in India. The major highlight of this concern however, has been on the direct uptake of toxic water. Rarely, its impact on plants and the entry of toxins in human and animals through the food chain is talked about.

COMBINATIONS OF PARAMETERS TO ASSESS WATER QUALITY

In addition to the problems created by individual parameters as discussed above, attempts have been made to categorize the waters using combinations of two or more parameters as given below.

- EC and SAR
- EC, SAR, RSC

USSL Diagram

The United States Salinity Laboratory (Richards, 1954) published a semi-logarithmic diagram, which rated waters considering EC and SAR of irrigation waters. Four classes each of salinity and SAR were devised (Table 10.11, Fig. 10.2). Each salinity class was associated with all the four sodium classes and vice versa, thus giving a total of sixteen classes for the purpose of evaluating the quality of irrigation water. It has been concluded that low salinity water (C1) can be used for irrigation with most crops on most soils, with little likelihood that soil salinity will develop, and that very high salinity water (C4) are not suitable for irrigation under ordinary conditions, but might be used occasionally under very special circumstances. The soil must be permeable, drainage must be adequate, irrigation water must be applied in excess to provide considerable leaching besides growing of high salt tolerant crops. Several limitations of the USSL diagram have been pointed out by various workers and users, some of them may not be tenable. Still the USSL diagram may not give clear-cut assessment of the quality of irrigation waters in the monsoon climatic conditions of India.

Table 10.11 EC and SAR values for different classes of salinity and sodium hazard of irrigation water

Salinity hazard	Class	EC (dS/m)	Sodium hazard	Class	SAR
Low	C1	0.1-0.25	Low	S1	<10
Medium	C2	0.25-0.75	Medium	S2	10-18
High	C3	0.75-2.25	High	S3	18-26
Very high	C4	2.25-5.0	Very high	S4	

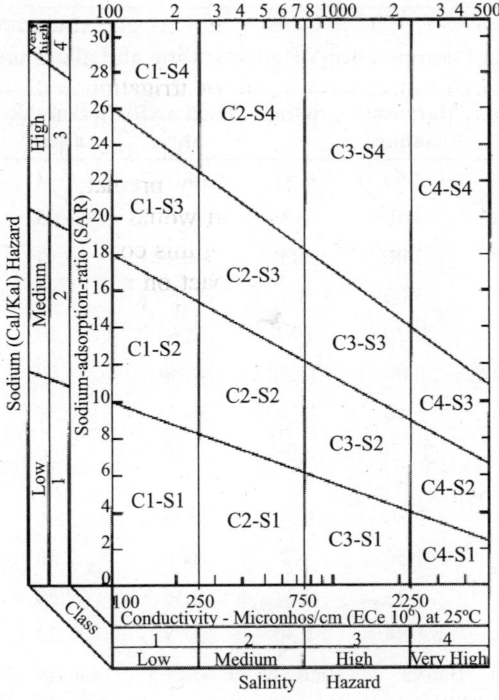

Fig. 10.2 USSL diagram to categorize water based on SAR and EC

EC, SAR and RSC

A Committee of Experts grouped the water into three main classes depending upon the EC, SAR and RSC of the water (Table 10.12). Saline and alkali water were further sub-grouped into three classes each to grade the severity of the problems. Like soils, alkali waters are also classified as sodic waters and the two terms are used synonymously. A glimpse of the chemical constituents in waters of various kinds is given in Table 10.13.

Table 10.12 Grouping of low-quality groundwater for irrigation in India

Water quality	EC_{iw} (dS/m)	SAR_{iw} $(mmol/L)^{1/2}$	RSC (meq/L)
A. Good	< 2	< 10	< 2.5
B. Saline			
i. Marginally saline	2-4	< 10	< 2.5
ii. Saline	> 4	< 10	< 2.5
iii. High-SAR saline	> 4	> 10	< 2.5
C. Alkali water			
i. Marginally alkali	< 4	< 10	2.5 – 4.0
ii. Alkali	< 4	< 10	> 4.0
iii. High-SAR alkali	variable	> 10	> 4.0
D. Toxic Water	The toxic water has variable salinity, SAR and RSC but has excess of specific ions such as chloride, sodium, nitrate, boron, fluoride, or heavy metals such as selenium, cadmium, lead and arsenic etc.		

Table 10.13 Chemical composition of good, saline and alkali water in Haryana

Parameter	Good	Marginally saline	Saline	High SAR saline	Marginally alkali	Alkali	High SAR alkali
EC (dS/m)	0.61	2.56	6.52	14.04	1.04	1.83	2.53
pH	8.60	8.30	8.3	8.55	8.05	8.20	8.50
Na (meq/L)	2.20	7.00	30.8	103.60	6.10	12.20	21.70
Ca (meq/L)	2.10	3.00	8.3	22.64	1.50	2.70	2.10
Mg (meq/L)	1.70	9.70	19.7	12.60	3.30	3.10	1.70
Cl (meq/L)	2.20	6.00	33.00	102.00	3.60	3.80	6.80
SO_4 (meq/L)	0.90	3.10	25.40	36.75	0.50	1.70	9.80
CO_3+HCO_3 (meq/L)	4.20	13.20	6.00	2.40	7.60	13.00	10.60
SAR $(meq/L)^{1/2}$	1.60	2.89	8.24	24.80	3.90	7.20	15.72
RSC (meq/L)	0.40	1.50	Nil	Nil	2.80	7.20	6.80
Mg/Ca ratio	0.81	3.23	2.37	0.56	2.20	1.15	0.81
Cl/SO_4 ratio	2.44	1.94	1.30	2.78	7.20	2.24	0.69
Village	Allipur	Barota	Bindrala	Durjanpur	Alipur	Amunpur	Kabulpur
Block	Nissing	Nissing	Assandh	B-Khera	Nissing	Nissing	Assandh
District	Karnal	Karnal	Karnal	Bhiwani	Karnal	Karnal	Karnal

INFILTRATION PROBLEMS

Rhoades Diagram

Because the effects of exchangeable sodium on swelling and dispersion are counteracted by high electrolyte concentration, the soil sodicity (permeability/ crusting) hazard cannot be assessed independently of electrolyte concentration. To keep the permeability of the soil stable, it is necessary to maintain threshold concentration of electrolytes in the infiltrating water. The threshold concentration of electrolytes increases with increase in ESP of the soil. Rhoades (1983) pointed out that because the soil surface usually limits infiltration, SAR of this layer should be estimated and the effect of electrolyte concentration of the irrigation water on the infiltration rate should be assessed. He devised a diagram by which permeability hazard is assessed by ascertaining whether the SAR-EC combination lies to the left (problem likely), or to the right (no problem likely), of the threshold line (Fig. 10.3). Since SAR of the top soil is in near equilibrium with SAR of irrigation water when EC_e is equal to EC_{iw}, it can be concluded that the irrigation waters having SAR, 10, 20 and 30 should not have TCC less than 10, 20 and 30 meq/L (or EC not less than 1, 2 and 3 dS/m, respectively).

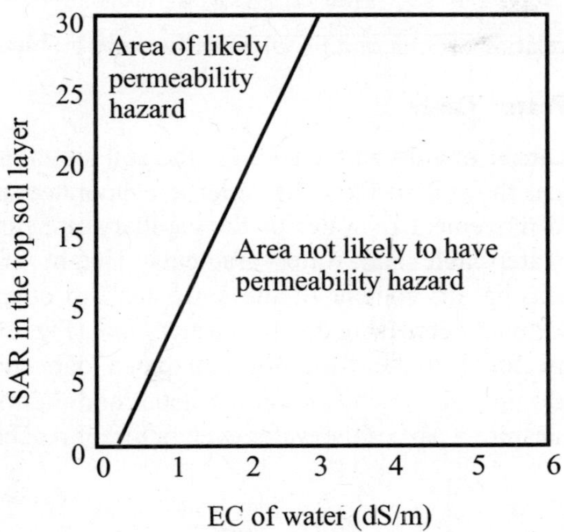

Fig. 10.3 Relative water infiltration rate as affected by salinity and SAR

FAO Guidelines

According to these guidelines, irrigation waters having SAR < 3, 3-6, 6-12, 12-20 and 20-40 should have EC greater than 0.7, 1.2, 1.9, 2.9 and 5.0 dS/m, respectively (Fig. 10.4). At lower than suggested EC, infiltration rate of water into the soil will be adversely affected.

SOIL SALINIZATION

A reference to soil salinization/sodication processes have been made in Chapter 7. However, basic principles of the two processes of soil salinization namely upward movement of groundwater through capillary rise and accumulation of salts applied with irrigation water are briefly discussed in the following sections.

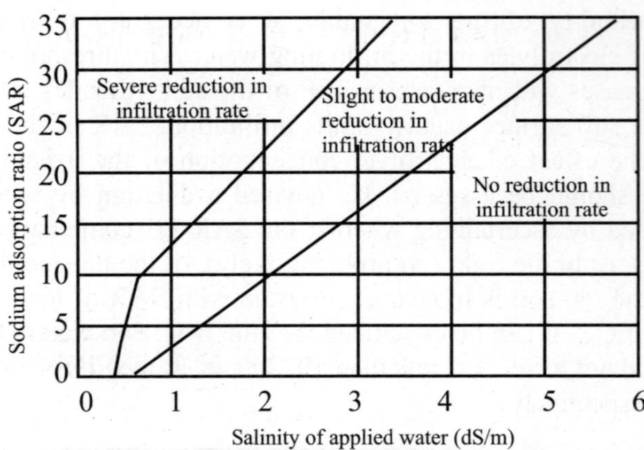

Fig. 10.4 Interrelation of SAR and EC of water to avoid problem of infiltration

Critical Depth to Water Table

Water is the main carrier of salts in the soil. As the soil solution reaches the site of evaporation at or near the soil surface, the water is evaporated and the salts are left behind. The upward movement of water or the capillary rise/flow is related to soil texture, depth to water table and water gradients. Height of the capillary rise increases with increasing silt content of the soils. On the other hand, the rate of evaporation increases with decreasing depth to water table (Fig. 10.5). Clearly for the same depth, capillary flow is higher for a silt loam than a loam soil. In cropped lands, the zone of salt build-up also depends upon the depth and the distribution of active roots, which may transpire a part of the water even before it reaches the soil surface.

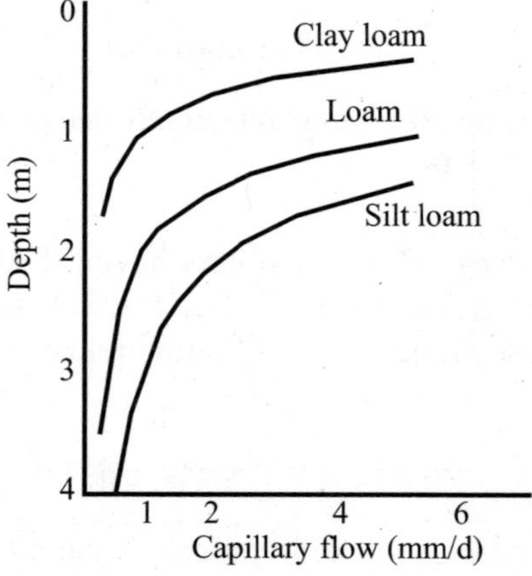

Fig. 10.5 Capillary flow velocity and depth of water table in 3 texture soil groups

The concept of 'Critical Depth to Water Table', first advocated by Talsma (1963), has been in vogue since then. According to this concept, the process of secondary salinization is quite active once the water table is above this critical depth. The critical depth, for the purpose of this concept, is defined *'as the depth to water table at which relative evaporation is 0.1 or the rate is about 10% of its value when the water table is at or near the soil surface'*. Soils with a large proportion of smaller pore sizes have higher values of critical depth to water table (e.g. silty loam). Coarse sand has a small value, whereas well-structured medium to fine soils have intermediate values. Thus, salinity in the profile increases with decreasing depth to groundwater table but increases with increasing salinity of the groundwater at the same depth to water table (van Hooren, 1979). For favorable water and salt balance, understanding and knowledge of the maximum flux assumes significance. For good quality groundwater, conventionally 1 mm/d is taken as the limiting value but for poor quality groundwater, the value is reduced to minimize soil salinization because salt movement is the product of flow rate and salt concentration'. As such, depth to water table values increase with increasing groundwater salinity (Fig. 10.6). For practical standpoint, the maximum allowable flux to maintain favourable water and salt balance should not exceed 0.5 mm/d. On this basis, critical depth may vary from a meter in light textured soils to a few meters in fine textured soils.

Recent studies have revealed the limitations of this concept and arguments contesting the applicability of this concept have been put forth. It is argued that if the flux at the bottom of the root zone is kept downward, there would be no opportunity for soil salinization even under shallow water table situation. In the Indian context, it may be noted that subsurface drains installed at shallow depths performed well and maintained water and salt balance because flux of the water at critical periods is downward (monsoon season, when the water table is at shallow depths). During the summer season, when evaporative demand is high, water table is deep enough to allow excessive upward flux. Thus, applicability of the concept of critical depth to water table needs to be examined before it is applied to field situations. Otherwise, we may end up in costly solutions in land reclamation programs.

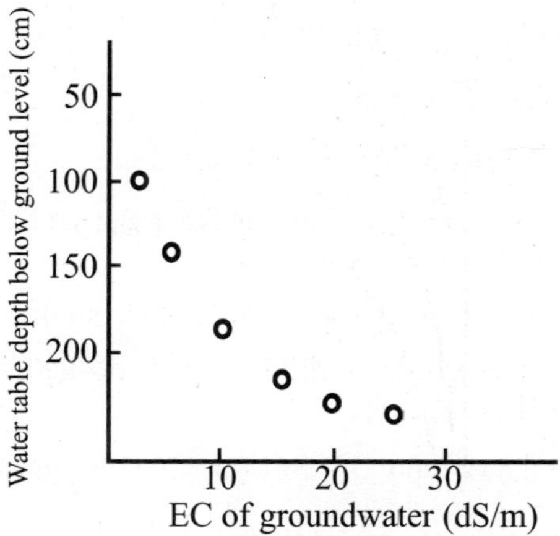

Fig. 10.6 Interaction between depth to water table and salt concentration of groundwater for good crops yields

Assessing Soil Salinization through Capillary Rise

Whenever the groundwater is within the capillary fringe, evaporation creates a gradient for water to move upward through soil capillaries. The total quantity of water brought into the root zone and the salt concentration of the groundwater determine the extent of soil salinization especially during the fallow period. The change in soil EC_e in terms of water entering the root zone in depth units and EC of groundwater (EC_g), can be given by the equation:

$$\Delta EC_e = D_g/D_s * \rho_w/\rho_b * EC_g/SP * 100 \quad (10.9)$$

Here ΔEC_e is the change in salt concentration of the soil depth, D_s, cm; D_g is the amount of water that enters the given depth of the soil, cm; SP is the water content in mass percentage at saturation, ρ_b and ρ_w are the bulk density of soil, g/cm^3 and density of water respectively usually taken as 1g/cm^3.

If salt concentration of groundwater is known in g/L or kg/m^3 such that one can calculate the salt added to the root zone in kg/ha then the above equation for the change in EC_e of the soil due to added salts is written as:

$$\Delta EC_e = (\Delta S/0.7) * 1/w_e * \rho_w/\rho_b * 1/D \quad (10.10)$$

Such that ΔS is the amount of salt added through capillary rise in kg/ha. The factor 0.7 is used to convert the salt in terms of EC. It may be noted that this factor varies from 0.64 to 0.80 and therefore, an average value of 0.7 is used in these calculations. For accuracy one may change this value accordingly.

Example 10.5

Determine the change in soil salinity in 30 cm soil depth if during this period of 3 months 15 cm of water gets evaporated from the soil surface. EC of groundwater is 5 dS/m. The bulk density of the soil is 1.5 g/cm^3 and saturation percent is 60. Assume any other appropriate value required for calculations.

We have all the values for use in the Eq. (10.9).

$$\Delta EC_e = D_g/D_s * \rho_w / \rho_b * EC_g/SP * 100$$

Assuming ρ_w as 1.0 g/cm^3, we have:

$$\Delta EC_e = 15/30 * 1.0/1.5 * 5/60 * 100 = 2.8 \text{ dS/m.}$$

Since the EC of the water is 5.0 dS/m or the water contains approximately 3.5 g/L (kg/m^3) of salts.

The amount of water entering the soil profile is 15 cm or 1500 m^3.

Thus, salt added to the profile is 1500*3.5 = 5250 kg/ha. Assuming ρ_w as 1.0 g/cm^3 and using Eq. (10.10) we have:

$$\Delta EC_e = (5250/0.64) * 1/60 * 1/1.5 * 1/30 = 2.77 \text{ dS/m.}$$

Salinization Due to Saline Water Irrigations

All irrigation waters contain some salts especially the groundwater in arid and semiarid regions. Every irrigation event, howsoever small the salt content may be, brings a certain amount of salt into the root zone. For example, if 10 cm of water having EC = 0.5 dS/m is applied to irrigate an area, it will add about 320 kg/ha of salts. If the EC is 1 dS/m, the addition will be 640 kg/ha. As the water is lost through evapotranspiration, the salts remain behind in the root zone/evaporation zone and accumulate as more and more salts are added with irrigation. In irrigated lands when there is no drainage and there are no other sources of salts, the salt build-up can be determined using the following relation:

$$\Delta EC_e = D_i/D_s{*}\ \rho_w/\rho_b{*}EC_i\ {*}100/SP \qquad (10.11)$$

Equation (10.11) has the same form as the Eq. (10.9). Eq. (10.11) has limited application to assess soil salinity development in soil profiles as some water always percolates below the zone of interest. Nonetheless, a very useful application of this equation can be made in the preparation of saline soils for conducting pot/lysimeter experiments for which saline soils need to be prepared. The amount of saline water required to salinize a soil to a pre-decided level is given as:

$$D_i = D_s{*}\ \rho_b/\rho_w{*}\Delta EC_e\ /EC_i\ {*}SP/100 \qquad (10.12)$$

Example 10.6

Determine the change in EC of the top 30 cm soil profile if 50 cm depth of saline water of 2 dS/m is applied and the whole water remains within this depth without any deep percolation. The bulk density of soil may be taken as 1.3 g/cc and the density of water as 1 g/cc. The soil saturation percentage is 42. What would be the depth of water applied if the salinity of the soil needs to be restricted to 4.0 dS/m?

Incorporating all the known values in Eq. (10.11)

$$\Delta EC_e = D_i/D_s{*}\ \rho_w/\rho_b{*}EC_i\ {*}100/SP$$

$$\Delta EC_e = 50/30{*}\ 1/1.3{*}2.0{*}100/40 = 6.4\ dS/m$$

Incorporating all the known values in Eq. (10.12)

$$D_i = D_s{*}\ \rho_b/\rho_w{*}\Delta EC_e\ /EC_i\ {*}SP/100$$

$$D_i = 30{*}\ 1.3/1.0{*}4.0/2.0\ {*}40/100 = 31.2\ cm$$

Since the salt uptake by the crops is negligible, the only way to avoid soil salinization is that an equivalent amount of salt added should be removed from the root zone by the percolating water. Whenever salt influx to the root zone exceeds the salt flux out of the root zone, soil salinization is a foregone conclusion,. The distribution of salts in the root zone is entirely different in a saline land formed due to saline water irrigation and a land salinized due to capillary rise. In the former, salt concentration increases with depth while in the later salt concentration decreases with depth being in equilibrium with the subsoil water salinity near the water table.

WATER AND SALT BALANCE OF THE ROOT ZONE

Some of the factors that influence the water and salt balance of the root zone are the amount of water applied, the salinity of the irrigation water, depth to water table, soil texture and other soil characteristic which affect its permeability, evapotranspiration, total rainfall and its distribution, irrigation methods employed, setting of the land in respect of the adjoining area (topography) and above all the overall management inputs that goes with any agricultural activity. The general salt balance equation irrespective of the scale is written as:

$$\text{Salt influx} - \text{Salt out flux} = \text{Change in salt storage} \qquad (10.13)$$

The salt influx or out flux are calculated by multiplying the water inputs times their concentrations and water outputs times their respective salt concentrations in a specific period of time. If water inflows into and outflows from a region are denoted by Q_i and Q_d, respectively and their salt concentrations are denoted by C_i and C_d, the regional salt balance equation can be written as follows:

$$\sum Q_i * C_i - \sum Q_d * C_d = \Delta S \quad (10.14)$$

The repeated use of equation on spatial and time scales is made to assess the salt balance.

The salt storage or leaching from the root zone is mathematically put into the form of an equation as follows (Fig. 10.7):

$$\pm S = R\,C_r + D_i\,C_i + G\,C_g + S_1 - D_d\,C_d - ET\,(0) - S_2 \quad (10.15)$$

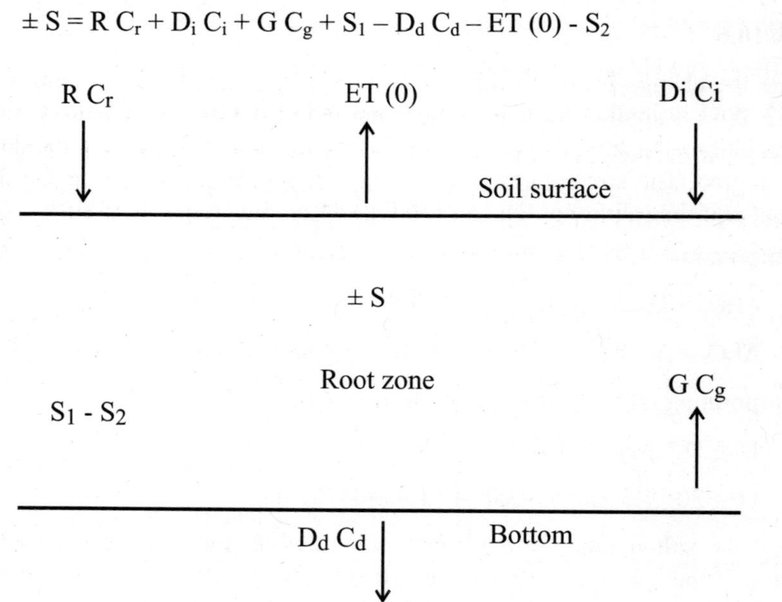

Fig. 10.7 Water balance components of the root zone of irrigated lands

Here salt storage (+) or leaching (-) S, in the root zone is given by the components which add salt to the root zone (+) or those which remove the salt (-). These components are added together with their respective signs. In this equation salt are added to the root zone through rainfall (RC_r), application of irrigation water (D_iC_i), capillary rise of groundwater (GC_g) and direct addition of salts such as through the application of amendments/fertilizer or generated in the root zone through mineral dissolution (S_1). On the minus side are the salt removed through drainage or deep percolation below the root zone (D_dC_d), salt removed by ET (being zero as the salt content is zero) and the salt precipitation (S_2). In most cases, the terms S_1 and S_2 are neglected leaving only four terms on the right hand side of eq. (10.15). It may be noted that in many situations irrigation induced mineral dissolution is significant and it may not be possible to neglect this component in irrigated lands without causing significant differences in the calculated and actual salt balance. It is clear that to control salinity build-up in the root zone one has to minimize the salt addition to the root zone or induce drainage from rainfall and applied irrigation water. Since rain water is a good quality water, salt additions can only be controlled either by reducing the salts applied with irrigation water or through the control of groundwater which might add substantial amounts of salts in the root zone if groundwater is at shallow depths and/or is saline in nature. Moreover, unknown parameters in Eq. (10.15) add another dimension to the complexity of the equation. For example G and C_d are not exactly known. Therefore, it is always useful to monitor the salt status of irrigated lands over a few years' time to work out the salt regime. It will not only help to diagnose the problem but also help to identify factors likely to accentuate salt accumulation or leaching of the salts.

The following salt balance scenarios are anticipated:

- Balance in favour of leaching (S in Eq. (10.15) is negative)
- Stable salt balance (S in Eq. (10.15) is zero)
- Balance in favour of salt accumulation (S in Eq. (10.15) is positive)

Besides, salt balance needs to be worked out for short durations (say between irrigations or a month), medium durations (say a season) and long durations (say annual) over a few years' time. Similarly, salt balance needs to be established over small terrestrial units (say a farm) or over an irrigation command or on a regional basis. Usually, larger areas are chosen for salt balance studies, since salt move from one area to another. Salt transport from an irrigated area to an unintended area is schematically shown in Fig. 10.8. Such processes operating on regional scale generally justify investments on assessing regional salt balance along with smaller units. While preparing the salt balance sheet, causes, rate and degree of accumulation or leaching needs to be established.

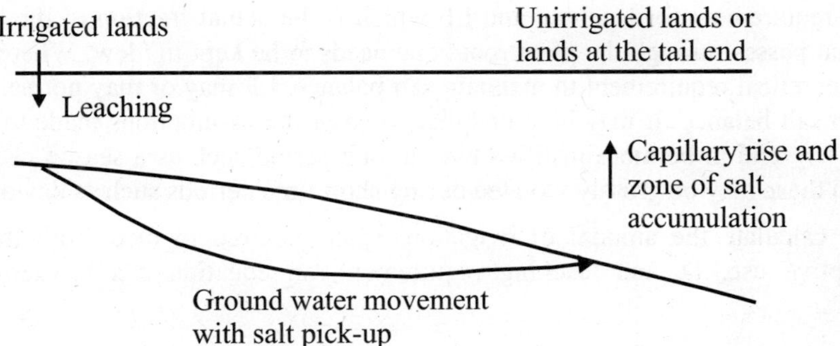

Fig. 10.8 Salt movement in irrigated lands showing areas of salt leaching and accumulation

LEACHING REQUIREMENT

For a stable salt balance (S is zero), Eq. (10.15) can be simplified as follows:

- Since rainfall is of very good quality and has negligible salts, the term can be ignored.

- For irrigated lands with deep water table, groundwater contribution to the root zone can be neglected.
- Salt application amounts being negligible and assuming no dissolution of salts, term S_1 can be neglected.
- For little uptake of salts by plants and assuming no precipitation etc. of the salts term S_2 can also be neglected.

With the above assumptions, Eq. (10.15) is written as:

$$D_i C_i - D_d C_d = 0 \tag{10.16}$$

From Eq. (10.16), it immediately follows that

$$\frac{D_d}{D_i} = \frac{C_i}{C_d} \tag{10.17}$$

From this equation, leaching requirement, LR, is defined as the *'fraction of the irrigation water that must pass through the root zone to maintain soil salinity at some predefined level'*. It is mathematically written as:

$$LR = D_d/D_i \text{ or } C_i/C_d \tag{10.18}$$

As electrical conductivity (EC) of the solution is a good index of the solution concentration, solution concentration is often replaced by EC such that Eq. (10.20) is written as:

$$LR = EC_i/EC_d \tag{10.19}$$

Here, EC_i and EC_d are the electrolyte conductivities of the irrigation and drainage water respectively. The distinction between LR which is the theoretically average requirement for leaching and LF which is the actual fraction of the applied water that passes through the plant root zone needs to be kept in view. Whereas, LR is the theoretical requirement to maintain salt balance, LF may or may not be able to maintain salt balance. It may be noted that some of the assumptions made to derive Eq. (10.18) tend to be more justified over a long period such as a season or a year, although these may be grossly violated during short time periods such as day or week

To calculate the amount of irrigation water required to meet both the crop consumptive use, D_c and leaching requirement, an equation can be derived as follows:

$$D_i = D_c + D_d \tag{10.20}$$

Since D_d can be written as Di* LR (from Eq. 10.18), Eq. (10.20) can be written as:

$$D_i = D_c + D_i\,LR \tag{10.21}$$

Or

$$D_i = D_c / (1-LR) \tag{10.22}$$

If $LR = EC_d/EC_i$ is incorporated in Eq. (10.22), D_i can be written as:

$$D_i = [EC_i/ (EC_d - EC_i)] * D_c \tag{10.23}$$

It may be noted that under field situations D_i is also calculated with the help of the following equation:

$$D_i = D_c/\text{application efficiency} \qquad (10.24)$$

Under such a situation, it is desirable to compare the D_i. If the D_i calculated by Eq. (10.24) is more than that calculated with Eq. (10.23), no additional water to meet LR is required. For example if the seasonal crop consumptive use is 250 mm and the LR is 0.20 then D_i with the help of Eq. (10.23) is:

$$D_i = 250 / (1\text{-}0.2) = 312.5 \text{ mm}$$

If the irrigation application efficiency is 75%, the total amount of water to be applied to satisfy crop water requirement is (Eq. 10.24):

$$D_i = 250 / 0.75 = 333.3 \text{ mm}$$

Since the excess water application due to low application efficiency will eventually go as deep percolation beyond the root zone, the application of 333.3 mm water will take care of both the irrigation and leaching requirements. However, one needs to take care of distribution efficiency. In the monsoon climatic conditions of India, usually it holds true as the excess rainfall and deep percolation losses together takes care of the leaching requirement especially when salt balance is made on annual basis.

If there is effective rainfall and it does not contribute to leaching, then effective rainfall D_r is subtracted from D_c and eq. (10.23) can be written as:

$$D_i = [EC_d/ (EC_d\text{-} EC_i)] * (D_c - D_r) = (1/1\text{-}LR) * (D_c - D_r) \qquad (10.25)$$

Here D_r is the effective rainfall.

Estimating Leaching Requirement for Salt Balance

Since EC_i in the LR equation is known, one must know C_d or EC_d so that a good estimate of LR is made. Several alternative procedures have been proposed by various workers. Bernstein (1961) suggested the use of average EC_e at which 50% yield reduction occurs as the value of EC_d in case of forage and vegetable crops. For fruit crops, EC_e at which 10% yield reduction occurs is taken as EC_d. Since EC_s= 2 EC_e, it may be reasonable to use twice the EC_e values at 50% or 10% reduction in yield as an approximation to EC_d. Bernstein and Francois (1973) on the other hand proposed that yields are nearly insensitive to the salinity of the lower root zone and therefore, the EC_e values at 50% or 10% reduction in yield can be increased by 4 times before use in Eqs. (10.25 or 10.27). Schilfgaarde *et al.* (1974) while commenting upon several equations notably the one proposed by Bernstein (1964) suggested that EC_d can be assumed twice of the EC_e at which yield reduction is 100%. A more logical semi empirical model of Rhoades (1974) states that average salt concentration in the profile can be estimated from the salinity at the top and bottom of the root zone such that:

$$C_S = K\,(C_t + C_l)/2 \qquad (10.26)$$

Here, C_l is the salt concentration at the bottom, which is the concentration of drainage water, C_d and C_t at the top of the root zone and is equal to C_i. K is a constant and its value is governed by the shape of the salt profile in the root zone. Equating C_S to the average root zone salinity for a pre-determined yield reduction, LR in terms of electrical conductivity can be written as:

$$LR = (EC_i/(2EC_s/K - EC_i) \quad (10.27)$$

This equation has been used by Ayers & Westcott (1976) assuming K=0.8 and EC_s= 2 EC_e. The proposed eq. (10.27) overestimates LR except at low leaching requirements.

Since large variations are noticed in the suggested procedures, it is advisable that values of EC_d should be determined experimentally by sampling the soil below the root zone. In its absence, depending upon the salt tolerance of the crops EC_d = 2 EC_e 25% for sensitive and EC_d = 4 EC_e 50% for tolerant crops may be taken. The value of EC_e 25% and EC_e 50% are the values at which yields reduce by 25% and 50% respectively. These values can be obtained from the table given by Maas & Hoffman (1977) and FAO (1985).

Example 10.7

Calculate LR and total irrigation requirement (D_i) of the season if irrigation water has EC_i = 1.2 dS/m and EC_e 50% for a sensitive crop is 8 dS/m. Crop consumptive use during the season is 250 mm with an effective rainfall of 20 mm. If the application efficiency is 80%, would it be necessary to apply additional irrigation water to meet the LR.

Since the EC_e 50% is 8, EC_d is taken as EC_e times 2 = 16 dS/m

Thus, LR from Eq. (10.19) is given as:

$$LR = 1.2/16 = 0.075 \text{ or } 7.5\%$$

This shows that to maintain average root zone salinity within desired limits LR of 0.075 is sufficient.

Using Eq. (10.25), we have:

$$D_i = [1/(1-0.075)] * (250-20) = 248.6 \text{ mm}$$

From application efficiency point of view, and taking care of effective rainfall, D_i is given by Eq. (10.24) as follows:

$$D_i = (250-20)/0.8 = 287.5 \text{ mm}$$

Since the total irrigation water requirement from application efficiency point of view is larger, there is no need to apply additional water to meet LR.

Salinization-Desalinization Cycles

Under field situations saline water irrigated lands experience salinization-desalinization cycles. While salts build-up during irrigation cycle or dry season, desalinization occur during the monsoon season as shown in Fig. 10.9. It shows:

- Period of soil salinity build-up (Zone A) during November to April till equilibrium salinity is reached

- Period of salt redistribution (Zone B) within the root zone from April to mid-June or till the onset of monsoon. During this period, soil salinity is usually maximum in the surface layer (0-15 cm).
- Period of desalinization (Zone C) during monsoon season when rainfall is in excess over the potential evapotranspiration.

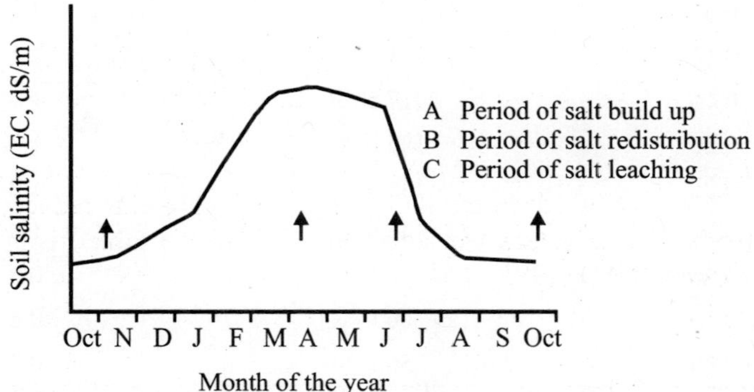

Fig. 10.9 Salinization-desalinization cycle in a saline water irrigated land

LEACHING FRACTION AND SOIL SALINIZATION

Several factors control the extent of salinity development in the soil profile, the most important being: (a) quality of irrigation water, (b) number of irrigations, (c) soil texture (d) leaching fraction and (e) rainfall.

In general soil salinity is related to the salinity of irrigation water being high with high salinity water and low with low salinity water. The soil salinity and salinity of irrigation water are related as per the following equation:

$$EC_e = EC_i * X \tag{10.28}$$

Here X is the concentration factor, which depends upon the leaching fraction. The relevant concentration factors reported in Table 10.14 are for a water uptake pattern of 40:30:20:10. These factors show that at low leaching fraction, steady state salinity is about 3.2 times the salinity of the irrigation water while at high leaching fraction it is even less than the salinity of the irrigation water as is evident from leaching fraction exceeding 0.30. The steady state root zone soil salinity (EC_e) likely to develop upon long-term use of saline water at various leaching fractions (LF) and a water uptake pattern of 40:30:20:10 is shown in Fig. 10.10. It is apparently clear that there is ample scope to control the mean root zone salinity at a pre-decided level by controlling leaching fraction. Moreover, for the same amount of water applied, the leaching fractions is more in a light than a heavy textured soil resulting in higher salinity build-up in the later. Clearly, the effect of amount of irrigation water applied and the soil texture can be lumped together in leaching fraction. It is again reiterated that LF should not be confused with LR. Whereas, LR is theoretically calculated value to maintain salt balance, LF is the actual fraction of the irrigation water that leaves the root zone. If rainfall contribution is neglected, LF can be easily related to the average water application efficiency. An application efficiency of 70% means that leaching fraction is around 0.30.

Table 10.14 Concentration factor (X) for use in Eq. (10.28)

Leaching fraction (LF)	Applied water needed (% of ET)	Concentration factor (X)
0.05	105.3	3.2
0.10	111.1	2.1
0.15	117.6	1.6
0.20	125.0	1.3
0.25	133.3	1.2
0.30	142.9	1.0
0.40	166.7	0.9
0.50	200.0	0.8
0.60	250.0	0.7

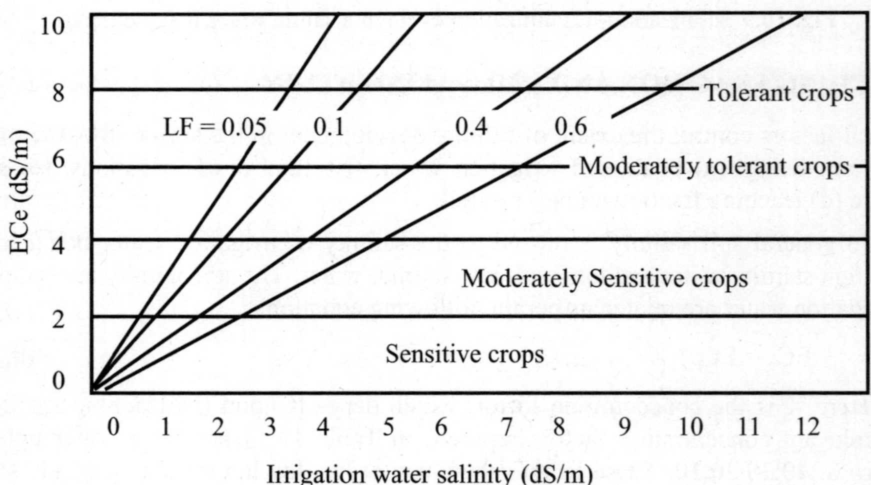

Fig. 10.10 Steady state root zone soil salinity (EC_e) likely to develop with applied irrigation water salinity (EC_i) at various leaching fractions (LF)

Soil Salinization with Known LF and Water Uptake Pattern

If the root zone is divided into four quarters, the plant roots will withdraw water from each of the four quarters. A plant water uptake pattern of 40:30:20:10 means that 40% of the total water uptake by the plant roots is from the top quarters of the root zone followed by 30% from the next quarter and 20 and 10% from the next two quarters from top to bottom (Fig. 10.11). The uptake pattern can vary from one crop to another and/or due to soil layering.

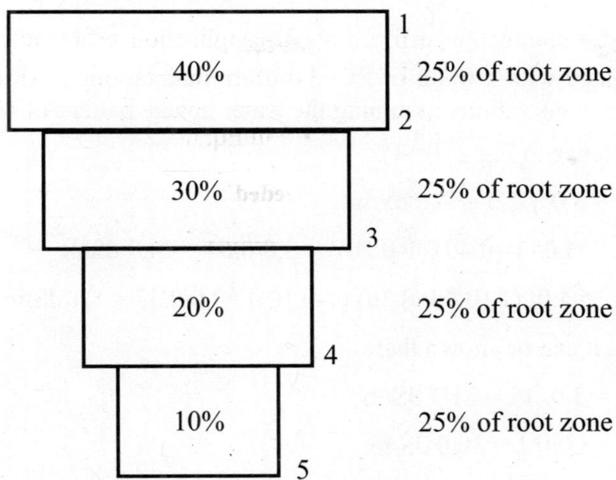

Fig. 10.11 Water uptake pattern 40:30:20:10 by the plants

If EC_i and LF are known, one can proceed to calculate the soil salinization in the following manner.

Since the leaching fraction at the top of the soil profile is 1.0, EC_1 at the soil surface is equal to EC_i. At the bottom of the first quarter the EC_2 is calculated as

$$EC_2 = EC_i/ [(1\text{-}0.40 (1\text{-}LF)] \tag{10.29}$$

The denominator shows the amount of water applied that goes below the 1st quarter i.e. 40% of the applied water less the leaching fraction desired is consumed in the top 25% of the root zone and the remaining moves out of the profile. Similarly, EC_3 at the top of the third and EC_4 at the top of the fourth quarter are calculated as follows:

$$EC_3 = EC_i/ [(1\text{-}(0.40+0.30) (1\text{-}LF)] \tag{10.30}$$

$$EC_4 = EC_i/ [(1\text{-}(0.40+0.30+0.20) (1\text{-}LF)] \tag{10.31}$$

The EC_5 at the bottom of the fourth quarter is equal to EC_i divided by the LF.

$$EC_5 = EC_i/ (1\text{-}1 (1\text{-}LF) = EC_i/LF \tag{10.32}$$

The average soil water concentration is taken as the average of the 5 values as follows:

$$EC_{sw} = (EC_1+EC_2+EC_3+EC_4+EC_5)/5 \tag{10.33}$$

$$EC_e = EC_{sw}/2 \tag{10.34}$$

$$\text{Concentration factor} = EC_e/EC_i \tag{10.35}$$

Thus, concentration factors given in Table 10.14 can be calculated in this manner. The examples given below illustrate the calculation of the concentration factor, X for a leaching fraction of 0.2 for two water uptake patterns i.e. 40:30:20:10 and 60:25:10:05. Concentration factors for other water uptake patterns can be calculated following the same procedure as outlined in the text and the following examples.

Example 10.8

A crop is irrigated with a water of EC_i 3.0 dS/m maintaining a LF of 0.1. Calculate the average root zone salinity assuming the water uptake pattern of 40:30:20:10.

With reference to Fig. (10.11)

$EC_1 = 3.0 \ / \ 1.0 = \ 3.0$ dS/m

$EC_2 = 3.0/(1-(0.40)\ (1-0.10)) = 3.0/0.64 \ = 4.7$ dS/m

$EC_3 = 4.0/\ (1-(0.40+0.30)\ (1-0.10)) = 3.0/0.37 = 8.1$ dS/m

Similarly, it can be shown that

$EC_4 = 3.0/.19 = 21.1$ dS/m

$EC_5 = 3.0/0.1 = 30.0$ dS/m

Thus average EC_{sw} is calculated as

$EC_{sw} = (3.0+4.7+8.1+21.05+30.0)/5 = 13.4$ dS/m

or $EC_e = EC_{sw}/2 = 13.4/2 = 6.7$ dS/m

Concentration factor = $EC_e/EC_i = 6.7/3.0 = 2.2$ times of EC_i

In Table (10.14), value at this LR = 2.1

EC of drainage water, $EC_{dw} = EC_5 = EC_i/LF = 3.0/0.1 = 30.0$ dS/m

Example 10.9

For the above example calculate the concentration factor for a water uptake pattern of 60:25:10:5.

EC_1 (Fig. 10.11) $= 3.0 \ / \ 1.0 = \ 3.0$ dS/m

$EC_2 = 3.0/(1-(0.60)\ (1-0.10)) = 3.0/0.46 \ = 6.5$ dS/m

$EC_3 = 4.0/\ (1-(0.60+0.25)\ (1-0.10)) = 3.0/0.235 = 12.8$ dS/m

Similarly,

$EC_4 = 3.0/.145 = 20.7$ dS/m

$EC_5 = 3.0/0.1 = 30.0$ dS/m

Thus, EC_{sw} can be calculated as

$EC_{sw} = (3.0+6.5+12,8+20.7+30.0)/5 = 14.6$ dS/m

or $EC_e = EC_{sw}/2 = 14.6/2 = 7.3$ dS/m

Concentration factor = $EC_e/EC_i = 7.3/3.0 = 2.4$ times of EC_i

EC of drainage water, $EC_{dw} = EC_5 = EC_i/LF = 30.0$ dS/m

Concentration factors for this water uptake patterns is 2.4 instead of 2.2.

The salt profiles with the two uptake patterns are shown in Fig. 10.12, which reveals a higher accumulation with the 60:25:10:05 pattern than the 40:30:20:10 pattern of water uptake.

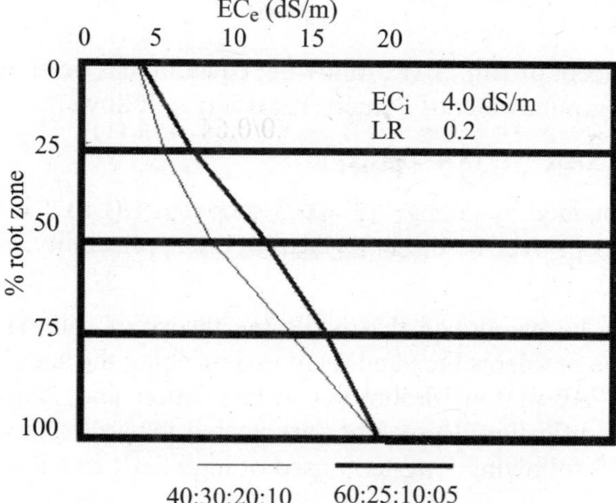

Fig. 10.12 Steady state salt profile in a saline water irrigated land with two water uptake patterns

It may be noted that in monsoon climatic conditions, steady state conditions do not prevail and hence the concept of leaching fraction has little relevance under the Indian monsoon conditions. With the expansion of sprinkler and drip irrigation, the concept might have limited applications.

Soil Salinization and Sodification

Sodification process never occurs in isolation of the soil salinization process. Primarily, sodification is also a salinization process but additionally involves the replacement of other cations on the soil exchange complex by sodium. Significant replacement can only occur when Na^+ becomes the dominant soluble cation in the soil solution. This may occur when the salinizing source is Na^+ rich or when the soil solution becomes more concentrated in Na^+ or when for other reasons the salinization processes favour the accumulation of Na^+. Sodium adsorption ratio (SAR), which determines the sodium hazard of irrigation water, increases within the soil solution as soil water concentration increases causing SAR to increase in proportion to the square root of the concentration factor as per the following reaction:

$$SAR_{\text{soil water}} = SAR_i * \sqrt{n} \quad (10.36)$$

Here n is soil water concentration factor. The presence of CO_3^{--} and HCO_3^- in the soil solution aids this process as these anions react and form less soluble salts with Ca^{++} as against the corresponding highly soluble Na^+ salts. Their presence leads to a relative enrichment of the soil solution with Na^+ as the overall salt concentration increases because the Ca^{++} salts precipitate.

A major difficulty in the soil sodification is to assess/predict the soil ESP when high SAR waters are used so that corrective measures are initiated against the build-up well in time. Several equations relate soil ESP with SAR of the water. According to Richard (1954), the relationship between SAR of the irrigation water and soil ESP is given by eq. (8.8). Since SAR of the soil water can be 1 to 3 times more than that of the irrigation water, higher ESP build-up may be expected than given by eq. (10.36).

Using the concept of Adj. SAR, following equation has been used to predict the sodium hazards of groundwater of Punjab (Pakistan) as follows:

$$ESP = 2\ SAR\ \{1+(8.4-pH_c)\} \tag{10.37}$$

Factor 2 is obtained by taking LF = 0.25 such that $1/(LF)^{1/2}$ is equal to 2. Since the Adj. SAR concept over predicts the hazard, the applicability of this equation is questionable.

It may finally be mentioned that while the theory of soil salinization is quite clear there are some problems here and there in describing the sodification processes. It is attributed to adsorption, desorption, precipitation and dissolution processes operative in soil sodification. Researches are geared towards this goal as well as to predict the soil ESP following long-term used of high SAR or alkali waters using the R_{Na} concept.

MANAGEMENT STRATEGIES FOR SALINE/ALKALI WATER

The quality of water used for irrigation plays a significant role in sustainable agriculture. The injudicious and indiscriminate use of poor quality water has often led to the problem of soil salinization and sodification, thereby, affecting crop yields. In view of the large quantum of poor quality groundwater, scarcity of fresh irrigation water and development of water logging and soil salinity due to rising water table in irrigation commands, it is of utmost importance that such waters are used judiciously. Any use of such waters must consider the interacting influence of the amount and type of salts in water and soils, type of soil to be irrigated, nature of crop to be grown, frequency and intensity of rainfall, and other climatic factors of the area. A number of management options have been perfected in India to manage saline/alkali waters. Although the basic principles of the technologies for saline and alkali water are similar, amendment application is needed only to reclaim/manage alkali waters. Many technologies discussed in Chapter 8 in relation to saline and alkali soils will also help to manage use of saline and alkali water respectively. Herein the technologies are discussed in a combined manner highlighting the differences wherever required.

Crop Management

Plants differ widely in salt tolerance from sensitive ones i.e. glycophytes, whose normal growth is inhibited at low concentration of salts to the most resistant halophytes, which grow profusely on saline habitats. Though most crop plants are glycophytes, there is rather wide range of salt tolerance amongst them. Beets and date palms, having halophytic ancestors, can be irrigated with saline water without much loss of yield. The relative crop salt tolerance can be assessed through the information reported in Table 10.15 (Maas, 1984). Most of the tables of relative tolerance report tolerance to soil EC_e and as such principles of soil salinization with saline water need to be applied to use these tables for crop selection. Based on the actual field testing of field crops at various AICRP centers on Management of Salt Affected Soils and Use of Saline Water, Minhas and Gupta (1992), prepared a relative salt tolerance table with respect to quality of water, which is reproduced as Annexure III. It may be mentioned that a crop tolerant to salinity may be sensitive or semi-tolerant to alkali conditions. As such, selection of a crop should be made to suit the irrigation water quality i.e. salinity, alkalinity or toxicity. It is also useful to select a cropping sequence such that all the crops in the sequence are able to tolerate the irrigation water salinity. The general recommendations are that semi-tolerant to tolerant crops and those with low water requirement should be grown while those requiring liberal water should be avoided.

Table 10.15 Relative crop salinity tolerance rating

Relative crop salinity tolerance	Soil salinity (EC_e) at which yield loss begins (dS/m)
Sensitive	<1.3
Moderately sensitive	1.3-3
Moderately tolerant	3-6
Tolerant	6-10
Unsuitable unless reduced yield acceptable	>10

Besides intra-genic variations in salt tolerance of crops, wide inter-genic variations in the salt tolerance of crop cultivars have been documented. Usually there is a negative correlation between salt tolerance of varieties and their potential yields. For example; 'Damodar' in rice and 'Kharchia' in wheat are well documented for their salinity tolerance but have low yield potentials. Recent breeding efforts have changed this outlook as cultivars like ‘CSR 30’ for rice, 'HD-2560' of wheat, 'CS-52' of mustard and 'MESR-16' of cotton have been developed. Apparently, it is possible to breed cultivars for high yield as well as higher salt tolerances.

Crops do not tolerate salinity equally at all stages of their growth. For example, even a good salt tolerant crop sugar beet is sensitive to salts at the germination stage but is quite tolerant at later growth stages (Gupta & Gupta, 2014a). During the initial stages, the root zone being limited to surface few cm, most salts from the evaporating soils concentrate there. As such, germination and early seedling establishment have been identified as the most crucial stages. Therefore, strategies to minimize the salinity of the seeding zone are needed to establish a good plant stand. The other critical periods for crops are when a phase change takes place say from vegetative to reproductive or heading and flowering to seed setting.

Soil Management

Soil texture

Accumulation of salts vis-à-vis tolerance limits to saline waters gets modified with soil texture. As a 'thumb rule' accumulation of salts in saline water irrigated soils is nearly one half the salinity of irrigation water in coarse textured soils (loamy sand and sand), equal to that of irrigation water in medium textured sandy loam to loam soils and more than two times in fine textured soils (clay and clay loam). As such, saline water that can be applied to grow crops vary from coarse to fine textured soils. Even the leaching of salts through rainfall is governed to a great extent by soil texture. Because of low infiltration rates of fine textured soils (having high clay content) rainwater either tends to runoff or evaporates from stagnated water at the surface. Adoption of measures such as tillage to open up heavy textured soils for better intake to conserve rainwater is recommended during monsoon.

Soil and land forming strategies

Distribution of water and salts in soils vary with land form. Land forming and seeding practices help in better germination and growth as spots with low accumulation of salts can be used to grow the crop. Seed bed design and seed placement should be done to minimize exposure of seeds and roots to salt particularly in the early sensitive stages of growth. Several of the land forming and irrigation strategy are discussed in Chapter 8. Readers may refer the chapter to utilize these strategy in planning the use of saline water.

Irrigation Water Management

On-farm irrigation management with saline and alkali waters should adopt irrigation schedules that minimize irrigations, eliminate salinity build-up and also ensures optimal crop production. Since the distribution of water and salts in soils vary with the method of irrigation, any irrigation method that creates and maintains favourable salt and water regimes in the root zone should be preferred. Shallow depth high frequency irrigations are preferred in saline environment. Irrigations with saline water should be applied more frequently to reduce the cumulative water deficits (matric and osmotic) between the irrigation cycles (Fig. 10.13). The shallow depth high frequency irrigations are possible by using improved irrigation techniques. Even in surface irrigation methods, it can be ensured through proper land leveling, laser land leveling being preferred to achieve better results.

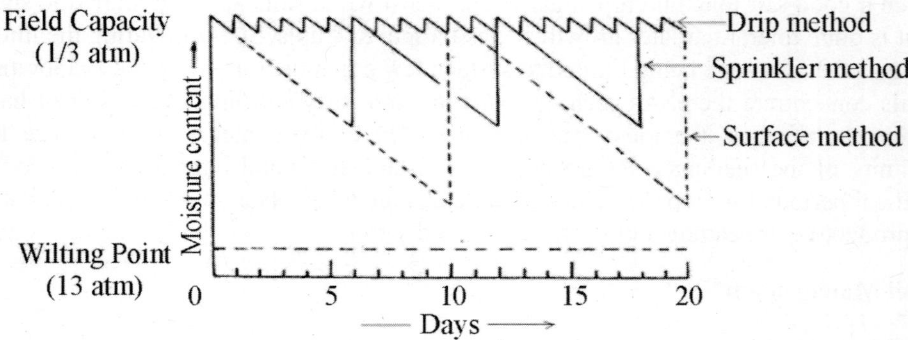

Fig. 10.13 Moisture availability in crops under different irrigation techniques, note low matric stress in frequently irrigated drip irrigation

Furrow irrigation seems to be one of the effective methods of irrigation to provide salt free zone to the plants in the furrows since most salts either present in the soil or applied through irrigation move towards the ridges of the furrows. Besides, by changing the size of the beds (broad bed and furrows) and/or the irrigation furrows (All furrows or alternate furrows) and placement of seeds, one can have several options as per the crop requirement discussed in Chapter 8.

The pressurized irrigation methods such as sprinklers and drip are more efficient as the quantity of water applied can be controlled. Sprinkler irrigation is an ideal method to apply irrigations at frequent intervals with shallow depth of water. Leaching of soluble salts is also more efficient with sprinkler as the water application rates can be controlled below the infiltration capacity of the soil. Sprinkler irrigation also helps in achieving uniform water application and salt leaching even under local topo-sequential differences in the field levels. The use of sprinkler irrigation helps in utilizing relatively more saline waters than surface irrigation methods. Compared to surface irrigation, sprinkler irrigation results in higher yields. Yet these advantages get nullified at higher salinity levels. It is attributed to evaporation and consequent salt build-up in the water droplets falling on the leaves and harming the crop adversely including crop burning effect with highly saline waters. Drip irrigation is still more beneficial as it also allows relatively more saline water use in crop production. As an example, good information on the relative merits of drip and sprinkler systems was made available way back in 1970, which in its modified form including surface irrigation is shown in Fig. 10.14 clearly revealing the superiority of drip over the sprinkler and surface irrigation method at all salinity levels for all the tested crops.

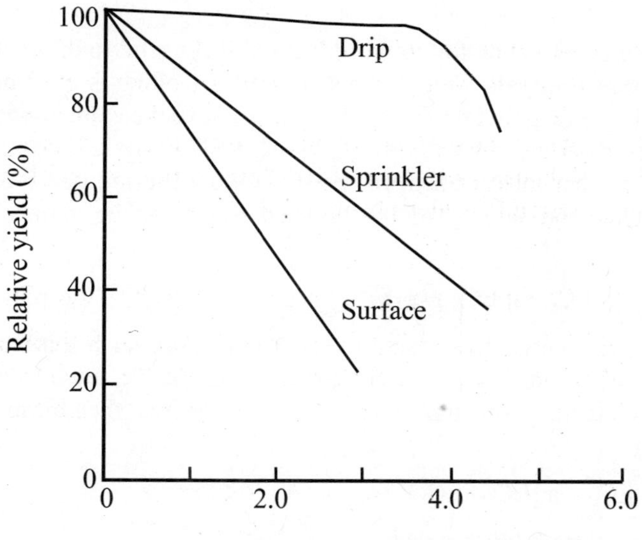

Fig. 10.14 Variation in yield with saline water in drip, sprinkler and surface irrigation

Conjunctive Use

The conjunctive use of multi-source/multi-quality waters has been defined as *'a strategy by which waters from various sources are used in a combined manner such that the net output is more compared to the net output when each source/quality of water is used separately'* (Gupta & Gupta, 2001). In conjunctive use, fresh and saline water are used in a manner that salinity of the soil does not exceed a pre-decided level to grow a desired crop. A number of alternatives strategies are shown in the following chart (Fig. 10.15).

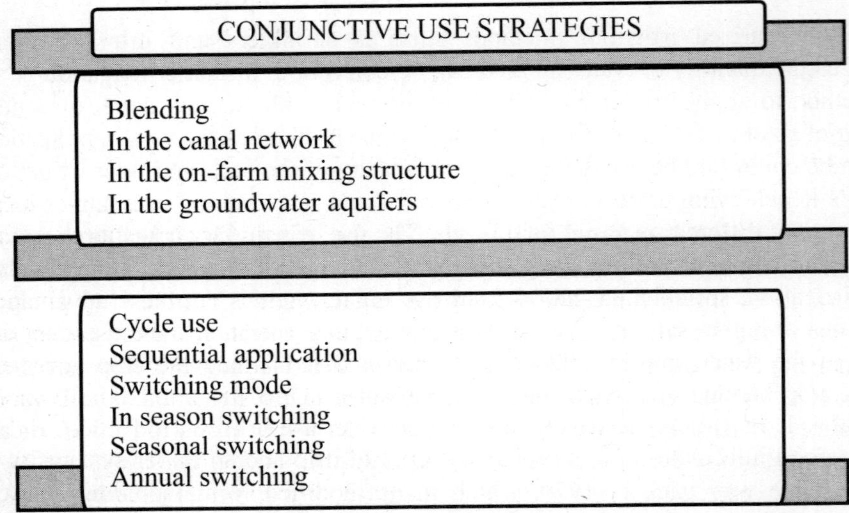

Fig. 10.15 Conjunctive use options for fresh and saline waters

Blending

When groundwater is too saline for crop production, it is useful to dilute it with good quality water to reduce its salt content to a safe limit. Blending is most promising in areas where fresh water is available on demand in sufficient quantities. Blending needs to be assessed on the basis of the EC, SAR, RSC or toxic constituents of the blended water such that most critical of them is below the pre-decided level. The equation to calculate the salinity of the blended water can be simply derived or written as:

$$EC_m = (a\,EC_a + b\,EC_b)/(a+b) \tag{10.38}$$

Here EC is the electrical conductivity of water, dS/m such that subscripts m, designate the EC of blended water; a, b represent the relative fraction of fresh and saline water in the blend. If one unit of mix water is prepared, then b can be replaced by (1-a) such that

$$EC_m = EC_b + a\,[EC_a - EC_b] \tag{10.39}$$

From this equation, it follows that:

$$a = \frac{EC_m - EC_b}{EC_a - EC_b} \tag{10.40}$$

From crop stand point of view, EC_m can be replaced by EC of the water that can be tolerated by the crops `EC_c' such that:

$$a = \frac{EC_c - EC_b}{EC_a - EC_b} \tag{10.41}$$

The value of b then is calculated by b = (1-a). Data on EC_c values are available in the literature and a good value is selected either for no or a pre-determined yield reduction (Annexure III).

Equation (10.38) through Eq. (10.41) are also valid to calculate RSC or toxic ion concentrations of the mix waters simply by replacing EC by RSC or the relevant toxic ion. However for SAR, these equations have to be used three times to find out the concentration of Ca, Mg and Na in the mixed water and then calculate the SAR of the mixed water. A mix for which all the components i.e. EC, SAR, RSC and toxic ion are below the critical value is chosen.

Blending can be accomplished either in the canal network where saline groundwater is pumped so that the salinity of the canal water does not exceed a pre-decided value. A major limitation is that the canal water is supplied as drinking water, which must be kept in view while planning such a strategy. The blending can also be done at the farm level where saline water is mixed in the water course or field channels (Fig. 10.16 left). An equally good way is to construct a mixing tank in which saline and fresh water can be mixed and used for irrigation (Fig. 10.16 right). In recent years, groundwater recharge to reduce salinity, sodicity and even toxicity of water has emerged as a viable option to blend low quality groundwater in the aquifers itself. Recharged water dilutes the groundwater and the diluted groundwater is used for crop production. Studies in Mohindergarh district in Haryana and Agra-Bharatpur region in Uttar Pradesh and Rajasthan have proved the adequacy of the recharge technologies to dilute groundwater.

Fig. 10.16 Blending saline water in canal network (left) and in a farm tank (right)

Example 10.10

The EC of fresh and saline water are 0.5 dS/m and 6 dS/m respectively. What fraction of saline water should be mixed in fresh water so that the salinity of the blended water does not exceed 1.5 dS/m ?

Since the fraction of fresh water source is given by a, we have from Eq. (10.41)

$$a = \frac{EC_c - EC_b}{EC_a - EC_b}$$

a= (1.5- 6.0) / (0.5- 6.0) = -4.5/-5.5 = 0.81

b = (1-a) =0.19

So 19% of saline water should be mixed in fresh water to get the desired quality of water.

Example 10.11

The EC of fresh water of a water course is 0.5 dS/m and has a discharge of 1 cumecs. The salinity of ground water is 5 dS/m respectively. If the discharge of one tube well is 0.01 cumecs, discharge of how many tube wells can be mixed in the water course such that salinity of the blended water does not exceed 1.2 dS/m ?

Using Eq. (10.39) as such, we have

$$EC_m = (a\,EC_a + b\,EC_b)/(a+b)$$

Since a =1.0 and b= 0.01*n

1.2 = (1*0.5 + 0.01*n *5)/ (1+0.01*n)

Here n is the number of tube wells

Solving for n

1.2 + 0.012 n = 0.5 + 0.05 n

0.7 = 0.038 n

n = 18.4 or to be on safer side it is 18 tube wells

Cyclic use

Cyclic use, also known as sequential application or rotational mode, is a technique in which canal water is replaced with saline water in a pre-decided sequence/cycle. Saline water is substituted for good quality waters either when fresh water is unavailable and/or at times when the crop is at a more tolerant growth stage. The timing and the amount of water depends upon the quality of water, the cropping pattern, the climate and the type of irrigation system. Cyclic mode is a more flexible technique for the conjunctive use of saline and fresh waters as both types of waters are independently available to the farmers as against the mixing technique where only one quality of water is available. Thus, depending upon the time when fresh water is available and on the basis of crop characteristics, number of alternative schemes have been proposed and tested. The following alternative strategies are adopted.

Sequential Application: In this technique saline water is applied in a pre-decided sequence. For example, alternate irrigation with fresh and saline water can be one sequence.

Switching Mode (in season switching): Many crops being more sensitive at the germination or initial establishment stage, switching mode technique has been found quite useful. In this mode after the pre-sowing or initial few irrigations, rest of the irrigations are applied with saline water.

Seasonal Switching: Seasonal switching is yet another alternate practice. The practice is based on the two basic premises that fresh water during *rabi*/summer season is limited and crops differ in their tolerance to soil salinity. Thus, a crop rotation is selected in such a manner that more tolerant crop is grown during *rabi*/summer season while a sensitive crop is grown during the *kharif* season. While fresh water is applied to the sensitive crop, it is replaced by saline water during the *rabi* season. Yet, in many cases annual switching is practiced. Here in one year fresh water is used whereas in other year only saline water is used. Such a switching mode helps in areas where both kinds of water are limited and one can switchover from one area to another in subsequent year.

Pre-sowing Irrigation with Fresh Water

Primary objectives of pre-sowing irrigation is to create optimal soil moisture conditions to facilitate tillage and seed bed preparation and to recharge the root zone with water for germination and to meet the ET needs of crops. In saline soils previously irrigated with saline water, to leach soluble salts below the seeding zone is also one of the objectives. Since failure of the crop at the germination and seedling emergence stages scales up to poor crop stand and low yields, root zone environment at these stages should be conducive for germination and seedling establishment stages. It has been proved that relative yield of various crops with pre-sowing irrigation with fresh water followed by saline water throughout was always higher in comparison to when no fresh water was applied at pre-sowing irrigation, salt load being the same.

Nutrient Management

The salt accumulation in saline water irrigated soils can affect the nutrient availability to plants in several ways such as: by changing the forms in which the nutrients are present in soils; by increasing the losses through leaching when saline soils are leached or as in nitrogen through denitrification or by precipitation in soils, through interactive effects of cations and anions, and through the effects of complementary (non-nutrient) ions on nutrient uptake. Besides, salt affected soils or soils irrigated with saline water are often poor in most of the essential plant nutrients owing to lack or limited vegetation and low organic matter content. Nitrogen deficiency is widespread in saline soils and a large fraction of the applied nitrogen is lost in gaseous forms under high soil salinity (Gupta & Gupta, 1997). It is useful to apply 25% more nitrogen than the soils irrigated with normal water. Availability of phosphorus increases up to a moderate level of salinity but thereafter it decreases. For chloride rich waters, additional dose of phosphorus may be required if these soils test low in phosphorus. For irrigation with alkali water, Zn @ 20 kg/ha per annum either during *rabi* season or to both the crops is desirable.

When saline water is used for irrigation, balanced use of essential nutrients is very important to achieve optimum productivity. Bulky organic materials such as FYM, green manure or even residue management have not only the nutritive value, but are important in structural improvements. These changes influence leaching of salts and minimize salt accumulation in the root zone. The other advantages of these materials in saline water irrigated soils are in terms of reducing volatilization losses, enhancing nitrogen-use efficiency and retention of nutrients in organic forms, which guards the plant nutrients from leaching and other losses.

CHEMICAL MANAGEMENT

Chemical amendments are essentially required to manage alkali waters having high SAR or RSC. If the water contains excessive sodium [>70% soluble sodium; Soluble Sodium Percentage, SSP = (Na * 100)/ (Ca + Mg + Na)], bicarbonate ions (> 8 meq/L) and/or Mg/Ca ratio (> 4.0), such waters may be low in calcium. Since calcium below a critical limit is harmful to soils and plants, the chemical composition of such waters is improved by adding some cheap and soluble calcium salt to raise the calcium content of the water. Gypsum is by far the best source of adding calcium. Besides reducing the SAR, it decreases the Mg/Ca ratio and helps to precipitate excessive bicarbonate ions as per the following reactions.

$$2\ NaHCO_3 + CaSO_4 = Ca(HCO_3)_2 + Na_2SO_4 \tag{10.42}$$

$$Ca(HCO_3)_2 = CaCO_3 + H_2O + CO_2 \tag{10.43}$$

Gypsum Application

Equivalent weight of gypsum ($CaSO_4.2H_2O$) is 86.09 g. Thus, 86.09 mg of gypsum fully dissolved in 1 L of water, constitutes 1 meq/L of calcium. In other words, 86 g of gypsum needs to be dissolved in 1 m^3 of water or 8.6 kg of pure gypsum needs to be dissolved in 1 ha-cm of water to add 1 meq/L of calcium. Accordingly, gypsum requirement per cropping season or per annum can be calculated by the following steps:

Calcium to be added (meq/L) = RSC to be neutralized (Initial RSC – Desired RSC) = R

Depth of irrigation = Y cm

No. of irrigations in the season = n

Desired RSC is usually taken as 2.5 meq/L or the actual value if known depending upon the soil type and crop to be irrigated. These steps can then be written in the form of a mathematical equation as follows:

Total pure gypsum required per ha= 8.6*R*Y*n/1000 t/ha

Total gypsum required (70% purity) = 8.6*R*Y*n/(1000*0.7) t/ha (10.44)

Example 10.12

Alkali water has RSC of 8.6 meq/L. The water is to be used to irrigate a crop, which needs 4 irrigations of 5 cm each. Calculate the gypsum requirement to treat the soil. Assume any value, if needed for completing the calculations.

Assuming the desired RSC as 2.5 meq/L

RSC to be neutralized = R =8.6-2.5 = 6.1

Depth of Irrigation = I = 5

Number of irrigation, n = 4

Incorporating the above parameters in eq. (10.44) and assuming gypsum purity as 75%, we get

8.6*6.1*5*4/(1000*0.75) = 1.4 t/ha

Since the efficacy of gypsum application depends upon its dissolution in the water, some guidelines for application of gypsum are as follows:

- Gypsum purity should not be less than 65%
- Gypsum needs to be powdered to pass through 30 mesh sieve
- Sometimes, application of gypsum as per requirement of the irrigation water alone may not be sufficient. If the soil has been irrigated with alkali water previously or for some other reasons it is alkali in nature, gypsum should be added to take care of soil requirement as well as to neutralize the RSC of the irrigation water. Subsequently, gypsum should be applied on the basis of irrigation water only.
- Unlike alkali land reclamation, application of gypsum in this case needs to be repeated every season or year. It is because irrigation water is used year after year resulting in regular alkalization of the soil.
- The best time for gypsum application is after the harvest of *rabi* crops, preferably in the month of May or June after some rains have occurred. After proper soil moisture conditions are attained, the soil should be ploughed to mix the applied gypsum in a shallow depth.
- Gypsum is added to the soil, being easier to accomplish. Evidences are now available to reveal that it will be more appropriate to treat the water through gypsum beds.

Gypsum Beds

Response of gypsum application is far better when applied after each irrigation than applying it on seasonal or annual basis. Since soil application after each irrigation poses several problems, technology to treat the irrigation water with gypsum has been developed so that there is no need to add gypsum to the soil. Earlier, the farmers used this technology by placing the gypsum lumps in the irrigation channel so that water picks up calcium from the gypsum clods while passing through the channel. The dissolution of gypsum in this manner was not very effective as it depended upon the channel dimensions (Length and width), RSC of irrigation water, tube well discharge and size distribution of gypsum fragments. The most practical way to treat irrigation water with gypsum is to construct a gypsum bed (Fig 10.17, left). In this method, the water structure at the outlet of the tube well is modified in such a way that water initially goes down and then rises to dissolve the gypsum placed in the chamber (Fig. 10.17, left). A constructed structure is shown in Fig. (10.17, right). If pyrites is to be applied, a similar bed can also be used for this purpose. To minimize the cost and construction difficulties, farmers also use PVC storage tanks to treat the water. In this case gypsum clods are filled in the tank. Water enters from an inlet provided at the bottom and comes out of the top, which serves as an outlet in this case.

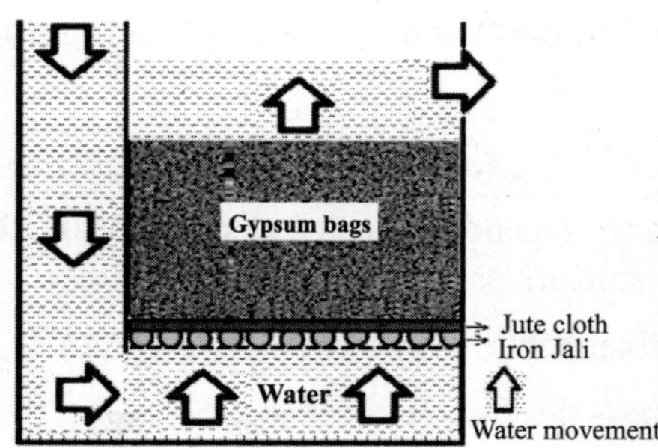

Fig. 10.17 Schematic structure of the gypsum bed (left) and a view of the constructed gypsum bed as it looks in the field (right)

Other Amendments

Besides gypsum, one can use pyrites and several industrial byproducts to treat the soil/water. Distillery spent wash and press mud (a byproduct of those sugar industries which use sulphitation method for purification of cane juice) have the potential for use to reclaim alkali water irrigated lands. On an average 1 L of distillery spent wash can reclaim about 250 L of water although exact quantum may need to be worked out for specific situations (Gupta and Gupta, 2014b).

Rainwater Management

One of the fortunate situations with continental monsoon climate of India is the concentration of rains in a short span of 2-3 months (July-September). Thus, if water penetrating into soils during this period exceeds evapotranspiration demand of crops, it induces leaching of salts added through saline irrigation. The amount and frequency of rains and soil texture basically determine salt leaching during the monsoon season. For removal of 80% of salts present in the soil, it would require 1.85, 0.95 and 0.76 cm of rainwater per cm soil depth in fine, medium and coarse textured soils respectively. Thus, annual rainfall of 250 mm can leach 80% of accumulated salts from the surface 33 cm depth of coarse-textured soils. In arid regions or where rainfall is limited, farmers prefer to grow a rain fed crop during the monsoon season to leach down the salts. Many farmers even leave the land fallow during monsoon to achieve favourable salt balance especially in the arid regions. In arid parts of Rajasthan with scanty rainfall, cultivation every alternate year is adopted as a common practice. It means a *rabi* crop is taken during alternate years, land remaining fallow during other times.

WATER QUALITY GUIDELINES FOR IRRIGATION

Based on the field experience and experimental evidences emerging from all over India, guidelines for using poor quality groundwater have been framed. A team of experts drawn from AICRP Saline Water; CSSRI, CCSHAU and PAU was constituted to frame these guidelines (Table 10.16 and Table 10.17). While guidelines for saline water are given in Table 10.16, guidelines for alkali water are given in Table 10.17.

Table 10.16 Guidelines for using poor quality irrigation water

Saline waters (RSC < 2.5 meq/L)

Soil texture (% clay)	**Crop tolerance**	**Upper limits of EC_{iw} (dS/m) in rainfall regions (mm)**		
		< 350	**350-550**	**550-750**
Fine	S	1.0	1.0	1.5
(<30)	ST	1.5	2.0	3.0
	T	2.0	3.0	4.5
Moderately fine	S	1.5	2.0	2.5
(20 – 30)	ST	2.0	3.0	4.5
	T	4.0	6.0	8.0
Moderately coarse	S	2.0	2.5	3.0
(10 – 20)	ST	4.0	6.0	8.0
	T	6.0	8.0	10.0
Coarse	S	-	3.0	3.0
(< 10)	ST	6.0	7.5	9.0
	T	8.0	10.0	12.5

S, ST and T denote sensitive, semi-tolerant and tolerant crops to salinity of irrigation water.

Special Considerations for Table 10.18

- Use gypsum when saline water has SAR >20 and/or Mg/Ca ratio >3 or is rich in silica or causes surface stagnation of water during rainy season and crops grown are sensitive to water stagnation.
- In conjunctive use mode, canal water should be used at early growth stages including pre-sowing irrigation.
- If saline water is used for pre-sowing irrigation, 20% extra seed rate and a quick post-sowing irrigation will ensure better germination.
- In low rainfall areas, it would be appropriate to leave the fields fallow during the rainy season when SAR > 20 and waters of high salinity are used to irrigate the crops during the *rabi* season.
- Additional phosphorus fertilization is beneficial when Cl/SO_4 ratio is >2.0.
- For soils having (i) shallow water table (within 1.5 m during *kharif*) and (ii) hard sub-soil layer, the next lower EC_{iw} /alternate mode of irrigation (canal/saline) is applicable.
- Use of organic materials in saline environment enhances yields.

Comments

(i) Textural criteria is applicable for all soil layers down to at least 1.5 m depth.

(ii) In areas where groundwater table reaches within 1.5 m at any time of the year or a hard sub-soil layer is present in the root zone, the limit of the next finer textural class should be used.

(iii) If in any textural class, SAR and RSC values are above the upper limits, gypsum should be used as per the requirement.

Table 10.17 Alkali waters (RSC > 2.5 meq/L and EC_{iw} < 4 dS/m)

Soil texture (% clay)	Upper limits of		Remarks
	SAR	RSC	
Fine (> 30)	10	2.5 – 3.5	Limits pertain to *kharif* fallow/*rabi* crop rotation when annual rainfall is 350-550 mm.
Moderately fine (20-30)	10	3.5 – 5.0	When the water have Na < 75% (Ca + Mg > 25%) or rainfall is < 550 mm, the upper limit of the RSC range become safe.
Moderately coarse (10-20)	15	5.0 – 7.5	For double cropping, neutralization of RSC with gypsum is essential
Coarse (< 10)	20	7.5 –10.0	

Source: Minhas and Gupta (1992)

DISPOSAL OF SALINE DRAINAGE EFFLUENTS

Disposal of saline drainage effluents has globally emerged as one of the major environmental issue. Considering the slow pace of land drainage in India it seems that such issues are not likely to crop up in near future. Nonetheless, it may be prudent to list some of the possible drainage disposal options (Sharma, 2014):

- Reuse for irrigation in and around the disposal structures or man holes (Fig. 10.18).
- Disposal in surface irrigation networks considering the network discharge and using appropriate mixing ratios.
- Disposal in rivers.
- Disposal through evaporation tanks in land locked areas.
- Disposal through irrigation of salt tolerant forest plantations or halophytic plants.
- Salt extraction.
- Disposal to the sea by constructing special drainage systems.

Fig. 10.18 Subsurface drainage water reuse for irrigation

MANAGEMENT OF TOXIC WATER

Overexploitation of groundwater in India is exerting tremendous stress on our aquifers. The quality of groundwater, which until sometime back was always considered benign, is deteriorating. Current data on water quality reveal that groundwater in many parts of India is contaminated by pollutants including heavy metals. Direct intake of such contaminated waters is causing severe health hazards to the populace. However, recent reports also reveal that the use of such contaminated water for irrigation is injecting these toxic elements into the food chain. While the drinking water essentially affects the local populations, the contaminated food may become a diffuse source of contamination since the food grains are transported to other regions. Adverse effects on crops and multiplying effect of toxicity on health due to uptake through agriculture produce have come into limelight only recently when higher than permissible arsenic has been detected in many people having no history of drinking arsenic contaminated water. It may be difficult to discuss the management of each kind of toxic element at this stage, yet few generic solutions that have emerged in the use of toxic waters are listed below:

- Selection of crops that require less water, crops that are tolerant to the toxic element and crops that absorb less toxic metal in their economic part should be cultivated. For example, if irrigation water is contaminated with fluoride, leafy vegetable and/or forage crops that accumulate fluoride should be avoided. Fluoride can enter the food chain even through cattle feeding on such forage crops.
- As far as possible, crops that do not enter the food chain such as flowers or trees of commercial and industrial importance should be preferred for cultivation.
- Conjunctive use of contaminated and fresh water appears to be a universal prescription that fits well in the safe use of all kinds of low quality waters such as saline, alkali, toxic or wastewater from other sectors of economy. Conjunctive use of fresh and toxic water blunts the adverse effects of contaminated water due to reduced load of contaminant as a result of the dilution effect. Groundwater recharge (sometimes referred as aquifer mixing) to reduce salinity, sodicity and even toxicity of water has emerged as a viable option to manage low quality groundwater.
- Application of gypsum to soils where boron rich water has been used for irrigation is a good option. Gypsum converts the sodium borate (soluble) to calcium borate (Insoluble form) making it unavailable to plants. Similarly, rice cultivation should be avoided if arsenic rich water is used for irrigation. Under the submerged conditions that result in low redox potential, arsenate is reduced to more soluble and mobile arsenite. Irrigation with sprinkler irrigation may be avoided if fluoride rich waters are used for irrigation as relatively excessive fluoride may enter the economic parts than with surface irrigation where it is retained by the roots and in other plant parts.
- Application of organic manures helps to manage fluoride or heavy metals contaminated waters due to increased adsorption potential of the soil.

- Some weed species accumulate toxic elements several times their biomass. Such species can be used for cost effective phyto-remediation of contaminated soils. *Ludwigia parviflora*, *Filmbristylis sp.*, *Ageratum conyzoides*, *Eleusine indica* are able to accumulate arsenic as much as 50 times their biomass.

MANAGEMENT OF WASTE WATER

In rural and peri-urban areas of most developing countries, the use of sewage and waste water for irrigation is a common practice. Even in areas where other water sources exist, small farmers often prefer waste water because of several reasons including its high nutrient content. Application of waste water reduces or even eliminates the need for expensive chemical fertilizers (Table 10.18). Although, volumes have been written on the pros and cons of this strategy, there is no denying the fact that this technology suits developed and developing countries the most. Across Asia, Africa and Latin America the wastewater micro-economies support countless poor people. Stopping or over-regulating these practices can remove the only livelihood source for many landless people around the world. Therefore, without disturbing or curbing the current use, efforts should be made to sensitize the stakeholders so as to minimize the health hazards to both the growers and the consumers.

Some guidelines to expand this technology with minimal impacts on the health of the growers or the consumers are as follows (Gupta & Gupta, 2014b).

- The irrigators should wear shoes and gloves while working with wastewater.
- Regularly treat the farmers and their families with antihelmintic drugs to prevent worm infections.

Table 10.18 Positive and negative effects of waste water irrigation

Positives	Negatives
It conserves water resource and provides a more reliable water supply to the cultivator	It is a health risk to the agricultural field workers and their families, crop handlers, consumers (of crops, meat and milk) and those living near affected fields
It is low-cost method for the disposal of municipal/industrial waste water	Likely contamination of groundwater
It helps to reduce pollution of rivers, canals and other surface water bodies	Build-up of chemical pollutants (heavy metals) in soils
It helps to conserve nutrients and reduces the need for fertilizer application	Excessive growth of algae and vegetation in drains/canals carrying waste water
Crop yields are increased	

- Crop restrictions should be enforced avoiding crops that are eaten raw. Although crops that are eaten after cooking can be grown but the preference be given to crops that are not consumed by humans/animals. Cultivate crops such as forest plantations/flowers.

- Practice conjunctive use of waste and fresh water.
- Withholding periods prior to harvest can reduce contamination of the products or allow sufficient time for pathogen die off.
- Adopt a suitable irrigation method such as drip irrigation. Even in drip irrigation, go for subsurface than surface drip.
- Even when surface irrigation cannot be dispensed with, broad bed-furrow method should be adopted. While plants are sown in or near the irrigation furrow, plants are spread on the bed so that the fruits don't come in direct contact with the wastewater.
- Sensitize the farmers about washing/rinsing/disinfecting the products using fresh water before it is marketed.
- Provide better information on hygiene behavior and risks of wastewater irrigation to farmers, consumers and policy planners.

QUESTIONS (THEORY)

1. What are the sources of salts in water?
2. Discuss various kinds of hazards of irrigation with saline/alkali water describing the various parameters used to assess the hazard.
3. Describe the United States Salinity Laboratory classification of irrigation water.
4. Describe the residual sodium carbonate and its role in assessing the water quality.
5. Water analysis of a given water sample is Ca =2.70 meq/L, Mg= 3.10 meq/L, Na=12.20 meq/L, CO_3+ HCO_3 =4.21 meq/L, Cl+SO_4 = 5.50 meq/L. Calculate the EC, RSC and SAR of the water.
6. Describe the infiltration related problems of soils irrigated with poor quality water. Which diagrams are commonly used to assess the likely problems?
7. Discuss the concept of critical depth to water table and its role in soil salinization.
8. Determine the change in soil salinity in 30 cm soil depth if during a period of 2 months 15 cm of water gets evaporated from the soil surface. EC of groundwater is 5 dS/m. The bulk density of the soil is 1.6 g/cm^3 and saturation percent is 50. Assume any other appropriate value required for calculations.
9. Determine the change in EC of the top 30 cm soil profile if 50 cm depth of saline water of 3 dS/m is applied and the whole water remains within this depth without any deep percolation. The bulk density of soil may be taken as 1.4 g/cc and the density of water as 1 g/cc. The soil saturation percentage is 45. What would be the depth of water applied if the salinity of the soil needs to be restricted to 3.0 dS/m?

10. Show by a neat sketch the water balance components of the root zone. Based on these components and any other salt inputs, write down the salt balance equation.

11. Simplify the salt balance equation derived in question 10 to arrive at the simplified equation of the leaching requirement.

12. What do you understand by salinization and desalinization cycle in the monsoon climatic conditions?

13. Crop consumptive requirement of the wheat crop is 250 mm. If the irrigation water has an EC of 2 dS/m and drainage water 8 dS/m, what is the depth of irrigation for this period to take care of leaching requirement? If there is 25 mm of rainfall during the period and it does not contribute to leaching, then what will be the depth of irrigation for this period?

14. Describe the method to find soil salinity profile in the root zone if salinity of irrigation water, leaching fraction and crop water uptake pattern is known.

15. What do you understand by the conjunctive use of water of different sources. Describe the various strategies of conjunctive use.

16. Name and describe various strategies for managing salt affected waters in crop production.

17. Why a chemical amendment is required to manage alkali water? Derive the equation to determine the gypsum requirement to manage irrigation water.

18. Write short notes on gypsum bed, management of toxic water and management of waste water

QUESTIONS (OBJECTIVE)

1. Fill in the blanks

1. Electrical conductivity of rain water is generally ______ than the groundwater.
2. Matric and osmotic stresses together determine the availability of ______ to the plants.
3. Adjusted SAR is always ________ than the SAR.
4. USSL diagram uses EC and ________ of the water for their characterization.
5. Residual sodium carbonate is assessed by subtracting sum of (Ca +Mg) from the sum of ________ .
6. Rhoades diagram is used to assess the problems related to __________ rate of the soils as a result of application of poor quality irrigation water.
7. The depth to water table at which relative evaporation is 0.1 or the rate is about 10% of its value when the water table is at or near the soil surface is termed as ______ depth to water table.

8.. Salt balance equation states that salt input minus salt output is equal to _______ in salt storage.

9. LR = EC_i/_______.

10. Under similar conditions, soil salinization is ________ with decreasing leaching fractions.

11. Water uptake pattern is usually taken as 40 : 30 : 20 : ______.

12. During initial years of reclamation, it is better to cultivate salt __________ crops.

13. For conjunctive use of saline and fresh water, two commonly used methods are _____ and cyclic use

14. Quantity of gypsum application to manage alkali water besides the alkalinity of the water depends upon the amount of _____________water applied

15. About 86 g of gypsum needs to be dissolved in 1 m^3 of water to add 1 meq/L of _______ .

16. Conjunctive use is also used to manage specific ion ________.

17. If 5 cm of irrigation water is to be applied and leaching requirement is 20%, then ______ of water should be applied to meet both the requirements.

18. When waste water is used for irrigation, products that are eaten ______should not be cultivated.

19. Irrigation with __________irrigation may be avoided if irrigation water contains excess of fluoride.

20 _______ is used as an amendment to manage high RSC irrigation water.

2. Tick mark correct or most appropriate answer (A, B, C or D) from the following

1 Quality of water is defined by some

(A) Physical parameters (B) Chemical parameters

(C) Biological parameters (D) All of these

2 Sodicity hazard of irrigation water is evaluated using

(A) Sodium Adsorption Ratio (SAR)

(B) Adjusted SAR

(C) Adj. R_{Na}.

(D) All of these

3 One dS/m is equal to

(A) 1 mmhos/cm (B) 1 μmhos/cm

(C) 1 mmhos/m (D) None of these

4 Constituent of water in meq/L is calculated as

(A) mg/L divided by equivalent weight

(B) mg/L divided by atomic weight

(C) Equal to mg/L

(D) None of these

5 Sum of the cations present in irrigation water is 45 meq/L, its approximate EC is

(A) 45 (B) 4.5

(C) 5.4 (D) None of these

6 Equivalent weight of calcium and sulphate are 20.04 and 48.03, the equivalent weight of calcium sulphate is

(A) 20.04 (B) 28.01

(C) 68.07 (D) None of these

7 The soil water having EC of 5 dS/m will exert an osmotic pressure of

(A) 1.8 atm (B) 55.44 atm

(C) 1.78 bar (D) Both A and C

8 4 g of calcium chloride is dissolved in 2 L of water. What is the approximate calcium concentration in mg/L

(A) 2000 (B) 4000

(C) 500 (D) None of these

9 4 g of calcium chloride is dissolved in 2 L of water. What is the calcium concentration in meq/L

(A) 72.08 (B) 36.04

(C) 99.8 (D) None of these

10 The sodicity hazard assessed by using Adj. SAR in general is

(A) Underestimate of the hazard

(B) Overestimate of the hazard

(C) Neither over nor underestimate

(D) None of these

11 As per United States Salinity laboratory, water quality is classified on the basis of

(A) EC (B) SAR

(C) Both A and B (D) RSC

12 As per United States Salinity laboratory, water quality is classified in

(A) 4 categories (B) 16 categories

(C) 8 categories (D) None of these

13 As per an Indian classification proposed by a committee of experts from CSSRI and other organizations, water quality is categorized on the basis of

(A) EC (B) SAR

(C) RSC (D) All of these

14 The formula to calculate adjusted R_{Na} is same as that of sodium Adsorption Ratio except that

(A) Ca content in water is modified

(B) Na content in water is modified

(C) Mg content in water is modified

(D) None of these

15 Rate of evaporation from groundwater increases

(A) With decreasing depth to water table from the soil surface

(B) With increasing depth to water table from the soil surface

(C) Does not change irrespective of the depth to water table from the soil surface

(D) None of these

16 The permeability hazard of irrigation water will be severe in case of

(A) Increasing SAR

(B) Increasing EC

(C) Increasing SAR and decreasing EC

(D) Decreasing SAR and increasing EC

17 if 10 cm of water having EC = 0.5 dS/m is applied, how much salts in kg/ha will be added to the soils

(A) 320 kg/ha (B) 640 kg/ha

(C) 120 kg/ha (D) None of these

18 Leaching requirement is given as

(A) Ratio of the EC of irrigation water to EC of drainage water

(B) Ratio of the EC of drainage water to EC of irrigation water

(C) Ratio of the EC of rain water to EC of drainage water

(D) None of these

19 EC of the soil solution at field capacity is nearly

(A) Same as the EC of the soil saturation extract

(B) 1.22 times the EC of the soil saturation extract

(C) 2 times the EC of the soil saturation extract

(D) None of these

20 The water uptake pattern of 40:30:20:10 means

(A) 20% of water is taken up by the crop from top 25% of the root zone

(B) 40% of water is taken up by the crop from top 25% of the root zone

(C) 30% of water is taken up by the crop from top 25% of the root zone

(D) 10% of water is taken up by the crop from top 25% of the root zone

21 The water uptake pattern of 40:30:20:10 means

(A) 40% of water is taken up by the crop from top 25% of the root zone

(B) 30% of water is taken up by the crop from the next 25% of the root zone

(C) 10% of water is taken up by the crop from the bottom 25% of the root zone

(D) All of these

22 Application of amendments is necessary for the management of

(A) Saline water (B) Alkali water

(C) Both A and B (D) None of these

23 Fresh water of 1 dS/m is to be mixed with saline water of 5 dS/m. If the salinity of blended water is 2 dS/m, in what ratio the two qualities of water should be mixed.

(A) 30:70 (Fresh to saline water)

(B) 70:30

(C) 50:50

(D) None of these

24 A crop requires 4 irrigations of 5 cm each to mature. How much gypsum of 70% purity (t/ha) should be added to soil after the season if the RSC of water is 8.5 meq/L? Assume desired RSC as 2.5 meq/L.

(A) 0.72 (B) 2.70

(C) 1.47 (D) None of these

25 To add 1 meq/L of calcium in 1 ha-cm of water one needs to add _____ kg of 100% pure gypsum

(A) 86 (B) 8.6

(C) 0.86 (D) None of these

26 Conjunctive use technologies comprise of

(A) Blending (B) Cyclic use

(C) Seasonal cyclic use (D) All of these

27 From use of saline water point of view, the most preferred method of irrigation is

(A) Drip (B) Sprinkler

(C) Surface (D) Any of these

28 Blending of different sources of water can be done

(A) In canal network (B) On-farm structure for blending

(C) Aquifer blending (D) All of these

29 To manage alkali water which of the following amendments is least effective?

(A) Gypsum (B) Pyrites

(C) Distillery spent wash (D) Lime

30 Under similar agro-climatic and management conditions, relatively higher EC water compared to others can be used in

(A) Light textured soil (B) Medium textured soil

(C) Heavy texture soil (D) Both B and C

31 Residual sodium carbonate (RSC) of water is calculated by

(A) RSC = $(CO_3 + HCO_3) – (Ca + Mg)$

(B) RSC = $(Ca + Mg) - (CO_3 + HCO_3)$

(C) RSC = $(Ca + CO_3) - (Mg + HCO_3)$

(D) None of these

32 If all ions are expressed in meq/L then Residual Sodium Carbonate (RSC) has the units of

(A) meq/L (B) mg/L

(C) A number with no units (D) None of these

33 If the sum of the cations in a water is 28 meq/L, the approximate EC of water in dS/m would be

(A) 0.28 (B) 28

(C) 280 (D) 2.8

34 Sum of the anions and cations when expressed in meq/L, should be

(A) Anions > cations (B) Anions = cations

(C) Cations > anions (D) None of these

35 Leaching requirement is given as (D and EC are depth and electrical conductivity of water, subscripts i and d denoting irrigation and drainage water)

(A) D_d/Di (B) EC_i/EC_d

(C) D_i/D_d (D) Both A and B

■■■■

Chapter 11

ECONOMIC ANALYSIS AND SOCIO-ENVIRONMENTAL ISSUES

Land drainage is practiced to enhance productivity of the cultivated lands or to bring new lands or lands gone out of cultivation back into production. Since, investments on land drainage are being made with an intention to receive future benefits, a sound economic analysis is essential to judge whether the investment decision is economically and financially sound (Fig. 11.1). This chapter describes the information that must be assembled to conduct a worthwhile economic or financial analysis. It also describes the procedure(s) to conduct the economic and financial analysis and discusses some of the social and environmental issues that impact the economics of drainage projects.

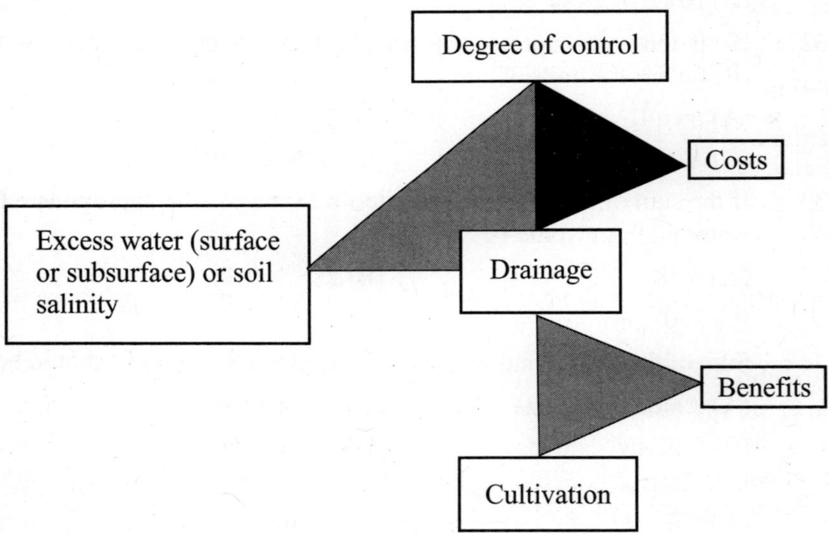

Fig. 11.1 Inter-relationship between drainage intensity, costs and benefits

COSTS

The eight major components of costs of a subsurface drainage system are:

1. Field surveys and investigations for DPR preparation including tendering
2. Drainage materials
3. Installation

4. Supervision of material and installation quality
5. Operation and maintenance
6. Contingencies
7. Farmers' awareness training including field visits
8. Monitoring and evaluation (M&E)

Preparation of Detailed Project Report (DPR) and Tendering

Assuming fairly well acquaintance of the team to the drainage area, cost of drainage investigations, detailed design, layout and budgeting including tender document preparation and evaluation of the tenders for final award of the work may be taken @ Rs. 2,500/ha.

Drainage Materials including Structures and Pump Set

Costs of drainage materials and installation of drain pipes and structures constitute major components of overall cost of a SSD project. Drainage materials include lateral and collector pipes, envelopes, manholes and sumps, pump set and materials for outlet pitching, etc. Structures comprise of manholes (inspection chambers/ junction boxes), sump/outlet structure and pump house (Table 11.1). In large projects, other cross drainage structures may also be required but these are not included in this chapter.

Installation

Installation of SSD system is a joint work between the contractor or installation unit, manufacturers/suppliers of quality drainage materials, and supervisors (quality control) under the overall supervision of the implementing authority. The task of the contractor/ installation unit is to install the drainage system in accordance with design and specifications (BIS, 1999). The supervisors have to ascertain that the installation (including the drainage materials) is done strictly in compliance with the design and specifications.

Different types of drainage machineries *viz.* tractor mounted trencher, hydraulic excavator, self-propelled trencher with automatic laser control, and self-propelled trenchless machines with V or L plough have been used to install subsurface drain pipes (Nijland *et al.,* 2005; Broughton and Fouss, 1999). The one time investments required to arrange the machinery is discussed in the next section.

Supervision of Material and Installation Quality

A well installed subsurface drainage system can function properly for a period of 50 years or even longer although economic life is usually taken as 30 years. Therefore, a quality assurance system for projects is to be put in place to ensure total quality for its effective installation and functioning (Nijland *et al.,* 2005). Provision of supervision charges for material and installation quality @ Rs. 4,000 per hectare is made in the project.

Operation and Maintenance

The common belief that a subsurface drainage does not require any operation and maintenance is untenable. In a pumped drainage system, the operation of pumps is essentially required at least in first few years of installation. Pumping charges for dewatering and first year leaching @ Rs. 500/- per ha may be allocated in the project. Pumping charges for subsequent years and maintenance charges should be borne by the farmers' drainage society. Maintenance of subsurface drainage systems mainly requires removal of sediment from the manholes, sump and pipes, and repairing, if necessary by replacing these pipes, manholes and outlet pipe (Nijland *et al.,* 2005). In controlled drainage systems, the operations can also involve opening and closing of gates for reuse of drainage water for crop irrigation.

Table 11.1 Quantities of drainage materials for 40 ha model drainage block with pumped/gravity outlet

S No	Drainage material/ installation	Material Quantity (metres /No) under different lateral drain spacing								
		20 m	30 m	40 m	50 m	60 m	67 m	80 m	90 m	100m
1	Lateral pipe, 80 mm	20,000	13,600	10,000	8,000	6,800	6,000	5,200	4,800	4,000
2	Collector pipe, 160 mm	400	390	360	354	360	360	324	270	300
	Collector pipe, 200 mm	330	330	324	354	300	300	324	360	346
	Collector pipe, 250/294 mm	270	282	324	300	342	342	360	372	354
3	Non-woven geotextile envelope for 80 mm	20,000	13,600	10,000	8,000	6,800	6,000	5,200	4,800	4,000
4	Woven nylon sock for 160 mm pipe	400	390	360	354	360	360	324	270	300
	Woven nylon sock for 200 mm pipe	330	330	324	354	300	300	324	360	346
	Woven nylon sock for 250/294 mm pipe	270	282	324	300	342	342	360	372	354
5	Manholes with ladder	3	3	3	3	3	3	3	3	3
6	Sump well	1	1	1	1	1	1	1	1	1
7	Pump house	1	1	1	1	1	1	1	1	1
8	Dewatering diesel pump set (6 HP)	1	1	1	1	1	1	1	1	1
9	Outlet with masonry structure and guard	1	1	1	1	1	1	1	1	1
10	Pipe installation length by trencher	21,000	14,602	11,008	9,008	7,802	7,002	6,208	5,802	5,000
11.	Dewatering hours	360	360	360	360	360	360	360	360	360

Note: For pumped outlet system, item 9 is not required and for gravity outlet, items 7, 8 and 11 are not needed.

Contingencies

The cost estimate is approximate and may vary within 5-8% either way when compared to actual implementation is undertaken. Therefore, provision of contingency cost @ 5% of drainage material and installation cost is made to take care of unforeseen expenses in accordance with the Central Public Works Department (CPWD) norms. It ensures that the project construction is not delayed due to lack of minor funds required for some unforeseen activities.

Farmers' Awareness Training

Farmers' awareness training including field visits to drainage area is required to sensitize them on operation and maintenance of SSD systems so that they manage the system themselves. Training charges @ Rs. 750/- per ha (~one per cent of total project cost) may be allocated for this purpose.

Monitoring and Evaluation (M&E)

It is not necessary that all SSD projects should be monitored and evaluated. Nonetheless, in the initial stages representative segments of all projects with area more than 250 ha should be independently monitored and evaluated using standard criteria by a third party (Nijland *et al.,* 2005; BIS, 1999). A national/state level organization having experience in agricultural land drainage works may be roped in for this purpose. Most SAUs have departments that have such capabilities and could be entrusted with the tasks of monitoring and evaluation. The charges @ Rs. 2,000/- per ha (~2% of total cost of the system) should be allocated for this purpose.

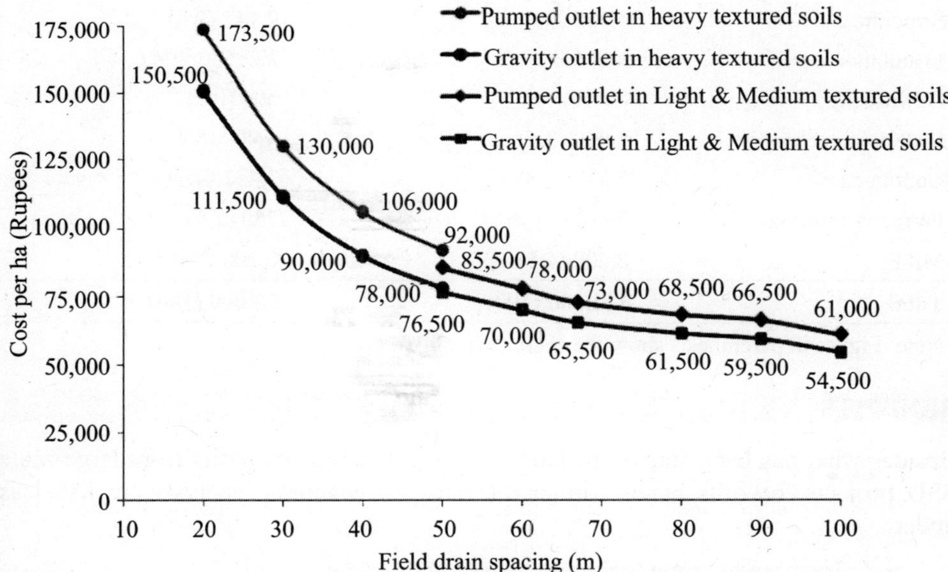

Fig. 11.2 Cost of SSD with M&E for pumped and gravity outlets in light and medium and heavy textured soils

The estimated cost on the basis of the above suggested norms is given in Table 11.2 (Bundela *et al.*, 2016). The contractor's margin of profit is also included in the cost. A complete range of costs for heavy and light textured soils with M & E charges for pumped and gravity outlets is shown in Fig. 11.2 (Bundela *et al.*, 2016) showing wide variations mainly depending upon drain spacing.

INVESTMENT ON MACHINERY

Subsurface drainage, an engineering intervention, is accomplished manually or by semi-mechanical operation or by fully mechanical operation with drain trenching machines. While minimum investments are needed in manual construction, large investments are needed in a fully mechanical construction. The items of investments needed for the fully automatic pipe laying operations using one trencher machine and semi-mechanical means is given in Table 11.3. It may be noted that besides these investments, relatively more investments are also needed in the mechanical laying because sufficient quantities of drainage materials and spares need to be stored.

Table 11.2 Cost of SSD for a heavy soils with gravity outlet (30 m spacing) and medium textured soils with pumped outlet (67 m spacing)

SSD Component	Cost (Rs./ha)	
	Heavy texture soils (30 m spacing) with gravity outlet	**Medium texture soils (67 m spacing with pumped outlet)**
DPR	2,500 (2.2)	2,500 (3.7)
Pipes & fittings	41,989 (37.6)	21,750 (29.7)
Filters	14,776 (13.2)	8,334 (11.4)
Structures	3,785 (3.4)	7,142 (9.8)
Installation cost	36,766 (33.0)	22,953 (31.4)
Dewatering	0 (0.0)	500 (0.7)
Supervision charges	4,000 (3.5)	4,000 (5.4)
Contingency	5,352 (4.8)	3,431 (4.7)
Farmer's training	750 (0.7)	750 (1.0)
M&E	2,000 (1.8)	2,000 (2.7)
Total	**111,500** (100)	**73,000** (100)

Note: Figure in parenthesis show the share (%) of total cost

BENEFITS

Besides what has been stated in Chapter 1 on the possible benefits from large-scale SSD projects, benefits having direct relevance to economic analysis are listed as under:

- Increase in crop yield to the extent of 40-50%
- Improved product quality and higher profits
- Reduced cost of farm operations and inputs
- Higher cropping intensity
- Advancement of farming operations for sowing of post monsoon crops resulting in higher yield
- A change in cropping pattern with introduction of more paying crops
- A source of irrigation

Table 11.3 Investments needed in mechanized pipe drain installation with one trencher and semi-mechanical means

Mechanized installation		**Semi-mechanical installation**	
Type of equipment	**Units**	**Type of equipment**	**Units**
Trencher (27 tonnes)	1	Hydraulic crawler excavator (16 tonnes)	2
Hydraulic crawler excavator (16 tonnes)	1	Hydraulic truck crane	1
Bulldozer (10 tonnes)	1	Wheel tractor	2
Low-loader for trencher, bulldozer and hydraulic excavator for transport	1	Trailer (4 wheel, 4 tonnes)	1
Wheel tractor (35 HP)	1	Bulldozer (10 tonnes)	1
Wheel tractor (45 HP)	1	Truck (7.5 tonnes)	1
Pick-up truck (4 wheel drive 1 tonne)	1	Total Station	2
Truck (5 tonnes)	1	Car	1
Truck (7.5 tonnes)	1	Jeep	1
Mobile workshop van (1.5 tonnes)	1	Dewatering pumps	4
Fuel trailer (4 wheel 2 tonnes)	1		
Diesel fuel tank (800 L)	1		
Trailer (4 wheel box size 2 m x 4 m)	1		
Trailer (4 wheel 4 tonnes)	1		
Portable generators	2		
Dewatering pumps	4		
Laser grade control* units	1		
Two-way radios	2		
Total station	2		
Car/Jeep/Bus (20 seated)	1 each		
Spare parts (20% of the cost)			

Like indirect costs, all indirect benefits must be evaluated. Some of the direct and indirect benefits from land drainage projects are given in Table 1.5 of Chapter 1. Usually, it is difficult to assign numbers (benefits in terms of economic value) to the indirect benefits. Both indirect costs and benefits could be left out from initial financial analysis, if numbers are difficult to assign (Gupta, 2014).

The principal benefit of land drainage is reflected in yield increase (Table 11.4). The total income resulting from direct increase in yields can be calculated by the following relation:

$$\sum_{i=1}^{n} (\text{Ao Yo Po} - \text{Ao Co}) / \sum_{i=1}^{n} \text{Ao} - \sum_{i=1}^{n} (\text{Ai Yi Pi} - \text{Ai Ci}) / \sum_{i=1}^{n} \text{Ai} \quad (11.1)$$

Table 11.4 Effect of subsurface drainage system on grain and straw yield of rice in Andhra Pradesh

Year	Grain yield (t/ha)	Straw yield (t/ha)
Pre-installation	1.80	2.10
2002 – 2003	4.10	5.50
2003 – 2004	4.24	5.90
2004 – 2005	6.40	7.40
2005 – 2006	5.78	7.32
2006 - 2007*	2.09	4.94
2007 – 2008	6.64	7.43
2008 – 2009	6.75	7.50

*Lower values are due to Oogni cyclone

Where Ai is the area under i_{th} crop, Yi is the yield of the i_{th} crop, Pi is the market price of the i_{th} crop and Ci is the cost of cultivation of the i_{th} crop per ha before land drainage. Ao, Yo, Po and Co have the same meaning but after land drainage.

Besides this direct increase, incremental benefits are also obtained in the yield through reduction in fallow land thereby resulting in increased cropping intensity and change in crop mix (Table 11.5). This fact can also be reflected easily in Eq. (11.1). In a recent study, Raju *et al.* (2016) reported an overall increase in cropping intensity by 20% and reduction in fallow land by 26.6%. Although, it is difficult to assign numbers to the benefits on year to year basis yet the most appropriate way is to evaluate the project on the basis of average benefits on account of average increase in crop yield. It may be noted that the full anticipated benefits cannot be obtained in the first year of the project. Usually, it takes 3-4 years before the full benefits of the project emerge (Fig. 11.3), although in few cases full benefits have been reported to emerge in 2 years. Under this situation and to be practical, the anticipated benefits should be spread over 2-4 years as per the expected pace of reclamation. After this period, it would be safe to assume that the benefits achieved are sustained for the remaining period (IPTRID, 2001).

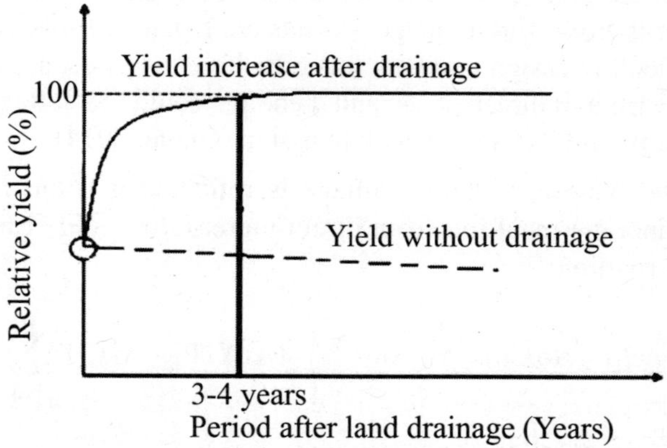

Fig. 11.3 Likely post-implementation benefits of drainage

Table 11.5 Increase in cropping intensity at various project sites

Place (State)	Cropping intensity (%)		
	Before drainage	**After drainage**	**Increase over pre-drainage**
Sampla (Haryana)	0	200	-#
Ismaila (Haryana)	73	148	103
Gohana (Haryana)	117	175	50
Konanaki (Andhra Pradesh)	70	90*	29
Uppugunduru (Andhra Pradesh)	130	165*	27
Islampur (Karnataka)			
ORP	0	200**	-#
Phase II	88	156***	77

*Cropping intensity did not increase as envisaged because of limited water resources; **In third year; ***In second year; # could not be calculated because initially area was not cropped

Time Period

When considering time period, one must distinguish between the economic life and the financial life of a project. The actual life of a well maintained drainage project is anywhere between 50 and 100 years (Gupta *et al.*, 1995). In economic terms, future costs and benefits occurring say after 30 years are insignificant. For example; the discounted benefit for the 12% discount rate in the 30th year would be 0.033 times the actual benefit. It would further go down thereafter. As such, the economic life of drainage projects is generally taken as 30 years.

Another important issue in an economic evaluation is the financial life of the project often imposed by the financing organizations. A farmer having obtained the loan from a financial institution must pay back the loan in the pre-decided duration say in 10 years. In that event, cash flow must exceed the costs during the financial life. Although the project has a terminal value in this case, yet it has little meaning to the farmer who is concerned with cash flow during the financial life of the project. In projects where farmers' contribution is limited or they are not supposed to contribute to the project, one may forego the financial evaluation part. The economic analysis is an indirect indicator of the financial viability of the project.

ECONOMIC ANALYSIS

Discounting

The costs and benefits of a land drainage project occur at different times during the project period. The amount available today is worth more than the same sum of money received 10 years after. It means present worth of benefits say after 10 years would be less than the ones received today. Same is true for costs although major component of cost is made in the first year. Such variation is taken into account by following the discounted cash flow method of analysis. The discount factors (DF) are calculated for a given discount rate as per the following relation:

$$DF = 1/\,(1+Dr/100)^t \tag{11.2}$$

Where Dr is the discount rate (%) and t is the year say 1, 2, 3 …..30. For t=1 and Dr = 6%, DF would be:

$$DF = \frac{1}{1+0.6} = 0.943$$

Discount factors for a discount rate of 8% over a period of 30 years are given in Table 11.6. Usually discount rates vary from a low of 6% to about 12%. Current bank rates on fixed deposits are good indicators of the discount rates to be adopted in economic evaluation. Since bank deposits rates fluctuate, it may be worthwhile to make the analysis using 3-4 discount rates (Less and more) to derive most appropriate information. For example; for the current rate of interest of 7% on fixed deposits, one can carry out the economic analysis for discount rates of 5.0%, 7.0% and 9.0%.

Table 11.6 Discount factors at a discount rate of 8%

Year	Discount factor	Year	Discount factor	Year	Discount factor	Year	Discount factor
1	0.926	9	0.501	17	0.270	25	0.146
2	0.857	10	0.464	18	0.251	26	0.135
3	0.784	11	0.429	19	0.232	27	0.125
4	0.735	12	0.397	20	0.215	28	0.116
5	0.607	13	0.368	21	0.199	29	0.108
6	0.681	14	0.341	22	0.184	30	0.099
7	0.584	15	0.316	23	0.171		
8	0.541	16	0.292	24	0.158		

Inflation

Since the future costs and benefits are usually calculated in terms of present prices, in many cases, it may be desirable to take care of inflation. Although inflation in most calculations is not accounted yet it should be possible to account inflation by increasing the costs and benefits by an assumed inflation rate.

EVALUATION INDICES

The costs and benefits are evaluated on the basis of following indices.

Simple Rate of Return (SRR)

SRR of a project is used to compare with the rate that has been fixed as cut off criterion or the minimum rate of return. The SRR is calculated by the following formula:

$$r = \frac{E-D}{I} * \frac{100}{n} \tag{11.3}$$

Where, r is the SRR, E is the annual net benefit, D is the annual depreciation including O & M costs, I is the initial capital investment and n are the number of years.

Net Present Worth (NPW)

It is also called Net Present Value (NPV). It is the difference between the present values of benefits over the present values of costs. To estimate the NPW, cost involved and benefits realized on any future dates are assigned equivalent present values using discount factors. Both net present worth of benefits and net present worth of costs are calculated. The difference gives the NPW of the project.

$$\text{NPW of benefits} = \sum_{t=1}^{n} \frac{B_t}{(1+i)^t} = \sum_{t=1}^{n} B_t * DF_t \tag{11.4}$$

Here B_t is the benefit received in t^{th} year; n is the number of years; DF_t is the discount factor in t^{th} year (Eq. 11.2). Other factors are as described previously.

$$\text{NPW of costs} = \sum_{t=1}^{n} \frac{C_t}{(1+i)^t} = \sum_{t=1}^{n} Ct * DF_t \tag{11.5}$$

Here C_t is the costs incurred in t^{th} year. Other factors are as described previously.

Subtracting NPW of costs from NPW of benefits, we get NPW as:

$$\text{NPW} = \sum_{t=1}^{n} \frac{B_t - C_t}{(1+i)^t} \tag{11.6}$$

If the NPW is positive, it indicates the economic viability of the project.

Benefit-Cost Ratio (B:C ratio)

It is the ratio of the present value of benefits (Eq. 11.4) divided by the present value of costs (Eq. 11.5). For an economically viable project, the ratio should be more than one.

Payback Period (PBP)

'Payback period is the time period for an investment to generate sufficient incremental cash to recover its full initial capital outlay'. Thus, the PBP measures number of years it will take for the net undiscounted benefits to repay the investment. If the cash flow is uniform over the years, the payback period, *P* is given by

$$P = (I/E) \tag{11.7}$$

Where *I* is the initial investments and *E* is the annual net benefits.

Internal Rate of Return (IRR)

'IRR is the rate of return which a project is likely to earn over its useful life'. Practically, 'IRR is the discount rate at which NPW is zero and B-C ratio is one'. It

means it is the rate at which project just breaks even. To determine IRR, a trial and error procedure is used. An IRR is assumed. If the NPW of the project is positive at a given discount rate, its IRR will be more than the assumed discount rate. If the NPW is negative, the discount rate will be less than the assumed discount rate. Since it is difficult to pinpoint the rate at which NPW is zero, a guess is made once two discount rates at which NPW is positive and negative are known. IRR is then approximated by the following relation (Tripathi, 2007):

$$\mathrm{IRR} = D_l + (D_h - D_l) * \frac{N_l}{|(N_h - N_l)|} \tag{11.8}$$

Here, D_l is the lower discount rate; D_h is high rate; and N_l and N_h are the NPWs at the lower and higher discount rate. A check on NPW is made with the assessed IRR using eq. (11.6). Minor adjustments might be needed in the assessed value.

It is not possible to give a fixed value of IRR at which the project is viable. The calculated IRR is compared with the minimum acceptable rate of return (say prevailing market rate). If the IRR is higher or equal to the minimum acceptable rate of return, the project is worthy of investment. In most drainage projects, all the three indices NPW, B-C ratio and IRR are determined to assess the economic viability of the project.

Other Parameters

In few situations, it may be difficult to assess the benefits. In that case it may be useful to approach the problem from the other end. Here, the minimum benefits that must be obtained to cover the investments are calculated. For this purpose following two parameters are commonly used.

Break Even Benefits (BEB)

Break even benefit is the minimal benefit required to meet the costs. It is calculated as follows (Smedema & Rycroft, 1983):

$$\mathrm{BEB} = \text{NPW of costs} / \sum_{i=1}^{n} \text{ discount factors} \tag{11.9}$$

For a project to be viable, the real undiscounted average benefits per year should exceed the BEB.

Equated Seasonal Installment (ESI)

Cost of the borrowed capital can be readily estimated from the known interest rate and the period with in which the loan has to be repaid. The equated seasonal/annual installment, (ESI) is calculated by the following formula.

$$\mathrm{ESI} = (L * I) * [(1+I)^M / \{(1+I)^M - 1\}] \tag{11.10}$$

Here, M is the loan period in seasons (Years * 2); I is the interest rate (Interest rate per annum /2) /100; L is the borrowed capital in Rs.

If the capital is invested from self-savings, the cost might be higher depending upon the alternative investment opportunities commonly known as opportunity costs of investment. It should be assessed from the likely returns from the alternative use of the capital. Sometimes, an average interest rate of the borrowed and estimated interest rate of self-funds is taken as the interest rate on the total capital to calculate the virtual ESI. For a project to be financially viable, the seasonal/annual benefits must exceed the seasonal/annual installment.

Example 11.1

A farmer obtains a soft loan of Rs. 110,000 at an interest rate of 6% per annum. He has to pay ESI for 10 years. What should be the minimum additional earnings per season so that he is able to repay the loan installments?

Incorporating the values in Eq. (11.10), we have

$$ESI = (110000 *.03) * [(1+0.03)^{20} / \{(1+0.03)^{20} -1\}]$$

$$ESI = \text{Rs. } 7394/-$$

It means the farmer must earn a minimum additional income of Rs. 7394/- from the land drainage.

CALCULATION OF INDICES

Various indices mentioned in the previous sections can be calculated either manually (calculator) or with the help of a computer. In the latter case financial functions of the Microsoft Excel are used to arrive at few of the indices. A full calculation procedure is shown in Table 11.7. Various steps used in preparing this table are listed as follows:

- Column 1 shows the number of years varying from 1 onwards. First year may also be written as zero year (See Table 11.8)
- In Column 2, all the costs incurred are entered year wise, although the major investment is made in the year 1. This investment is assigned to year 1. O & M costs are entered in Column 3. The sum of the costs (sum of values in columns 2 and 3) are written in column 4
- An appropriate discount rate (DR) is used and entered in column 5. Year wise discount factors (DF) are calculated as per eq. (11.2) and entered in column 6.
- Discounted cost for each year is calculated by multiplying total costs (column 4) by DF (column 6 and entered in column 7.
- All benefits accruing due to projects are entered in column 8.
- Discounted benefits in column 9 are the values obtained by multiplying values in column 8 and DF in column 6.
- Cash flow for each year is calculated by subtracting the total costs (Column 4) from net benefits (column 8). Note the first value as negative since initial investments in first year are higher than the net benefits.

Table 11.7 Calculations of B:C ratio, NPW, IRR and payback period for a drainage project

Year	Cost	O& M cost	Total cost	DR	DF	DC	Net benefits	DB	Cash flow
1	2	3	4	5	6	7	8	9	10
1	50000	0	50000	10	0.9091	45455	14000	12727	-36000
2	0	2250	2250	10	0.8264	1860	15000	12396	12750
3	0	2250	2250	10	0.7513	1690	16500	12396	14250
4	0	2250	2250	10	0.6830	1537	16500	11269	14250
5	0	2250	2250	10	0.6209	1397	16500	10245	14250
6	0	2250	2250	10	0.5645	1270	16500	9314	14250
7	0	2250	2250	10	0.5132	1155	16500	8467	14250
8	0	2250	2250	10	0.4665	1050	16500	7697	14250
9	0	2250	2250	10	0.4241	954	16500	6998	14250
10	0	2250	2250	10	0.3855	867	16500	6361	14250
11	2500	2250	4750	10	0.3505	1665	16500	5783	11750
12	0	2250	2250	10	0.3186	716.92	16500	5257	14250
13	0	2250	2250	10	0.2897	651.74	16500	4779	14250
14	0	2250	2250	10	0.2633	592.50	16500	4345	14250
15	0	2250	2250	10	0.2394	538.63	16500	3950	14250
16	0	2250	2250	10	0.2176	489.67	16500	3591	14250
17	0	2250	2250	10	0.1978	445.15	16500	3264	14250
18	0	2250	2250	10	0.1799	404.68	16500	2968	14250
19	0	2250	2250	10	0.1635	367.89	16500	2698	14250
20	0	2250	2250	10	0.1486	334.45	16500	2453	14250
21	2500	2250	4750	10	0.1352	641.87	16500	2230	11750
22	0	2250	2250	10	0.1228	276.40	16500	2027	14250
23	0	2250	2250	10	0.1117	251.28	16500	1843	14250
24	0	2250	2250	10	0.1015	228.43	16500	1675	14250
25	0	2250	2250	10	0.0923	207.67	16500	1523	14250
26	0	2250	2250	10	0.0839	188.79	16500	1384	14250
27	0	2250	2250	10	0.0763	171.62	16500	1259	14250
28	0	2250	2250	10	0.0693	156.02	16500	1144	14250
29	0	2250	2250	10	0.0630	141.84	16500	1040	14250
30	0	2250	2250	10	0.0573	128.94	16500	946	14250
31	0	2250	2250	10	0.0521	117.22	16500	860	14250
			122500				507500		385000
						65950.93		152891	
									IRR=38%

O & M: Operation and maintenance, DR, DF, DC and DB are the discount rate, discount factors discounted costs and discounted benefits

Following calculations are then made.

Benefit: Cost ratio = ∑Discounted benefits/∑Discounted costs =

∑Column 9/∑column 7 = 152891/65950 = 2.32

NPW = Discounted benefits – Discounted costs = 86941

Payback period = 3 years

IRR = 38%

The values of NPW and B-C ratio for short-term say for 10 years can also be calculated in the same fashion by truncated the calculation at 10 years.

The average annual break even benefit is calculated as

Present value of costs for 10 years = 57235

$$\sum_{i=1}^{10} \text{Discount factor} = 6.14$$

Annual break even benefit = 57235/6.14 = 9321.7

Since the real average benefits during the 10 years are 16,100 (sum in column 8), more than the break even benefits, the project is financially viable.

CASE STUDIES OF ECONOMIC ANALYSIS

Several studies on financial viability of SSD projects in India have been conducted. At all places, it has emerged that drainage has positive impacts and is the socio-economically viable technology. For the first ever large-scale drainage project in Rajasthan (RAJAD), India B-C ratio was found to be 3.2 (Table 11.8). It may be noted that in this case lands were not barren and farmers were already cultivating these lands albeit low productivity resulting in minimal benefits and posing several other limitations. Therefore, benefits that occurred due to subsurface drainage are only considered in calculating the various economic parameters (RAJAD, 1995). From Table 11.8, it emerges that

Benefit – Cost Ratio = 114325/35400 = 3.23

Net Present Worth (NPW) = 114325 – 35400 = 78925 Rs./ha

The economic parameters for few other projects in various states are given in Table 11.9. Following observations can be made:

- Benefit : Cost ratio is more than 1.0 at all places and varies from 1.2 to 4.0
- The payback period in most cases is 3 years but may range from 3-4 years
- Internal rate of return is high to very high making this intervention an economically viable option to reclaim waterlogged saline lands.

Later projects implemented under various agro-climatic conditions more or less prove these observations.

Table 11.8 Benefit-cost analysis for the RAJAD project

Year	Discount factor at 15%	Total cost (Rs.)	Present worth of total cost (Rs.)	Net benefit without SSD (Rs./ha)	Net benefit with SSD (Rs./ha)	Net benefit due to SSD	Present worth of net benefit (Rs./ha)
0	1	34250	34250	6120	6120	0	0
1	0.87	175	152	6059	10270	4211	3664
2	0.76	175	133	5998	15208	9210	7000
3	0.66	175	116	5938	20935	14997	9898
4	0.57	175	100	5879	27449	21570	12295
5	0.5	175	88	5820	27449	21629	10815
6	0.43	175	75	5762	27449	21687	9325
7	0.38	175	67	5704	27449	21745	8263
8	0.33	175	58	5647	27449	21802	7195
9	0.28	175	49	5591	27449	21858	6120
10	0.25	175	44	5535	27449	21914	5479
11	0.21	175	37	5480	27449	21969	4613
12	0.19	175	33	5425	27449	22024	4185
13	0.16	175	28	5371	27449	22078	3532
14	0.14	175	25	5317	27449	22132	3098
15	0.12	175	21	5264	27449	22185	2662
16	0.11	175	19	5211	27449	22238	2446
17	0.09	175	16	5159	27449	22290	2006
18	0.08	175	14	5107	27449	2342	1787
19	0.07	175	12	5056	27449	22393	1568
20	0.06	175	11	5005	27449	22444	1347
21	0.05	175	9	4955	27449	22494	1125
22	0.05	175	9	4905	27449	22544	1127
23	0.04	175	7	4856	27449	22593	904
24	0.03	175	5	4807	27449	22642	679
25	0.03	175	5	4759	27449	22690	681
26	0.03	175	5	4711	27449	22738	682
27	0.02	175	4	4664	27449	22785	456
28	0.02	175	4	4617	27449	22832	457
29	0.02	175	4	4571	27449	22878	458
30	0.02	175	4	4525	27449	22924	458
TOTAL		**39500**	**35400**	**163818**	**793656**	**629838**	**114325**

Source: Garg and Bhargava (2001)

Table 11.9 Economic viability of the composite subsurface drainage systems

Economic parameter	Segwa (Gujarat)	Konanki (Andhra Pradesh)	Uppugunduru (Andhra Pradesh)	Lakhuwali (Rajasthan)	Islampur (Karnataka)	Sampla (Haryana)
Net present worth (Rs.)	74957	36655	50872	51500	14192	7500
Benefit-cost ratio	1.7	2.8	2.5	3.2	1.2	3.99
Internal rate of return (%)	58	32	36	39	20	52.6
Payback period (Years)	3	3	3	4	3-4	3

Source: Anonymous (2002) and Ritzema (2009)

Sensitivity Analysis

The sensitivity analysis is carried out to determine how some change in cost factors or benefits are likely to alter the project economic/financial viability. It is usually done by assessing the economic viability with increased cost of drainage materials say by 100 or 200% or with decreased benefits due to unforeseen circumstances say by 50 or 100%. The other way to do a similar analysis is shown in Fig. 11.4, which reveals that B-C ratio increases linearly with increase in yield and is equal to one if yield increases due to drainage is 0.41 t/ha. Apparently, if the increase is less than this value, project is not financially viable. The benefit-cost ratio is 2.45 with an additional yield of 1.0 t/ha. These calculations have been carried out for a project life of 20 years and a discount rate of 12%. The costs are held at a constant level, while benefits (yield in this case) is increased to calculate the B:C ratios (Mathew, 2003).

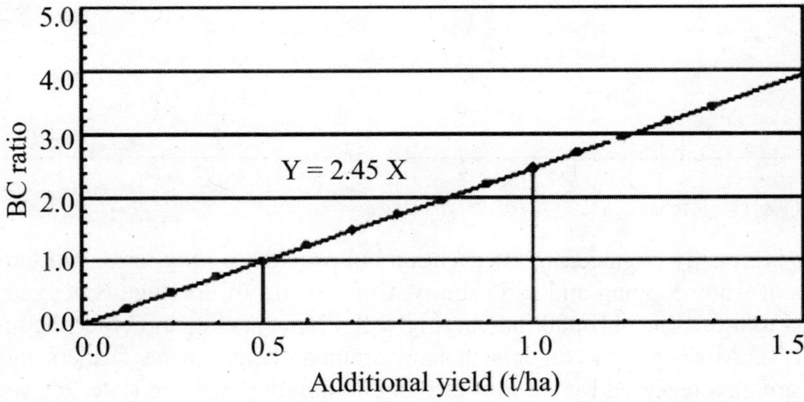

Fig. 11.4 Sensitivity analysis of a land drainage project in Kerala

COMPARISON OF DRAINAGE SYSTEMS

We have several options to decide the kind of drainage system to be implemented in an area. Similarly, we have few alternate designs to choose one from these alternatives. The investments required as well as benefits emerging from each of them may vary. The question, therefore, is to compare the relative economics of various options. An economic analysis of the options is made. However, in this case only discounted costs are compared to take a decision although, only complete analysis will yield the optimal answer.

Let us consider a situation, where a subsurface (System A) and a surface (System B) drainage system are employed to solve the drainage problem. The initial cost of SSD is high whereas its operation and maintenance cost is less. On the contrary, the initial cost of surface drainage is less but its operation and maintenance cost is high. The costs over a thirty years period are assessed and the discounted costs of the two systems are calculated, summed and compared. For example, the notional data in Table 11.10 reveal that over a period of 30 years, the cost of SSD is Rs. (A1-B1)/ha more than system B if A1>B1. Therefore, the SSD system in this case to be financially attractive must result in higher income by more than Rs. (A1-B1)/30 per ha. However, if benefits from the two systems can be assessed than one must go for full economic analysis as in Table 11.8. It may be noted that data in this table represents a comparison between a subsurface drainage system and the no drainage system, the former being economically attractive than leaving the land without a drainage system.

Table 11.10 Method to compare two systems or designs

Year	Undiscounted costs		Discounted costs	
	System A	**System B**	**System A**	**System B**
1	50000	12000	45455	10909
2				
30	20	500	1.15	28.65
			A1	B1

COST RECOVERY

Drainage is a costly engineering intervention but provides a long-term solution to the problems of water logging and soil salinity. Current cost of drainage is pegged at Rs. 70,000/ha to more than 110000/ ha varying with drain spacing and other factors (See Table 11.2). Although investments in land drainage benefits the farmers most yet benefits are also received by the non-farming community and the state. It may be in the form of employment generation, reduced expenditure on health services and increased production leading to regional food self-sufficiency. The reduced maintenance cost of public property can also be taken into account while assessing the benefits to the state. It is therefore, unreasonable to recover 100% of the drainage cost from the farmers. The minimum recoveries should cover the operation and maintenance costs. The cost to be recovered should be uniformly distributed in proportion to the land holdings. The Egyptian policy is to recover the whole cost. No recovery is made during the first 3 years. Thereafter, the actual cost is recovered in installments over a period of 20 years.

Subsidy

There is ample justification to grant subsidy to the farmers who wish to undertake land drainage activity by themselves. The following factors justify the grant of subsidy:

- The damage to the lands is attributed to canal irrigation for which farmers pay revenue to the state exchequer. They should not be penalized for technical faults or neglect of drainage in the project.
- Drainage projects have many indirect social benefits which are shared by all the village communities and the state. Since the communities who do not own lands will be sharing the benefits but not the cost, suitable subsidy to the farming community is justifiable.

The subsidy at the current prices can be up to 80% of the total cost.

INDIRECT BENEFITS

Besides the direct benefits, indirect socio-economic impacts of the project should be monitored. Some of the indirect impacts are as follows.

Increase in Land Value

The increase in land value alone may sometime be sufficient to compensate the investments. The data monitored at four places Konanki and Uppugunduru in Andhra Pradesh, Segwa in Gujarat and Islampur in Karnataka reveal that land reclamation projects enhanced the land value at all the places. This alone was much more than the investments made in land drainage (Gupta, 2011). This fact must be used to sensitize the farmers to adopt this technology.

Enhanced Status of the Poor in Rural Setting

In a rural setting, farmers having barren salt affected lands are ill-treated mainly because of their poor economic status. They are forced to do all kinds of works. With reclamation, these farmers become the proud owners of lush green fields. They are able to engage their family members at their own farm and in some cases even provide employment to few landless labourers. With this transformation, they can play an independent role in socio-politico-economic functions of the village. They need not migrate from the village for livelihood especially during lean periods.

Employment Generation

Employment generation is another important indirect benefit of this technology. The employment generation for various drains spacing is given in Table 11.11. It is estimated that a person gets employment for a full year upon reclamation of 3-5 ha of barren agricultural land. Impact of employment generation in reducing migration has been documented in a survey in the Islampur project in Karnataka (Anonymous, 2002).

Table 11.11 Employment generation in saline land reclamation projects involving pipe drainage

Spacing (m)	Installation of laterals and collectors	Construction of		
		Junction boxes	Sump	Total
25	290	5	14	309
50	170	3	14	187
75	112	2	14	128
100	103	2	14	119

Source: Joshi *et al.* (1987)

Inter-sectoral Terms of Trade

Resource mobilization from one sector to another is an indication of strengthening inter-sectoral terms of trade and economic development. Total cost involved in constructing a subsurface drainage system is associated with three different sectors. The share of the industrial sector dominates the total cost of drainage being as high as 60%. The share of the rural based non-farm resources occupies the second place with 20.8%. The data suggest that land drainage projects should strengthen both the industrial and the agriculture sector. As such, one can anticipate more employment opportunities and overall infrastructure development of the region.

Source of Irrigation

Subsurface drainage besides reclaiming saline soils provides a reliable source of water. Although, drainage effluent is generally saline, yet drainage water quality improves with time. Within 2-3 years, drainage effluent is good enough to be reused within the reclaimed land for supplemental irrigations (See Fig. 10.18). At one of the sites in Haryana, it was observed that water quality is better than the tube well water being used by the farmers. The EC of the drainage water at this site was 2.6 dS/m compared to the shallow tube well, which had an EC of 5.7 dS/m.

Environmental Improvement

Canal irrigation is often associated with increased human health problems due to increased incidences of water borne diseases. Increased incidences of malaria have been reported from waterlogged areas as Anolpheles mosquitoes breed in the standing water. Improved environment, reduced health problems and ecological improvements due to land drainage have been documented by various organizations. Documented evidence of such a situation is a report on migration of people living in and around Amritsar in the first decade of the last century. The vertical drainage scheme designed to halt the migration of people from Amritsar (India) in 1917 is often quoted in this context. The main aim of the scheme was to improve the general hygienic conditions of the area by lowering the water table. As per IWMI reports, government expenditure on health services can be curtailed in the same order of magnitude as the investments made in the land reclamation programs.

ENVIRONMENTAL IMPACT ASSESSMENT

An environmental monitoring and mitigation program should be an integral part of economic evaluation of drainage systems. ICID has developed a comprehensive 'Environmental Checklist to Identify Environmental Impacts of Irrigation, Drainage and Flood Control Projects' (ICID, 1993). The World Bank has also prepared an Environmental Assessment Sourcebook (World Bank, 1991) while FAO has produced a paper on the steps in the EIA process and the major environmental impacts of irrigation and drainage projects (FAO, 1995). These documents enable detailed environmental impact assessments of irrigation and drainage projects, the ICID checklist being the most comprehensive. The environment Impact Assessment (EIA) of drainage projects comprises of the following main steps:

- Defining objectives
- Selecting the evaluation method
- Assessing the environmental effects
- Formulating alternative options

Potential environmental impacts should be assessed in terms of their significance i.e. insignificant, low, moderate and high. Impacts of high significance mean that proper precautions need to be taken against environmental or economic damage. In that case, alternative options, alternative designs or phasing of the project should be explored. The identified alternative options should also be evaluated to arrive at the best option.

To assess the environmental impacts, an environmental baseline assessment (EBA) must precede a drainage project. In this assessment existing and potential environmental issues are identified and assessed. It helps to prepare an environmental Management Plan (EMP). Some of the environmental issues that are included in EBA are:

- Land losses
- Soil salinity
- Water logging
- Discharge
- Surface water quality of river/drain
- Fertilizer use
- Pesticides use
- Trace elements
- Socio-economic impacts

Detailed analysis of the useable land losses due to permanent land acquisition by the project components should be made. The economic and social impacts of such land loss as well as resettlement and compensation issues should be addressed. Detailed analysis of epidemiology of the diseases should be made with particular emphasis on water-based and water-related diseases such as malaria, schistosomiasis and other contagious diseases. In drainage projects, this issue may prove to be insignificant.

Although the main purpose of a drainage project is to reduce water logging and soil salinization, it should be ascertained that the drainage project is not just transporting the salts and water from one place to another. Since the drainage effluent takes away a part of the water of the project area, the effect of the project on other water use patterns, its quality, availability particularly in surroundings of the project area should be studied. Discharge in the surface water bodies and specifically of the main drain is likely to increase because of the project. Thus, the carrying capacities of all the drains leading to main drain should be assessed for their sufficiency. Besides what has been stated above, the water in water bodies might be used by the inhabitants in the following day to day activities.

- Domestic and livestock consumption
- Washing of clothes
- Human and livestock bathing

In all these activities, human and animals are likely to come in contact with the water that has come from the drainage system. Therefore its impact on a long-term basis must be evaluated.

The drainage effluent in arid and semiarid regions is likely to be salty. As such, the discharge from the drainage system might affect the quality of water bodies depending upon the quantum of discharge. It may be relevant to assess types of biocides (insecticides, herbicides, fungicides and rodenticides etc.) currently used and likely increase in their application rate following the project. Their likely impact on aquatic ecologies of the recipient water bodies should be assessed. The water quality both at the upstream and downstream of the project should be measured and assessed. The parameters of interest are: EC, pH, NO_3, PO_4 and trace elements. It may also impact the terrestrial and aquatic ecology (Flora and fauna). Appropriate studies should be made to assess the damage.

Application of chemicals particularly fertilizers is likely to go up after the project implementation. Nitrate being highly soluble and mobile, nitrate-N should be monitored to evaluate water quality from a subsurface drainage system. Phosphate although less mobile can still find its way through the surface drainage system. These nutrients may contribute to eutrophication of drainage channels to reduce their flow efficiency and capacity.

Trace elements in effluents have emerged as an environmental issue for drainage projects in the San Joaquin, Sacramento and Imperial valleys of California (Westcot *et al.,* 1989 and 1993). Various trace elements such as arsenic, aluminium, lead, manganese, boron, cadmium, selenium may adversely impact the crops/plants as well as aquatic and terrestrial wildlife. While many heavy metals are not readily mobile, irrigation/drainage may transport these contaminants to receiving waters.

All socio-economic activities in and around the proposed site should be monitored. These may impact the project outcome or get adversely impacted during the project implementation.

Reuse of Drainage Water

The main focus of the environmental impact assessment should revolve around options for the environmentally sound disposal/reuse of drainage effluent from the project site (See Chapter 10). The same may be upscaled for large-scale drainage schemes that may succeed the project. It is commonly believed that reuse is only a temporarily solution as salts will go on circulating and therefore, in the long run a more durable solution has to be found. In spite of this limitation, reuse seems to be the most viable alternative to meet the water deficit commonly encountered in the arid and semiarid regions as well as to buy time till some viable alternative of disposal is worked out (DRI, 1995). In spite of what is said, the issue is not as simple as stated as supplemental irrigation with drainage effluent might result in uptake of some contaminants by crops that may finally enter into the food chain. Such issue should be focused in the environmental Impact Assessment. For the construction of evaporation ponds, a separate EIA (Environmental Impact Assessment) may be sponsored.

FARMERS' PARTICIPATION

Definition and Objectives

Farmers' participation is a continuous, two ways communication process. It involves sharing the status, progress and implications of project formulation and evaluation activities with the farmers to keep them fully informed. It also includes soliciting their opinion and perception of objectives and needs and their preferences regarding resource use and alternative development of management strategies. It also seeks to derive any other information and assistance related to project formulation and evaluation to make it more effective.

For drainage projects, farmers' participation is sought with the following objectives.

- To inform and education and for liaison
- To identify problems and needs
- For idea generation and problems solving
- To receive reaction and feedback on the proposals
- To seek information on other alternatives and evaluation of these alternatives
- Conflict resolution
- Arrive at consensus decisions

Need for Farmers' Participation

It is now increasingly realized that success and sustainability of development projects besides technical solution depends largely on social dimensions of the project communities. Some of the reasons for farmers' participation in drainage projects are:

- Public sector organizations can no longer afford the burden of subsidies Farmers are now expected to assume a greater share of the cost of rural development activities. In this context, one important objective would be to explore the capacity of farmers to contribute to the capital as well as O & M expenses in post-drainage development.

- Irrigation and drainage works are by nature collective goods which, generally speaking are not divisible. In other words, the benefits of pipe drainage systems are shared more or less by all the farmers in the drainage unit. If a group of farmers objects to the installation of the system, the entire project is jeopardized. Therefore, some basis of farmers' cooperation and consensus must be created at the planning and design stage.
- Construction, especially mechanized, creates a flurry of activities involving men, women, and machines and inevitably destroys a fraction of the standing crops. While the technical dimensions of design are largely an engineering exercise, farmers can contribute to design by providing input regarding social constraints and opportunities for workable layouts with respect to land tenure patterns, land acquisition for outlets/sumps, and acceptable compensation for land required by project authorities or for crop losses during the construction.
- Farmer's involvement nurtured from the planning and design stage can contribute to reduced conflicts during the construction phase. Similarly, to the extent farmers can assume a participatory role; the process of association from the beginning will accelerate a sense of ownership in the system.

Institutional Set-up

Farmers' participation in drainage projects cannot be ensured through a top down approach. To ensure effective farmer participation, one should employ a bottom-up, democratic, decentralized strategy in which farmers have a sense of ownership of the project. A line section within the organizational setup of the project to deal with these activities and issues should be created.

Farmer Participation Section should make a list specifying the actual and potential activities, membership criteria, funding considerations, possible sources of subsidies, expenditures and training requirements for the farmers' drainage societies (FDS). The farmer's participatory section should also develop agreement forms for: farmers' permission to undertake SSD construction (Annexure IV) including construction of pump houses, sumps and effluent discharge channels and farmers' agreement to become members of the FDS. The agreement forms should be translated into local language.

Type of the Farmers Organizations and Mandate

Any decision on the kind of organization should emerge only after understanding the functioning of site specific social organizational arrangements of the beneficiaries. Exact nature regarding the structure and responsibilities of the appropriate farmer organizational might emerge in time. The Farmer's Participation Section (FPS) should plan the farmer mobilization process and should organize the farmers in a society even before the construction work in the project area begins. This will enable the farmers to actively participate in the construction/implementation phase of the project. It will inculcate a feeling of ownership from the very beginning and it will generate willingness to take over the project upon its completion.

The general body of the committee shall constitute all the stakeholders of the project and should be got registered under the relevant act of a cooperative or a similar societies act. An executive committee of 7-10 members should be elected from the stakeholders in accordance with the established by-laws of the association. The functions of the committee should be included in the memorandum of Association (Annexure V). The generalized mandate of the society besides what has been stated in the previous section should include one or more of the following tasks:

- To fix cash contributions by the members
- To act as intermediary between the farmers and the drainage agency to resolve difficulties in operation and maintenance
- Arrange water management training and facilitating extension activities
- Arranging farm inputs and implements
- Identifying and prioritizing local needs.

The Farmer's Participation: the Pitfalls

Major causes of failure to ensure farmers' involvement in agricultural development projects are:

- Insufficient funding
- Time pressure
- A preoccupation with physical structures
- Projects unresponsiveness to farmers
- Expecting unrealistic inputs from farmer without project delivery
- Organizational structures that lack accountability to farmers
- Rapid staff changes in the FPS
- Unskilled or untrained community organizers.

Handing Over the Project

Once it is decided to withdraw from the project area, the attention is needed to handover the system to the community. Sometimes, one needs to experiment till a successful socially compatible solution is in place. The handing over of the system should be in association with the Land Development and M&E Section so that they can finalize the land development and M & E activities in the project as per agreement in Annexure VI.

LAND DRAINAGE: CURRENT STATUS AND ISSUES

The science of drainage has moved forward on a worldwide basis, India also having made tremendous progress in the field. More than 60,000 ha area has been treated with subsurface drainage in different irrigation commands including lift irrigation commands (Table 11.12). Based on these experiences, salient achievements are listed as:

- While the productivity of agricultural lands depends on irrigation but drainage is essential to sustain high yield irrigated production system.
- We are now conscious and aware that too much water and salts adversely affects crop yields and that such adverse conditions can only be remedied through land drainage.
- The technical developments in the last 30 years have opened up new vistas making it possible to install drainage systems over large areas in relatively short time over a wide range of agro-climatic conditions.
- The design guidelines, construction and operational methods are in place.
- Evidences of economic, social and environmental benefits of land drainage have been documented.

In spite of the recognition that drainage has a vital role to play to restore the productivity and to ensure sustainability of irrigated agriculture, so far it has not received the due attention of the planners and the farmers (Gupta, 2002). It can be attributed to the fact that secondary benefits that make this technology attractive have not been recognized. As such nation is bearing huge economic losses year after year. A time bound action program to put vulnerable lands under drainage is needed duly recognizing the role of drainage as an instrument for rural and agricultural development. It calls for development of national and state level integrated Master Drainage Plans. A summary from the proposed drainage Master Plan of Haryana is given in Table 11.13. It illustrates that only a mix of drainage alternatives discussed in this book can resolve the problems.

Table 11.12 Extent of the area treated with subsurface drainage

State	Canal command	Area covered (ha)
Andhra Pradesh and Telangana	Lower Krishna, Lower Godavari, Nagarjuna Sagar and Krishna Western Delta	500
Assam	Tea gardens	50
Gujarat	Mahi-Kadana, Ukai-Kakrapar	1,300
Haryana	Western Yamuna canal, Bhakra canal command, Agra canal	10,584
Karnataka	Upper Krishna, Tungabhadra, Malprabha, Ghatprabah	25,000
Kerala	Acid sulphate soils	200
Madhya Pradesh	Barna, Tawa, Chambal	950
Maharashtra	Lift irrigation schemes in Krishna river basin and Neera canal command	3,500
Punjab	Sirhind and Sirhind feeder canals	2,000
Rajasthan	Chambal and Indira Gandhi *Nahar Pariyojana*	16,500
Total		**61,084**

Table 11.13 Summary sheet of drainage plan of Haryana

S.N.	Project	Quantity	Cost (Rs. million)
1.	Surface drainage	1432 km length	12310
2.	Horizontal sub surface drainage	47000 ha area	2570
3.	Vertical tube well drainage	1200 nos	430
4.	Conjunctive use tube wells	1 lakh nos. (2 lakh ha area)	2500
5.	Bio-drainage	2 lakh ha area	3850
6.	Repair of canal water courses, water supply tanks	-	440
7.	Road raising and cross drainage	-	680
Total			**22780**

Apart from what is stated in the preceding paragraphs, there is an urgent need for a better understanding of the socio-economic issues and environmental consequences of land drainage. It should determine the new research agenda for a more harmonious planning of drainage activities. Policy and action research agenda needs to be upgraded. Integrated drainage plans need to be developed for cost-effective and eco-friendly solutions (Gupta, 2014).

Current state of institutional arrangements for land drainage is seriously inadequate. There are number of central and state organizations dealing with land drainage, many of these duplicating the work or having no focused assignments. Therefore, the most challenging task is to establish a strong, specialized institution fully authorized to carry out drainage works. There is a need for rationalizing and streamlining the respective roles of the government, the communities and private sector in drainage development and management. The need for such rationalization has to be seen in the context of required investments on drainage assessed at Rs. 4000 billion (Chakraborty *et al.*, 2012). With its strong human resource and basic infrastructure, it should not be difficult to build-up a desired structure to accomplish such a challenging task that would impact future agricultural more than anything else.

QUESTIONS (THEORY)

1. What are the cost components of subsurface drainage system?
2. Explain the terms Net Present Worth, B:C ratio and internal rate of return.
3. What do you understand by payback period? How is it calculated?
4. What do you understand by breakeven benefits? How it is calculated?
5. Explain how you will proceed to conduct economic analysis of a drainage project?
6. Differentiate between economic and financial analysis of a project.
7. Briefly describe some of indirect benefits associated with drainage projects.

8. What do you understand by stakeholder's participation in land drainage? What are its benefits?

QUESTIONS (OBJECTIVE)

1. Fill in the blanks

1. Funds available today are worth ________ than the same sum of money available say after 10 years.
2. Net present worth is calculated as present worth of benefits ________ present worth of costs.
3. Ratio of the present value of benefits divided by the present value of costs is termed as ________.
4. ________ period is the number of years it will take for the net undiscounted benefits to repay the full investment.
5. Internal rate of return is the discount rate at which NPW is ________ and B-C ratio is ______.
6. __________benefit is the minimal benefit required to meet the costs
7. Success and sustainability of development projects besides technical solutions depends largely on ________ issues of the project communities.

2. Tick mark the correct or most appropriate answer (A, B, C or D) from the following

1. Net present worth of a project is calculated as

   (A) Difference between the present values of benefits over the present values of costs

   (B) Present values of costs over the period

   (C) Present values of benefits over the period

   (D) None of these

2. Time period for an investment to generate sufficient incremental cash to recover its initial capital outlay in full is known as

   (A) Cost overrun period (B) Payback period

   (C) Breaking period (D) None of these

3. IRR is that rate of discount which makes

   (A) Net present worth as zero (B) B:C ratio one

   (C) Both A and B (D) None of A and B

4. The economic life of a drainage project for economic analysis is usually taken as

   (A) 50 years (B) 30 years

   (C) 10 years (D) None of these

5. Discount factor, DF for any discount rate, Dr (%) for a particular time, t (years) is given as

   (A) $DF = 1/(1+Dr/100)^t$ (B) $DF = 1/((1+Dr)/100)^t$

   (C) $DF = (1+Dr/100)^t$ (D) None of these

6. What is the discount factor for the second year of a drainage project, if discount rate is 10%

   (A) 0.9 (B) 0.83

   (C) 0.0049 (D) None of these

7. Which is not the economic evaluation parameter

   (A) Internal rate of return (B) Benefit cost ratio

   (C) Net present worth or value (D) Discount rate

8. The Internal rate of return is the discount rate at which

   (A) B:C ratio is equal to one (B) B:C ratio is equal to zero

   (C) B:C ratio is negative (D) None of these

9. The Internal rate of return is the discount rate at which

   (A) Net present worth positive

   (B) Net present worth is zero

   (C) Net present worth is negative

   (D) None of these

10. The Internal rate of return is the discount rate at which

    (A) Net present worth is zero (B) B:C ratio is one

    (C) Both A and B (D) None of these

11. Payback period is the period in years given as (If the net profits and gross income are equal in each year)

    (A) Investments divided by net profit per year

    (B) Investments divided by gross income per year

    (C) Investments divided by discounted net profit per year

    (D) None of these

12. Net present worth is the difference of

    (A) Gross income and gross costs during the period

    (B) Discounted income and discounted costs during the period

    (C) Discounted income and discounted benefits during the period

    (D) None of these

13. For a drained area of 40 ha, the length of the lateral pipes required will be approximately _____m if the lateral drain spacing is 40 m

    (A) 40000 (B) 10000

    (C) 20000 (D) None of these

14. Maximum initial investments are needed when construction of the system is to be done

    (A) Manually (B) Semi-mechanical means

    (C) Fully mechanical means (D) Equal investments in A, B or C

15. In calculating break-even benefits, we use

    (A) NPW of costs (B) NPW of benefits

    (C) NPW of the project (D) None of these

■■■■

Annexure I

BILL OF QUANTITIES FOR SURFACE AND SUBSURFACE DRAINAGE SYSTEMS

S.No.	Description of Work	Unit	Qty.	Unit cost	Total cost	Remark
1	**River Course Treatment**					
	1.1. Excavation work of 1 m of main drain to a depth of 2/2.5 m with embankment	m^3				
	1.2. Disposal of excess excavated soil at a distance of 500 m or more					
2	**For Surface Drainage Systems (Open Drain Systems)**					
	2.1. Excavation work of 100 m of surface drains with embankments, road and appropriate inlets	m^3				
	2.2. Disposal of excess excavated materials with in initial lead of 500m	m^3				
3	**Excavation, laying, and back filling of trenches**					
	3.1. For Lateral					
	3.1.1. Excavation of earth work (for trench for laying 100 m of pipe) pipe in a hard clay soil 1.2 m deep and 0.60m width. The top 30 cm soil would be kept on one side while the remaining on the other side of the trench	m^3				
	3.1.2. Laying of 72/80 mm P.V.C pipe on the desired grade including connections as per drawings	m	100			
	3.1.3. Backfilling of trenches immediately following placing of the pipes. The back fill composed of soil excavated from trenches shall be placed carefully up to 40 cm initial depth with soil excavated below 30cm depth from the ground surface. Large stones, Clods, pieces of concrete or other materials likely to damage the pipes and farm equipment shall be removed from the backfill. Back fill and manual compaction after first 40 cm. Top 30 cm soil shall be used to fill the top 30 cm of the trench	m^3				

S.No.	Description of Work	Unit	Qty.	Unit cost	Total cost	Remark
	3.1.4. Compaction of earth work by Tractor / manual after complete fill of trench.	m^2	60			
	Sub total					
4	**Excavation, laying, and backfilling of trenches**					
	4.1. For Collector					
	4.1.1. Excavation of earth work (trench for laying 100 m pipe) in a hard clay soil 1.5 m deep and 0.60 m width	m3				
	4.1.2. Laying of 88/100, 144/160 mm, 178/200 mm corrugated and 258/294 mm PVC Collector pipes on the desired grade including connections as per drawings	m	100			
	4.1.3. Backfilling of trenches immediately following placing of the pipes. The backfill composed of soil excavated from trenches shall be placed carefully up to 40 cm initial depth with soil excavated below 30 cm depth from the ground surface. Large stones, clods, pieces of concrete or other materials likely to damage the pipes and farm equipment shall be removed from the backfill. Backfill and manual compaction after first 40 cm. Top 30 cm soil shall be used to fill the top 30 cm of the trench	m^3				
	4.1.4. Compaction of earth work by tractor/ manual after complete fill of trench	m^2	60			
	Sub total					
5	**Supply of Corrugated perforated/non-perforated PVC pipes**					
	5.1. 72/80 mm dia perforated pipe corrugated	m	100			
	5.2. 88/100 mm dia non- perforated pipe	m	75			
	5.3. 144/160 mm dia of non-perforated	m	30			
	5.4. 178/200 mm dia of non-perforated	m	6			
	5.5. 258/294 mm dia non-perforated PVC pipe	m	6			
	Sub total					
6	**Supply of PVC Fittings**					
	6.1. Couplers 72/80 mm dia pipe	No	1			
	6.2. Couplers for 88/100 mm dia pipe corrugated	No	1			
	6.3. Couplers 144/160 mm dia pipe corrugated	No	1			
	6.4. Couplers 178/200 mm dia pipe corrugated	No	1			
	6.5. Couplers for 254/294 mm dia pipe plain	No	1			

S.No.	Description of Work	Unit	Qty.	Unit cost	Total cost	Remark
7	**Supply of PVC bend**					
	7.1. 88/100 mm dia pipe bend	No	1			
	7.2. 144/160 mm pipe bend	No	1			
	7.3. 178/ 200 mm dia pipe bend	No	1			
	7.4. 254/294 mm dia pipe bend	No	1			
8	**Supply of PVC Tee**					
	8.1. 100*80*100 (O.D)	No	1			
	8.2. 160*80 *160 (O.D)	No	1			
	8.3. 200* 80*200 (O.D)	No	1			
	8.4. 294*100 *294 (O.D)	No	1			
9	**Supply of end caps**					
	9.1. End caps for 72/80 mm dia pipes		1			
	9.2. End caps for 80/100mm dia pipes		1			
	9.3. End caps for 144/160mm dia pipes		1			
10	**Supply of reducers**					
	144-88		1			
	178-144		1			
	Sub-total					
11	**Providing and fixing of synthetic envelope materials 240gm/m^2 made from polypropylene fibre/ Geo synthetic materials with opening size (O^{90}) more than 250 micron for 72/80 mm dia of pipe**	m				
	sub total					
12	**Construction of 1 No. of manholes for collectors as per the Drawings**					
	12.1. Excavation of earth materials dry or moist including dressing and disposal (1.75 m * 1.75m *2.0) m^3	m^3	6.2			
	12.2. Providing and laying of lean concrete (0.25m)		0.72			
	12.3. Providing, laying reinforced cement concrete (1:4:8) with mix aggregates 40mm	m^3	1.63			
	12.4. Provide, laying and fixing of man hole cover at the top (1.35m* 1.35*0.05m)	No	1			
	12.5. Providing and laying of variable dia of (100-300 mm) RCC pipe of 2 m long including joint filling with cement mortar (1:3)	No	2			
	Sub total					

S.No.	Description of Work	Unit	Qty.	Unit cost	Total cost	Remark
13	**Construction of 1 No. of junction boxes for collectors as per the drawings**					
	13.1. Excavation of earth materials dry or moist including dressing and disposal (1.20 m * 1.20 m *1.5) m^3	m^3	2.16			
	13.2. Providing and laying of lean concrete (0.25m)	m^3	0.36			
	13.3. Providing, laying reinforced cement concrete (1:4:8) with mix aggregates 40 mm		0.62			
	13.4. Providing, laying and fixing of RCC junction box cover at the top (0.90m* 0.90*0.05m)	m^3	0.01			
	13.5. Providing and laying of variable dia of (200-300 mm) RCC pipe of 2 m long including joint filling with cement mortar (1:3)	No.	2			
	Sub total					
14	**Construction of 1 outlet structures for singular laterals**					
	14.1. Excavation of earth materials dry or moist including dressing and disposal of excavated materials with in initial lead of 30 m and lift of 1.5m	m^3	2.34			
	14.2. Providing and laying of 100 mm dia RCC pipe of 2 m long including joint filling with cement mortar (1:3)	m	3			
	14.3. Cart away of excavated materials with in initial lead of 30m and lift of 1.5 m excluding dressing	-	-			
	14.4. Providing and fixing of rodent controls and identification structures as per drawings	No	1			
	Sub total					
15	**Construction of 1 outlet structures for collectors**					
	15.1. Excavation of earth materials dry or moist including dressing and disposal of excavated materials within initial lead of 30m and lift of 1.5m	m^3	7.28			
	15.2. Provide lean concrete of 0.15m thickness	m^3	0.11			
	15.3. Provide and laying of reinforced cement concrete 1:4:8 including curing and finishing	m^3	0.65			
	15.4. Providing and laying of variable dia of (200-300 mm) RCC pipe of 2 m long 4 pipes including joint filling with cement mortar (1:3)	No.	4			
	15.5. Cement Plaster 20 mm thick (1:4)	m^2	1.1			
	15.6. Cart away of excavated materials with in initial lead of 30m and lift of 1.5 m excluding dressing	m^3	7.28			

S.No.	Description of Work	Unit	Qty.	Unit cost	Total cost	Remark
	15.7. Providing and fixing of rodent controls structures as per drawings	No	1			
	Sub total					
16	**Construction of one road crossing structure**	No				
	16.1. Excavation of earth materials dry or moist including dressing and disposal of excavated materials with in initial lead of 30m and lift of 1.5 m (7*0.6*1.5)	m^3	6.3			
	16.2. Providing of variable dia RCC pipe (200-300mm) 2.0 m long, 3 No., as per drawings including joint filling, etc.	m	6			
	16.3. Reshaping of excavated materials with in initial lead of 30 m and lift of 1.5 m and making the road to its original shape including dressing	m^2	6.3			
	Sub total					
17	**Construction of one drain crossing structures**					
	17.1. Excavation of earthen materials dry or moist including dressing and disposal of excavated materials with in initial lead of 30 m and lift of 1.5 m (14.0*0.60*1.5)	m^3	12.6			
	17.2. Provide and laying and fixing of 250 mm diameter RCC pipe 2 m long with collar in required slope, including joint filling in CM 1:3	m	10			
	17.3. Rehandling of excavated materials with in initial lead of 30 m and lift of 1.5 m and reshaping the drain to its original shape including dressing	m^3	12.6			
18	**Construction of /reshaping of the surface drains (20 m/ha)**					
	18.1. Excavation of earthen materials for the main drains (20 m length, 2 m deep) dry or moist including reshaping and disposal of excavated materials with in initial lead of 30 m	m^3	-			
	Sub total					
	Total					
	Contingency					
	Grand total					

dia = diameter

■■■■

Annexure II

GROUNDWATER BALANCE

Groundwater balance is one of the most important criteria to determine the intensity of drainage. Water balance describes the gains and losses of water in a given area or soil layer over a certain time period. In general a water balance equation is written as:

$$I - O = S_g \qquad (A3.1)$$

Here I is the inflows; O is the outflows; and S_g is the change in storage, all defined in the same units. All quantities are measured over a given time period. Based on the value of S_g the groundwater of an area is described as stabilized ($S_g = O$), non-stabilized positive ($+ S_g$) and non-stabilized negative ($- S_g$). A negatively non-stabilized groundwater regime may become stabilized in due course of time because of increasing recharge. This regime is generally preferred in waterlogged areas as it helps to decrease degree of water logging, decreases the intensity of soil salinization and improves the general productivity of the land. A generalized groundwater balance equation of an area can be mathematically written as:

$$(R + T_i + C_s) - (E+M) + L_i - L_0 = S_g \qquad (A3.2)$$

Where, R, T_i and C_s are the recharge components from rainfall, irrigation and seepage from surface water sources in the area, E and M are the draft component due to capillary rise and evaporation and draft through minor irrigation/domestic wells respectively. L_i and L_0 are lateral inflow and outflow across the boundaries of the area. If the net inflow minus outflow is a positive, it means the groundwater is receiving foreign water which might also be the source of water logging in the area.

Recharge from Rainfall

Only a part of the rainfall percolates to groundwater which can be estimated by empirical formulae. A formula to estimate rainfall recharge for alluvial zones of Indo-Gangetic plains is given by Chaturvedi as:

$$R_G = 2\,(R_1 - 15)^{0.45} \qquad (A3.3)$$

where R_1 is the annual rainfall, cm and R_G is the groundwater recharge due to rainfall, cm. On the other hand Groundwater Estimation Committee of the CGWB (2009) recommended that groundwater level fluctuation and specific yield approach should be adopted to estimate groundwater recharge. Specific yield values for different geological formations are given in Table A3.1. Only at places where adequate groundwater fluctuation data are not available, ad-hoc norms given by the Committee for rainfall infiltration method should be followed (Table A3.2). The amount of recharge is obtained by measuring the net accretion to groundwater and by multiplying the same to the specific yield of geological formation in the zone of water table fluctuation. The average rainfall during the period is used to evaluate the net recharge for normal monsoon rainfall by the principle of linear proportionality.

Table A3.1 Specific yield for various hydrogeological settings

Hydrogeological situation	Specific yield (%)
Sandy alluvial area	12-18
Valley fills	10-14
Silty/clayey alluvial area	5-12
Granites	2-4
Basalts	1-3
Laterite	2-4
Weathered phyllite, shales, schists and associated rocks	1-3
Sandstone	1-8
Limestone	3
Highly karstified limestone	7

Table A3.2 Ad-hoc norms for rainfall infiltration

Hydrogeological situation	% of normal rainfall
Alluvial areas in sandy areas	20-25
Areas with higher clay content	10-20
Semi-consolidated sandstones friable highly porous	10-15
Hard rock areas: Granitic terrain	
(a) Weathered and fractured	10-15
(b) Unweathered	5-10
Basaltic terrain -	
(a) Vesicular and jointed basalt	10-15
(b) Weathered	4-10
Phyllites, lime stones (sand stones, quartzites, shales etc.)	3-10

To calculate the annual recharge during monsoon, the following formula is adopted.

$$\text{Recharge during monsoon} = (S + DW - R_s - R_{igw} - R_{is}) \times (\text{Normal Monsoon RF/ Annual Monsoon RF}) + R_s + R_{is} \qquad (A3.4)$$

Where S is the change in groundwater storage volume during pre and post-monsoon period (April/May to November), DW is the gross groundwater draft during monsoon, R_s is the recharge from canal seepage during monsoon, R_{igw} is the recharge from recycled water from groundwater irrigation during monsoon and R_{is} is the recharge from recycled water from surface water irrigation during monsoon, RF is the rainfall.

To eliminate the effects of drought or surplus rainfall years, recharge is estimated for periods ranging from 3 to 5 years and an average value is adopted. Recharge from winter rainfall may also be estimated in a similar manner.

Recharge due to Canal Irrigation System

Numerous investigations have been carried out to evaluate the total losses in the canal irrigation system. No set values have been reported because of the location specific nature of losses and wide variations in the soil types. Irrigation efficiencies in conventional irrigation methods are low such that as much as 60-70% of the water is lost in transit at many places. These losses occur as seepage from the distribution network and during field application. The seepage losses from the canals and distributaries can be estimated by the relation

$$\text{Seepage loss} = \text{Length of canal distributory} \times \text{Wetted perimeter} \times \text{Number of running days} \times \text{Seepage factor} \quad \text{(A3.5)}$$

Seepage factor is a function of soil type as well as depth to water table. It varies from 3.2 – 3.5 cumec/10^6 sq. m of wetted area for sandy soils to 1.8 – 2.8 cumecs/10^6 sq. m of wetted area for soils with some clay content. The CGWB guidelines are:

- For unlined canals in normal type of soil with some clay content along with sand 15-20 ha-m/d-10^6 sq m
- For unlined canals in sandy soils 25 to 30 ha-m/d-10^6 m^2 of wetted areas
- For lined canals the seepage losses may be taken as 20% of the above values.

For field channels and field water application, losses are taken as a fraction of the total water delivered at the outlet. For major irrigation sources, the losses are taken as 35% of water delivered at the outlet for all crops except rice for which a value of 40% is taken. For minor irrigation sources a value of 30% is used for all crops except paddy where it is taken as 35%. In all the above cases, return seepage figures include losses in the field channel also and therefore, should not be accounted for separately. The effect of groundwater table on recharge can be gauged from the data given in Table A3.3. Since for drainage point of view, shallow water table situation would prevail, the values in the 3rd column would be more appropriate.

Table A3.3 Recharge as percentage of application

Source of irrigation	Type of crop	Water table below ground level		
		<10 m	10-25 m	>25 m
Groundwater	Non-paddy	25	15	5
Surface water	Non-paddy	30	20	10
Groundwater	Paddy	45	35	20
Surface water	Paddy	50	40	25

Seepage from Other Sources

Drains, surface water bodies and flooded areas during monsoon are some other important sources of recharge to groundwater. The seepage from these sources is calculated as per the equation and norms listed below.

$$\text{Seepage} = \frac{\text{Area} \times \text{days of flooding}}{365} * \text{Selected norm value} \qquad \text{(A3.6)}$$

For tanks, a value of 1.4 mm/d is used for the period tank contains water and based on the average area of water spread. If data on the average area of water spread is not available, 60% of the maximum water spread area may be used instead of average area of water spread. Other guidelines are as follows:

- 9 to 20% of the live storage or 40 to 60 cm/year over the total water spread.
- Seepage from percolation tanks is 50% of its gross storage times the number of times the tank is filled. Half of this recharge is assumed to occur in the monsoon season while the balance in the non-monsoon season. Same norm may be used for check dams and *nala* (drain) bunds.

The potential recharge from flood plains may be estimated on the same norms as for ponds and lakes, i.e., 40 to 60 cm/year over the water spread area for period equal to the retention period.

Inflow-Outflow Measurements

Lateral groundwater inflow and outflow can be calculated either by Darcy's law or flow net analysis. Data on water table gradients of the aquifer across the project area's boundaries are calculated by drawing regional water table contours including the area under study. The transmissivity is estimated by pump test analysis. In case other components of water balance equation are measured accurately, the water balance itself can be used to estimate net inflow-outflow from the area.

Total Annual Recharge

Total Annual Recharge = Recharge during monsoon + Non monsoon rainfall recharge + Seepage from canals + Return flow from irrigation + Inflow from influent rivers- etc. + recharge from submerged lands, lakes etc.

Net recoverable recharge: Net recoverable recharge is taken as 70- 90% of the total gross recharge to the ground water.

Mining of Groundwater

In order to avoid water logging and soil salinity, it is necessary to maintain water table sufficiently below the soil surface. Diverse opinions have emerged whether fresh groundwater should be continuously pumped to greater depths. It is argued that when groundwater table is lowered through mining to greater depths, the cost of vertical drainage increases on account of following factors:

- Investments on tube wells will increase in terms of number of tube wells and in terms of deeper depth of tube wells
- The operational cost of pumping from deeper depths will increase
- Following factors will also limit the extent of mining.
- Possibility of saline water intrusion from surrounding saline water aquifers
- Possibility of groundwater salinity increasing with increasing depths

There are two schools of thoughts. According to the one school, it is necessary to lower the water table to a desired depth normally 3 to 4.5 m and then to restrict the abstractions from the aquifer to an extent of average annual recharge. It is not necessary to balance year to year recharge and discharge. In the opinion of the other school of thought, there is no harm in mining the groundwater over a longer period of time provided there are no technical difficulties. The arguments of this group are based on a comparison of the groundwater in the aquifer by the money in a bank. If the savings are not withdrawn in times of need; the purpose of savings is defeated. They further argue that in times to come pumping technologies will develop to an extent that there would be no limitations either technology-wise or cost-wise.

■■■■

Annexure - III

IRRIGATION WATER SALINITY LIMITS FOR SELECTED CROPS

Crop/ Location	Soil type	No. of years	Previous crop	EC_{iw} (dS/m) for yield reduction of		
				10 %	25 %	50 %
Wheat						
Agra	Sl	6	Pearl millet	6.6	10.4	16.8
Agra (late sown)	Sl	2	*Toria*	4.3	6.6	11.0
Dharwad	Scl	5	Sorghum	3.4	7.0	12.9
Hisar	Sl	5	Sorghum/fallow	6.1	8.7	13.0
Indore	Cl	8	Maize	4.7	8.7	15.2
Jobner	Ls	4	Fallow	8.3	11.7	17.5
Jodhpur	Scl	2	Fallow	8.3	10.2	13.4
Karnal	Sl	5	Fallow	9.1	10.8	13.7
"	S	5	Fallow	14.0	16.1	19.5
Barley						
Agra	Sl	2	Fallow	7.2	11.3	18.0
Jobner	Ls	5	Fallow	13.0	21.1	-
Rice						
Agra	Sl	1	*Berseem*	2.3	4.6	8.6
Bapatla	Scl	6	*Kharif* rice	2.2	3.9	6.8
		3	*Rabi* rice	1.8	2.9	4.8
Maize						
Dharwad	Scl	5	Sorghum	3.7	7.8	14.5
Indore	Cl	7	Wheat	2.2	4.7	8.8
Pearl millet						
Jobner	Ls	5	Fenugreek	11.9	22.7	-
Agra	Sl	4	Wheat	5.4	9.0	15.0
Italian-millet						
Bapatla	S	5	Sunflower (K)	2.4	4.6	8.2
	S	4	Italian-millet (R)	2.5	4.9	8.7
Sorghum						

Crop/ Location	Soil type	No. of years	Previous crop	EC_{iw} (dS/m) for yield reduction of		
				10 %	25 %	50 %
Dharwad	Scl	6	Wheat	2.6	5.1	9.2
Agra	Sl	3	Mustard	7.0	11.2	18.1
Sorghum (fodder)						
Agra	Sl	2	*Berseem*	5.2	10.2	18.4
Hisar	Sl	2	Wheat	2.5	5.7	11.0
Mustard						
Agra	Sl	6	Sorghum	6.6	8.8	12.3
Bapatla	Scl	5	Soybean	3.8	7.9	14.7
Jobner	Ls	2	Cluster bean	6.6	13.5	-
Toria						
Agra	Sl	1	Wheat	4.7	5.1	5.9
Safflower						
Dharwad	Scl	5	Maize	3.3	6.8	12.6
Sunflower						
Bapatla	Sl	3	Mustard	3.5	7.2	13.4
Hisar	Sl	1	Wheat	2.5	5.7	11.0
Groundnut						
Bapatla	S	5	Italian millet	1.8	3.1	5.3
Soybean						
Bapatla	Scl	3	Mustard	2.0	3.1	5.0
Pigeon pea						
Agra	Sl	6	Onion	1.3	2.3	3.9
Cluster bean						
Bapatla	Sl	3	Variable	3.2	4.5	6.8
Jobner	Ls	2	Mustard	3.9	6.6	11.1
Cowpea						
Jobner	Ls	2	Variable	8.1	13.2	-
Berseem						
Agra	Sl	5	Rice/sorghum	2.5	3.2	4.4
Onion						
Agra	Sl	5	Variable	5.1	6.0	7.5
Bapatla	S	5	Pigeon pea	1.8	2.3	3.3
Potato						
Agra	Sl	5	Okra	2.1	4.3	7.8
Tomato						
Bapatla	S	3	Variable	2.4	4.1	6.9

Crop/ Location	Soil type	No. of years	Previous crop	EC_{iw} (dS/m) for yield reduction of		
				10 %	25 %	50 %
Okra						
Agra	Sl	2	Variable	2.1	3.9	6.7
Bapatla	S	5	Potato	2.7	5.6	10.5
Brinjal						
Bapatla	S	2	Variable	2.3	4.1	7.1
Fenugreek						
Jobner	Ls	3	Pearl millet	3.1	4.8	7.6
Chillies						
Bapatla	S	2	Variable	1.8	2.9	4.9
Jobner	Ls	2	”	4.5	7.5	12.5
Coriander						
Bapatla	S	3	Variable	2.9	5.8	10.7
Jobner	Ls	2	”	9.8	15.4	-
Bittergourd						
Bapatla	S	3	Variable	2.0	3.4	5.8
Bottlegourd						
Bapatla	S	3	Variable	3.2	4.5	6.8

S, sandy; ls, loamy sand; sl, sandy loam; l, loam; scl, sandy clay loam; cl, clay loam

The annual rainfall in Agra, Bapatla, Dharwad, Indore, Jobner, Hisar, and Karnal was 660, 803, 778, 750, 500, 450 and 850 mm, respectively.

Source: Minhas and Gupta (1992)

■■■■

Annexure IV

FARMER'S CONSENT / AGREEMENT

I _______________________son/daughter/wife of ____________________ resident of Village_______________P.O.___________Tehsil______________ District_________________, hereby give my consent and I have no objection with the Project authorities in executing surface/subsurface/other works (reclamation of waterlogged soils) through subsurface drainage (SSD) system on my agricultural land (A*____ K*____ M*____) in *Khasra* No. ___________________.

I hereby agree to abide by the following:

1 That I will become a permanent member of the farmers' drainage society to be constituted to take up the above mentioned works

2 That I will not put any hindrance in the execution of works

3 That I will not claim any compensation for any damage to my crops etc. during the operations

4 That I will extend full cooperation during the installation of surface/SSD system and its operation and maintenance

5 That I will share the operation and maintenance costs of the SSD system according to my holdings as decided by the Farmers Drainage Society

6 I will follow the recommendations of executing agency for adopting appropriate farm practices.

I have read carefully the above mentioned agreement.

Farmer's signature

Date : ....
Place : ....

Witnesses

1. *Sarpanch/ Numberdar*

2. Community Organizer (CO)

Signature of the Sociologist

Countersigned

Project Director of the Agency

*AKM are acres, *Kanals* and *Marlas*

■■■■

Annexure V

MOA FOR FARMER DRAINAGE SOCIETIES (FDS)

A.1. Memorandum of Association

1. Name of the Society:
2. Location of the Registered Office:
3. Area of Operation:
4. Aims and Objectives: As mentioned under para 4 of Rules and Regulations.
5. Conditions

a. The income and property of the society shall be solely towards the promotion of the objectives of the society as set forth in the Memorandum of Association and no portion thereof shall be paid or transferred directly or indirectly to the members of the society.

b. No member of the Governing Body of the society shall be appointed to any salaried office of the society or any office of the society paid by fees, that no remuneration shall be given by the society to any member of such Governing Body except repayment of out of pocket expenses and interest on money lent or rent for premises to the society.

c. The society by its constitution is required to apply its profits, if any, or other income in promoting its objectives.

d. If upon the winding up or dissolution of the society there remains, after satisfaction of all its debts and liabilities, any property whatsoever, the same shall not be paid to or distributed amongst the members of the society, but shall be given or transferred to some other institution having objectives similar to the objectives of the society to be determined by the members of the society at or before the time of dissolution.

6. The management of the affairs of the society is entrusted in accordance with the Rules and Regulations of the society to a Governing Body of which the first members are:

Sr. No.	Name	Address	Occupation	Designation

-----				

7. We the following, whose names and addresses are hereunder subscribed are desirous of being formed into a society in pursuance of this Memorandum of Association:

Sr. No.	Name	Address	Occupation	Designation

-------.				

Witness:
I know the above persons &
they have signed in my presence.

Station:
Signature: (*Sarpanch*)

Place:
Other than member of the society

Name:

Address:

Annexure VI

HANDING OVER THE BLOCK TO DRAINAGE SOCIETY

AGREEMENT

This agreement is entered into among the following parties regarding the following subject matter:

Parties

1. ______________________________ herein after called -----------.
2. Farmers Drainage Society (FDS), --------, hereinafter called the **FDS**.

Subject Matter

Handing over the Subsurface Drainage (**SSD**) system, constructed by the -------- on the land belonging to the farmers of the FDS.

On being handed over the above-mentioned SSD system (without any cost) in proper running and good maintenance condition to the FDS, the following responsibilities/ obligations are taken over by each party to the agreement:

- The system is handed over without charging any cost in perpetuity to the FDS subject to the society keeping its side of the obligations/responsibilities.
- In case of violation of the responsibilities/obligations by the FDS, the first party would be free to levy any damages/ cost of the SSD system from the FDS.
- The SSD system is handed over in a properly functional manner and in good maintenance. A final as built map has been supplied to the society.
- The FDS undertakes the following:

Besides what is stated in the objectives of the FDS in the registration deed the society undertakes the following:

- Proper security of the SSD system.
- Proper repair and maintenance of the SSD system.
- Payment of all future dues which would accrue for running/maintenance of the SSD system.

Project Director

Chairman
Farmer Drainage Society

Witnesses

1.
2.

Dated:

Place:

GLOSSARY

Acidification: Process whereby soil becomes acidic (pH < 7) because acid parent material is present or in regions with high rainfall, where soil leaching occurs. Acidification can be accelerated by human activities (use of fertilizers, deposition of industrial and vehicular pollutants).

Acid Sodic Soils: Soils having pH less than 7 but exchangeable sodium percentage greater than 15.

Acid Soils: A soil with a preponderance of hydrogen ions, and probably aluminum, in proportion to hydroxyl ions. Specifically, soil with a pH value of less than 7.0; for most practical purposes, a soil with a pH value of less than 6.5.

Acid Sulphate Soils: A soil with pH < 4, acid sulphate soils are saturated with water, almost oxygen-free and contain microscopic crystals of iron sulphide minerals (commonly pyrite).

Active Acidity: Acidity caused by hydrogen and aluminum ions in the soil solution or free H^+ ions in the soil solution

Aerated Soil: An aerated soil is a soil with a good movement of air through the soil structure. The opposite is a wet waterlogged soil.

Aeration (Soil): The process by which soil air is replaced by atmospheric air at low airflow rates. In a well-aerated soil, the soil air is very similar in composition to the atmosphere above the soil. Poorly aerated soils, on the other hand contain a much higher percentage of carbon dioxide.

Aerial Photograph: A photograph of the Earth's surface taken from an aero plane or some other type of airborne equipment.

Agricultural Lime: Carbonates, oxides and hydroxides of calcium and magnesium are referred as agricultural lime.

Agro-climatic Zone: A land unit defined in terms of major climate and growing period which is climatically suitable for a certain range of crops and cultivars.

Air Porosity: The proportion of the bulk volume of soil that is filled with air at any given time or under a given condition such as a specified moisture tension. Usually the large pores drained by a tension of less than approximately 100 cm of water.

Alkali Soil: A soil that contains sufficient exchangeable sodium to interfere with the growth of most crop plants, either with or without appreciable quantities of soluble salts. A soil having ESP more than 15, and pH_s more than 8.2. It is also called sodic soil. May not be confused with alkaline soil having alkali reaction (See alkaline soil)

Alkali Water: Water having RSC more than 2.5. EC may or may not be high.

Alkaline Soil: A soil that has an alkaline reaction, i.e. a soil for which the pH reading of the saturated paste is more than 7.0.

Alkalization: The process whereby the exchangeable sodium content of the soil is increased. Sometimes also referred as sodication or sodification.

Amendment Soil: Substances that are added to the soil such as lime, gypsum, sawdust, etc. to make them more suitable for crop production. Strictly speaking even fertilizers constitute a special group of soil amendments.

Anaerobic Soils: Anaerobic soils having very little oxygen present e.g. the wet, waterlogged, marshy soils in a bog. While anaerobic conditions are important for some processes, such as bacterial reduction of nitrate to nitrogen, these conditions can also produce hydrogen sulphide, methane and other undesirable substances.

Anisotropic: A soil having physical properties that depend on direction.

Aquifer: A formation that contains sufficient saturated permeable material to yield and transmit significant quantity of water to wells and springs.

Aquifer Properties: The properties of an aquifer that determine its hydraulic behaviour and its response to abstraction.

Artificial Drainage: Drainage constructed artificially by men in contrast to natural drainage which develops naturally in a region.

Auger: A tool used for boring into the soil and withdrawing small samples for field or laboratory examination.

Auger Hole Method: A technique to determine saturated hydraulic conductivity of a soil at a certain depth in the presence of water table.

Available Soil Moisture Capacity: Total amount of available soil moisture held by a soil for use by plants. Usually the moisture held between field capacity and permanent wilting point. Also known as available water holding capacity.

Backfilling: Refilling the trench with earth after the tiles or pipes have been laid.

Base Saturation Percentage: The proportion of the soil's cation exchange capacity occupied by basic cations namely sodium, potassium, magnesium and calcium.

Black Alkali: Alkali soils also called black soils because black patches appear on these soils due to the deposition of organic matter that gets dissolved at high pH and accumulates at the soil surface.

Benefit: Cost Ratio: The discounted values of benefits divided by the discounted values of costs (adjusted to a common time basis).

Bio-drainage: Use of water from the shallow water table by quick growing plants to cause lowering of the water table; drainage with the help of biological material, the plants.

Bio- interceptor Drain: Interception of canal seepage by growing plants on both sides of the canal.

Bioremediation: The use of biological agents to remove substances hazardous to human health or the environment from contaminated soil or water.

Blending: Mixing; term used in conjunctive use of multi quality waters when two or more qualities of waters are mixed to grow a crop.

Blind Drain: An excavated trench refilled with pervious materials such as coarse sand, gravel or crushed stone, through which water percolates and flows towards an outlet. Often referred to as a French drain because of its initial development and widespread use in France.

Blind Inlet: Surface water inlet to subsurface drain in which water enters through percolation rather than open channels flow. Most commonly used where quantity of water is small and/or water contains high sediment content.

Blinding Material: Material placed on top of and around a subsurface drain to improve the flow of water to the drain as well as to prevent displacement of the drain during back-filling of the trench.

Blinding: The process of placing the blinding material.

Borehole: A hole, usually vertical, bored to determine subsurface conditions, for extraction of water or measurement of groundwater level.

Buffer: A substance that prevents a rapid change in pH when acids or alkalis are added to the soil, these include clay, humus and carbonates.

Buffering Capacity: Buffering capacity is the ability of the soil to resist high alkalinity or acidity levels.

Bulk Density: The ratio of the mass of oven dried soil to its bulk volume. When expressed in g/cm^3, bulk density is numerically equal to apparent specific gravity or volume weight. The bulk density of a soil is always smaller than its particle density. The bulk density of sandy soil is about 1.6 g/cm^3, whereas that of organic matter is about 0.5. Bulk density normally decreases, as mineral soils become finer in texture.

Bullet: Round-nosed cylindrical plough point of a mole drain plough, which forms a rounded cavity as the plough is drawn through the soil. Also called a torpedo.

Calcareous Soil: A soil that contains enough calcium carbonate so that it effervesces when treated with hydrochloric acid. Acidifying materials can be used to reclaim calcareous alkali soils.

Calcitic Lime: Limestone consisting of $CaCO_3$ - based material with very low magnesium content.

Calcium Carbonate Equivalent (CCE): The liming potential of a material as compared to $CaCO_3$.

Canal Seepage: Water that leaves a canal or any water conveying structure through capillary action and percolation through its wet perimeter, usually a concern when it results in rise in water table in the adjoining area.

Capillary Fringe: Soil zone above the water table where water rises due to surface tension and remains almost saturated. It may consist solely of capillary water or it may be combined with gravity water in transit to the water table.

Capillary Rise: The upward movement of water from a free water table due to adhesion of water to the soil capillaries and the cohesion of water molecules.

Cased Well: A well usually constructed to observe water table, provided with a casing to control caving in of the soil and for long life in unstable soils.

Cation Exchange Capacity: The total quantity of cations which a soil can absorb through cation exchange, usually expressed as c mol(p^+)/kg soil or milliequivalents per 100 g of soil.

Chisel, Subsoil: A tillage implement with one or more cultivator-type feet to which are attached strong knife like units used to shatter or loosen hard, compact subsoil layers, to depths below normal plough depth.

Clay Tile: Pipe made from shales or clays, commonly of short lengths, used as conduits for subsurface drains.

Coefficient of Permeability: See Hydraulic Conductivity.

Coefficient of Runoff: Factor in the rational runoff formula expressing the ratio of the peak runoff rate to rainfall intensity.

Coefficient of Storage: The amount of water in cubic units, discharged from each vertical column of aquifer with a basal unit area as the water level falls through a unit depth. In an unconfined aquifer the coefficient of storage is essentially equal to the specific yield.

Coefficient of Transmissibility: Coefficient of transmissibility of an aquifer is the product of the field coefficient of permeability multiplied by the thickness of the saturated portion of the aquifer.

Collector Drain: For surface drainage system, it is an open drain into which a number of field drains discharge their water. For subsurface drainage system, it is usually a closed conduit, larger in size than the lateral drains into which a number of laterals discharge their water.

Composite System: A combination of drainage systems, for example, the herringbone and gridiron system. In subsurface drainage, a system where laterals discharge into collectors before the drainage water is disposed of.

Conduit: Any open or closed channel intended for the conveyance of water.

Conjunctive Use: Use of multi-quality, multi-sources water in conjunction to get better returns in comparison to when these waters are used separately.

Consolidation: The gradual slow compression of a cohesive soil due to weight acting on it as water and air are driven out of the voids in a soil, commonly encountered in clays or slowly permeable soils.

Controlled Drainage: Regulation of the water table by means of pumps, control dams or check dams, or a combination of these for maintaining the water level at a depth favourable to boost crop growth.

Conveyance Efficiency: Ratio between water received at the inlet to a block of fields and that released at the project's headwork's; usually expressed in fraction or percentage.

Conveyance Loss: Loss of water from a channel or pipe during conveyance, including losses due to seepage, evaporation leakage and transpiration by plants growing in or near the channel.

Corrugations: The design of the profile on the pipe surface having ridge and valley

Cost Recovery: Fee structures to cover the cost of providing a service.

Cover Drain: A closed drain placed below the surface of the soil.

Critical depth: (i) The depth of flow in a channel of specified dimensions at which specific energy is minimum for a given discharge. (ii) The depth to water table at which rate of evaporation through capillary rise is reduced to almost zero. In practice, it is 1/10th of the rate which occurs when the water table is at or near the soil surface.

Crop Tolerance: Tolerance of crops to adverse abiotic stresses such as water logging, soil salinity, alkalinity, heat and cold waves etc.; it may be due to the biotic stresses as well

Crop Tolerance Period: It is the duration up to which a specific crop can withstand adverse stresses such as water logging condition without significant damage to the quality and quantity of yield.

Cross-slope Ditch System or Drainage-type Terraces: A system of ditches to drain sloping land. These ditches usually function both for surface drainage and for erosion control. When specifically designed for erosion control, these are called 'terraces'.

Cumulative Infiltration: Total sum of the depth of water absorbed by a soil in a specified elapsed time in reference to the time of initial water application.

Cut-off Drain: A tile drain system normally placed near the upper edge of a wet area to intercept water before it gets down the slope.

Cyclic Use: A term used in conjunctive use of multi quality waters when the two or more qualities of waters are used to irrigate the crop in some cycle or sequence, to avoid stress at salt sensitive stages.

Deep Percolation: Water which percolates downward below the root zone and cannot be used by plants.

Degradation: The process of soil change from one type to another resulting in increased production constraints. Soils being salinized or the process of alkalization or even the process of replacement of sodium by hydrogen during leaching.

Degrees of Limitation: The scaling of a single factor (land use requirement, land quality or land characteristic) according to its adverse effects on a specified land utilization type. In drainage it is the degree of limitation exerted by water logging, salinity or alkalinity of the soil or their consequential impacts.

Desalinization: Removal of excessive soluble salts from the saline soil, usually by leaching.

Design discharge: Specific value of flow rate selected for the design of the dimensions of a structure or a parts of a system.

Diagnostic Criterion: A variable (land quality, land characteristic etc.) that has an influence on the outputs and inputs of a specified kind of land use, and which serves as a basis for assessing the suitability of the land for that use.

Dike: An embankment constructed of earth or other suitable materials to protect land against overflow from streams, lakes, and tidal influences.

Discounted Benefits: Actual benefits occurring in future adjusted through the use of discount rates to a common time base.

Discounted Costs: Actual costs occurring in future adjusted through the use of discount rates to a common time base.

Discounting: The use of discount factors to adjust cash flows to a common time basis.

Dispersed soil: Soil in which clay readily forms a colloidal solution, usually have low permeability, tend to shrink, crack and become hard on drying and become plastic on wetting.

Distribution Loss: See conveyance loss.

Distribution System: (a) System of ditches, and their appurtenances, which conveys irrigation water from the main canal to the farm units. (b) Any system which distributes water to users.

Ditch: Constructed open channel for collecting and/or conveying water. A ditch may carry flow from both surface and subsurface drainage.

Diversion Ditch: A graded channel constructed across the land slope to intercept and divert water to a suitable outlet; the system comprising ditches and dikes which diverts water away from a low-lying area or prevents water from flooding land is known as diversion system.

Dolomitic Lime: A naturally occurring liming material composed chiefly of carbonates of magnesium and calcium

Double Ditch or Drain: An area of land, which is drained by both the laterals and a main tile.

Double-main system: A system of drains used where seeps appear on both sides of a broad flat, also where a stream separates an open flat into two parts.

Drain: (a) A channel, such as open ditches or drain tile, to remove excess water by surface or by internal flow. (b) To lose water (from the soil) by percolation.

Drain Depth: For open surface drains, it is the depth from the ground surface to the drain bottom. For subsurface drains, it is the distance between the ground surface and the drain tube centre.

Drain Spacing: The horizontal distance between the centrelines of adjacent parallel drains.

Drain Tile: Concrete or ceramic pipe used to conduct water from the soil.

Drain Well: Vertical opening to a permeable substratum into which surface and subsurface drainage water is discharged.

Drainable Pore Space: The ratio of the change in soil-water content in the profile above the water table to the corresponding rise/fall of the water table in the absence of evaporation. Also see specific yield and effective porosity.

Drainage: Process of removing groundwater or surface water by artificial or natural means.

Drainage Channels: An open channel either artificial or natural for carrying off surplus ground or surface water.

Drainage Coefficient: It is the amount of water to be removed in 24 hours to provide favourable conditions for crop growth. It is usually expressed in mm/d. It is also known as drainage modulus.

Drainage-Density: The average length of streams, natural or artificial, per unit area within the basin.

Drainage Design: It is process of arriving at an appropriate combination of depth and spacing of drains, cross-sectional area and slope such that the drainage system is economically viable for any or a combination of water table control, disposal of excess surface water, salinity control and alkalinity control.

Drainage, Excessive: Too much or too rapid loss of water from soils, either by percolation or by surface flow.

Drainage Material: Any material that can be laid below the soil surface to provide an easy passage of flow to the subsurface water. Examples are clay tiles (with or without perforations), rigid or corrugated flexible perforated PVC pipes, etc. In real sense of term, it also includes envelope materials.

Drainage Modulus: See Drainage coefficient

Drainage Pumping Plant: Pump's power unit and other appurtenances for lifting drainage water from a collecting basin to a gravity outlet.

Drainage System: A system comprising a main drain, its branches and subsidiary drains, used to collect and dispose of excess surface or subsurface water.

Drawdown: The lowering of the water table or piezometric surface caused by the discharge of groundwater through wells, springs or other openings.

Economic Analysis: Assessment of an investment's profitability to a whole society or economy regardless of who in the society contributes to the costs and receives the benefits, conventionally measured in 'opportunity' rather than 'market' prices.

Ecosystem: A complex system formed by the interaction of a community of organisms with its environment.

Edaphic: Of or pertaining to the soils is the condition resulting from or influenced by factors inherent in the soil or other substrata, rather than by climatic factors.

Effective Calcium Carbonate Equivalent (ECCE): A measure of the effectiveness of liming materials calculated as the product of the purity value (CCE) and the fineness value divided by 100.

Effective Porosity: The portion of pore space in saturated porous media in which movement of water takes place.

Electrical Conductivity: It is the property of a medium expressing its capacity to carry electrical charge from one point to another point. It is the reciprocal of the electrical resistivity. The resistivity is the resistance in ohms of a conductor which is 1cm long and has a cross-sectional area of $1cm^2$. This property, usually expressed in milli or micro ohm/cm, is used to express the salinity of the soil extracts, drainage water, irrigation water, etc. International units of EC are deci-Simens per meter or dS/m. One dS/m is equal to 1 mmhos/cm.

Entrance Head: Extra head required to overcome the entrance resistance.

Entrance Resistance: The extra resistance to water flow in the vicinity of a drain pipe either due to decreased permeability of the material around the drain and/or due to contraction of flow lines resulting from drain openings.

Environmental Impact: The effect of human interference on the environment. It could be both favourable and unfavourable.

Envelop Material: A generic name for both envelops and filters. Envelop acts as permeable constraint to impede entry of excess quantities of soil particles into drains pipes whereas a filter improves the permeability around the pipe.

Equivalent Depth: It is fictitious reduced depth of effective flow zone due to increased length of streamlines of flow when the tile or open drains do not reach down to the impermeable layer.

Evapotranspiration: The rate of water loss through transpiration from vegetation plus evaporation from the soil surface or from standing water on the soil surface - mm/day or m^3/day.

Evaporation pond: A pond for storage and evaporation of low quality drainage effluent, bottom may be impervious to minimize seepage of water in the area.

Exchange Complex: The surface active constituents of soils (both inorganic and organic) that are capable of cation exchange.

Exchangeable Sodium Percentage (ESP): This term indicates the degree of saturation of the soil exchange complex with sodium and is defined as follows:

$$ESP = \frac{\text{Exchangeable sodium (meq/100 g soil)}}{\text{Cation exchange capacity (meq/100 g soil)}} \times 100$$

Exchangeable-cation Percentage (ECP): The extent to which the adsorption complex of a soil is occupied by a particular cation. See Exchangeable Sodium Percentage.

Farm Drainage: The disposal and removal of excess water from farm lands to prevent or to relieve water logging, to prevent accumulation of harmful amounts of salts and deterioration of soil structure, and to provide healthy environment for crop growth.

Farm Financial Viability: Analysis to ascertain whether the farmer is sufficiently better-off with the project than without the project using farm gate prices and actual cash flow.

Field Application Efficiency: The ratio of water available to the crop to that received at the field head; expressed either in fraction or percentage.

Field Capacity (Field Capillary Moisture Capacity and Specific Retention): The amount of water retained in a soil against the force of gravity at any specified time (about 24 to 36 h) after flooding. It is approximated by moisture equivalent and one-third atmosphere percentage.

Field Drain: It is usually a shallow drain provided in the field to facilitate the removal of excess accumulated surface water from different parts of the field.

Filter (drain): A material placed around the subsurface drains to prevent the soil entering into the drains and to provide a zone of higher hydraulic conductivity (than the soil) around the drain tubes.

Financial Analysis: Assessment of an investment's profitability to those people or entities contributing capital and directly benefitting from the returns, conventionally measured at market prices.

Final Infiltration Rate: Rate at which water is absorbed by soil when the infiltration has become essentially constant. Also called final intake rate (See infiltration)

Fineness Factor: A factor used to calculate the effective CCE of lime materials and controlled by the fineness of the lime material.

Flushing: A process to remove salts from the top few cm by impounding and draining the water through surface drainage. Used for heavy textured soils where leaching may pose problems. See leaching.

Free Groundwater: Groundwater in aquifers not bounded or confined by impervious strata.

Free Water: Water, which drains through a soil under the influence of gravity. It is also known as gravitational water.

French drain: See blind drain

Gauge: Device for measuring water level, discharge, velocity, pressure, etc.

GIS (Geographical Information System): A term for a computer-based system for storing, accessing and representing maps and other data in electronic form.

Gravel Filter: Graded sand and gravel aggregate placed around a drain or well screen to prevent the infiltration of fine materials from the aquifer into the drain or well. To provide a high hydraulic conductivity zone around the drain pipe. Sometimes it is also known as gravel pack.

Gravel Mole: A mole drain filled with gravel

Gravel Pack: See gravel filter.

Gravitational Water: see free water

Gravity Outlet: Outlet for a drain line or system where water flows by gravity.

Grid-iron System: A system in which parallel laterals enter into the branch or main drain only from one side.

Gross Irrigated Area. The area of land irrigated in a year (land having two irrigation seasons is counted twice). See Net Irrigated Area.

Groundwater: Water in soil beneath the soil surface, usually under conditions where the pressure in the water is greater than the atmospheric pressure, and the soil voids are substantially filled with the water.

Groundwater Balance: It is a concept which states that all inputs of water in a defined space and time are equal to the sum of all outputs of water, and the changes of water storage, in the same space and time.

Groundwater Flow: Flow of water in an aquifer or soil. That portion of the discharge of a stream which is derived from groundwater under hydraulic gradient.

Groundwater Quality: Quality of groundwater in relation to various purposes. For the purpose of this book purpose is irrigation.

Groundwater Runoff: That part of the groundwater which is discharged into a stream channel as spring or seepage water.

Groundwater Mining: Pumping groundwater from a basin beyond the safe yield, thereby extracting groundwater which had accumulated over a long period of time.

Groundwater Table: The level of water storage (above mean sea level) in an aquifer; hence, the point at which the soil is fully saturated with water.

Gypsum: Calcium sulphate ($CaSO_4.2H_2O$). Used for the replacement of sodium by calcium in alkali soils/water.

Gypsum Requirement (GR): It is the quantity of gypsum required to lower down the ESP of a soil from an initial high value to the desired value.

Halophyte: A plant capable of growing in salty soil; i.e. a salt tolerant plant.

Heavy Metals: Trace elements regulated because of their potential for human, plant, or animal toxicity, including cadmium (Cd), copper (Cu), chromium (Cr), mercury (Hg), nickel (Ni), lead (Pb) and Zinc (Zn). In chemistry, these have a high atomic number, atomic weight and a specific gravity greater than 5.0

Herringbone System: A drainage layout in which drain laterals enter the collector from both sides at an angle to ensure adequate slope.

High Water Mark: A distinct mark made on vegetation, buildings or rocks that shows the extent of water rise.

Horizontal Drainage: A method of drainage to control water table through tile or pipes placed horizontally beneath the soil surface.

Hydraulic Conductivity: The rate of flow of water through a porous material of a unit cross-sectional area under a unit hydraulic gradient, at a specified temperature. It is also called unit of permeability, transmission constant or coefficient of transmission. It is the coefficient, K in the Darcy's equation v = -Ki, such that v is the flow velocity of a specified liquid under saturated conditions, and i is the hydraulic gradient. Thus, unit of K is L/T.

Hydraulic Gradient: The decrease in hydraulic head per unit distance in the soil in the direction of flow.

Hydraulic Head: It is the sum of pressure head and gravitational head at the point under consideration.

Hydraulic Radius: It is ratio of cross sectional area of flow to the wetted perimeter.

Hydraulic Soil Properties: Soil properties that affect the flow of water in the soil profile.

Hydrograph (Groundwater): Groundwater graph which shows water level in a well, plotted against time.

Hydro-isobaths, or Isobaths of Water Table: Contour of similar depth of the subsoil water table below the ground surface.

Hydro-isopleth Map: A map showing fluctuation of water table with respect to time and space.

Hydromorphic Soils: Formed under conditions of poor drainage in marshes, swamps, seepage areas or flats.

Hydrophytes: Plants that grow in water or in wet soils partially or wholly immersed.

Hydrostatic Pressure: Force per unit of area exerted by a liquid due to its elevation above a point.

Ideal drain: A drain having no entrance resistance.

Impeded Drainage: Condition in which the downward movement of gravitational water in soil is restricted.

Impermeable Layer: In drainage, a layer of sub-soil which does not easily permit water flow through it. A layer is usually called impermeable if its hydraulic conductivity is one-tenth or less than that of the zone in which the subsurface drains are proposed to be laid.

Infiltration: The process of entry of water into the soil profile through the ground surface.

Infiltration Rate: The maximum rate at which a soil, in a given condition at a given time, can absorb water impounded on the surface when adequate precautions are taken regarding border or fringe effects. Defined as the volume of water passing into the soil per unit of area per unit of time, it has the dimensions of velocity (L/T).

Infiltrometer: A device to measure infiltration rate. A double ring infiltrometer is commonly used to measure the infiltration rate.

Inlet: Surface connection to a subsurface drain; structure at the diversion end of a conduit; upstream end of any structure through which water may flow.

Inspection Chamber: Structure provided in the form of a manhole at the junction of a lateral drain and the collector drain, to act as a silt trap and to inspect the functioning of the drainage system.

Intangible Benefits: All the benefits (direct and indirect) that cannot be expressed in monetary terms.

Interceptor Drain: (i) Surface/subsurface drains or their combination, installed to collect subsurface flow before it resurfaces (ii) Surface/subsurface drains to intercept seepage from canals.

Internal Drainage: Drainage of the soil profile; may be either natural or artificial.

Internal Rate of Return: The rate of discounting at which the present value of benefits becomes equal to the present value of costs.

Inverse Auger Hole Method: A method similar to auger hole method to determine hydraulic conductivity above the water table.

Inverted Well: Disposal of drainage water through wells into a porous layer of earth or an open-rock formation.

Isograms, or Isohaline Lines: The locus of points of equal salt concentration.

Isopachs: Lines of equal groundwater rise or fall during a specified period of time.

Isotropic: Soils having the same physical properties in all directions.

Isobath Map: A map showing lines connecting all points on a land surface that have the same depth to water table.

Junction: Point of intersection of two or more closed drains; accessory used to effect a junction between two tile lines.

Junction Box: A rectangular manhole or other structure which serves to join one or more drains.

Land Forming: Changing the micro-topography of the land to meet the pre-decided requirements say of drainage or irrigation; in drainage to facilitate the movement of surface water over a field or a part of a field.

Land Grading: The operation of producing a plane land surface with a continuous slope; also referred as land leveling.

Land Improvement: An alteration in the quality of land to improve its potential for land use sometimes referred as major land improvement and minor land improvement.

Land Leveling: See land grading.

Land Reclamation: Changing general physico-chemical characteristics of the land to make it capable of more intensive use for agriculture.

Land Smoothing: Process of eliminating minor surface irregularities without changing the general contours of the land; the finishing operation in land levelling. Also called land smoothening.

Lateral: Secondary or side channel, ditch or conduit. Also called branch line or drain, spur, lateral ditch, group lateral, etc.

Lateral Drain: It is a field drain to remove excess water from root zone. It can be unlined open channel, mole drain, tile drain, perforated pipes, etc.

Leaching: The process of removal of soluble material from soil by the passage of water through the soil. A primary step in the reclamation of saline soils.

Leaching Efficiency: State of mixing of salts within the crop root zone depth when profile is leached with some irrigation water.

Leaching Fraction: The fraction of water applied on the soil that actually passes through the root zone as against the leaching requirement which must pass through the root zone to prevent soil salinity from exceeding a specified value.

Leaching Requirement: The fraction of the water entering the soil that must pass through the root zone in order to prevent soil salinity from exceeding a specified value. Leaching requirement is used primarily under steady state or large-time average conditions.

Lime: Compounds of calcium used to correct the acidity of soils.

Lime Fineness: The particle size of limestone determined by the fineness of grinding. The finer the grind, the more reactive the material and the more quickly the material neutralizes soil acidity.

Liming: The process of application of liming materials.

Liming Material: A material capable of neutralizing soil acidity.

Lime Purity: The measure of impurities in a given liming material, needed to estimate the material's neutralizing value.

Lime Requirement: The amount of liming material required to change the soil pH to a specific desired value.

Line-pan: A hardened layer cemented by calcium carbonate.

Main Ditch: Principal ditch, channel, or stream, natural, improved, or constructed, serving one or more drainage enterprises. Local names of streams, rivers or channels are often used for identification.

Main Drain: An existing natural or artificial drain into which the collector drains discharge their water to an outlet.

Map Scale: Relationship between a certain distance on the map and the corresponding distance on the ground (e.g. 1:10,000 means 1 cm on the map equals to 10,000 cm or 100 m on the ground).

Mechanical Analysis: To determine particle size distribution of a soil by screening, sieving and International pipette method or hydrometer analysis.

Mesophytic Vegetation: Plants which for their growth avoid both extremes of moisture and drought; plants which are intermediate between 'hydrophytes' and 'xerophytes'.

Mhos: Reciprocal of Ohms. A unit of Electrical Conductivity to measure the quality of water, See Electrical Conductivity.

Micro-relief: Small-scale, local differences in topography, including mounds, swales, or pits that are only a few meters in diameter and with elevation differences of up to 2 m.

Milliequivalent (meq): One thousandth of an equivalent: atomic weight divided by valence/1000.

Milliequivalent per 100 g soil (meq/100 g soil): Milligrams of a cation divided by its equivalent weight per 100 g soil.

Milliequivalent per Litre: A milliequivalent of an ion or a compound in one litre of solution.

Mole Drains: Cylindrical drains formed in the subsoil by pulling a bullet-shaped cylinder through it.

Mole Plough or Mole Plough: A plough by which a steel ball following a pointed-iron shoe makes an underground channel, resembling the track of a mole, that serves as a drain.

Natural Drainage: It refers to the natural topographic and/or subsoil conditions of an area for disposing of excess surface and/or subsurface water from it. Natural system of drains follows the natural depression in the land, and diverges like the frame.

Net Irrigated Area: The actual irrigated area by a system as against the gross irrigated area where an area is considered twice if irrigated in both the seasons.

Net Present Value: The present value of benefits minus the present value of costs.

Non-saline Alkali Soil: A soil that contains sufficient exchangeable sodium to interfere with the growth of most crop plants and contains variable quantities of soluble salts. The exchangeable sodium percentage is greater than 15, the conductivity of the saturation extract is usually less than 4 dS/m (at 25°C) and the pH of the saturated soil usually ranges between 8.2 to 10.0 or even more.

Observation Well: It is a perforated tube installed vertically into the soil profile going below the water table to monitor the water table fluctuation by taking measurement of the water level inside the tube. It may be cased or uncased.

Opportunity Cost of Capital: The opportunity cost of using investment resources in a project rather than in their next best alternative use, usually expressed in the form of an interest rate.

Organic Material: Organic matter, also called 'humus', forms from the decay of leaves, plants and other plant biological materials.

Organic Soil: Soils having high composed or decomposed organic carbon and a low mineral content.

Osmotic Pressure: The equivalent negative pressure that influences the rate of diffusion of water through a semi-permeable membrane or the pressure difference required to equalize the diffusion rates between the solution and pure water across a semi-permeable membrane. Osmotic pressure in atmospheres may be calculated from the freezing-point depression ΔT in °C by the formula $OP=12.06\ \Delta T- 0.021\ \Delta T^2$. If expressed per unit weight, it is also called osmotic head.

Outfall: Point where water flows from a conduit, stream or drain.

Outlet: The terminal point of the drainage system, ditch or drain under consideration; point of water disposal.

Overland Flow: The flow of storm water over the ground before becoming channelized.

Over-liming: Applying lime to a soil in excess of soil requirement.

Oxidation: Oxidation is the addition of oxygen, removal of hydrogen, or the removal of electrons from an element or compound. In the environment, organic matter is oxidized to more stable substances. Oxidation is the opposite of 'reduction'.

Parallel Field Ditch System: A system similar to bedding, except that the drains are spaced farther apart and have a greater capacity than the dead furrows. The minimum depth of the ditch is 22.5 cm and minimum cross-section area 0.4 sq m.

Parallel Lateral Open Ditch System: A system similar to field ditch system, except that the ditches are deeper and cannot be crossed with most machinery.

Parent Material: Mineral or rock material on and/or from which soils are formed during pedogenesis (soil formation process); parent material is one of the five major soil forming factors.

Particle Density: The weight per unit volume of the solid portion of soil is called particle density. Generally particle density of normal soils is 2.65 g/cm^3. The particle density is higher if large amount of heavy minerals such as magnetite; limonite and hematite are present in the soil. With increase in organic matter of the soil the particle density decreases. It is also termed as true density.

Particle-size Analysis: Determination of the various amounts of different separates in a soil sample, usually by sedimentation, sieving, micrometry, or combinations of these methods.

Particle-size Distribution: The amounts of the various soils separate in a soil sample, usually expressed as weight percentage.

Parts Per Million (ppm): Weight units of any given substance per one million equivalent weight units of oven-dry soil; or in the case of soil solution or other solutions, the weight units of solute per million weight units of solution.

Perched Water Table: A water table, usually of limited area, maintained above the normal groundwater level by the presence of an intervening, relatively impervious confining stratum.

Percolation: The gravity induced downward flow of water which passes below the effective root zone of crop.

Permeability: A soil property, which is a measure of the readiness with which the soil transmits water. The permeability is a velocity, and is expressed in cm/h. For most soils, both the chemical nature of the water used and the history and treatment of the sample greatly affect the permeability.

pF: The numerical measure of energy with which water is held in the soil, expressed as the common logarithm of the absolute value of the matric head in cm of water.

pH: The common logarithm of the reciprocal of the hydrogen ion concentration of a system. pH 7 is neutral, pH less than 7 is acidic and pH more than 7 is basic or alkaline.

Phreatic Divide: An underground divide that fixes the boundary of area that contributes groundwater to each stream system.

Phreatic Surface: The upper boundary of an unconfined groundwater body, at which the water pressure is equal to the atmospheric pressure.

Phreatophytes: Deep-rooted plants which obtain their water from the water table or the layer of soil just above the roots.

Piezometer: An instrument for measuring the pressure head of liquid. Under field conditions it is similar to the observation well but is without perforations. See observation well.

Pipe Deflection: The ratio of the deflection in pipe inside diameter to the initial inside diameter expressed as the percentage of the initial inside diameter

Pipe Stiffness (PS): The value obtained by dividing the force per unit length of specimen by the resulting deflection in the same units at the prescribed percentage deflection.

Pond: See Reservoir.

Porosity: The fraction of the soil volume not occupied by soil particles i.e. the ratio of the sum of the volume of the liquid and gas phases to the sum of the volumes of the solid, liquid and gas phases of soil.

Potential Acidity: Acidy caused by hydrogen and aluminum ion adsorbed on cation exchange capacity (CEC) of the soils.

Potential Evapotranspiration: The theoretical quantity of water that is lost as water vapour from the cropped surface under given meteorological conditions and when the water is not a limiting factor.

Precipitation: The total measurable supply of water of all forms of falling moisture, including dew, rain, mist, snow, hail and sleet; usually expressed as depth of liquid water on a horizontal surface in a day, month, or year, and designated so daily, monthly or annual precipitation.

Primary Benefits or Primary Effects: Benefits from project including the value of any reduction in costs due to the project and of any increase of farm products marketed or consumed by the farm family.

Project Justification: Analysis and testing through benefit: cost ratio, net present value or internal rate of return whether a project meets an acceptable or pre-determined standard of economic worth.

Radial Flow: Groundwater flow towards the wet perimeter of a drain such that flow lines converge towards the drain.

Radial Resistance: Resistance caused by radially converging flow lines.

Radius of Influence: It is the radial distance from the center of the well to the point where the cone of depression tangentially meets the static or unaffected water table.

Rainfall: 1. precipitation in the form of liquid water. 2. The amount of rain, usually expressed in millimeters (inches) depth of water on an area, that reaches the surface of the earth. The term is sometimes expanded to include other forms of atmospheric precipitation, such as snow, dew, but technically the term precipitation would be used in this broader scope.

Rainfall Frequency: The average time interval between the occurrences of rainfall event of an equal or greater intensity/magnitude.

Rainfall Intensity: The rate at which rainfall occurs expressed in depth units per unit time. It is the ratio of the total amount of rain to the length of the period in which the rain falls.

Reaction Factor: A factor in non-steady state drainage theory that expresses the rate at which a drainage system is able to lower the water table after a recharge event.

Recharge: Process by which water is added to the zone of saturation, as recharge of an aquifer.

Reconnaissance Survey: An exploratory study of the area to identify problems, degree and extent of the problem and even possible solutions although only qualitatively.

Recurrence Interval: The average time interval between occurrences of a hydrological event of a given or greater magnitude, e.g. in an annual flood series the average interval in which a flood of a given size recurs as an annual maximum.

Recycle: Using same water several times in the same sector i.e. using irrigation return flow or drainage water for irrigation

Reduction: Reduction is the addition of hydrogen, removal of oxygen, or the addition of electrons to an element or compound. Under anaerobic conditions (where there is no dissolved oxygen present). Sulphur compounds are reduced to hydrogen sulphide (H_2S) and other compounds. Opposite of oxidation.

Relief Drain: A drain to lower the groundwater where gradients of water table and subsurface strata do not permit sufficient lateral movement of groundwater. The principal source of groundwater is from the overlying land and the water table is relatively flat.

Remote Sensing: The collection and analysis of data from above the crop canopy, using sensors that respond to different frequencies of electromagnetic radiation.

Reservoir: A natural or constructed place in which water is collected and stored for use, also known as pond, tank etc.

Residual Acidity: Acidity due to aluminium and hydroxyl ions present in non–exchangeable form with organic matter and clay.

Residual Sodium Carbonate (RSC): It is defined as the excess of CO_3+HCO_3 ions over the Ca+Mg ions present in the water. The RSC is calculated with the following relation having units of meq/L.

$$RSC = (CO_3 + HCO_3) - (Ca + Mg)$$

Return Period: Strictly speaking, it is the period, as in a time series, for a value to return to the same value. Generally, it is used in the sense of the average time interval within which a given event will be equalled or exceeded. For example, where the interval between observations is a year, a return period of 100 years means that, on the average in the long run, an event of this magnitude, or greater, is expected to occur once in 100 years. It is inverse of frequency

Reuse: Use of same water several times often in other sectors such as use of domestic wastewater in agriculture. Often recycle and reuse is used synonymously.

River Basin: A geographical area determined by the watershed limits of a system of waters, including surface and groundwater flowing into a common terminus.

Roughness Coefficient: A dimensionless parameter appearing in Manning's equation related to surface irregularity, vegetation drag and retardance of the wetted perimeter.

Rubble Drain: A small shallow trench filled with coarse rubble, clinker or similar material, with or without field drain pipes.

Runoff: That portion of precipitation which is not absorbed by the deep strata but finds its way into the streams after meeting the demands of evapotranspiration including interception and other losses. It includes surface runoff received into the channels after rainfall, delayed runoff that enters the streams after passing through the earth and other delayed runoff that has been temporarily detained as a snow cover and stored in natural lakes or swamps.

Runoff Coefficient: i) The ratio of the peak rate of runoff to the uniform rate of rainfall with a duration equalling or exceeding the time of concentration which produced this rate of runoff. ii) Also used incorrectly for 'percentage runoff'.

Rupture: A crack or break extending entirely or partly through the pipe wall.

Safe Yield: The maximum rate at which water can be artificially withdrawn from an aquifer throughout the foreseeable future without impairing the aquifer.

Saline-Alkali Soil: A soil containing sufficient exchangeable sodium to interfere with the growth of most crop plants and containing appreciable quantities of soluble salts. The exchangeable sodium percentage is greater than 15 and the conductivity of the saturation extract is greater than 4 dS/m (at 25°C). The pH of the saturated soil is usually more than 8.2.

Saline Soil: A non-alkali soil containing soluble salts in such quantities that they interfere with the growth of most crop plants. The conductivity of the saturation extract is greater than 4 dS/m (at 25°C), the exchangeable sodium percentage is less than 15, and the pH of the saturated soil is less than 8.2.

Saline Water: A water having excessive amount of soluble salts such that if used for irrigation would interfere with plant growth. The EC is usually more than 2 dS/m. SAR or RSC may or may not be high.

Salinity: Concentration of salt in a given fluid or material.

Salinization: The process of accumulation of salts in the soil particularly in the root zone.

Salt Balance: The salt balance of an irrigated area is commonly defined as the difference between the total dissolved solids brought to the land in a pre-decided duration by the irrigation water and the total solids carried away by the drainage water during the same period.

Saturated Soil Paste: A mixture of soil and water, at saturation it reflects light, flows slightly when the container is tipped, and the paste slides freely and cleanly from a spatula for all soils at its saturation percentage.

Saturation Extract: The solution obtained by pressure or vacuum filtration of a soil paste that has been made up to a saturated condition by adding water while stirring.

Saturation Percentage: The moisture content of a sample of soil that has been brought to saturation by adding water while stirring. Expressed as grams of water per 100 g of dry soil.

Scraping: Removal of the salt incrustations that have accumulated at the surface due to excessive evaporation.

Seepage: The infiltration of water downwards or laterally into soil or substrata from a source of supply such as a reservoir, irrigation canal or channel.

Seepage Meter: Device for measuring seepage rate.

Sensitive Crops or Plants: A qualitative term used to designate the plants which are sensitive to water, salts or some toxic element. Also see tolerant.

Singular System: A drainage system in which each lateral directly discharge into an open collector/ main drain.

Sink-hole Drainage: A system of tile drains similar to the cut-off system and used for the same purpose, *viz.* to intercept seepage water, but has, in addition, wells dug at regular intervals to let the water come up from a lower stratum and enter the drain.

Slick Spots: Small areas in a field that are somewhat black alkali when wet due to high exchangeable sodium.

Slope: Degree of deviation of a surface from the horizontal, usually expressed in per cent or degrees.

Sodic Soil: (a) A soil containing sufficient exchangeable sodium to interfere with the growth of most crop plants. Used synonymously with alkali soils. See Alkali soils

Sodic water: A water with residual sodium carbonate more than 2.5. EC/SAR may or may not be high.

Sodication or Sodification: See Alkalization

Sodium Adsorption Ratio (SAR): A ratio for soil extracts and irrigation waters used to express the relative activity of sodium to calcium and magnesium. SAR is calculated as follows:

SAR= Na/((Ca+Mg)/2)$^{1/2}$ (all ions expressed in meq/L)

SAR= Na/(Ca+Mg)$^{1/2}$ (All ions expressed in mmol/L)

Expressed in units of (mmol/L)$^{1/2}$.

Soil Aeration: Movement of air into the soil as water is drained out. A process by which air and other gases in the soil are renewed.

Soil Auger: A tool for boring into the soil and withdrawing a small sample for field or laboratory determinations. Soil augers are classified into several types.

Soil Extract: The solution separated from a soil suspension or from a soil by filtration, centrifugation, suction or pressure (may or may not be heated prior to separation).

Soil Map: A map showing the distribution of soil types or other soil mapping units in relation to the prominent physical and cultural features of the earth's surface.

Soil Permeability: Capacity of the soil to transmit fluids, see permeability.

Soil Structure: The combination or arrangement of primary soil particles into secondary particles, units, or peds. These secondary units may be but usually are not, arranged in the profile in such a manner as to give a distinctive characteristic pattern. The secondary units are characterized and classified on the basis of size, shape, and degree of distinctness into classes, types and grades, respectively.

Soil Structure Classes: A grouping of soil structural units or peds on the basis of size.

Soil Survey: The systematic examination, description classification, and mapping of soils in an area. Soil surveys are classified according to the kind and intensity of field examination.

Soil Texture: The tèxture is the term indicating the coarseness or fineness of the soil. It is decided on the basis of the amount and quantity of each of the grain size group of particle that constitute the soil.

Soil Water Balance (Field Water Balance): The sum of all gains and losses of water over a given period of time; mm/per unit time.

Soluble Sodium Percentage: The proportion of sodium ions in solution in relation to the total cation concentration, defined as follows:

$$SSP = \frac{\text{Soluble sodium concentration (meq/L)}}{\text{Total cation concentration (meq/L)}}$$

Specific Ion Effect: Any effect of a salt constituent in the substrate on plant growth which is not due to the osmatic properties of the substrate.

Specific yield: The ratio of the volume of water which a rock or soil, after being saturated, will yield by gravity, to the total volume of rock or soil.

Spoil Bank: Rock waste, banks, and dumps, from the excavation of ditches.

Storage Coefficient: A dimensionless fraction, is defined as the volume of water that an aquifer releases from or takes into storage per unit area of the aquifer per unit change in the component of head normal to that surface. The storage coefficient for an unconfined aquifer corresponds to its specific yield.

Submerged Outlet: Outlet for discharging water below a free water surface.

Sub Soil: That part of the soil beneath the top soil, usually not having an appreciable organic matter content.

Sub-soiling: Tillage operation that is primarily concerned with altering the soil below plough depth. This includes chiselling or other means of breaking into subsoil.

Subsurface Drainage: The removal of surplus or excess groundwater accomplished by means of natural drainage (for gravelly subsoil) or by artificial means, such as drains placed under a surface.

Subsurface Drains: Subsurface channels used primarily to remove subsurface water from soil. Also known as tile drains or pipe drains.

Summation Curve, Particle Size: A curve showing the cumulative percentage by weight of particles within increasing (or decreasing) size limits as a function of diameter; the per cent by weight of each size fraction is plotted cumulatively on the ordinate as function of the total range of diameters represented in the sample plotted on the abscissa.

Sump: Pit, tank, or reservoir in which water is collected or stored for withdrawal by a pump.

Surface Drainage: The removal of surplus or excess surface water collected on land accomplished by natural or artificial means, such as open ditches and terracing.

Surface Inlet: Structure for diverting surface water into an open ditch or a tile line. May be called drain inlet structure intake, open inlet, surface well, direct inlet.

Surface Runoff: Overland runoff when it has reached a definite stream channel.

Surface Water: Water stored on or flowing over the surface of the soil.

Sustained Use: Sustainability of any measure over a long-term; continuing use of land without severe or permanent deterioration in the resources of the land after drainage and land reclamation.

Tangible Benefits or Tangible Effects: All the benefits that can be expressed in monetary terms.

Tank (Earth): An excavation for impounding water. Also see reservoir.

Tensiometer: A device to measure the tension with which water is held in the soil.

Tide Drainage: Removal of excess water by gravity from inland area to the river/sea which has periodic low water levels owing to tides.

Tile Drain: Concrete or ceramic pipe placed at suitable depths and spacing in soil or subsoil to provide water outlets from the soil.

Tile Drainage: Land drainage by means of a series of tile lines laid at a specified depth and grade.

Tile Drainage System: Drainage system consisting of a drainage outlet tile main, submain and laterals. The laterals remove free water from the soil and the submains and main convey water to the outlet.

Tile Joint: Opening between two drain tiles through which water from the surrounding soil flows.

Tile Outlet: Point of discharge of a tile drainage system.

Tilth: The physical condition of soil as related to its ease of tillage fitness as a seedbed, and its impedance to seeding emergence and root penetration.

Tolerant: Capable of withstanding a condition usually un-favourable. Some plants are tolerant to soil acidity and others are tolerant of soil salinity or alkalinity. **See** crop Tolerance.

Tolerant Crops: See tolerant.

Topographic Divide: A divide that demarcates the boundary of the area from which surface runoff is derived.

Torpedo: See Bullet.

Total Acidity: Sum of active, potential and residual acidity.

Toxic: Poisonous; tending to reduce the yield of crops below the normal.

Toxic water: A water containing one or more of toxic elements. Yield reduction is attributed to toxic effects of the ions rather than the osmotic potential of the soil solution.

Transmissibility: See Coefficient of transmissibility.

Trencher: A drainage machine usually self-propelled to dig trenches for laying drain pipes and envelopes.

Tube wells: Wells consisting of perforated tubes or pipes placed in holes bored into the ground to tap groundwater supplies from one or more aquifers.

Uncased Well: A well usually constructed to observe water table without any casing, used in stable soils.

Vadose Zone: The aerated region of soil above the permanent water table.

Variable Costs: The farming costs which can be assigned to specific farm enterprises as against the fixed costs.

Vertical Drain: The disposal of excess water through surface inlets or wells into pervious subsoil.

Vertical Drainage: (1) Disposal of drainage water through wells into a porous layer of earth or an open-rock formation. (2) The term vertical drainage is also used for a drainage system in which tube wells are used to lower the water table.

Wall Cracking: The occurrence of a break in the pipe wall visible to the naked eye.

Wasteland: Land not suitable for producing material or services of value. A miscellaneous land type.

Water Application Efficiency (Farm): The ratio of water stored in soil root zone to the amount of water delivered at the farm.

Water Conservation: The physical control, protection, management, and use of water resources in such a way so as to maintain crops, grazing and forest lands, vegetal cover, wildlife and water supplies for maximum sustained benefits to people, agriculture, industry, commerce and other segments of the national economy.

Water Disposal System: Complete system for removing excess water from land with minimum erosion. For sloping land it may include a terrace system, terrace outlet channels, dams, and grassed waterways. For level lands it may include only surface drains or both surface and subsurface drains.

Water Level: Water surface; also its elevation above any datum; gauge height; stage.

Water Logging: It refers to the condition of land in which it cannot be put to its normal use as a result of either or both of shallow water table and surface ponding. In arable lands depending upon the extent and duration of water logging, the crop yield may reduce substantially or even become zero.

Water Stable Aggregates: Soil aggregates which are stable to the action of water such as falling drops, or agitation as in wet sieving analysis.

Water Standards: Definitions of water quality established as a basis of control for various water use classification.

Water Stress: The stress of water caused to the plant due to inadequate supply of water through the soil. Also known as matric stress, usually expressed in bars. One bar is equal to 1023 cm of water.

Water Table: The upper surface of groundwater, locus of points in soil water at which the hydraulic pressure is equal to atmospheric pressure.

Water Table Contour: A line drawn on a map and passing through the points where the water table has the same elevation above a specified datum plane.

Water Yield: The total outflow of groundwater and surface runoff from a watershed.

Weep Holes: Openings left in retaining walls, aprons, livings or foundations to allow drainage for preventing build-up of pressure.

Well: An artificial excavation (pit, hole, tunnel) often walled sunk (drilled, dug, driven, bored or jetted) into the ground to such a depth so as to penetrate water yielding rock or soil and allow the water to flow or to be pumped to the surface.

Wetlands: Areas of marsh, fen, peat land or water that include natural, artificial, permanent or temporary areas with either static or flowing fresh, brackish or marine water.

Wetted Perimeter: Length of the wetted contact of a liquid, measured along a plane at right angles to the direction of flow.

Woodland Drainage: Drainage needed for the removal of excess surface water to permit the germination and production of forest seedlings.

Zigzag System: System in which field drains as well as mains are constructed zigzag to reduce high velocities.

■■■■

REFERENCES

Abrol, I.P., Yadav, J.S.P. and Massoud, F.I. 1988. Salt Affected Soils and Their Management. FAO Soils Bulletin, Soil Resources Management and Conservation Service, FAO Land and Water Development Division, 39: 131-139

ACIAR Project 9316. 1996. Tree growing on salt-affected land in Pakistan, Thailand and Australia – 3rd Annual Report (1996), Australian Centre for International Agricultural Research, Canberra.

Agarwal, R.R., Yadav, J.S.P. and Gupta, R.N. 1979. Saline and Alkali Soils of India, ICAR, New Delhi, 286 p.

Akilan, K., Marshall, J.K., Morgan, A.L. Farrell, R.C.C. and Bell, D.T. 1997. Restoration of catchment water balance: responses of clonal river red gum (*Eucalyptus camaldulensis*) to waterlogging. Restoration Ecology, 5: 1-9.

Aldabagh, A.S.Y. 1971. Effect of drainage on trafficability for agricultural equipment. Ph. D. Dissertation. Iowa State University, Ames., USA.

Ambast, S.K., Gupta, S.K. and Sharma, D.K. 2015. Laser Land Levelling. Central Soil Salinity Research Institute, Karnal, Technical Bulletin, CSSRI/Karnal/2015/1, 40 p.

Ami, S.R. 1987. Drainage Pipe Testing Manual SRO-KMA Consultants Inc. Kirkland. Quebec, Canada.

Anonymous. 2002. Recommendations on Water Logging and Salinity Control based on Pilot Area Drainage Research. ICAR/CSSRI/Alterra ILRI Indo-Dutch Network Publication, 100 p.

Arora, S., Singh, A.K., Mishra, V.K., Singh, Y.P. and Sharma, D.K. 2015. Kinetics of gypsum and native $CaCO_3$ dissolution and nutrient transformations mediated through organic amendments and microbial inoculants for crop production in sodic soils. Annual Report 2014-15. Central Soil Salinity Research Institute, Karnal, 95-100.

ASAE. 1979. Design and Construction of Surface Drainage Systems on Farms in Humid Areas. Engineering Practice EP 302.2, American Society of Agricultural Engineers (ASAE), Michigan.

ASAE. 1980. Design and Construction of Surface Drainage Systems on Farms in Humid Areas. ASAE Engineering Practice (ASAE EP 302-2), Agricultural Engineers Handbook 1980, American Society of Agricultural Engineers (ASAE), St. Joseph, MI, 453-460.

ASAE (American Society of Agricultural Engineers). 1984. Design and construction of subsurface drains in Humid Areas. EP260.4. St. Joseph, MI.

ASTM 2152-80. 1986. Standard Test Method for Degree of Fusion of Extruded Poly (Vinyl Chloride) (PVC) Pipe and Molded by Acetone Immersion. Am. Society for Testing and Materials, Philadelphia.

ASTM 2412-87. 1987. Standard Test Method for Determining the External Loading Characteristics of Plastic Pipe by Parallel-Plate Loading. American Society for Testing and Materials, Philadelphia.

ASTM 2444-93. 1984 Standard Test Method for Impact Resistance of thermoplastic Pipe and Fittings by Means of a Tub (Falling Weight). American Society for Testing and Materials, Philadelphia.

ASTM F800-89. 1989. Standard Specification for Corrugated Poly Vinyl chloride (UPVC) Tubing and Compatible Fittings: American Society for Testing and Materials, Philadelphia.

Attanandana, T. and Vacharotayan, S. 1986. Acid sulfate soils: their characteristics, genesis, amelioration and utilization. Southeast Asian Studies, 24: 154-180.

Ayers, R.S. and Westcot, D.W. 1976. Water Quality for Agriculture. FAO Irrigation and Drainage Paper 29. FAO, Rome, 97 p.

Ayers, R. S. and Westcot, D.W. 1985. Water Quality for Agriculture. Irrigation and Drainage paper 29, Rev. 1. FAO, Rome, 174 p.

Bear, J. 1972. Dynamics of Fluids in Porous Media. Dover Publications Inc. New York.

Beg, M.A. 2016. Performance of Various Crops under Acidic Soil Conditions of Northern Hills Climatic Agro-Climatic Zone of Chhattisgarh. A Farmers Perspective Study. M.Sc. Thesis. Indira Gandhi *Krishi Visvavidyalaya*, Raipur, Chhattisgarh, 104 p.

Behera, R.D. and Pattanayak, S.K. 2016. Influence of different sources of liming materials on growth, yield and productivity of the maize crop grown in acid soil. International Journal of Environment, Ecology, Family and Urban Studies, 6:117-124.

Bernstein, L. 1961. Tolerance of plants to salinity. Proc. Amer. Soc. Civil Eng., 87:1-12.

Bernstein, L. 1964. Salt Tolerance of Plants. U.S. Dept. Agr. Inf. Bull., 283. 23p.

Bernstein, L. and Francois, L.E. 1973. Leaching requirement studies: sensitivity of alfalfa to salinity of irrigation and drainage waters. Proc. Soil Sci. Soc. Am., 37: 931-943.

Bhandarkar, D.M. and Singh, R. 2000. Drainage problems and suitable drainage system for Vertisols. Proc. 8th ICID International Drainage Workshop: A Role of Drainage in 21st Century, Jan. 31 - 4th Feb. 2000. INCID, *Vigyan Bhavan*, New Delhi. Vol. II, 245-254.

BIS. 1979. Indian Standard: Specification for Farm Drainage Concrete Tiles IS : 8968 – 1978. Indian Standards Institution, New Delhi (Reaffirmed 2007). https://archive.org/stream /gov.in.is. 3597.1998/is.3597.1998_djvu.txt)

BIS 1980. IS: 9271. 1980. Indian Standard Specification for Farm Drainage Plastic Pipes. Indian Standard Institution, *Manak Bhawan*, New Delhi.

BIS. 1999. Farm Drainage System Performance Evaluation of Horizontal Subsurface Drainage-Guidelines. Indian Standard IS 14632: 1999. Bureau of Indian Standards, New Delhi, 18 p.

BIS. 2004. Unplasticized Polyvinyl Chloride (UPVC) Single Wall Corrugated Pipes for Drainage-Specification (first revision). IS: 9271, Indian Standard Institution, *Manak Bhawan*, New Delhi. 10 p

BIS. 2009. Specifications for Farm Drainage Clay Tiles. Par 2. Perforated Tiles with Collar Joints. IS 8967 Part 2. 1983 reaffirmed 2009. Indian Standard Institution, *Manak Bhawan*, New Delhi. 10p.

Bhaumik, H.D. and Roy, D.L. 1964. Soil Acidity and the Use of Lime in India. Farm Information Unit. Directorate of Extension, Ministry of Food and Agriculture, Government of India.

Bos, M.G. and Boers, Th.M. 1994. Land Drainage: Why and How? In: H.P. Ritzema (Editor-in-Chief), Drainage Principles and Applications. International Institute for Land Reclamation and Improvement (ILRI), ILRI Publication 16, Wageningen, The Netherlands, 23-31.

Bouman J. 1984. Studies for the Use of Saline Water in the Command Areas of Irrigation Projects (Haryana). UNDP/FAO/HSMITC Report, 290 p.

Boussinesq, J. 1903. Sue le debit, en temps de secheresse, d'une source alimentee par une nappe d'eaux d'infiltration. Comptd. Rend. Sceances Acad. Sci., 136: 1511-1517.

Bouwer, H. and van Schilfgaarde, J. 1963. Simplified method of predicting fall of water table in drained lands. Trans. ASAE, 6: 288-291, 296.

Bower, C.A. 1961. Prediction of the effect of irrigation water on soil. Proc. Tehran Symposium on Salinity Problems in Arid Zones,1-8.

Brown, J. R. and Cisco, J.R. 1984. An improved Woodruff buffer for estimation of lime requirement. Soil Science Society of America Journal, 48: 587-591.

Bowonder, B. and Ravi, C. 1984. Water logging from Irrigation Projects - An Environmental Management Problem. Centre of Energy and Tech., Administrative staff college of India, Hyderabad.

Broughton, R.S. and Fouss, J.L. 1999. Subsurface drainage installation machinery and methods. *In*: Skaggs, R.W. and Van Schilfgaarde, J. (Eds.) Agricultural Drainage. Agronomy Monograph 38, ASA, CSSA, and SSSA, Madison, WI, 967-1003.

Broughton, R.S., Fuller, B.W. and Bonnell, R.B. 1992. Water conveyance capacity of corrugated plastic drain pipe with internal diameters between 38 and 75 mm. Proceedings of the 6th International Drainage Symposium. Nashville Tennessee, American Society of Agricultural Engineers (ASAE), St. Joseph, 288-296.

Bundela, D.S., Kaledhonkar, M.J., Gupta, S.K., Mohan Lal, Kamra, S.K., Sharma, D.K., Sharma, P.C. and Chaudhari, S.K. 2016. Cost estimation of sub surface drainage systems for reclamation of waterlogged saline lands. J. Soil Salinity and water Quality, 8: 131-143.

Canada Department of Agriculture. 1977. Management of Saline Soils. Publication 1624, 31 p.

CGWB. 1997. Inland Groundwater Salinity in India. GOI, Ministry of Water Resources, Central Ground Water Board, Faridabad, 62p.

CGWB. 2010. Groundwater Quality in Shallow Aquifers of India. Central Ground Water Board, Faridabad, 113 p.

Chaitanya, A.K., Badole, S., Gupta, A.K. and Pal, B. 2017. Prevention, reclamation and management of acid and acid sulphate soils. In: Soil Salinity Management in Agriculture: Technological Advances and Applications (Gupta, S.K. and Goel, M.R. Eds.). Apple Academic Press, USA, 337-364.

Chakraborty, S.S., Mohan, S., Tyagi, N.K., Sonde, R.R. and Chander, S. 2012. Water Resources Management. Indian National Academy of Engineering, New Delhi, 170p.

Chauhan, H. S. 2000. Bio-drainage and other alternatives to tile drainage. Proc 8th ICID Int. Drainage Workshop, New Delhi, India. ICID, New Delhi, IV: 79-95.

Chauhan, H.S., Jaiswal, C.S. and Seth H.B.S. 1975. Analysis of flow to non-penetrating well in an artesian aquifer. J. Institution of Engineers (India), 56(C12): 45-47.

Chhabra, R. and Thakur N.P. 1998a. Bio-drainage using trees to control water logging and secondary salinization in canal irrigated areas. Proceedings National Conference on Salinity Management in Agriculture. Central Soil Salinity Research Institute, Karnal, India.

Chhabra, R. and Thakur, N.P. 1998b. Lysimeter study on the use of bio-drainage to control water logging and secondary salinization in (canal) irrigated arid/semi-arid environment. Irrigation and Drainage Systems, 12: 265-288.

Chossat, J.C. 1987. Relation between hydraulic conductivity and drainable porosity. In Fifth National Drainage Symposium, Proceedings, ASAE, Chicago. International ST Book, 403: 81–92.

Dagan, G. 1964. Spacing of drains by an approximate method. Journal Irrigation and Drainage Division. ASCE, 90: 41-46.

Dagar, J.C. 2011. Bio-drainage for amelioration of water logging. In: Salinity Management for Sustainable Agriculture in Canal command (Dey and Gupta Eds.). CADA, Haryana and CSSRI, Karnal, 54-58.

Das, G. 2012. Soil Conservation Engineering Including Watershed Management. PHI Learning Private Limited, New Delhi,539 p.

Denecke, H.W. 2000. Concept Note on Bio-drainage and Bio-Disposal systems. Available on http://www.fao.org/iptrid/drainage/bio_drainage/ concept_note _bio_ drainage.htm (Opened on 13.07.2017)

De Zeeuw, J.W. and Hellinga, F., 1958. Neerslag en afvoer. landbouwkundig Tijdschrift. 70: 405-422 (in Dutch with English summary).

Dhruvanaryana, V.V. 1980. Reclaiming Alkali Soils: Engineering Aspect. Bulletin No. 6, Central Soil Salinity Research Institute, Karnal, India.

Dieleman, P.J and Trafford, B.D. 1976. Drainage Testing. Irrigation and Drainage Paper, 28. FAO, Rome.

Dierickx, W. 1980. Electrolytic Analogue Study of the Effect of Opening and Surrounds of various Permeabilites on the Performance of Field Drainage Pipes. Report 77. National Institute for Agricultural Engineering. B-9220. Merelbeke, Belgium.

Dierickx, W. 1987. Choice of subsurface drainage materials. ASAE Paper #87-2589, ASAE, MI.

Dierickx, W. 1999. Opening size determination of technical textiles used in agricultural applications. Geotextiles and Geomembranes, 17: 231-245:

DIN-1187. 1987. Unplasticized Polyvinyl Chloride (PVC-U) Drain Pipes. Dimension Requirements and Testing. Germany.

Donnan,W.W. 1959. Drainage of agricultural land using interceptor lines. Journal of Irrigation and Drainage Division. ASCE: 13-23

DRI (Drainage Research Institute). 1995. Re-use of Drainage Water in the Nile Delta: Monitoring, Modeling and Analysis. Re-use Report No. 50, Cairo, Egypt.

Dumm, L.D. 1954. Drain spacing formula. Agricultural Engineering, 35: 726-730.

Dumm, L.D. 1964. Transient flow concept in subsurface drainage, its validity and use. Trans. ASAE, 7: 142-146, 151.

Eggelsmann, R. 1978. Subsurface Drainage Instructions. Paul Parey, Hamburg, Germany.

Ernst, L. F. 1950. A new formula for the calculation of the permeability factor with the auger holes. Proc. Soil. Sc. Soc. Am. J., 15: 90-96.

Ernst, L.F. 1956. Calculation of the steady flow of groundwater in vertical cross sections. The Netherlands. J. Agric. Sci., 4: 126-131.

Ernst, L.F. 1962. Grondwater stormingin in de verzadigde zone en hun berekening bij aanwezidheid van horizontal evenwigde open leidingen vers. Landbouwk. Onderz. 67-15 Pudoc, Wageningen. 189 p.

FAO. 1973. Drainage of Salty Soils. FAO Irrigation and Drainage Paper 16, Food and Agriculture Organization of the United Nations, Rome.

FAO. 1985. Water Quality for Agriculture. Irrigation and Drainage Paper 29 Revision-1. FAO, Rome. 174 p.

FAO. 1995. Environmental Impact Assessment of Irrigation and Drainage Projects. FAO Irrigation and Drainage Paper No. 53. FAO, Rome.

Faure, Y. 1991. Developments récents de la recherche sur les problèmes de filtration par géotextiles. In: Colmatage des drains etenrobages: état des connaissances et perspectives. Linard (Ed.). CEMAGREF, Groupement d'Antony, France. 83-97.

Fenemor, A.D., Bevier, B.R. and Schwab, G.O. 1979. Prediction of deflection for corrugated plastic tubing. Transactions ASAE, 22: 1338-1342.

Fraser, G.W., Thorburn, P.J. and Taylor, D.W.1996. Growth and water use of trees in saline land in southeast Queensland. In: Proc. 4th National Conference and Workshop on the Productive Use and Rehabilitation of Saline Lands, Albany, Western Australia, Promaco Conventions, 219-224.

Gardner, W.R. and Fireman, M. 1958. Laboratory studies of evaporation from soil columns in the presence of a water table. Soil Sci., 85: 244-249.

Garg, A.K. and Bhargava, K. 2001.Land and Water Management in Irrigated Areas. Himanshu Publications, Udaipur, 370 p.

Goel, A.C. and Gupta, S.K. 1986. Drainage requirement of wheat in relation to its cultivation in Haryana. Haryana Agric. Univ. J., 26: 30-36.

Grable, A.R. 1966. Soil aeration and plant growth. Advances in Agronomy, 18: 57-106.

Gupta, I.C. and Gupta, S.K. 2001. Use of Saline Water in Agriculture: A Study of Arid and Semi-arid Zones of India. Scientific Publishers (India), Jodhpur, 302p.

Gupta, I.C., Yaduvanshi, N.P.S. and Gupta, S.K. 2012.Standard Methods for Analysis of Soil Plant and Water. Scientific Publishers (India), Jodhpur, 224 p

Gupta, R.K., Singh, N.T. and Sethi, Madhurama. 1994. Ground Water Quality for Irrigation in India. CSSRI, Tech. Bull No. 19, 13p.

Gupta, S.K. 1984. Analysis of flood and rise in water table problem of Rohtak district. Curr. Agric., 8: 16-25.

Gupta, S.K. 1987. Non-penetrating wells for characterizing aquifer constants. In: Trends in Groundwater Research (Ramesam, V. Ed.). The Geological Society of India, Bangalore, 107-116.

Gupta, S. K. 2002. Future prospects of subsurface drainage in India. J. of Water Management, 10: 49-57.

Gupta, S.K. 2005. A critical analysis of bio-drainage with special reference to Indian experiences. Souvenir: All India Seminar Reclamation of Waterlogged Saline Soils through Drainage. Institution of Engineers (Kota Chapter), Kota, 81-96.

Gupta, S.K. 2005. Water logging, soil salinity and land drainage In: Souvenir National Symposium on Efficient Management for Eco-friendly Sustainable Agriculture. ISWAM, New Delhi, 36-46.

Gupta, S.K. 2011. Subsurface Drainage for Waterlogged Saline Lands. Report submitted to Agricultural Education Division, New Delhi. Central Soil Salinity Research Institute, Karnal (Unpublished), 353 p.

Gupta, S.K. 2011. Case studies on subsurface drainage systems and guidelines for design and evaluation. In: Salinity Management for Sustainable Agriculture in Canal Commands (Dey, P. and Gupta, S.K. Eds.). Central Soil Salinity Research Institute, Karnal, 11-22.

Gupta, S.K. (Ed.). 2014. Agricultural Land Drainage in India. Agrotech Publishing Academy, Udaipur, 308 p.

Gupta, S.K. 2014. Response of crops to surface stagnation and ameliorative strategies. In: Agricultural Land Drainage in India (Gupta, S.K. Ed.) Agrotech Publishing Academy Udaipur, 25-48.

Gupta, S.K., Dinkar, V.S. and Tyagi, N.K.1995. Reclamation and Management of Waterlogged and Salt Affected Areas in Irrigation Commands- A Manual. Ministry of Water Resources, New Delhi and CSSRI, Karnal, 144 p.

Gupta, S.K. and Dhruvanarayana, V.V. 1974. Design consideration for dual purpose ponds. Soil Conservation digest, 2: 1-7.

Gupta, S.K. and Gupta, I.C. 1997a. Crop Production in Saline Soils. Scientific Publishers, Jodhpur, 239 p.

Gupta, S.K. and Gupta, I.C. 1997b. Management of Saline Soils and Waters. Scientific Publishers, Jodhpur, 299 p.

Gupta, S.K. and Gupta, I.C. 2014a. Salt Affected Soils: Reclamation and Management. Scientific Publishers, Jodhpur, 321p.

Gupta, S.K. and Gupta, I.C. 2014b. Management of Saline and Waste Water in Agriculture. Scientific Publishers, Jodhpur, 316 p.

Gupta, S.K. and Pandey, R.N. 1979. Crop and water yield as affected by rain water storage in rice fields: a field evaluation. Field Crops Research, 2: 365-371.

Gupta, S. K., Sharma, D. P., Tyagi, N. K. and Dubey, S. K. 2002. *Lavaniya-Kshariya Mirdaon ka Sudhar aur Parbandh.* CSSRI, Karnal, 153 p.

Gupta, S.K., Sharma, D.P. and Swarup, A. 2004. Relative tolerance of crops to surface stagnation. J. Agril. Engg., 41: 44-48.

Gupta, S.K. and Sharma, S.K. 1990. Response of crops to high exchangeable sodium percentage. Irrigation Science, 11: 173-179.

Gupta, S.K., Tejwani, K.J. and Ram Babu. 1971. Drainage coefficient for surface drainage of agricultural land for different parts of the country. Journal of Irrigation and Power, January-1971, Issue, 53-60.

Gurunadharao, V.V.S. 2016. Personal communication.

Hamza, S., Srinivasan, V. and Dinesh, R. 2013. Management of acid soils under major spice crops. Indian Journal of Arecanut, Spices and Medicinal Plants, 15: 3-6.

Hanson, B.R., Gratten, S.R. and Fulton, A. 2006. Agricultural Salinity and Drainage. UC ANR Publication 3375 (anrcatalogue.ucdavis.edu).

Hardjoamidjojo, S., Skaggs, R.W. and Schwab, G.O. 1982. Corn yield response to excessive soil water conditions. Transactions of the ASAE, 25:922-927, 934.

Heuperman, A. F. 1999. Hydraulic gradient reversal by trees in shallow water table areas and repercussions for the sustainability of tree growing systems. Agricultural Water Management, 39: 153-167.

Heuperman, A.F., Kapoor, A.S. and Denecke, H.W. 2002. Bio-drainage - Principles, Experiences and Applications. IPTRID, Food and Agriculture Organization of the United Nations, Rome.

Hiler, E.A. 1977. Drainage requirements of crops. Proceedings of Third National Drainage Symposium. ASAE Publication No. 177, ASAE, St. Joseph, MI 49085, 127-129.

Hiler, E. A. and Clark, R.N. 1971. Stress day index to characterize effects of water stress on crop yields. Transactions of the ASAE, 14: 757-761.

Hirekhan, M., Gupta, S.K. and Mishra, K.L.2007. Application of WaSim to assess performance of subsurface drainage system under semi-arid monsoon climate. Agricultural Water Management, 88: 224-234.

Hooghoudt, S.B. 1940. General consideration of the problem of field drainage by parallel drains, ditches, watercourses, and channels. Publ. No.7. In the series of 'Contribution to the knowledge of Some Physical Parameters of the Soil (translated from Dutch)'. Bodemkundig Instituut, Groningen, The Netherlands.

Hooja, R., Srinivas, V. and Govind Sharma. 1994. Water logging and possible remedial measures in the Indira Gandhi canal command area development project. Nat. Seminar on Reclamation and Management of Waterlogged Saline Soils. CSSRI, Karnal.

HOPP. 1997. Mission Report of the Machine Drainage Specialist, TR-23, Haryana Operational Pilot Project (HOPP), Department of Agriculture, Haryana.

HOPP. 2004. Report on Benefits of Subsurface Drainage in Gohana, HOPP. Haryana Operational Pilot Project (HOPP), Department of Agriculture, Haryana.

http://www.kiran.nic.in/acid_soil.html.Management of Acid Soils of NE Region. Opened on 15.05.2016.

https://www.nabard.org/english/ld_acidsoils.aspx (Opened on 30.09.2015)

IARI. 2014. Drainage Technology for Land Improvement. Indian Agricultural Research Institute, New Delhi. 14 p. (www.iari.res.in opened on 14.03.2014)

ICID. 1979. Amendments to the Constitution, Agenda of the International Council Meeting at Rabat. International Commission on Irrigation and Drainage (ICID), Morocco, ICID, New Delhi, India, A-156-163.

ICID. 1982. ICID Standard 109, Construction of Surface Drains. ICID Committee on Irrigation and Drainage Construction Techniques, ICID Bulletin, 31: 47-57.

ICID. 1993. The ICID Environmental Check list to Identify Environmental Effects of Irrigation, Drainage, and Flood Control Projects. Oxfordshire, U.K.: HR Wallingford.

Indian Council of Agricultural Research. 2009. Handbook of Agriculture. Indian Council of Agricultural Research, New Delhi, 1617 p.

Inglish, G.C. and Gokhale, V.K. 1927. Report on Land Drainage Experiment at Baramati, Technical Paper No. 24, Bombay Public Works Department, Bombay.

IPTRID (International Programme for Technology and Research in Irrigation and Drainage). 2001. Drainage and sustainability. Issues Paper No. 3. Food and Agriculture Organization of the United Nations, Rome, 28p.

Irrigation Commission, 1972. Report of the Irrigation Commission. Ministry of Irrigation, Govt. of India, New Delhi.

Jena, D. 2008. Management of Acid Soils for Sustainable Crop Production. NAE-Acid soil Bulletin No. 4, Orissa University of Agriculture and Technology, Bhubaneswar.

Jena, D. 2013. Potential of industrial by-products in ameliorating acid soils for sustainable crop production. The Fourth Dr. G.S. Sekhon Memorial Lecture delivered on 2nd September 2013 at Birsa Agricultural University, Kanke, Ranchi. Ranchi Chapter of the Indian Society of Soil Science, Ranchi.

Joshi, P.K., Singh, O.P., Rao, K.V.G.K. and Singh, K.N. 1987. Subsurface drainage for salinity control- an economic analysis. Indian J. Agric. Economics, 42:198-206.

Kaledhonkar, M.J. 2014. Drainage investigations and design of subsurface drainage system. In: Agricultural Land Drainage in India (Gupta, S.K. Ed.). Agrotech Publishing Acad., Udaipur. 70-90.

Kaledhonkar, M. J., Sharma, D. P., Gupta, S.K., Bundela, D. S. and Singh, Gurbachan. 2008. Farmers' Response and Strategies to Manage Rainstorms and Low Temperatures during Winter. Tech. Bull CSSRI/Karnal/2008/5. CSSRI, Karnal. 12 p.

Kanwar, R.S., Baker, J.L. and. Mukhtar, S. 1988. Excessive soil water effects at various stages of development on the growth and yield of corn. Transactions of the ASAE, 31: 133-141.

Kelly, J. and Rengasamy, P. 2006. Diagnosis and management of soil constraints: transient salinity, sodicity and alkalinity. The University of Adelaide and Grain Research and Development Corporation, Australia. Available at http://www.arris.com.au/index.php?id=39

Kapoor, A.S. 1997. Bio-drainage of lands under irrigation in dry arid regions with particular reference to the Indira Gandhi *Nahar* project, Rajasthan.

Kar, G., Kumar, A. and Sahoo, N. 2012. Acid Soils Management for Higher Crop Productivity: An Integrated Approach. Directorate of Water Management, Bhubaneswar, Odisha, Leaflet, 8p.

Khanzada, A.N., Morris, J.D., Ansari, R., Slavich, P.G. and Collopy, J.J. 1998. Groundwater uptake and sustainability of *Acacia* and *Prosopis* plantations in southern Pakistan. Agricultural Water Management, 36: 121-139

Khosla, B.K., Gupta, R.K. and Abrol, I.P. 1979. Salt leaching and the effect of gypsum application in a saline-sodic soil. Agricultural Water Management, 2: 193-202.

Kirkham, D. 1958. Seepage of steady rainfall through soil into drains. Amer. Geophys. Union Transactions, 39: 892-908.

Klute, A. and Dirksen, C. 1986. Hydraulic conductivity and diffusivity: laboratory methods. In: Klute, A. (Ed.), Methods of Soil Analysis. Part 1. Physical and Mineralogical Methods, 2nd ed. Agron. Monogr. 9. ASA, Madison, WI, 687–734.

Kostiakov, A.N. 1932. On the dynamics of the coefficient of water percolation in soils and on the necessity of studying it from dynamic point of view for purposes of amelioration. Trans. 6th Comm. Soil Sci. Soc. Moscow. Part A: 17-21.

Kraijenhoff, Van de Leur, D.A. 1958. A study of non-steady groundwater flow with special reference to a reservoir coefficient. I. De. Ingeneiue, 70: B87-B94.

Kumar R., Chatterjee, D., Kumawat, N., Pandey, A., Roy, A. and Kumar M. 2014. Productivity, quality and soil health as influenced by lime in rice bean cultivars in foothills of north-eastern India. The Crop Journal, 2: 338-344.

Kumar, R. and Singh, J. 2005. Textbook of Drainage Engineering. Indian Council of Agricultural Research, New Delhi. 342 p.

Lagacé, R. 1983. Predicting Drain Silting Potential. Ph.D. Thesis, North Carolina State University, Raleigh, NC, United States.

Lal, Cheddi. 2005. Develop design criteria of interceptor drain for preventing/minimizing water logging in areas along major canals. Progress during 204-05 (Personal communication)

Lal, Cheddi. 2006. Control of canal seepage through conventional and bio-interceptor drains. In: Sustainable Irrigated Agriculture through Command Area Development (Ambast, Gupta and Singh Eds.). CAD and WM, New Delhi, GoI; and Central Soil Salinity Research Institute, Karnal. 93-105.

Lok Sabha Secretariat. 1996. Fourth Report of the Standing Committee on Agriculture of the 11th Lok Sabha. New Delhi, India.

Lovas, L. 1972. Salinity control trials by sub-soil drainage at Digod. Proc. All India Symp. on Soil Salinity. CSA University of Agricultural Sciences, Kanpur. 156-160.

Luthin, J.N., Taylor, G.S. and Prieto, C. 1968. Exit gradients into subsurface drains. Hilgardia, 39: 419-428.

Maas E.V. 1980. Crop tolerance to saline soil and water. In: Prospectus for Bio-saline Research. Proc. US –Pak Bio-saline Res. Workshop, Karachi, Pakistan. 205- 211.

Maas, E.V. 1984. Crop tolerance. California Agric., 38: 20-21.

Maas, E.V. 1987. Salt tolerance of plants. In: Handbook of Plant Science in Agriculture. (Christie, B.R. (Ed)). CRC Press Inc. Vol. 2. 57-75.

Maas, E.V. and Hoffman, G.J. 1977. Crop tolerance assessment. *J. Irrigation and Drainage Div. Proc. Am. Soc. Civil Engineering* 103: 115-134.

Marshall, J.K., Morgan, A.L., Akilan, K., Farrell, R.C.C. and Bell, D.T. 1997. Water uptake by two river red gum clones (*Eucalyptus camaldulensis*) in a discharge site plantation in the western Australia wheat belt. *J. of Hydrology,* 200: 136-148.

Massland, M. 1959. Water table fluctuation induced by internal recharge. J. Geophysical Research, 64: 549-559.

Mathew, E.K. 2003. Adaptability constraints of a technically and economically feasible subsurface drainage system in the low-lying acid sulphate soils of Kerala, India. Paper No 116. 9th International Drainage Workshop, September 10 – 13, 2003, Utrecht, The Netherlands.

Mathew, E.K., Nair, M., Raju, T.D. and Jai Kumaran, U. 2003. Drainage Digest: A Report. AICRP on Agricultural Drainage. Kerala Agricultural University, Thrissur, Kerala. India. 200 p.

Mehlich A. 1976. New buffer method for rapid estimation of exchangeable acidity and lime requirement. Communications in Soil Science and Plant Analysis, 7: 637–52.

Minhas, P.S. and Gupta, R.K. 1992. Quality of Irrigation Water – Assessment and Management. Indian Council of Agricultural Research, New Delhi, 123p.

Ministry of Agriculture. 1984-85. Statistics on Land Degradation. Min. of Agriculture. New Delhi.

Ministry of Agriculture. 1994. Draft Report on Status of Land Degradation in India. Ministry of Agriculture. New Delhi. 102 p.

Ministry of Agriculture. 2011. http://www.nfsm.gov.in/Presentations/Soil_Rice_ 22may 2010.ppt (Opened on 07.04.2014).

Misra, B. 1996. Highlights of Research on Crops and Varieties for Salt Affected Soils. Research Bulletin. Central Soil Salinity Research Institute, Karnal. 28 p.

Mohan Lal, 2011. Impact of Subsurface Drainage on Land Quality, Ground Water Quality and Crops Yield in Haryana. Ph.D. Dissertation submitted to National Institute of Technology, Kurukshetra.

Mohan, Lal, Arora, V.K. and Gupta, S.K. 2012. Application and performance of geo-synthetic filter materials as drain envelope in land reclamation in Haryana, Indian Journal of Agricultural Engineering, 49:14-19.

Mohan, Lal, Arora, V.K. and Gupta, S.K. 2014. Subsurface drainage, construction, monitoring and evaluation. In: Agricultural Land Drainage in India. Agrotech Publishing Academy, Udaipur. 133-163.

Mohan Lal, Arora, V.K., Gupta, S.K. and Singh, R.V. 2006. Design considerations and guidelines for large-scale subsurface drainage projects using pipe-laying machine (Trencher) at farmer's field. J. of Water Management, 14: 48-56.

Mondal, A.K., Obi Reddy, G.P. and Ravisankar, T. 2011. Digital database of salt affected soils in India using Geographic Information System. Journal of Soil Salinity and Water Quality, 3:16-29.

Murthy, V.V.N. 1985. Land and Water Management Engineering. Kalyani Publishers, Delhi. 586 p.

Murty, V.V.N. and Jha, M.K. 2009. Land and Water Management Engineering. Fifth Edition, Kalyani Publishers, Ludhiana, India, pp. 374-387.

Murty, V.V.N. and Jha, M.K. 2011. Land and Water Management Engineering. Sixth Edition, Kalyani Publishers, Ludhiana, India.

NAAS (National Academy of Agricultural Sciences). 2015. Bio-drainage: an Eco-friendly Tool for Combating Water Logging. Policy paper 74. National Academy of Agricultural Sciences, New Delhi, 27p.

National Commission on Agriculture, 1976. Ministry of Agriculture and Irrigation, Government. of India, New Delhi.

National Institute of Hydrology (NIH). 1996. Drainage Manual. NIH, *Jal Vigyan Bhawan*, Roorkee. 216 p.

Nijland, H.J., Croon, F.W. and Ritzema, H.P. 2005. Subsurface Drainage Practices: Guidelines for the Implementation, Operation and Maintenance of Subsurface Pipe Drainage Systems. ILRI Publication No. 60, Wageningen, Alterra, 608 p.

Pai, A.A. and Hukkeri, S.B. 1979. Manual on Irrigation Water Management. Department of Agriculture, Krishi Bhavan, New Delhi, India.

Panda, N. and Mishra, U.K. 1970. Utilization of partially acidulated rock phosphates in acid laterite soil. Plant Soil, 33: 225-234

Panda, N., Sarkar, A.K. and Chamuah, G.C. 2009. Soil acidity. In: Fundamentals of Soil Science, Indian Society of Soil Science. 2nd ed., New Delhi, India. 317-328.

Pandey, R.S. and Gupta, S.K. 1983. A Comparative evaluation of steady state drain spacing formula for application in field drainage design. J. Institution of Engineers, 64(ND): 27-31.

Patil, B.N., Patil, S.G., Hebbara, M., Manjunatha, M.V., Gupta, R.K. and Minhas, P.S. 2005. Bio-ameliorative role of tree species in salt-affected Vertisols of India. Journal of Tropical Forest Science, 17: 346-354.

Patiram. 2007. Management and future research strategies for enhancing productivity of crops on the acid soils. The 14th Dr. D.P. Motiramani Memorial Lecture. Journal of the Indian Society of Soil Science, 55: 411-420.

Planning Commission. 2013. Report of the High Level Expert Group on Water Logging in Punjab. Planning Commission, New Delhi, 95 p.

Ponnamperuma, F. N. 1972. The Chemistry of Submerged Soils. Advances in Agronomy. 24: 29-96.

Poysa, V.W., Tan C.S. and Stone, J.A. 1987. Flooding stress and the root development of several tomato genotypes. Hort. Science, 22: 24-26.

RAJAD, 1995. Analysis of Subsurface Drainage Design Criteria. Chambal Command Area Development & Canadian International Development Agency. Kota, Rajasthan.

Raju, R., Thimmappa, K., Kumar, P., Kumar, S. and Tripathi, R.S. 2016. Reclamation of saline soils through subsurface drainage technology in Haryana – An economic impact analysis. Journal of Soil Salinity and Water Quality, 8: 194-201.

Ram, Sewa. 2008. Drainage Training Manual. Unpublished.

Rao, K.V.G.K., Singh, O.P., Gupta, R.K., Kamra, S.K., Pandey, R.S., Kumbhare, P.S. and Abrol, I. P. 1986. Drainage Investigations for Salinity Control in Haryana. Bull. No. 10. CSSRI, Karnal. 95p.

Rao, K.V.R., Ravi, K. and Singh, R. 2009a. Mole drainage to enhance soybean production in waterlogged Vertisols. J. of Agricultural Engineering, 46: 54-58.

Rao, K.V.R., Ravi, K. and Singh, R. 2009b. Mole drainage a viable alternate to pipe drainage-a case study. 60th Int. Executive Council and 5th Asian Regional Conference, New Delhi. www.tndholkia.com/data/delhi/2.c.d.

Ravelo, C.J., Reddell, D.L., Hiler, E.A. and Skaggs, R.W. 1982. Incorporation of crop needs into drainage system design. Transactions of the ASAE, 25:623-629, 637.

Rawat, K.S., Tripathy, V.K., Gupta, S.K. and Rao, K.V.G.K. 2001. Effect of drain placement in a low hydraulic conductivity zone in a two-layered soil profile on transmissivity of shallow aquifers. J. Ind. Soc. Soil Sci., 49: 399-406.

Rengasamy, P. 2010. Soil processes affecting crop production in salt-affected soils. Functional Plant Biology, 37: 613–620

Rhoades, J. D. 1974. Drainage for salinity control. Agron. Journal, 17: 433:470

Rhoades, J.D. 1977. Potential for using saline drainage waters for irrigation. Proc. Water Management for Irrigation and Drainage. ASCE, Reno, Nevada, USA, 85-116.

Rhoades, J.D. 1983. Using saline water for irrigation. Proc. Intern. Workshop Salt Affected Soils of Latin America, Maracay, Venezuela, 233-264.

Richards, L.A. 1954. Diagnosis and Improvement of Saline and Alkali Soils. Handbook No. 60. United States Department of Agriculture, Washington DC, 160 p.

Ritzema, H.P. 1994. Drainage Principles and Applications. ILRI Pub. 16. International Institute for Land Reclamation and Management (ILRI), Wageningen, The Netherlands. 1125 p.

Ritzema, H. 2009. Drain for Gain: Making Water Management Worth its Salt. PhD Thesis. Wageningen University, Wageningen, The Netherlands.

Ritzema, H., Wolters, W., Bhutta, M., Gupta, S. K. and Abdel-Dayem, Safwat. 2007. The added value of research on drainage in irrigated agriculture. Irrigation and Drainage, 56: 205-215

Royal Commission on Agriculture, 1928. Report of the Royal Commission on Agriculture. New Delhi, India.

Rycroft, D.W. and Alcock, M. 1974. The Assessment of Suitability of Soils for Moling. Technical Bulletin 75/5. Field Drainage Experimental Unit, Cambridge.

Rycroft, D.W and Amer, M.H .1995. Prospect of the Drainage of Clay Soils. FAO Irrigation and Drainage Paper 55. Food and Agriculture Organization, Rome. 135 p.

Salama, R.B., Bartle, G.A., Farrington, P. 1994. Water use of plantation *Eucalyptus camaldulensis* estimated by groundwater hydrograph separation techniques and head pulse method. J. of Hydrology, 156: 163-180.

Samani, J.A. and Willardson, L.S. 1981. Soil hydraulic stability in a subsurface drainage system. Trans. ASAE, 24: 666-669.

Sanjay Kumar, Gupta, S.K. and Sewa Ram. 1994. Inverse techniques for estimating transmissivity and drainable pore space utilizing data from subsurface drainage experiments. Agric. Water Management, 26: 41-58

Sarkar, A.K. 2005. Annual Report, ICAR Network Project on Soil Characterization and Resource Management of Acid Soil Regions for Increasing Productivity. Birsa Agricultural University, Ranchi.

Schilfgaarde van J., Bernstein L., Rhoades J.D. and Rawlins S.L. 1974 Irrigation management for salt control. J. Irrigation and Drainage Division, ASCE, 100:321–338.

Schoonover, W.R. 1952. Examination of Soils for Alkali. Univ. Of California. Extension Service, Berkley, California.

Schwab, G.O., Fangmeier, D.D., Elliot, W.J. and Frevert, R.K. 2005. Surface Drainage and Land Forming. Fourth Edition, John Wiley and Sons (Asia) Pvt. Ltd., Singapore, 245-264.

Schwab, G., Maheswari, K.M., Gupta, S.K. and Johri, G.B. (Eds.) .1987. Handbook for Drainage of Irrigated Areas in India. Tech. Report No. 5. IMTP (LBII and WAPCO), New Delhi.

Sen, H. S. 2003. Problem soils in India and their management: Prospect and retrospect. Journal of the Indian Society of Soil Science, 51: 388- 408.

Sehgal, J.C. and Abrol, I.P. 1994. Soil Degradation in India: Status and the Impact. Oxford and IBH Publ. Co. Pvt. Ltd., New Delhi.

Shakya, S. K., Gupta, P. K. and Kumar, D. 1995. Innovative Drainage Techniques for Waterlogged Sodic Lands. Tech, Bull. 3. Punjab Agricultural University, Ludhiana. 5 p.

Sharma, D.K. and Gupta, S.K. 2013. State of the Art Report: Increasing Irrigation Water Use Efficiencies. Indian National Committee on Hydrology. Ministry of water Resources, Roorkee (UP). 51p.

Sharma, D.K., Mishra, V.K. Singh, Y.P., Nayak, A.K., Singh, Ranbir , Kaledhonkar, M. J., Varshney, R.K. and Singh, Gurbachan. 2008. Integrated Farming System Model for Waterlogged Sodic Soils. Technical Bulletin: CSSRI/Lukhnow/01/2008, CSSRI, Regional Research Station, Lucknow.12 p.

Sharma, D.P. 2014. Reuse and disposal strategies for saline drainage effluents. In: Agricultural Land Drainage in India (Gupta, S.K. Ed.). Agrotech Publishing Academy, Udaipur. 253-285.

Sharma, P., Rishi, M.S. and Thakur, T. 2013. Assessment of waterlogged and salt affected area in Muktsar district of southwest Punjab. International Conference on Advances in Water Resources Development and Management (AWRDM-2013), 23rd – 27th October, 2013, Chandigarh. Paper No. R089, 10 p.

Sharma, P.D. and Sarkar, A.K. 2005. Managing Acid Soils for Enhancing Productivity. Technical Bulletin. 23. ICAR, NRM, Division, KAB II, New Delhi. 24p.

Sharma, S.K. and Bhargava, G.P. 1982. Ground water behavior and soil salinity in Haryana. Annual report. CSSRI, Karnal.

Shiratori, Y., Watanabe, H., Furukawa, Y., Tsuruta, H. and Inubushi, K. 2007. Effectiveness of a subsurface drainage system in poorly drained paddy fields on reduction of methane emissions. Soil Science & Plant Nutrition, 53: 387-400.

Shoemaker, H.E., McLean, E.O. and Pratt, P.F. 1961. Buffer methods for determining the lime requirement of soils with appreciable amounts of extractable aluminum. Soil Science Society of America Proceedings, 25: 274-277.

Sieben, W.H. 1964. Het verban tussen ontwatering en opbrengst bi j de jonge zavelgronden in de Noordoostpolder. Van Zee tot Land. 40, Tjeenk Willink V. Zwolle, The Netherlands, (as cited by Wesseling, 1974).

Singh, Gurbachan, Kumar, A., Kundu, S.S., Tripathi, R.S., Yaduvanshi, N.P.S., Batra, L., Sharma, P.C. and Pandey, R.S. 2007. Multi-enterprise Agriculture Model for Livelihood Security in Reclaimed Sodic Soils. Tech. Bull 21/2007. CSSRI, Karnal. 12 p.

Singh Gurmel, Ram Babu, Pratap Narain, Bhushan, L.S. and Abrol, I.P.1992. Soil erosion rates in India. Journal of Soil and Water Conservation, 47: 97- 99.

Singh, M. 2007. Large-scale implementation of subsurface land drainage technology in canal irrigated commands and coastal tracts. In: Agricultural Land Drainage: Reclamation of Waterlogged Saline Lands (Eds. Ambast, Gupta and Singh). Central Soil Salinity Research Institute, Karnal and CAD&WM, New Delhi. 143-155.

Singh, N.T. 1992. Salt affected soils in India. In: Land and Soils (Khoshoo, T.N. and Deekshatulu, B.L. Eds.). Indian National Science Academy, New Delhi. 65-100.

Singh, P. Drainage problems in India. http://citeseerx.ist.psu.edu/viewdoc/download?doi= 10.1.1.630.6881&rep=rep1&type=pdf. 278-285.

Singh, R. and Ramana Rao, K.V. 2014. Agricultural drainage technology for temporary waterlogged Vertisols' In: Agricultural Land Drainage in India (Gupta, S.K. Ed.). Agrotech Publishers, Udaipur. 207-225.

Smedema, B. 1997. Biological drainage: myth or opportunity? GRID, 9: 3.

Smedema, L.K. and Rycroft, D.W. 1983. Land Drainage: Planning and Design of Agricultural Drainage Systems. Cornell University Press, Ithaca, New York. 375p.

Smith, R.M. and Browning, D.R. 1946. Some suggested laboratory standards of subsoil permeability. Soil Sci. Soc. Amer. Proc., 11:21-26.

Spangler, M.G. 1957. Underground conduits- an appraisal of modern research. Proc. Am. Soc. Civil Engineers, 73: 855-884.

Spangler, M.G. and Handy, R.L. 1973. Soil Engineering. Intex Education Publishers, New York.

Somani, L.L., Totawat, K.L. and Sharma, R.A. 1996. Liming Technology for Acid Soils. Agrotech. Publishing Academy, Udaipur. 240 p.

Sondhi, S.K. 2007. Subsurface drainage: a case study of southwest Punjab. In: Agricultural Land Drainage-Reclamation of Waterlogged Saline Lands (Eds. Ambast, Gupta and Singh). Command Area Development and Water Management, New Delhi and CSSRI, Karnal. 116-127.

Stephens, J.C. and Mills, W. C. 1965. Using the Cypress Creek Formula to Estimate Run-off Rates in the Southern Coastal Plain and Adjacent Flatwoods Land Resource Areas. U. S, Dept. of Agriculture, ARS-41-95.

Stirzaker, R.J., Cook, F.J. and Knight, J.H. 1997. How to space trees in a paddock for control of dry land salinity: a first approximation. In: Proc. Agroforestry for Sustainable Land-use: Fundamental Research and Modelling, Temperate and Mediterranean Applications, Montepellier. 169-173.

Stolzy, L. H. and Letey, J. 1964. Measurement of oxygen diffusion with platinum- micro-electrodes: correlation of plant response to soil oxygen diffusion rates. Hilgardia, 35: 567-576.

Stuyt, L.C.P.M. 1992, The water Acceptance of Wrapped Subsurface Drains. Ph.D Thesis. Land bouw Universiteit Wageningen, Agricultural University, Wageningen, the Netherlands. 305p.

Stuyt, L.C.P.M., W. Dierickx and J.M. Beltran. 2000. Materials for Subsurface Land Drainage Systems. FAO , Irrigation and Drainage Paper 60. FAO, Rome.

Stuyt, L.C.P.M., Dierickx, W. and Beltrán, J.M. 2005. Materials for Subsurface Land Drainage Systems. FAO Irrigation and Drainage Paper 60 (Rev. 1). FAO, Rome. 131 p.

Tarez, D.L. 1981. Relationship between pH and SAR and an alternative method for estimating SAR of soil or drainage water. Soil Sci. Soc. Am. J., 45:469-475.

Talsma, T. 1963. The control of saline groundwater. Meded Landb Hoogesch, Wageningen. 63:1-68.

Tezidis, G. 1968. Discussion: falling water table between tile drains. Amer. Soc. Civil Engg., 94: 159.

Toksoz, S. and Kirkham, D. 1961. Graphical solution and interpretation of a new drain spacing formula. J. Geophysical Research, 66: 509-516.

Toky, O.P., Angrish, R., Datta, K.S., Arora, V., Rani, C., Vasudevan, P. and Harrish, P.J.C. 2011. Bio-drainage for preventing water logging and concomitant wood yield in the arid eco-system in northwestern India. J. Scientific and Industrial Research, 70: 639-644.

Tripathi, R.S. 2007. Economic analysis of subsurface drainage projects. In: Agricultural Land Drainage: Reclamation of Waterlogged Saline Lands (Eds.: Ambast, Gupta, Gurbachan (Eds.). Central Soil Salinity Research Institute, Karnal, India, 222-231.

Uppal, H.L. and Sehgal, S.R. 1965. Drainage Requirements of the Punjab State. International Commission on Irrigation and Drainage (ICID), Bulletin, New Delhi, India. 23-31.

U.S. Army Corps of Engineering 1977. Civil Works: Construction Guidelines. Specification for Plastic Filter Fabrics. Specification No. CW-02215. Office Chief of Engineers, Washington, US.

USBR (U S Bureau of Reclamation), SCS. 1971. Soil Conservation Service. National Engineering Handbook, Section 16: Drainage of Agricultural Lands, USDA, Washington, D.C.

USBR, 1978. Drainage Manual: A Water Resources Technical Publication. US department of Interior, Bureau of Reclamation, USA. 321 p.

USDA, 1951. Soil Survey Manual. Agricultural Research Administration, U.S. Dept. of Agriculture, Washington, D.C., USA.

USDA, 2008. Drainage Guide for New York State. Natural Resource Conservation Service, USDA, New York. 83 p.

USDA-SCS. 1985a. National Engineering Handbook, Section 4 - Hydrology. Washington, D.C.: USDA-SCS.

USDA-SCS. 1985b. National Engineering Handbook, Section 16 - Drainage of Agricultural Land. Washington, D.C.: USDA-SCS.

US SCS. 1988. Standards and Specifications for Conservation Practices – Standard on Subsurface Drain No. 606. USDA-SCS, Washington, DC, USA. 606/1 -606/7.

Van der Molen W.H. and Wesseling J. 1991. A solution in closed form and a series solution to replace the tables for the thickness of the equivalent layer in Hooghoudt's drain spacing formula. Agricultural Water Management, 19: 1–16.

van Hooren, J.W. 1979. The effect of capillary flow on salinization and the concept of critical depth for determining drain depth. In: Proc. of the Int. Drainage Workshop (Wesseling, Ed.) ILRI Pub. 25, Wageningen. 686-700.

Van Schilfgaarde, J. 1971. Drainage yesterday, today, and tomorrow. Proc. ASAE National Drain. Symposium, St. Joseph, MI, United States.

Van Schilfgaarde, J. (Ed.). 1974. Drainage for Agriculture. American Society of Agronomy, Madison, USA. Agronomy no. 17. 700 p.

Verma, A.K., Gupta, S.K. and Isaac, R.K. 2010. Application of SWAP to assess performance of subsurface drainage system under semi-arid monsoon climate. CIGR XVII World Congress – Québec City, Canada.

Vlotman, W.F. 1998. Agricultural drain envelope design and laboratory testing. In Proc. Conf. on Filtration and Drainage in Geotechnical/ Geo- environmental Engineering.

Vlotman, W.F., L.S. Willardson and W. Dierickx. 2000. Envelope Design for Subsurface Drains. International Land Reclamation and Improvement Institute, Wageningen. Publication 56, 356 p.

Werner, P.W. 1957. Some problems in non-artesian groundwater flow. Amer. Geophysical Union Transactions, 38: 511-518.

Wesseling, J. 1964. A comparison of the steady state drain spacing formulae of Hooghoudt and Kirkham in connection with design practice. J. Hydrology, 2:25- 32.

Wesseling, J. 1973. Subsurface flow into drains. Drainage Principles and Applications Vol. II: Theories of Field Drainage and Watershed Runoff. Publ. 16. ILRI, Wageningen. 2-56.

Wesseling, J. 1974. Crop growth and wet soils. In: Drainage of Agricultural Lands. J. van Schilfgaarde (Ed). Agronomy, 17:7-32.

Wesseling, J. and Homma, F. 1967. Hydraulic resistance of drain pipes. Netherlands Journal of Agricultural Sciences, 15: 183-197.

Williamsan, W. E. and Kriz, G.J.1970. Response of agricultural crops to flooding, depth of water table and soil gaseous composition. Transactions ASAE, 13: 216-220.

Working Group. 1991. Report of the Working Group on Problem Identification in Irrigated Areas with Suggested Remedial Measures. Ministry of Water Resources, Government of India, New Delhi.

World Bank. 1991. Environmental Assessment Sourcebook. Vol. II. Sectoral Guidelines. World Bank Technical Paper No. 140, World Bank, Washington, D.C.

Zangar, C.N. 1953. Theory and Problems of Water Percolation. U.S. Bureau of Reclamation. Engineering Monograph No. 8, Denver, 76 p.

■■■■

ANSWERS TO OBJECTIVE TYPE QUESTIONS

Chapter 1

Q:1

1.	Aeration	2.	Infiltration rate	3.	Irrigation	4.	Perched	5.	Cooler
6.	Redox	7.	Oxygen	8.	Coastal	9.	Lecahing	10.	Seepage

Q:2

1.	B	2.	C	3.	A	4.	B	5.	B
6.	C	7.	A	8.	A	9.	C	10.	B
11	B								

Chapter 2

Q:1

1.	Sensitive	2.	Marks	3.	mm/d	4.	Concentration	5.	Low
6.	Wilting	7.	Lower	8.	Area	9.	Capacity	10.	Groundwater
11.	Manning	12.	Depth	13.	Top width	14.	Hydraulic	15.	Peak

Q:2

1.	A	2.	D	3.	C	4.	B	5.	A
6.	C	7.	C	8.	A	9.	A	10.	B
11.	C	12.	D	13.	C	14.	C	15.	B

Chapter 3

Q:1

1.	Reconnaissance	2.	Total	3.	Feel	4.	4	5.	2
6.	Geographical	7.	Uncased	8.	water table	9.	Hydraulic	10.	Slope
11.	Piezometers and Terminating	12.	Hook/ point	13.	Vertical	14.	Hetero- geneous	15.	Isotropic
16.	Inverse	17.	Volume	18.	Hydraulic conductivity	19.	Hanging	20.	Kostiakov

Q:2

1.	C	2.	A	3.	D	4.	B	5.	A
6.	D	7.	A	8.	B	9.	D	10.	A
11.	A	12.	B	13.	C	14.	C	15.	B
16.	A	17.	A	18.	B	19.	C	20.	A

Chapter 4

Q:1

1.	Heringbone	2.	Singular	3.	Depth	4.	Horizontal	5.	Less
6.	Bottom	7.	Less	8.	Equi-valent	9.	Non-uniform	10.	Fluctu-ating
11.		12.		13.		14.		15.	
16.		17.		18.		19.		20.	
21.		22.		23.		24.		25.	
26.		27.		28.		29.		30.	
31.		32.		33.		34.		35.	

Q:2

1.	A	2.	D	3.	B	4.	C	5.	D
6.	C	7.	D	8.	D	9.	C	10.	C
11.	C	12.	B	13.	D	14.	A	15.	A
16.	C	17.	C	18.	D	19.	A	20.	B
21.	C	22.	B	23.	C	24.	B	25.	A
26.	D	27.	A	28.	B	29.	C	30.	B
31.	A	32.	D	33.	A	34.	C	35.	
36.	B	37.	B						

Chapter 5

Q:1

1.	Density	2.	16	3.	Coils	4.	Ultraviolet	5.	Higher/ more
6.	Higher/more	7.	0.9	8.	Acetone	9.	Concentrated	10.	Narrow
11.	Load	12.	Organic	13.	Silting	14.	Liquid	15.	Day/ meter
16.	Hydraulic conductivity	17.	O_{90}	18.	O_{90}	19.	Failure	20.	Plasticity

Q:2

1.	B	2.	A	3.	C	4.	A	5.	A
6.	B	7.	A	8.	C	9.	D	10.	D
11.	C	12.	A	13.	A	14.	B	15.	D
16.	B	17.	C	18.	A	19.	C	20.	B
21.	B	22.	C	23.	A				

Chapter 6

Q:1

1.	Blinding	2.	Quicksand	3.	50	4.	Subsurface	5.	Neagtive
6.	Rodding	7.	Above	8.	Y-axis	9.	Improves	10.	Rodent

Q:2

1.	C	2.	A	3.	C	4.	D	5.	B
6.	D	7.	D	8.	D	9.	D	10.	B
11.	A	12.	C	13.	B				

Chapter 7

Q:1

1.	Pipe less	2.	Unlined	3.	Compact	4.	Tube wells	5.	Unconfined
6.	Interference	7.	*Eucalyptus*	8.	Bio-harvested	9.	Pipe	10.	Chiesling

Q:2

1.	A	2.	D	3.	B	4.	C	5.	A
6.	B	7.	D	8.	A	9.	B	10.	C
11.	A	12.	C	13.	D	14.	B	15.	A
16.	D	17.	D	18.	A	19.	C	20.	D

Chapter 8

Q:1

1.	1	2.	Equivalent	3.	EC	4.	More	5.	SAR
6.	Alkali Alkali	7.	0.36	8.	Acid	9.	$CaSO_4.2H_20$	10.	Acid
11.	More	12.	Continuous	13.	Wetting	14.	High/more	15.	More

Q:2

1.	D	2.	B	3.	C	4.	A	5.	B
6.	A	7.	A	8.	A	9.	B	10.	A
11.	C	12.	A	13.	C	14.	A	15.	B
16.	A	17.	B	18.	B	19.	D	20.	A
21.	C	22.	D	23.	C	24.	A	25.	C

Chapter 9

Q:1

1.	Less	2.	Humid/sub-humid	3.	Active	4.	Potential	5.	Lime, $CaCO_3$
6.	CaO	7.	Increases	8.	More	9.	Kerala	10.	More

Q:1

1.	A	2.	D	3.	C	4.	B	5.	D
6.	A	7.	A	8.	B	9.	C	10.	C
11.	A								

Chapter 10

Q:1

1.	Less	2.	Water	3.	More/higher	4.	SAR	5.	(CO_3+HCO_3)
6.	Infiltration	7.	Critical	8.	Change	9.	EC_d	10.	Higher
11.	10	12.	Tolerant	13.	Blending or mixing	14.	Irrigation	15.	Calcium
16.	Toxicity	17.	6.25	18.	Raw	19.	Sprinkler	20.	Gypsum

Q:2

1.	D	2.	D	3.	A	4.	A	5.	B
6.	C	7.	D	8.	A	9.	B	10.	B
11.	C	12.	B	13.	D	14.	A	15.	A
16.	C	17.	A	18.	A	19.	C	20.	B
21.	D	22.	B	23.	A	24.	C	25.	B
26.	D	27.	A	28.	D	29.	D	30.	A
31.	A	32.	A	33.	D	34.	B	35.	D

Chapter 11

Q:1

1.	More	2.	Minus	3.	B:C ratio	4.	Payback	5.	Zero and one
6.	Break even	7.	Social or sociological	8.		9.		10.	

Q:2

1.	A	2.	B	3.	C	4.	B	5.	A
6.	B	7.	D	8.	A	9.	B	10.	C
11.	A	12.	B	13.	B	14.	C	15.	A

■■■■

CONVERSION TABLES

Length (Linear)

To convert from	To	Multiply by
micrometers	millimeters	1.0×10^{-3}
	meters	1.0×10^{-6}
	angstrom units (A)	1.0×10^{4}
	mils	0.03937
	inches	$3.937\ 01 \times 10^{-5}$
millimeters	micrometers	1.0 x 103
	centimeters (cm)	0.1
	meters	1.0×10^{-3}
	mils	39.370 08
	inches	0.039 37
	feet (ft)	$3.280\ 84 \times 10^{-3}$
centimeters	millimeters	10.0
	meters	0.01
	mils	0.3937×10^{3}
	inches	0.3937
	feet	0.032 81
inches	millimeters	25.40
	meters	0.0254
	mils	1.0×10^{3}
	feet	0.083 33
feet	millimeters	304.8
	meters	0.3048
	inches	12.0
	yards (yd)	0.333 33
yards	meters	0.9144
	inches	36.0
	feet	3.0
meters	millimeters	1.0×10^{3}
	kilometers (km)	1.0×10^{-3}
	inches	39.370 08
	yards	1.093 61
	miles (mi)	$6.213\ 71 \times 10^{-4}$
kilometers	meters	1.0×10^{3}
	feet	$3.280\ 84 \times 10^{3}$
	miles	0.621 37

To convert from	To	Multiply by
miles	meters	$1.609\ 34 \times 10^{3}$
	kilometers	1.609 34
	feet	5280.0
	yards	1760.0

Area

To convert from	To	Multiply by
square millimeters	square centimeters (cm^2)	0.01
	square inches (in^2)	1.550×10^{-3}
square centimeters	square millimeters (mm^2)	100.0
	square meters (m^2)	1.0×10^{-4}
	square inches	0.1550
	square feet (ft^2)	1.076 39 x 10-3
square inches	square millimeters	645.16
	square centimeters	6.4516
	square meters	6.4516×10^{-4}
	square feet	69.444×10^{-4}
square feet	square meters	0.0929
	hectares (ha)	9.2903×10^{-6}
	square inches	144.0
	acres	$2.295\ 68 \times 10^{-5}$
square yards	square meters	0.836 13
	hectares	8.3613×10^{-5}
	square feet	9.0
	acres	$2.066\ 12 \times 10^{-4}$
square meters	hectares	1.0×10^{-4}
	square feet	10.763 91
	acres	2.471×1^{-4}
	square yards (yd^2)	1.195 99
acres	square meters	4046.8564
	hectares	0.404 69
	square feet	4.356×10^{4}
hectares	square meters	1.0×10^{6}
	acres	2.471
square kilometers	square meters	1.0×10^{6}
	hectares	100.0
	square feet	107.6391×10^{5}
	acres	247.105 38
	square miles (mi^2)	0.3861
square miles	square meters	$258.998\ 81 \times 10^{4}$
	hectares	258.998 81
	square kilometers (km^2)	2.589 99
	square feet	$2.787\ 84 \times 10^{7}$
	acres	640.0

Volume-capacity

To convert from	To	Multiply by
cubic millimeters	cubic centimeters (cm^3)	1.0×10^{-3}
	liters (L)	1.0×10^{-6}
	cubic inches (in^3)	$61.023\ 74 \times 10^{-6}$
cubic centimeters	liters	1.0×10^{-3}
	milliliters (ml)	1.0
	cubic inches	$61.023\ 74 \times 10^{-3}$
	fluid ounces (fl oz)	33.814×10^{-3}
milliliters	liters	1.0×10^{-3}
	cubic centimeters	1.0
cubic inches	milliliters	16.387 06
	cubic feet (ft^3)	$57.870\ 37 \times 10^{-5}$
liters	cubic meters (m^3)	1.0×10^{-3}
	cubic feet	0.035 31
	gallons	0.264 17
	fluid ounces	33.814
gallons	liters	3.785 41
	cubic meters	$3.785\ 41 \times 10^{-3}$
	fluid ounces	128.0
	cubic feet	0.133 68
cubic feet	liter	28.316 85
	cubic meters	$28.316\ 85 \times 10^{-3}$
	cubic decameters (dam^3)	$28.316\ 85 \times 10^{-3}$
	cubic inches	1728.0
	cubic yards (yd^3)	$37.037\ 04 \times 10^{-3}$
	gallons (gal)	7.480 52
	acre-feet (acre-ft)	$22.956\ 84 \times 10^{-6}$
cubic miles	cubic decameters	$4.168\ 18 \times 10^{6}$
	cubic kilometers (km^2)	4.168 18
	acre-feet	3.3792×10^{6}
cubic yards	cubic meters	0.764 55
	cubic feet	27.0
cubic meters	liters	1.0×10^{3}
	cubic decameters	1.0×10^{-3}
	gallons	264.1721
	cubic yards	1.307 95
	acre-feet	8.107×10^{-4}

To convert from	To	Multiply by
acre-feet	cubic meters	1233.482
	cubic decameters	1.233 48
	cubic feet	43.560×10^3
	gallons	325.8514×10^3
cubic dekameters	cubic meters	1.0×10^3
	cubic feet	$35.314\ 67 \times 10^3$
	acre-feet	0.810 71
	gallons	$26.417\ 21 \times 10^4$
cubic kilometers	cubic decameters	1.0×10^6
	acre-feet	$0.810\ 71 \times 10^6$
	cubic miles (mi^3)	0.239 91

Velocity

To convert from	To	Multiply by
feet per second	meters per second (m/s)	0.3048
	kilometers per hour (km/h)	1.097 28
	miles per hour (mi/h)	0.681 82
meters per second	kilometers per hour	3.60
	feet per second (ft/s)	3.280 81
	miles per hour	2.236 94
kilometers per hour	meters per second	0.277 78
	feet per second	0.911 34
	miles per hour	0.621 47
miles per hour	kilometers per hour	1.609 34
	meters per second	0.447 04
	feet per second	1.466 67
feet per year (ft/yr)	millimeters per second	$9.665\ 14 \times 10^{-6}$

Volume per unit Time Flow

To convert from	To	Multiply by
cubic feet per second	liters per second (L/s)	28.316.85
	cubic meters per second (m^3/s)	0.028 32
	cubic decameters/day (dam^3/d)	2.446 57
	gallons per minute (gal/min)	448.831 17
	acre-feet per day (acre-ft/d)	1.983 47
	cubic feet per minute (ft^3/min)	60.0

To convert from	To	Multiply by
gallons per minute	cubic meters per second	0.631×10^{-4}
	liters per second	0.0631
	cubic decameters per day	5.451×10^{-3}
	cubic feet per second (ft^3/s)	2.228×10^{-3}
	acre-feet per day	4.4192×10^{-3}
acre-feet per day	cubic meters per second	0.014 28
	cubic decameters per day	1.233 48
	cubic feet per second	0.504 17
cubic decameters per day	cubic meters per second	0.011 57
	cubic feet per second	0.408 74
	acre-feet per day	0.810 71

Volume per Area per Unit Time

To convert from	To	Multiply by
cubic feet per square foot per day	cubic meters per square meter per day ($m^3/m^2.d$)	0.3048
	cubic feet per square foot per minute ($ft^3/ft^2.min$)	0.6944×10^{-3}
	liters per square meter per day ($L/m^2.d$)	304.8
	gallons per square foot per day ($gal/(ft^2.d)$)	7.480 52
	cubic millimeters per square millimeter per day ($mm^3/mm^2.d$)	304.8
	cubic millimeter per square millimeter per hour ($mm^3/mm^2.h$)	25.4
	cubic inches per square inch per hour ($in^3/in^2.h$)	0.5
gallons per square foot per day	cubic meter per square meter per day ($m^3/m^2.d$)	40.7458×10^{-3}
	liters per square meter per day ($L/m^2.d$)	40.7458
	cubic per square foot per day ($ft^3/ft^2.d$)	0.133 68

Volume per length per Unit Time

To convert from	To	Multiply by
Cubic feet per foot per day	cubic meters per meter per day ($m^3/m.d$)	0.0929
	gallons per foot per day (gal/ft.d)	7.480 52
	liters per meter per day (L/m.d)	92.903

Salt Concentration

To convert from	To	Multiply by
TDS (mg/L)	TDS (ppm)	1.0
Salinity (dS/m)	salinity (mmho/cm)	1.0
Salinity (mmho/cm)	salinity (μmho/cm)	1000
Salinity (dS/m) (3-30)	osmotic pressure	0.36
Salinity (dS/m) (0.1-5.0)	ppm	0.64
Parts of salt per million parts of irrigation water	tons/ ha.cm	0.001323
Salinity (dS/m) (< 5 dS/m)	TDS (mg /L)	640
Salinity (dS/m) (>5 dS/m)	TDS (mg/L)	800
Water, NO_3-N, SO_4-S, B (ppm)	kg/ha.cm of water applied	0.1015

TDS = Total dissolved salts

Pressure

To convert from	To	Multiply by
Atmosphere	psi	14.7
	Feet (water)	34
	cm (mercury)	76
	kg/cm^2	1.0322
	millibar	1013.25
	Atmosphere (metric)	1.033
Atmosphere (metric)	0.9678	Atmosphere
	m (water)	10
	kg/cm^2	1
	feet (water)	32.8
	millibar	98.2
	cm (mercury)	73.5

Power

To convert from	To	Multiply by
HP (horse power), British	ft-lb/s	550
	HP (metric)	1.018
	watts	746
	kg-m/s	76.04
HP (metric)	kg-m/s	75
	HP (British)	0.986
	watts	736

Useful Approximations

Value	Approximation
1 in	2.5 cm
1 ft	30 cm
33 ft	10 m
1 ha	2.5 ac
1 ac	0.4 ha
1 in^2	6.5 cm^2
1 m^2	11 ft^2
1m^3	35 ft^3
1 in^3	16.4 cm^3

■■■■

SUBJECT INDEX

A

Acetone test 195
Acid soils 356
Acid-sodic soils 379
Acid-sulphate soils 376,378
Active acidity 358
Adj. R_{Na} 392
Adjusted SAR 390
Aeration, soil 9,11
Agronomic practices 373
Alkali soils 2,5,86,321,322,338
Alkali water 386,394
Amendment application 328,422
    Gypsum 329,422
    Gypsum requirement 330,422
    Lime 362
    Sulphur 329
    Sulphuric acid 329
Anisotropic 101,113
Antecedent moisture conditions 35,37
Apparent opening size, O_{90} 214
Aquifer conditions 290
Artesian conditions 93
Artificial drainage 19
Asbestos pipes 191
Auger hole method 104

B

Back filling 229,233
Base saturation 365
Bedding system 47
Bending test 195,196
Benefit: cost ratio 445
Benefits, drainage 20,64,440
Bill of quantities 465
Bio-drainage 23,300
    Case studies 307
    Design 303
    Limitations 305
Bio- interceptor Drainage 23,304
Blending 418
Blind inlet 243
Blinding material 229
Blinding 229,233
Boston Society formula 42
Break even benefits 446
Bullet 279

C

Calcium carbonate equivalent 366
Calcium oxide 363
Calibrated rainfall -discharge 32
Capillary rise 401
Cavity wells 293
Chemical deposits 250
Chemical management 422
Chiselling, subsoil See sub-soiling
Clay tile 188
Cleaning drain line 248
Clogging, roots 251
Coefficient of curvature 83,84,211
Coefficient of uniformity 83,207
Collector drain 130,173
Composite system 132
Concrete tiles 190
Conduits
    Flexible 188

Rigid 188
Conjunctive use 417
Construction methods
Inspection 251
Manual 226
Mechanical 229
Sequence 226
Tender document 252
Work organization 253
Construction surveys 73,95
Contour map (Water table) 91
Controlled drainage 246
Cost, drainage 436
Cost recovery 452
Critical depth 138,400
Crop cutting 268
Crop management 414
Crop selection 335
Crop tolerance
Acid soils 374
Alkali 335
ESP 336
Salinity 343,347
Shallow water table 136
Cross slope drains 50
Cultural practices 349
Curve number method 34
Cyclic use 420
Cypress creek formula 41

D

Dagan equation 151
Darcy´s law 100
Depth-duration-frequency curves 40
Desalinization 408
Design (collectors) 41
Detailed project report 437
De Zeeuw-Hellinga equation 166
Discounted benefits 447
Discounted costs 447
Discounting 443
Disposal, drainage effluents 426
Canals/rivers 426
Evaporation tanks 426
Irrigation, halophytic plants 426
Salt extraction 426
Sea 426
Diversion drain 51
Dolomite 363
Double main 131
Drain depth 136
Drain materials 187
Drain pipes 187
Drain outflow method 119,261
Drain size/capacity 171
Collector 173
Laterals 171
Maximum area drained 173
Drain spacing 141
Steady state 141,261
Unsteady state 157,158,267
Drainable pore space 114
Correlation method 115
Field methods 119
Hanging water column 116
Laboratory method 116
Drainage
Artificial 19
Benefits 20
Construction 61
Criteria 134
Non-conventional 65
Paddy fields 178
Problems 16,17
Types 20
Drainage coefficient
Non-steady state124
Subsurface 120
Surface 32
Drainage criteria
Steady state 134
Non-steady state 135
Fluctuating water table 135
Salinity control 135

Trafficability 135
Drainage investigations 31,71,72
Drainage, organization 253
Drainage structures
Blind inlet 243
Breather 235,243
Closing devices 235,245
Crossing 235,244
Drain bridges 235,244
Drops 235,239
Junction box 235,239
Manholes 235,239
Outlets, Gravity 235,236
Outlets, sump 237
Regulating sluices 245
Relief well 235,243
Surface water inlets 242
Vent pipes 244
Drainage vertical: types 289
Drainage water (quality) 270
Drainage wells
Interference 294
Theory 292
Drawdown 292
Dupuit-Forchheimer (D-F) assumptions 142

E

Economic analysis 436,443
Case studies 449
Effective calcium carbonate eq. 368
Electrical conductivity, EC 315,387
Ellipse equation 144
Elongation 195
Empirical formulae 41
Employment 453
Engineering measures (Acid soils) 373
Entrance head loss 208
Entrance resistance 208
Envelope 203
Bedding function 204
Design 211,219
Filter function 204
Granular 205,217
Selection 205
Synthetic 205,214
Thickness 213
Envelop material 203,204
Environmental improvement 455
Equated installments 446
Equivalent depth 145,162
Equivalent drainage area 44
Ernst equation 152
Eucalyptus 301
Evapotranspiration, plants 302
Exchangeable acidity 358
Exchangeable Sodium Percentage 319

F

Fallow 424
Farmers' drainage societies 458,479
Farmers' participation 256,457
Farmers' organizations 458
Field drain 56
Filter, drain 204
Financial viability 443
Fineness (lime) 368
First wetting 229
Fittings (tests) 197
Flexible conduits 188
Load 202
Tests 194
Fluctuating water table 166
Flushing 334
Fly ash 372
Foliar injury 395
Formation (Acid soils) 357
Freeboard 55

G

Geological surveys 77
Glover Dumm equation 158
Gradient ratio 218
Gravel filter
Amount 213

Shortcomings 213
Gravel mole 282
Gravity outlet 236
Ground water
Balance 470
Contour maps 91
Inflow/outflow 470,473
Investigations 87
Mining 473
Guidelines/recommendations
Drain depth 141
Drain layout 133
Drain spacing 169
Lime application 368
Use of saline/alkali water 424
Gypsum beds 423
Gypsum requirement, water 422

H

Herringbone system 131
Heterogeneous soil 101
History (Water logging) 7
Homogeneous soil 101
Horizontal drainage 21
Hooghoudt equation 142
Auger hole 105
Integrated equation 165
Hydraulic conductivity 47,99
Auger hole method 104
Constant head method 102
Definition 99
Falling head method 103
Inverse auger hole 109,111
Laboratory methods 101
Soil texture-HC 101
Spatial variation 112
Vertical 99,113
Two layered profile 108
Hydraulic criteria 218
Hydraulic failure gradient 218
Hydraulic radius 51
Hydrograph 91
Hydrological soil group 35

I

Impact test 195
Impermeable layer 85
Indirect benefits 433
Industrial wastes 370
Infiltration (Problem) 398
Infiltration rate 95
Infiltrometer 98
Inflation 444
Inspection chamber: see man hole
Integrated drainage system 308
Integrated nutrient management 337,374
Interceptor drainage 23,51
Interceptor drainage, sloping lands 169
Interceptor drain, canal 284
Interception-on farm ponds 287
Interference (Wells) 294
Internal rate of return 445
Inverse anger hole method 109
Investment, machinery 439
Irrigation (source) 454
Frequency 349
Irrigation techniques
Drip 416
Sprinkler 416
Surface irrigation 416
Irrigation versus drainage wells 298
Irrigation water management 416
Isobath map 92

J

Junction box 239

K

Kostiakov equation 95

L

Land forming 45
Land grading 45
Land modification 346

Land smoothening 45
Land value 453
Laser land leveling 45
Layout 131
  Guidelines 133
  Singular 131
  Composite 131
Leaching 334,340
  Continuous 345
  Intermittent 345
  Water requirement 341
Leaching curves 343
Leaching efficiency 345
Leaching fraction 342,409
Leaching requirement 122,405,407
Levees 50
Lime requirement 363
Liming 361
  Impact 369
Load on conduits 197
  Concentrated load 201

M

Maintenance
  Open drains 62
  Subsurface drains 247,257
Management 18
  Acid soils 361
Manning equation 51
Marriott arrangement 97
Mixing 418
Mole drainage 278
Mole drains
  Depth 280
  Experiences 281
  Economics 282
  Life 281
  Failure and remedies282
  Spacing 280
Mole plough 279
Monitoring and evaluation 257
Muck soil 234
Multiple well point system 300

N

Net present worth 445
Non steady state See unsteady state
Nutrient management 337,373,421

O

Observation well 88.259
  Hydrograph 91
Ochre formation 249
On-farm ponds 65,287
Organic envelopes 205,219
Organic manures 421
Organizational set-up 253
Outlet 61
Oxygen diffusion rate11

P

Paper mill sludge 372
Parallel system 131
  Field drain 48
  Lateral open ditch 49
Particle-size analysis 85,207
Payback period 445
Peak runoff 33,37
Peat soils 234
Perched water table 7
Perforations 194
Permeable back fills 311
pH 318,389
Photogrammetric survey 86
Phyto-remediation 338
Piezometer 89
Pipe drainage 21
Plastic pipes 192
  PVC 192,194
  PE 192,194
  PP 192
  Tests 194
Plasticity index 207
Potential acidity 358

Potential theory (Drainage) 142
Pre-installation activities 225
Pre-sowing irrigation 421
Press mud 372
Prevention 18
Puddling 337
Puncture test 233

Q

Quality control, construction 251
Quick lime 363
Quicksand conditions 233

R

Rainfall 33
Rainwater management 424
Random system 49,132
Rating curve 33
Rational formula 33
Recharge-groundwater 66,470
Reclamation 18
    Alkali soil 328
    Saline soil 339
    Saline alkali soil 350
Reconnaissance survey 72,74
Redox potential 12,13
Residual acidity 358
Residual Sodium Carbonate (RSC) 391
Rhoades diagram 398
Rigid conduits 188
Rodent guard 237
Runoff 33

S

Saline soil 2,322,323,339,350
Saline alkali soil 322,325,350
Saline water 382,383
Salinization 400,402
Salt affected soils 321,325
    Extent 326
Salt balance 272,306,403,407
Saturation extract 320
Scraping 339
Seasonal switching 421
Sedimentation 62,270
Seedling (age) 337
Seed rate 336
Seepage (interception) 283,287
Sensitivity analysis 451
Sequential application 420
SEW_{30} 168
Shaft-recharge 66,67
Shallow water table 6
Sieve analysis 82
Silting tendency 206
Simple rate of return 444
Singular system 132
Skimming (well drainage) 300
Slags 362,370
Sodic soil 321
Sodic water: See alkali water
Sodium Adsorption Ratio (SAR) 318,389
Soil degradation 1
Soil extract 320
Soil management 415
Soil parameters (Drain outflow) 261
Soil salinity 15,269
Soil salinization 13,137,339,401,409,413
Soil sample 78
Soil sodification 413
Soil structure 116
Soil survey 77
Soil texture 81
    Feel and touch method 81
Specific ion toxicity 386
Spoil bank 56
Stakeholders 63
Steady state equations (Summary) 157,262
Steel slag 370
Stiffness test 195,196
Subsidy 453
Sub-soiling 309
Subsurface drainage 22,71
    Design 130

Sump 239
Surface drainage 22,28
  Design 31
  Factors affecting 30
  Requirement (Crops) 29
Surface inlet 242
Surface stagnation 4
Switching mode 420
Synthetic envelope (tests) 219

T

TDS 315
Tender document 252
Time of concentration 33
Tolerant crops 335,343,374,475
Topographic survey 74
Total station 75,76
Toxic effect (Low pH) 359
Toxicity hazard 394
Toxic water 394,427
Trafficability 135
Trench digging 227
Trencher 230
Tube wells drainage
  Number of tube wells 295

U

Uncalibrated rainfall-runoff 33
Uncased well 88
Unsteady state equations 157,262
USSL diagram 396

V

Velocity (Permissible) 52
Vertical drainage 22,288
  Case studies 299
  Cost 298
  Design information 290
  Number of wells 295
  Theory of flow 291
  Types 289
Vertical hydraulic conductivity 99,113
Visual observations 31

W

Waste water 428
Water balance 122,123,272,290,403
Waterlogged salt affected soils 325
Water logging 2,5,8
Water quality 94,269
Water samples 80
Water table control, bio-drainage 305
Water table, perched 7
Water transmission 95
Water uptake pattern 410
Well versus tile drainage 298
Wetted perimeter 148
Wrapping machine 231